D1607504

FUNDAMENTALS OF III-V DEVICES

FUNDAMENTALS OF III-V DEVICES

HBTs, MESFETs, and HFETs/HEMTs

WILLIAM LIU
Texas Instruments

A WILEY-INTERSCIENCE PUBLICATION
JOHN WILEY & SONS, INC.
New York / Chichester / Weinheim / Brisbane / Singapore / Toronto

This book is printed on acid-free paper. ♾

For ordering and customer service, call 1-800-CALLWILEY.

Library of Congress Cataloging-in-Publication Data:

Liu, William.
Fundamentals of III-V devices: HBTs, MESFETs, and HFETs/HEMTs/William Liu.
p. cm.
"A Wiley-Interscience publication."
Includes index.
ISBN 0-471-29700-3 (cloth: alk.paper)
1. Bipolar transistors. 2. Field-effect transistors. 3. Metal semiconductor field-effect transistors. 4. Modulation-doped field-effect transistors. I. Title.
TK7871.96.B55L567 1999
621.3815'28—dc21 98-38918

Printed in the United States of America.

10 9 8 7 6 5 4 3 2 1

To: *my wife*, Lee-Ping Chong
my daughter, Yiling Ashley Liu
my parents, Chien-Shun and Li-Yue Liu
the Liu family
the coach, Prof. James Harris

and ...
my idiosyncratic chess pals who master both

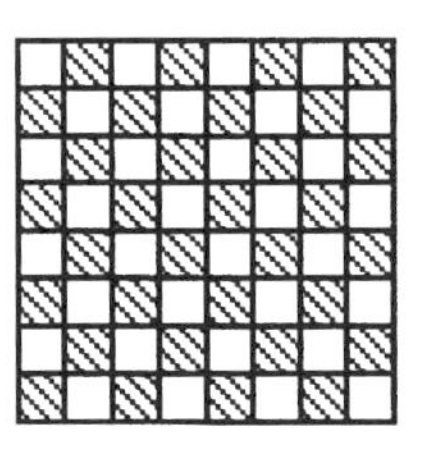 and 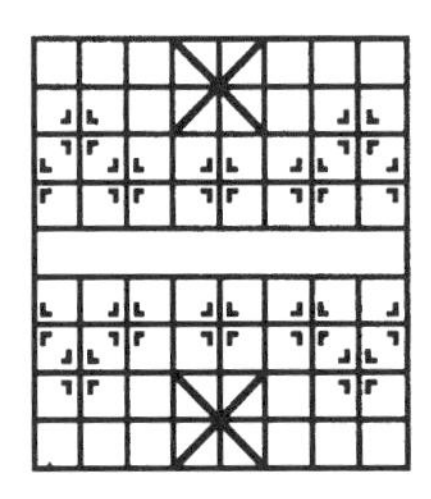

CONTENTS

PREFACE

III-V transistors, such as heterojunction bipolar transistors HBT, metal-semiconductor field-effect transistors (MESFETs), and heterojunction field-effect transistors (HFETs), which include high electron mobility transistors (HEMTs), extend the advantages of silicon counterparts to significantly higher frequencies. They have been natural choices of device for wireless communication and military applications operating between 1 and 100 GHz. Despite their popularity, existing books relevant to III-V devices have been directed to practitioners by delineating the state of the art. They tend to be "survey" books, consisting of chapters contributed by several authors working in various fields. They also tend to emphasize one particular III-V device, concerned with either HBT or FET exclusively.

It is the purpose of this book to present the III-V devices systematically and consistently. This book is written in a manner so that an elementary course in device physics is not a prerequisite. In many ways it is intentionally made to mimic the popular undergraduate textbook on device physics by Muller & Kamins, except that this book focuses on III-V materials instead of silicon. Both books begin with introductory chapters on semiconductor basics and two-terminal junctions, continue with two chapters on bipolar devices, and end with two chapters on FET devices. Plenty of worked-out examples are given to facilitate design of a device for a given application. The problem sets at the ends of the chapters are created for students to reinforce device concepts as well as design considerations.

This book emphasizes physical understanding of the devices. Traditional textbooks tend to have a gap between the descriptions of the d.c. and the high-frequency characteristics of a device. The high-frequency model seems to be introduced in a hasty fashion, or given based on hand-waving arguments.

While this approach may at times be necessary to shorten the length of the book, it often leaves students with a nebulous understanding of high-frequency characteristics. In contrast, this book presents the logical steps leading to an equivalent circuit. In this manner, the physical meaning of each circuit element becomes clear.

The latter portion of this book has been inspired by the textbook written by Y. Tsividis in 1987, on the *Operation and Modeling of the MOS Transistor*. The author fully believes what Prof. Tsividis states in the preface of his book, while commenting on MOS modeling, that "once one has decided to fudge, things grow out of hand very quickly, and one ends up with a hodgepodge of careless derivations, conflicting models, and a lot of patchwork." While the author has hoped to follow the spirit of Prof. Tsividis' book in the treatment of field-effect transistors, there are some noticeable differences. For example, this book develops the drain/source charge expression without invoking the "Ward equation" directly, but instead uses the first-moment technique. This book develops the non-quasi-static y-parameters prior to the quasi-static y-parameters, rather than the other way around. Moreover, this book intentionally separates the wave equation governing the transistor dynamics into a drift equation and a continuity equation. The author believes that the approach used here is more appropriate for the simultaneous presentation of various III-V FETs such as MESFETs and HFETs, in contrast to just the silicon MOSFET alone.

The development of this book has been benefited by my colleagues at or formerly at Texas Instruments. I especially would like to thank Mi-Chang Chang, Ulvi Erdogan, Timothy Henderson, Darrell Hill, Ming-Yi Kao, Andrew Ketterson, and Larry Witkowsky for offering valuable comments. Suggestions made by Chris Bowen, Hin-Fai Chau, Steve Evans, Paul Saunier, Hwa-Quen Tserng, Jerold Seitchik, and Ron Yarborough, as well as general support from Damian Costa, Ali Khatibzadeh, David McQuiddy, and William Wisseman, are acknowledged. I would like to thank Jenn-Hwa Huang of Motorola for discussion on FETs. My special gratitude goes to Coach, Professor James Harris of Stanford University, who introduced me to III-V semiconductors. I would like to thank Wonill Ha, a Stanford graduate student, for kindly supplying me the device simulator problems. My past colleagues at Stanford, Chong-Hong Dai, Won-Seong Lee, Tony Ma, and Alan Massengale, also contributed to this work. The book editor at Wiley, George Telecki, as well as the anonymous reviewers, have all been extremely helpful in shaping this project. Finally, I thank my wife, Lee-Ping Chong, and my parents, Chien-Shun and Li-Yue Liu, for their encouragement and support.

WILLIAM LIU

Plano, Texas
January 1999

FUNDAMENTALS OF III-V DEVICES

CHAPTER 1

BASIC PROPERTIES AND DEVICE PHYSICS OF III-V MATERIALS

§ 1-1 SEMICONDUCTOR CRYSTALLINE PROPERTIES

There are several ways to classify a solid. Based on its electrical properties, a solid can be viewed as a conductor, an insulator, or a semiconductor; based on the chemical bonding between the atoms comprising the solid, it can be an ionic solid, a covalent solid, or a metal. We describe some of these terms later in this section. However, it is instructive first to classify a solid in accordance to its atomic arrangement. In this regard, solids are grouped into three broad categories: amorphous, crystalline, and polycrystalline.

A solid is *amorphous* when there is little, if any, geometric regularity or periodicity in the way in which the atoms are arranged in space. Its two-dimensional representation is shown schematically in Fig. 1-1. The irregularity of the atomic structure prevents an amorphous solid from being easily studied; consequently, much of the knowledge about amorphous materials is empirical. The previous definition does not mention the bonding between the atoms. Although the atoms of an amorphous material lack periodicity in their arrangement, they can be tightly bound to each other and remain stable. The counterpart of an amorphous solid is a *crystalline* solid, characterized by a perfect periodicity in the arrangement of the atoms. A crystal appears exactly the same at one point as it does at a series of other equivalent points. This regularity property allows the material to be analyzed with a simple mathematical construct. Semiconductor materials are nearly perfect crystalline solids with small amounts of imperfections, such as impurity atoms, lattice vacancies, or dislocations. These imperfections are intentionally (or sometimes inadvertently) introduced to alter the electrical characteristics of the semiconductors. The third category, the *polycrystalline* solid, has an atomic arrangement that is somewhat between the two extremes of amorphous and crystalline solids. It consists of

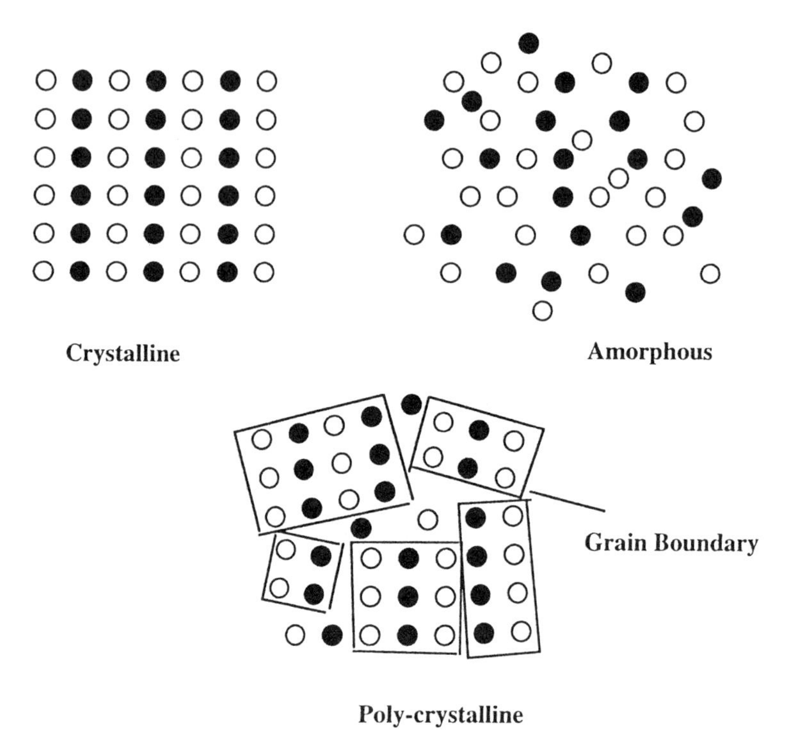

FIGURE 1-1. Two-dimensional representations of amorphous, crystalline, and poly-crystalline solids.

several regions of single-crystalline material bounded together in no particular order. The boundaries separating these distinct regions of periodicity are called the *grain boundaries*, as marked in Fig. 1-1.

The basic building block making up a crystalline solid is called a *unit cell.* Figure 1-2 illustrates the lattice of a two-dimensional crystal. The region bound by the points ABCD is a unit cell, whose replicas reproduce the crystal. This unit cell, however, is not unique. A parallelogram bound by A′B′C′D′ also forms a unit cell, since its replicas reproduce the whole crystal. The former choice for the unit cell is preferred because its characteristics lead naturally to the use of the Cartesian coordinate system. Although it is not immediately obvious, the region A″B″C″D″ also fits the definition of a unit cell. It is not, however, a *primitive* unit cell, which is defined as the smallest unit cell in volume that can be defined for a given lattice. The first two examples are both primitive unit cells for the lattice of Fig. 1-2.

Of the several unit cell structures existing in nature, the simplest is the *cubic* unit cell. It differs from the square primitive unit cell of Fig. 1-2 by its third-dimensional extension. As shown in Fig. 1-3*a*, each corner of the cubic lattice is occupied by an atom, which is shared by eight neighboring unit cells. Because there are eight corners, the total number of atoms residing in this unit cell is one.

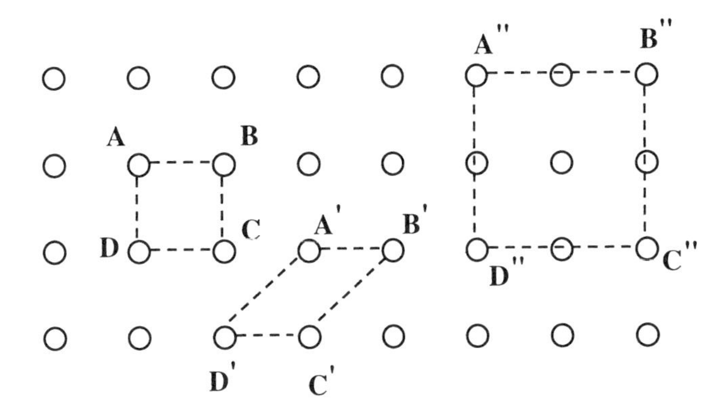

FIGURE 1-2. Two-dimensional crystalline structure showing the various unit cells.

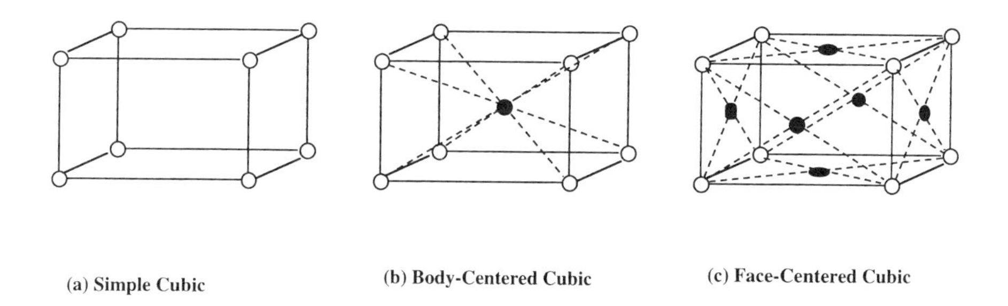

FIGURE 1-3. Some examples of three-dimensional unit cells: (*a*) simple cubic; (*b*) body-centered cubic; (*c*) face-centered cubic.

Hence, the simple cubic unit cell is a primitive unit cell. Although they are simplistic in structure, few crystals are characterized by this type of unit cell. Polonium is the only element that crystallizes in this form.

The next unit cell of interest is the *body-centered cubic* (bcc) structure, which differs from the simple cubic structure by an additional atom located in the center. This is shown in Fig. 1-3*b*. Because the total number of atoms enclosed in this unit cell is two, the bcc structure is not a primitive unit cell. The primitive unit cell can be constructed, in one way, by connecting four center atoms and four corner atoms of five bcc unit cells as shown in Fig. 1-4*a*. The primitive unit cell, though having the advantage of being the most fundamental building block, is awkward in a Cartesian coordinate system. It is easier to visualize the crystal with the bcc unit cell of Fig. 1-3*b*. The bcc structure is found in several metals, such as molybdenum, tungsten, and sodium.

The *face-centred cubic* (fcc) crystal contains one atom at each of the eight corners, as well as one atom at the center of each of the six faces. The atom at each face of the unit cell is shared with another unit cell. There are a total

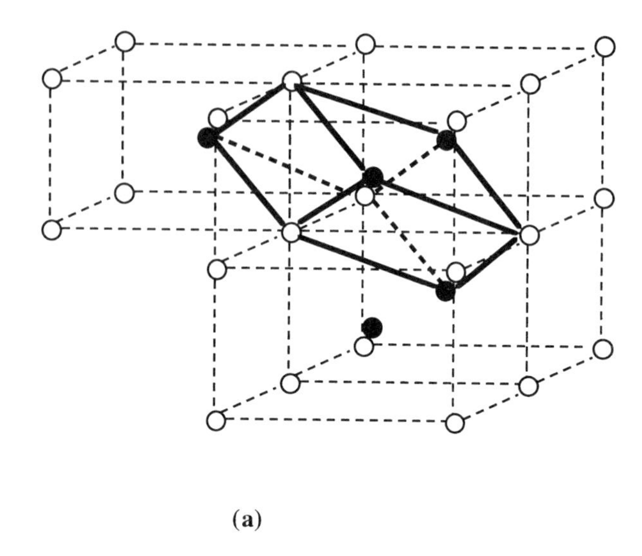

(a)

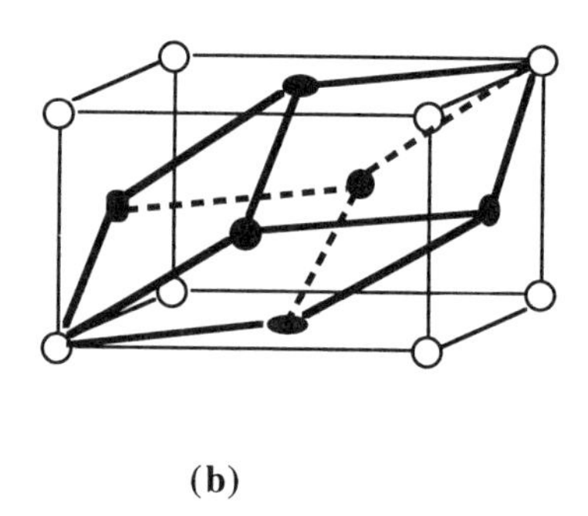

(b)

FIGURE 1-4. (*a*) The primitive unit cell of a bcc structure. (*b*) The primitive unit cell of a fcc structure.

of $8 \cdot \frac{1}{8} + 6 \cdot \frac{1}{2} = 4$ atoms in a fcc unit cell, suggesting that it is not a primitive unit cell. Nonetheless, just like the bcc unit cell, the fcc unit cell is the preferred choice for analysis because of its simple representation in the Cartesian coordinate system. The primitive unit cell can be formed by connecting the six face atoms with two corner atoms, as shown in Fig. 1-4*b*. The fcc unit cell is an important crystal form, which characterizes several elements such as aluminum, copper, gold, silver, nickle, and platinum.

Most ionic crystals resemble the structure of salt (NaCl), which has alternating Na and Cl atoms at the lattice points of a simple cubic lattice, as shown in Fig. 1-5. When examined separately, the Na atoms occupy the lattice points of

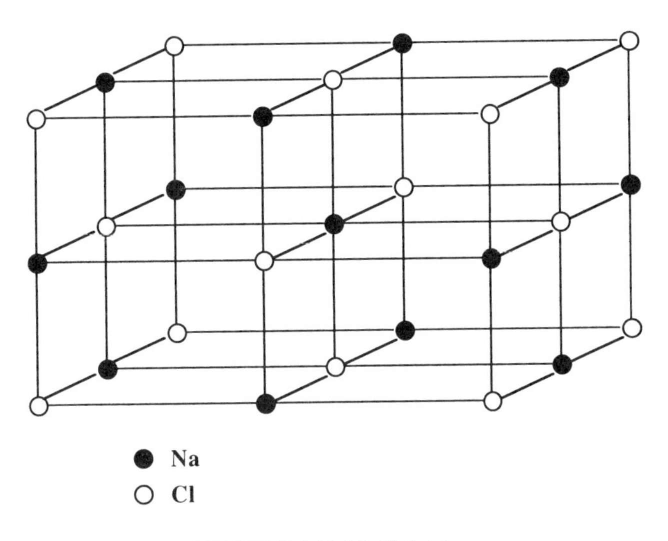

FIGURE 1-5. NaCl lattice.

a fcc structure, as do the Cl atoms of another fcc structure. Most semiconductors have another variation of the fcc structure. GaAs, the most-studied III-V compound, has the so-called *zincblende lattice*. The lattice can be regarded as two interpenetrating fcc lattices, displaced from one another along the cube diagonal by one-fourth the length of that diagonal. This is shown in Fig. 1-6, with the Ga atoms occupying some fcc lattice sites while the As atoms occupy the others. Although for clarity in picturing the zincblende structure we talked about two fcc lattices, the whole structure is regarded as one fcc unit cell. Each lattice point of the fcc unit cell is associated with two *basis atoms*, one Ga and one As. Similarly, the NaCl structure is also classified as a fcc structure, with one Na and one Cl atoms associated with each lattice point. Because multiple atoms can be associated with a given lattice point, the correct definition of a primitive unit cell is one that contains a single lattice point. The previously given definition that a primitive unit cell contains one atom is incorrect. Likewise, we were not careful when we said that there are four atoms in a fcc unit cell: We should have said that there are four lattice points.

III-V compounds such as InP and GaP also have the zincblende structure. The *lattice constant* (a) is defined as the distance at one edge of the fcc unit cell, as shown in Fig. 1-6. The value of a for GaAs is 5.6533 Å. The group IV semiconductors such as Si and Ge have the *diamond* structure. This is identical to the zincblende structure except that the two fcc lattice sites are occupied by the same atom.

A set of three integers enclosed in square brackets is used to specify direction in a crystal. Let $\hat{x}$, $\hat{y}$, and $\hat{z}$ be the unit vectors along the Cartesian coordinate axes. The integers a, b, and c in the notation $[abc]$ define a vector $a\hat{x} + b\hat{y} + c\hat{z}$,

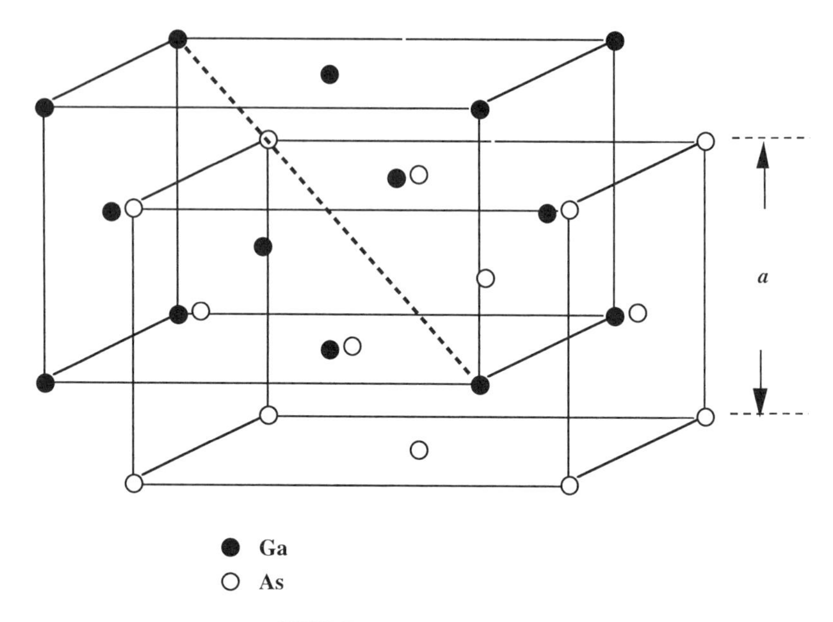

FIGURE 1-6. GaAs lattice.

which points along the given direction. The integers are chosen such that they do not have a common integral divisor. Some directions of practical importance, such as [100], [110], and [111], are shown in Fig. 1-7. If the direction points toward the negative direction, such as the opposite of the [100] direction, the symbol $\bar{1}$ is used instead of -1. For instance, the direction in the $-\hat{x}$ direction is written as [$\bar{1}$00]. The surface perpendicular to a direction [*abc*] is denoted by (*abc*), that is, with parentheses rather than square brackets. For example, the surfaces drawn in Fig. 1-7*a* are the (100) surfaces.

It is straightforward to visualize from Fig. 1-6 that Ga and As atoms form alternative layers along the [100] direction. In other words, in the [100] direction, one layer of pure Ga atoms is followed by a layer of pure As atoms, which is followed by another layer of pure Ga atoms. A GaAs monolayer grown in this direction has a thickness equal to half the lattice constant of the fcc unit cell. Although more difficult to visualize, the Ga and As atoms also form alternative layers along the [111] direction. This fact can be understood from examining the fcc structure along that direction, as shown in Fig. 1-8. Along the [110] direction, in contrast, each layer contains an equal number of Ga and As atoms. Because of the electrostatic force between the Ga and As layers, it is difficult to cleave a GaAs wafer along the [100] and [111] directions. The electrostatic force is weakest along the [110] direction and generally the GaAs wafer breaks along such a direction. Besides wafer cleavage, the distinction of various crystalline directions is important in explaining the etching properties of the wafer. This is elaborated in Chapter 7.

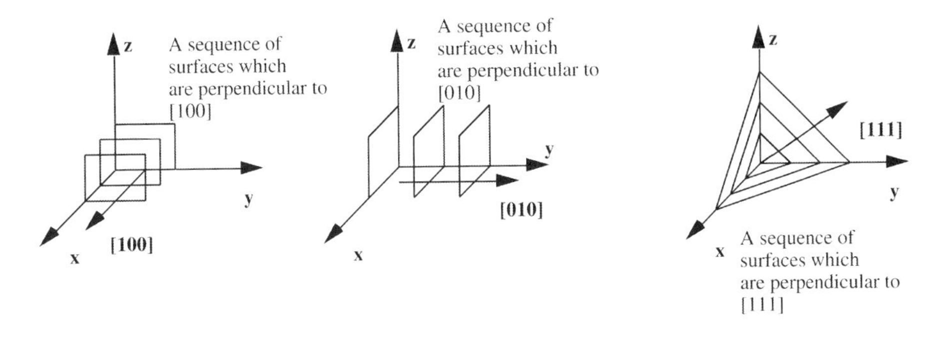

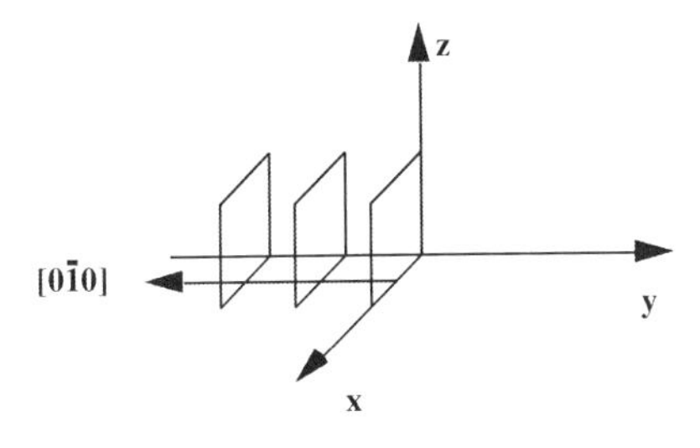

FIGURE 1-7. Some common crystalline directions.

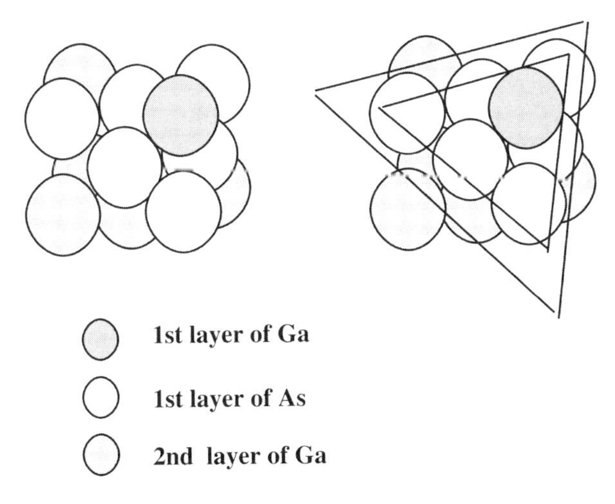

FIGURE 1-8. GaAs unit cell in the [111] direction.

Example 1-1:

The lattice constant of a silicon fcc lattice is $a = 5.431$ Å at 300 K. (1) What is the shortest distance between two Si atoms? (2) What is the maximum proportion of space that may be filled by hard spheres arranged in this crystal structure? (3) What is the volume of a primitive unit cell?

1. The shortest distance between two atoms is that between two atoms in the [111] direction. They are the two corner atoms of the two fcc unit cells that are interpenetrating, and displaced from one another along the cube diagonal by one-fourth the length of that diagonal. Since the distance between the two extreme corners along the [111] direction of the conventional unit cell is $\sqrt{3}a$, the shortest distance between two atoms is

$$\frac{\sqrt{3}a}{4} = \frac{\sqrt{3}\cdot 5.431}{4} = 2.35\ \text{Å}$$

2. The shortest distance between two atoms are $\sqrt{3}a/4$. Therefore, the radius of each atom that bumps against its nearest neighbor is equal to $r = \sqrt{3}a/8$. The volume of a sphere with radius r is equal to $4\pi r^3/3$. Finally, each fcc lattice has four atoms, implying that this conventional unit cell has a total of eight atoms. Therefore, the filling percentage is

$$\frac{8(4\pi/3)r^3}{a^3} = \frac{8(4\pi/3)(\sqrt{3}a/8)^3}{a^3} = 34\%$$

3. The diamond structure is a fcc unit cell with two basis atoms associated with each lattice point. In this case, each lattice point characterizes two atoms. We know that a fcc lattice contains four lattice points. However, a primitive unit cell has only one lattice point per cell (in this case, with two atoms again associated with such lattice point). Therefore, the volume of the primitive unit cell is

$$\frac{a^3}{4} = \frac{5.431^3}{4} = 40\ \text{Å}^3$$

Although the electrostatic force characteristic of an ionic bond was used to explain the crystalline direction along which the wafer breaks, the dominant bonding force between Ga and As atoms is the covalent bond. If we consider a nonpolar material made of a single element in Group IV (e.g., Si), then the covalent bond is the only bonding force. The ionic nature of a bonding force begins to surface in a III-V compound and becomes even more pronounced in a II-VI compound.

A covalent bond is formed when the outer electrons (the valence electrons) of the adjacent atoms are shared equally between them, rather than being transferred from one atom to another. A schematic bonding diagram is shown in Fig. 1-9. The outermost shell of each atom consists of eight electrons, with five of them contributed from an As atom (the Group V atom) and the remaining three from a Ga atom (the Group III atom). These electrons make up four bonds, and

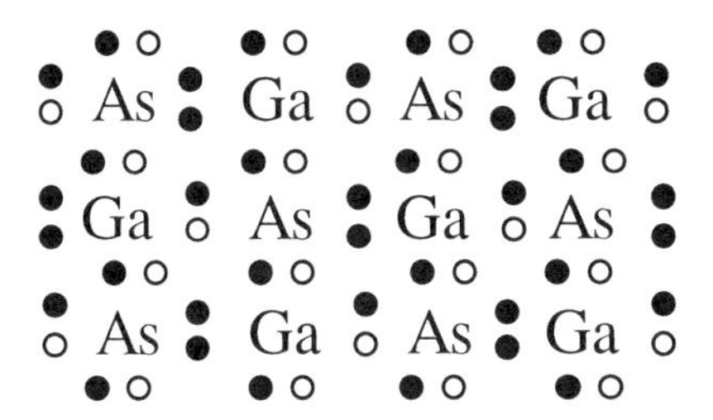

FIGURE 1-9. Schematic diagram showing the atomic bonding in GaAs.

each bond contains two electrons. The electrons in a given bond are indistinguishable from one another. They obey the Pauli exclusion principle: therefore, one electron has spin in one direction whereas the other has the opposite spin. The origin of the electron (Ga or As) is not identifiable.

At 0 K, all of the outer electrons participate in covalent bond formation, leaving no free electrons for conduction. At elevated temperature, thermal energy enables some outer electrons to be liberated from the bonding. Once they become free, they contribute to current conduction. The free electron concentration at a given temperature is called the *intrinsic carrier concentration.* As the free electron of a given bond moves away, it leaves behind a vacancy. This vacancy may be filled by a neighboring electron, an action that leaves behind another vacancy even though it fills the prior one. This process of vacancy filling and vacancy regeneration repeats as the electrons move. This motion is effectively vacancy movement, in which the vacancy hops around in the opposite direction of the actual electron movement. In this picture, the vacancy is a carrier just like the electron, and is referred to as a *hole.* A hole carries a positive charge rather than the negative charge associated with an electron. Because a vacancy is generated when an electron is liberated from the bonding, the hole concentration is equal to the intrinsic carrier concentration.

The free electrons released by thermal energy represent only a small fraction of the overall electrons, most of which remain bound to the crystal atoms. The typical intrinsic carrier concentration ranges from 10^6 to 10^{10} cm^{-3}. This is miniscule in comparison to the atomic concentration, which is on the order of 10^{23} cm^{-3}. Besides thermal excitation, there is another method to introduce mobile carriers to the semiconductor and thereby vary the conductivity of the material. Suppose that a beryllium atom is placed in an otherwise perfect GaAs lattice and occupies a Ga site. The Be atom, a member of Group II, contributes only two outer electrons to the bond with the adjacent atoms. If the Ga atom were still in place, it would have contributed three electrons. There is one less

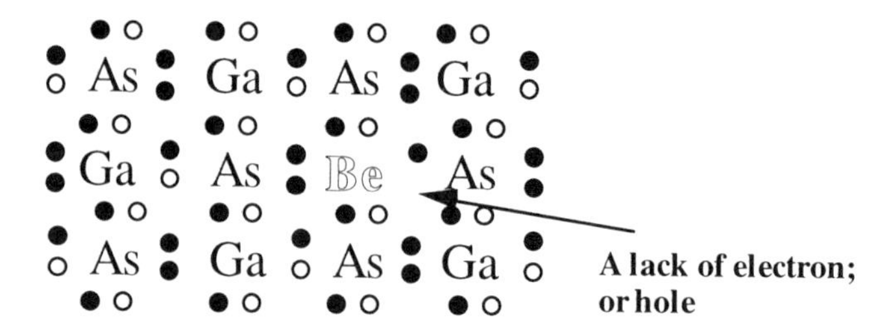

FIGURE 1-10. Atomic bonding in a *p*-type GaAs.

electron participating in the covalent bonding when a Be atom replaces a Ga atom. The deficiency of an electron can be viewed equivalently as the generation of a hole, as shown in Fig. 1-10. Therefore, with the use of an impurity such as Be, we are able to introduce holes into the semiconductor. By controlling the amount of "doping," the hole concentration is varied and the conduction property of the semiconductor is greatly modified. A dopant that gives arise to a mobile hole, such as Be in this case, is called a *p*-type dopant, or an acceptor. Conversely, a dopant that gives rise to a free electron is called a *n*-type dopant, or a donor. A doped material is said to be an extrinsic material, as opposed to an intrinsic material whose mobile carriers come entirely from thermal excitation.

Another popular dopant is silicon. When a Si atom occupies an As site, there is one less electron in the bonds connecting to the Si atom. This makes Si a *p*-type dopant, just like Be. If instead the silicon atom occupies a Ga site, there is one additional electron that is free to move in the crystal, as shown in Fig. 1-11. In this case, the Si atom is an *n*-type dopant. Whether a Si atom occupies a Ga or As site depends on the growth condition in which the dopant is incorporated into the lattice. A dopant that can act as either an acceptor or donor is said to be *amphoteric*. Under typical growth conditions, Si is an *n*-type dopant. Carbon, another Group IV element that has the potential of being amphoteric, acts predominantly as a *p*-type dopant.

When an electron moves in free space, the only potential it encounters is established by the applied electric field. In a crystal, a conducting electron encounters in addition a periodic potential brought about by the periodicity of the crystal atoms. It turns out that if we assume the electron to have an effective mass that differs from its mass in free space, we can neglect such a periodic potential. Essentially, the electron is treated as though it travels in free space and the relationship between its kinetic energy (E_k) and its momentum is identical to that encountered in elementary physics:

$$E_k = \frac{P^2}{2m_e^*} \tag{1-1}$$

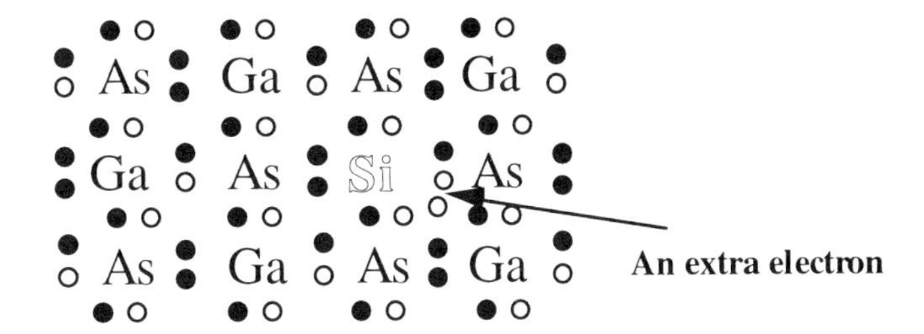

FIGURE 1-11. Atomic bonding in an *n*-type GaAs.

P is the momentum and m_e^* is the electron effective mass. The periodic crystal potential in GaAs is such that m_e^* is 0.067 times the free-space electron mass. The periodic crystal potential in Si, which differs from that of GaAs, results in an effective mass that is roughly the same as the free electron mass.

The effective mass property is just one of the differences between an electron traveling in a crystal potential and an electron traveling in free space. An electron in free space assumes a continuous range of energies, whereas an electron in a semiconductor can have only a certain values of energy. A semiconductor band diagram that illustrates the available bands of electron energy as a function of position is shown in Fig. 1-12. The valence electrons occupy the lower band of energy levels, called the *valence band* (E_v). The free electrons, in contrast, travel in the upper band, which is named the *conduction band* (E_c). Between the two bands there is a range of energy wherein no available electron state exists. The amount of this forbidden energy, the *energy gap* (E_g), is a key parameter of a semiconductor. Before a valence electron can move in the conduction band, it must acquire enough thermal energy to overcome the energy gap, which is 1.42 eV for GaAs and 1.1 eV for Si. A simple rule of thumb for the amount of thermal energy acquired by the valence electrons is kT, where k is the Boltzmann constant and T is the lattice temperature. At room temperature, $T = 300$ K, the thermal energy kT is on the order of 0.025 eV. Since the thermal energy is significantly smaller than the energy gap, we expect that only a small fraction of electrons in the valence band can acquire enough energy to jump to the conduction band.

As mentioned, another and more effective method of introducing electrons to the conduction band is to incorporate *n*-type impurities into the semiconductor. These donors introduce energy states in the forbidden energy gap, denoted as E_d in Fig. 1-12. Because the energy level E_d is within a few kT from the bottom of the conduction band edge, the electrons introduced by the dopants easily absorb enough thermal energy to jump into the conduction band at the room

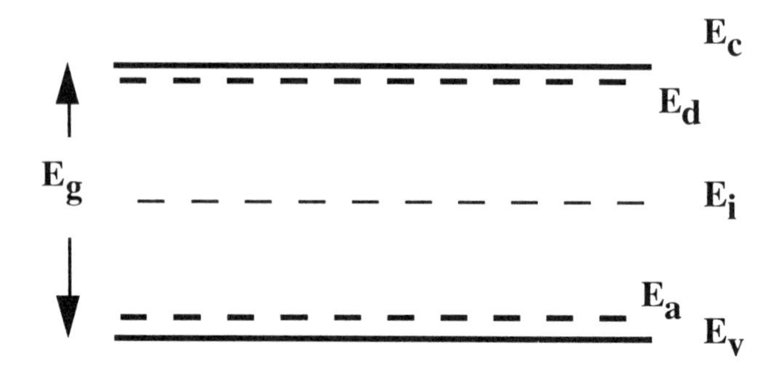

FIGURE 1-12. Semiconductor band diagram, showing the electron energy as a function of position.

temperature. However, if the ambient temperature is low enough (such as about 0 K), then the thermal energy is too small to allow the dopant electrons to make the transition from E_d to the conduction band. Under this condition, the electrons are said to be "frozen out," and they remain in the E_d energy states. Conversely, acceptors introduce an energy level E_a that is just above the top of the valence band edge. It is easy for a valence electron at room temperature to jump into such an energy state, thereby creating a hole in the valence band. This transition from the valence band to E_a, however, ceases when the temperature is low enough.

The above depiction of the donor and the acceptor levels is adequate for light and moderate doping levels. When the doping concentration is high, the dopant atoms are closely spaced and interact with each other. The interaction in *n*-type dopants, for example, causes the previously discrete energy level E_d to broaden into an impurity band with a continuous band of energy levels extending to the conduction band edge. This situation is illustrated in Fig. 1-13. The impurity level broadening produces one important result. Because E_d of Fig. 1-13 marks the lower edge of the states for the free electrons to travel, the energy gap of the material effectively decreases. The amount of energy gap narrowing depends on the amount of doping as well as on the semiconductor material itself. Another heavy doping effect relates to the carrier freeze-out at low temperatures. Since the energy separation between the original E_d state and E_c is replaced by a continuous band of states, the extra electrons introduced by the dopants are not bounded to E_d even as the temperature approaches 0 K. Hence, the carrier freeze-out disappears in heavily doped materials and current conduction persists even at low temperatures.

Conceptually, the energy diagram of an insulator is similar to that of a semiconductor. The insulator's energy states also consist of two bands separated by an energy gap. However, the magnitude of the energy gap is considerably larger than that of the semiconductor, being generally in the range of 8 eV as compared to 1 eV in semiconductors. Therefore, the amount of intrinsic electron concentration liberated from the valence band by thermal energy is practically zero. This results in the poor conductivity observed in insulators even at high temperatures. Metals, in contrast, can be viewed as materials whose conduction and valence bands overlap or are only partially filled. Because of the lack of an energy gap, the electron in the valence band moves freely in the conduction band. The conducting electrons are therefore large in number, resulting in the high conductivity associated with the metals. Schematic band diagrams for insulators and metals, as compared to that of a semiconductor, as shown in Fig. 1-14.

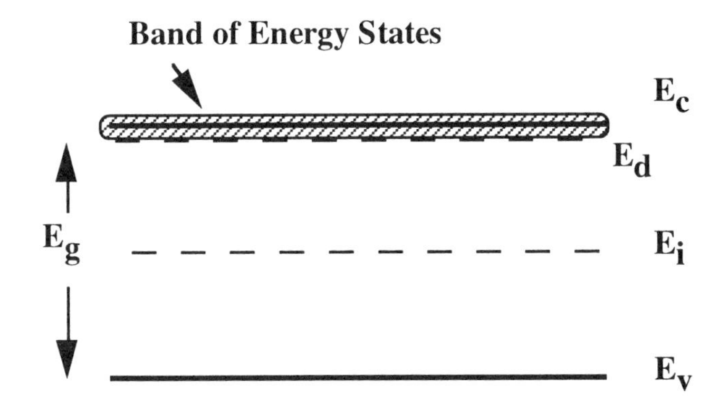

FIGURE 1-13. Band diagram of a heavily doped *n*-type semiconductor.

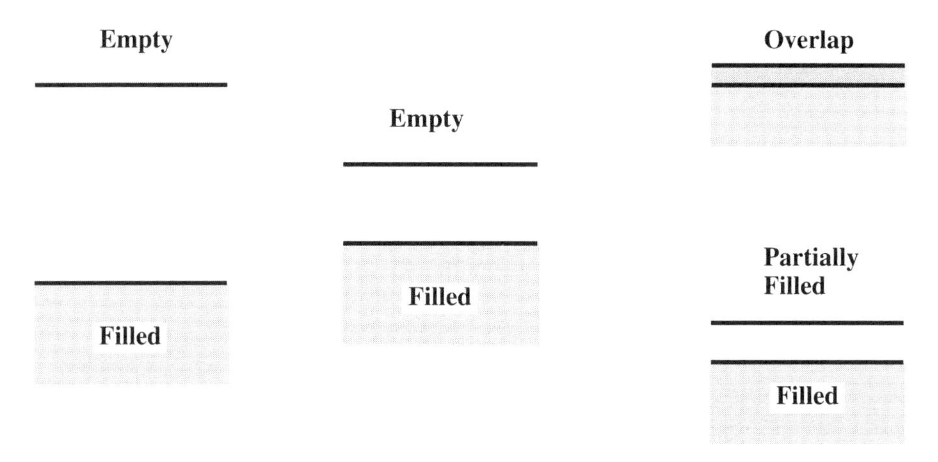

FIGURE 1-14. Band diagrams of insulator, semiconductor, and metal.

§ 1-2 MOLECULAR BEAM EPITAXY

Two growth techniques are pertinent to III-V devices. They are molecular beam epitaxy (MBE), the subject of this section, and metal-organic chemical vapor deposition (MOCVD), the subject of the next section.

MBE is a growth technique by which a specific number of atomic layers can be precisely grown. It allows the growth of abrupt interfaces, both in material composition and in intentional impurity doping profiles. The epitaxial growth takes place in an ultrahigh vacuum chamber whose base pressure is maintained at $\sim 10^{-8}$ to 10^{-10} torr. Figure 1-15 illustrates a schematic diagram of a high-vacuum chamber. A starting GaAs wafer is mounted on the substrate heater, which heats the substrate to approximately 600°C. A total of typically eight effusion cells (also known as the furnaces) containing different solid source materials face the target substrate. Each effusion cell is equipped with a shutter. After the effusion cell heats the source material to its melting temperature, the source material is emitted out of the effusion cell once the shutter is opened. Because the chamber is maintained at ultrahigh vacuum, the evaporated source

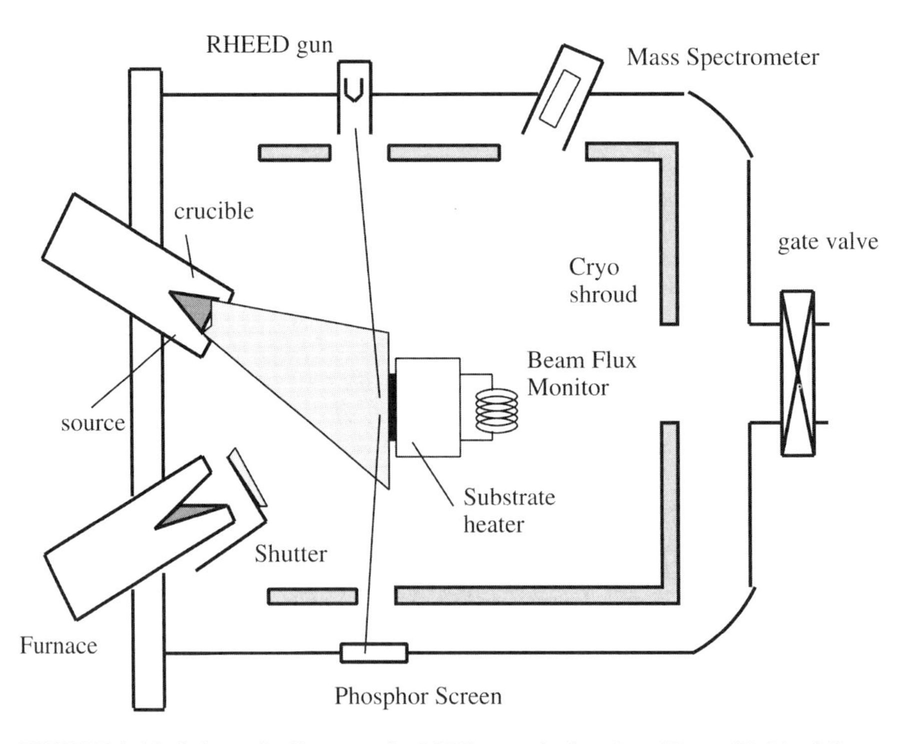

FIGURE 1-15. Schematic diagram of a MBE growth chamber. (From W. Liu, Microwave and d.c. Characterizations of *Npn* and *Pnp* HBTs. Ph.D. dissertation, Stanford University, Stanford, CA, 1991.)

molecules do not collide with each other, but form a molecular beam that impinges directly on the semiconductor substrate. The high substrate temperature promotes the movement (through diffusion) of the impinging molecules around the substrate surface, allowing them to arrange themselves into proper crystalline sites. The evaporation rate of the source material is determined by the furnace temperature, which is maintained at desirable values through a thermal couple tapped either at the bottom or around the center of the crucible. Since the switching time of the shutters is short compared to the growth time for a single atomic layer, the exact number of atomic layers to be grown is controllable. As shown in Fig. 1-15, the flux of each source material arrives at a different angle for different parts of the substrate. The substrate is rotated during growth to reduce this directional nonuniformity across the wafer.

The typical source materials used in a MBE growth are As, Ga, Al, In, Si (*n*-type dopant), and Be (*p*-type dopant). The source materials are contained in pyrolytic boron nitride (PBN) crucibles in the effusion cells. Because the temperature in parts of the effusion cells may exceed 1000°C, the effusion cells are constantly cooled by circulating a chilled water/isopropanol solution through a shroud surrounding the crucibles. The chilled coolant also provides thermal isolation among the various effusion cells. The PBN crucibles generally withstand high furnace temperatures without emitting foreign substances that might potentially contaminate the source materials. However, there has been investigation into the use other crucible materials, such as sapphire, which may be a better alternative in reducing contaminants.

The cryoshroud surrounding the interior of the ultrahigh-vacuum chamber is constantly cooled with a flow of liquid nitrogen. This cryoshroud is effective in trapping foreign molecules in the chamber, thus making the growth of the epitaxial layers less susceptible to impurities. It also adsorbs any source molecule that misses the substrate, preventing unwanted molecules from being subsequently incorporated in the epitaxial layer. Cryopumps and titanium sublimation pumps (TSPs) are also used to lower the chamber pressure. Whenever the MBE system is serviced, such as to replace the source materials, the stainless steel chamber are extensively baked and outgassed. The system is typically baked for 2–5 days at a temperature of $\sim$200°C after the source loading. A mass spectrometer is used to ensure that traces of foreign species such as H_2O, O_2, CO, and N_2 are within tolerable limits, and that the chamber is leak-tight. The use of metal-sealed gate valves eliminates leaks associated with simple O-ring-sealed valves. These practices ensure an ultrahigh vacuum in the MBE chamber, which is critical to high-quality material growth.

MBE growth takes advantage of the fact that Group V atoms impinging on the substrate surface immediately stick to the Group III atoms already there, until an atomic layer of Group V atoms is formed. Any additional Group V atoms that arrive after the formation of this atomic layer will not stick on the surface; they simply reevaporate from the surface. This is not true for Group III atoms. If there are excess Group III atoms such as Ga atoms, the excess Ga atoms remain on the surface and eventually form a "Ga-rich" surface. The crystalline structure of GaAs is then lost. Therefore, during growth of GaAs wherein both gallium and arsenic shutters remain open, an arsenic overpressure is intentionally used. Under this condition, the surface of the substrate is immediately covered with an As layer for the impinging Ga atoms to stick on. Before additional Ga atoms arrive and work to deteriorate the surface, the excess arsenic overpressure provides enough As atoms that the fresh Ga layer is immediately covered with another As layer. Since the adsorbing of As atoms on a fresh Ga surface takes negligible time, the Ga flux determines the growth rate. The growth of GaAs results in successive completion of a layer of Ga, then a layer of As, and so on. The substrate for growth is normally in the [100] orientation. In this direction, any particular single layer contains only either Ga or As atoms, consistent with the natural growth mechanism just described. Growth on [110] or [111] substrates is possible, but requires certain additional precautions.

Excessive As flux does not affect the growth rate. However, too much As is not desirable, since it unnecessarily introduces contaminants that might be present in the As source. However, experience has shown that the overpressure of the As flux needs to be at least 10 times the Ga flux to prevent the formation of a "Ga-rich" surface at a substrate temperature of 600°C. Generally, an As-to-Ga ratio of $\sim$20 is used, as measured by the beam equivalent pressure (BEP) of the flux gauge. The substrate temperature is read from a thermocouple located near the back side of the substrate. This location may not reflect the substrate's actual temperature, although it is reasonably consistent.

The growth of the $Al_\xi Ga_{1-\xi}As$ resembles the growth of GaAs. It also requires an appreciable As overpressure. The fluxes of Al and Ga are adjusted by varying the furnace temperatures such that a desired aluminum composition, ξ, is achieved. Let Gr_{GaAs} be the growth rate of GaAs when the Al shutter is closed and Gr_{AlAs} be the growth rate for AlAs when the Ga shutter is closed. The aluminum composition is approximated as

$$\xi = \frac{Gr_{AlAs}}{Gr_{AlAs} + Gr_{GaAs}} \tag{1-2}$$

and the growth rate of $Al_\xi Ga_{1-\xi}As$ is $Gr_{GaAs} + Gr_{AlAs}$. The growth of AlGaAs is especially susceptible to the incorporation of foreign species such as oxygen. This is because aluminum is quite reactive and easily reacts with oxygen to form aluminum oxide. The oxide degrades the AlGaAs crystalline quality and, consequently, the device performance. Compared to a field-effect transistor, this degradation is less important for a heterojunction bipolar transistor (HBT), since it primarily affects the d.c. current gain but not the high-frequency performance. If an AlGaAs layer of good quality is required (such as in laser diodes or in low-noise HBTs), the AlGaAs can be grown at a substrate temperature in excess of 700°C. This represents a temperature range in which significant Ga desorption occurs. That is, the Ga sticking coefficient on an As atomic layer is no longer unity as expected at lower substrate temperatures, such as 600°C. (The Al sticking coefficient can still be taken as unity near 700°C.) Therefore, the net growth rate and aluminum composition cannot be predicted with great precision from calibrations done for lower substrate temperatures. A separate calibration must be performed at the intended growth temperature.

A calibration of growth rate can be done with a reflection-high-energy-electron-diffraction (RHEED) assembly, which consists of an electron gun on one side of the chamber and a phosphorous screen on the opposite side. The central spot in the RHEED pattern is imaged onto a photodiode to observe the intensity variation of this spot. Figure 1-16 shows a typical RHEED intensity oscillation plot as monolayers of GaAs are grown. The oscillation pattern is explained as follows. Initially, when the substrate surface is fresh with no adatoms, a distinct RHEED pattern results because of the constructive interference of the diffracted electrons from the crystalline substrate surface. The intensity at the specular spot is at its maximum value. At the moment the Ga shutter is opened, Ga atoms begin to adhere to the surface. Before the completion of one Ga layer, the electrons diffracted from different parts of the substrate surface, having one layer of difference in height, interfere destructively with one another. The RHEED intensity decreases, reaching a minimum when half of a layer of Ga is deposited. Afterward, the intensity gradually recovers to its original peak value as a full layer of Ga is nearly completed. Since the deposition of the As layer takes a negligible amount of time, each oscillation period indicates a completion of monolayer of GaAs. It is ~2.83 Å for GaAs grown in the [100] direction.

One notable feature of Fig. 1-16 is that the maximum intensity decreases after each successive layer of GaAs is grown. This suggests increased roughening of the surface as more and more epitaxial layers are deposited. However, if the Ga shutter is closed for a short period of time, this peak intensity increases to its original maximum value. This finding forms the basis of the so-called *migration enhanced epitaxy* (MEE) growth technique, in which the Ga shutter alternately opens and closes so that near-perfect surface smoothness is maintained during the entire growth process. However, this rapid, continuous shuffling of shutters decreases the lifetime of the shutters and is generally not used in practice.

Another feature of the RHEED oscillation, which is not so obvious from Fig. 1-16, is that the initial period of the oscillation is slightly shorter than the later ones. This is due to the shutter transient effect. When the effusion cell of Ga is heated, for example, the Ga atoms are prevented from leaving the crucible because the shutter blocks their path. The shutter, in addition, reflects a significant amount of heat back to the crucible, which is sensed by the furnace thermocouple. The heat reflection causes the surface temperature, which controls the evaporation rate, to be elevated compared to the true furnace temperature. When the shutter is suddenly opened, the abrupt heat loss caused by the absence of heat reflection reduces the surface temperature. The thermocouple

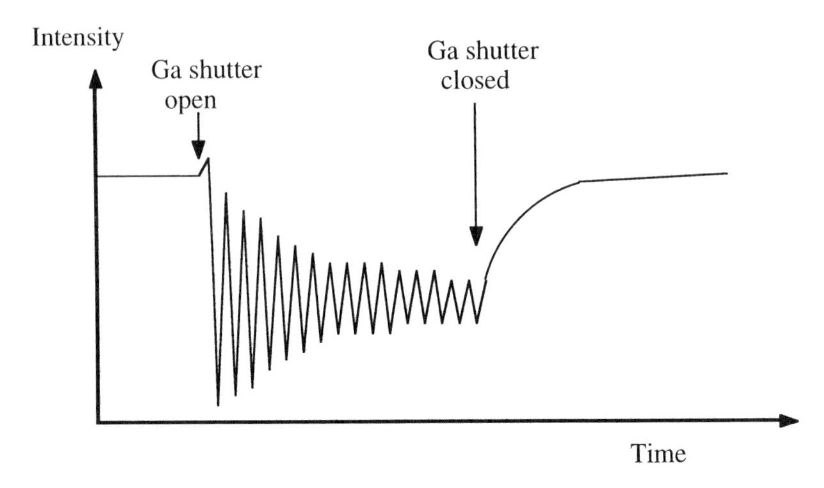

FIGURE 1-16. Schematic diagram of the RHEED oscillation waveform. (From W. Liu, Microwave and d.c. Characterizations of *Npn* and *Pnp* HBTs, Ph.D. dissertation, Stanford University, Stanford, CA, 1991.)

senses this lowered temperature, and the effusion cell heats rapidly, trying to maintain the temperature prior to the shutter opening. A large transient flux results, and the initial transient growth rate is larger than the steady growth rate. Since the initial growth rate is larger, it takes less time to grow one monolayer. Consequently, the initial oscillation period is shorter. This transient flux problem is reduced by adding a $\frac{3}{4}$-in. spacer, for instance, to the effusion cell, thereby increasing the distance between the shutter and the source crucible. This design lessens the heat reflected back by the shutters, and thus the transient flux problem.

Modern MBE growth is done with computer-controlled furnace temperatures and shutter switching. The computer control allows reproducible growth of precise layer structures. Just before each growth, the flux of each source is measured with a beam flux monitor, which is located behind the substrate, as shown in Fig. 1-15. The monitor is flipped externally into the position where the substrate is located during growth. The desired furnace temperatures are set in order to achieve the desired fluxes. This reduces the layer thickness error related to the fact that, as the source material is being expended with time, the flux of a given source material changes for a specific furnace temperature.

For the growth of epitaxial layers on the GaAs substrate, the substrate is typically etched by a $NH_4OH/H_2O_2/H_2O$ solution before being loaded into the MBE machine. When the wafer is transferred into the ultrahigh-vacuum chamber, the naturally occurring surface oxide is thermally desorbed by ramping the substrate temperature to around 700°C. Once the oxide is desorbed in an arsenic overpressured ambient, the substrate temperature is reduced and the actual epitaxial growth begins.

can be understood from Fig. 1-17. In regions where ξ is to be made small, the aluminum shutter is open for a short time; and in regions where ξ is to be made ~ 0.35, and Al shutter is open for a longer time. Grading is achieved by shuffling the Al shutter in the time domain, rather than by varying the aluminum furnace temperature directly. Since $Al_{\xi}Ga_{1-\xi}As$ has a higher energy gap than GaAs, the bottom part of the figure can be treated as the band diagram of the resulting chirped superlattice. The feasibility of chirp grading relies on the fact that carriers can tunnel through the narrow energy barriers. The resulting band diagram, which is formed after averaging out the thickness of the well and the height of the barrier, approaches the straight line shown in the top of the figure. For *Npn* transistors, each chirp period is generally 15 Å. For *Pnp* transistors, because of the hole's larger effective mass and thus greater difficulty to tunnel through energy barriers, each chirp period is roughly 5 Å.

The above discussions describes the chirping technique used to grade the aluminum composition in an $Al_{\xi}Ga_{1-\xi}As$ layer. The same technique is applicable to grade the indium composition in an $In_{\xi}Ga_{1-\xi}As$ layer, for example.

§ 1-3 METAL-ORGANIC CHEMICAL VAPOR DEPOSITION

Field effect transistor structures require precision thickness and doping controls since the device characteristics critically depend on the channel thickness and doping profile. These structures are generally grown by MBE, in which control of the shuttering actions allows the growth of abrupt interfaces. Although metal-organic chemical vapor deposition (MOCVD) can also be used to grow such abrupt interfaces, it would certainly be more difficult. For HBT structures, however, there is really no demand for any abrupt interface, at least not critical for the device operation. Even if by design an abrupt base-emitter junction is desired, a few monolayers of grading layers present at the interface have negligible consequences on the overall transistor operation. Therefore, both MBE and MOCVD are popular growth techniques for HBTs.

There seems to be a growing trend to use MOCVD to grow HBT epitaxial structures. A major reason favoring MOCVD relates to the dopant species used for the heavily doped *p*-type base layer. In MBE, the usual *p*-type dopant is beryllium, which has a high diffusion coefficient in GaAs. At a doping level above about $4 \times 10^{19} cm^{-3}$, Be tends to diffuse outward from the base layer into the adjacent emitter and collector layers. The movement into the emitter produces deleterious effects, since the emitter is made of a high-energy-gap material. If the base-emitter junction is abrupt, the Be diffusion easily compensates the emitter layer, whose doping level in HBT structures is about 100 times less than the base doping level. The overcompensation converts part of the emitter into *P*-AlGaAs. As the junction boundary moves far enough into the emitter, the *N*-*p* AlGaAs/GaAs base-emitter heterojunction gradually becomes an *N*-*P* AlGaAs/AlGaAs base-emitter homojunction. All the desirable properties associated with the heterojunction disappear and the transistor ceases to function properly. Even if the Be diffusion does not occur instantaneously during device operation (as in the case of a graded base-emitter junction), it poses a long-term reliability problem in that the device characteristics degrade over time. If MBE must be used, it is mandatory to apply special growth techniques (such as low-temperature growth) to migrate dopant diffusion problems. Success with suppression of dopant diffusion has been demonstrated.

In comparison, HBT structures grown by MOCVD are more immune to the reliability problems associated with dopant diffusion. In MOCVD, the predominant *p*-type dopant is carbon, whose diffusion coefficient is significantly lower than that of Be even at concentration levels approaching 1×10^{20} cm^{-3}. As a simple rule of thumb, the HBT high-frequency performance improves as the base doping concentration increases. This drive toward heavy base doping has led to the popularity of MOCVD, although MBE was the dominant growth technique in early HBT work and is still being used by some major companies.

It is difficult to use carbon as the dopant source in MBE. In MBE, all of the material sources Ga, As, and Al, as well as the dopant sources, are in solid form. This limits the choice of the carbon source material. Typically, if carbon is required for MBE growth, graphite is used as the source material. Since carbon's melting temperature is higher than those of other source materials, the design of the effusion cell is slightly different from that of other sources. The doping level achievable in conventional MBE growth is generally lower than in MOCVD.

The operation of MOCVD is more complicated than that of MBE, involving several chemical reactions that are often too complex to be modeled efficiently. To comprehend the basics of MOCVD, we first list the names of the chemicals involved in the reactions (see Table 1-1).

A basic chemical reaction in MOCVD is that between a metal-organic compound and a hydride. For example, when trimethylgallium vapor is mixed with arsine over a high-temperature substrate, GaAs is formed along with a gaseous by-product (methane):

$$(CH_3)_3Ga + AsH_3 \rightarrow GaAs + 3CH_4 \tag{1-3}$$

The substrate temperature in MOCVD is in the range 550 to 650°C, similar to that used in MBE. In the equation, $(CH_3)_3Ga$ is referred to as an alkyl-metal and AsH_3 as a hydride. All constituents inside the growth chamber are in the vapor phase; therefore, a mass flow controller is used to regulate the gas flow and thereby control the growth rate. Alkyl-metals such as trimethylgallium and trimethylaluminum are liquids at or near room temperature. They are contained in stainless steel bubblers, which are kept in well-controlled temperature baths to maintain certain vapor pressures of the sources over their liquids.

Purified hydrogen gas is passed through the bubbler to carry the alkyl-metal vapors to the growth chamber. A MOCVD assembly is schematically shown in Fig. 1-18.

Equation (1-3) represents the net result of the reaction between TMGa and arsine. Several subprocesses actually take place prior to producing the final

TABLE 1-1 Chemicals Involved in Metal-Organic Chemical Vapor Deposition

Source	Chemical Symbol	Abbreviation(s)
Trimethylgallium	$(CH_3)_3Ga$	TMGa, TMG
Trimethylaluminum	$(CH_3)_3Al$	TMAl, TMA
Trimethylindium	$(CH_3)_3In$	TMIn, TMI
Trimethylarsine (trimethylarsenic)	$(CH_3)_3As$	TMAs, sometimes TMA
tert-Butylarsine (tertiarybutylarsine)	$(C_4H_9)AsH_2$	*t*-BAs, TBA, TBAs
tert-Butylphosphine (tertiarybutylphosphine)	$(C_4H_9)PH_2$	*t*-BP, TBP
Triethylarsine	$(C_2H_5)_3As$	TEAs, TEA
Triethylphophine	$(C_2H_5)_3P$	TEP
Arsine	AsH_3	
Phosphine	PH_3	
Disilane	Si_2H_6	
Silane	SiH_4	
Carbon tetrachloride	CCl_4	
Diethylzinc	$(C_2H_5)_2Zn$	DEZn
Alkane	C_nH_{2n+2}	For example, methane = CH_4
Alkyl radical	C_nH_{2n+1}	For example, methyl = CH_3

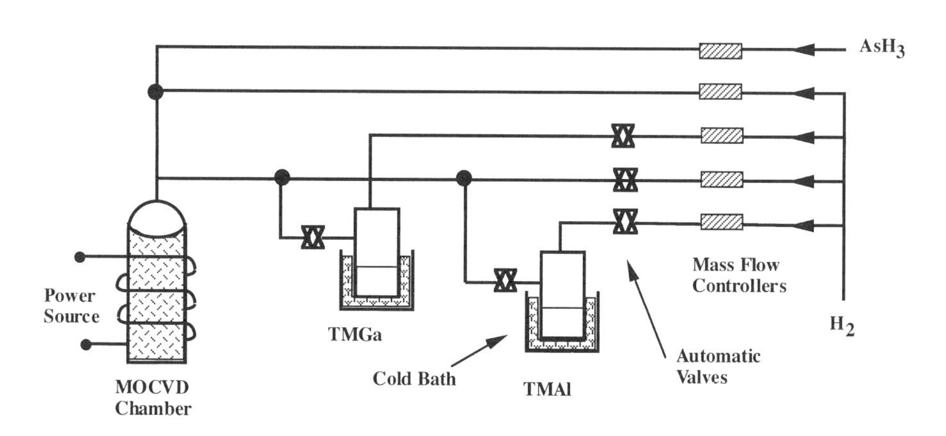

FIGURE 1-18. Schematic diagram of a MOCVD growth system.

products of GaAs and $3CH_4$. We describe some plausible subprocesses that can occur in the reactions.

$$AsH_3 \rightarrow AsH_2 + H \tag{1-4a}$$

$$(CH_3)_3Ga \rightarrow (CH_3)_2Ga + CH_3 \tag{1-4b}$$

$$(CH_3)_2Ga \rightarrow (CH_3)Ga + CH_3 \tag{1-4c}$$

$$(CH_3)Ga + AsH_2 \rightarrow GaAs + CH_4 + H \tag{1-4d}$$

It is readily verifiable that the net result of the above subprocesses is consistent with Eq. (1-3). These events in which a molecule is broken down into unstable elemental forms are brought about by the presence of heat energy; hence, they are referred to as *pyrolytic reactions*.

One observation from Eq. (1-4) is that there are a lot of hydrogen atoms present in the MOCVD growth. These hydrogen atoms, when incorporated in the GaAs epitaxial layer, usually do not introduce recombination sites as other impurities do. Instead, under certain growth conditions they can electrically inactivate the silicon dopants or the carbon dopants by forming complexes with them. The change of current gain after biasing the device with high current and voltage is a device phenomenon attributable to the hydrogen compensation. There have also been some links between HBT long-term reliability and the amount of these hydrogen-dopant complexes. The hydrogen compensation is an issue of concern when MOCVD is used to grow HBT epitaxial layers.

Sometimes *tert*-butylarsine replaces arsine as the arsenic source in MOCVD growth. All the reactants at the beginning of the reaction are then organometallic, without any hydride. The main advantage of *t*-BAs over AsH_3 is the reduced hazard. In addition, *t*-BAs, which is in liquid form, is easier to purify and maintain in the purified form than the gaseous AsH_3, thus leading to less impurities during growth. Although the reaction becomes more complicated with *t*-BAs, we may simplify it as follows. During the beginning stage of the reaction, AsH_2 is formed by the decomposition of *t*-BAs. With the production of AsH_2, the subsequent subprocesses are then the same as those outlined in Eqs. (1-4b)–(1-4d).

We mentioned previously that in MBE growth, the As atoms impinging on the GaAs surface immediately stick to the Ga atoms already there, until an atomic layer of As is formed. Any additional As atoms that arrive after the formation of the atomic layer will not stick but simply reevaporate from the surface. Having excess As fluxes ensures that MBE growth is determined solely by the Ga atom flux and prevents the growth surface from entering the Ga-enriched condition. Similarly, in MOCVD growth, the growth proceeds in an environment with excess AsH_3. The growth rate of GaAs within a specified temperature range is limited by the diffusion of TMGa to the wafer surface and becomes linearly dependent on the TMGa concentration. As a first-order approximation, we write the GaAs growth rate as

$$\mathrm{Gr}_{\mathrm{GaAs}} \propto p_{\mathrm{Ga}} \cdot F_{\mathrm{Ga}} \tag{1-5}$$

where P_{Ga} is the vapor pressure of TMGa and F_{Ga} is the flow rate (volume/time) of the hydrogen through the TMGa bubbler. These two parameters are the most important factors controlling the GaAs growth rate in MOCVD, unlike MBE growth in which gallium furnace temperature is the dominant factor. The growth rate of AlAs is similar, except with a multiplication factor of 2 to reflect the fact that TMA1 is a dimer under typical pressures and temperatures in the bubbler:

$$\mathrm{Gr}_{\mathrm{AlAs}} \propto 2p_{\mathrm{Al}} \cdot F_{\mathrm{Al}} \tag{1-6}$$

When both TMGa and TMAl are flowing into the reaction chamber, the resulting growth rate of $\mathrm{Al}_{\xi}\mathrm{Ga}_{1-\xi}\mathrm{As}$ is simply $\mathrm{Gr}_{\mathrm{AlAs}} + \mathrm{Gr}_{\mathrm{GaAs}}$. The aluminium mole fraction is in principle identical to Eq. (1-2) given for MBE growth:

$$\xi = \frac{\mathrm{Gr}_{\mathrm{AlAs}}}{\mathrm{Gr}_{\mathrm{AlAs}} + \mathrm{Gr}_{\mathrm{GaAs}}} \simeq \frac{2p_{\mathrm{Al}}F_{\mathrm{Al}}}{2p_{\mathrm{Al}}F_{\mathrm{Al}} + p_{\mathrm{Ga}}F_{\mathrm{Ga}}} \tag{1-7}$$

We have added the approximate sign because realistically, the proportional factors appearing in Eqs. (1-5) and (1-6) are not necessarily exact.

Forming a truly abrupt interface, such as an abrupt GaAs/AlGaAs heterojunction, is more difficult in MOCVD than in MBE. The key reason for the difference is that the chamber pressure is high in MOCVD. Any pressure imbalance between lines during switching of gases may cause a rush of gases. Although purging the gas line relieves the problem somewhat, it is not a panacea because gas purging does not occur instantaneously. Even after a certain gas flow is switched off, a small fraction of the gas remains in the line and the chamber for some time. The high growth rate associated with MOCVD makes this problem even more severe. (The MOCVD growth rate is ~ 10 μm/h, as opposed to 0.5–1 μm/h in MBE.) Therefore, in an attempt to grow an "abrupt" GaAs/AlGaAs heterojunction in MOCVD, some monolayers of grading are believed to exist despite earnest attempts to prevent them. This natural grading has no major device implications for HBTs, and the junction is practically an abrupt heterojunction from the device operation's point view. For resonant tunneling diodes, however, the finite grading that is not easily avoided in MOCVD can alter the device's characteristics deleteriously, since the tunneling probability has an exponential dependence on the thickness of the tunneling barrier.

Disilane ($\mathrm{Si_2H_6}$) and silane ($\mathrm{SiH_4}$) are the usual choices for the n-type dopant of GaAs in MOCVD growth. The silicon is incorporated into the lattice, occupying gallium sites and giving off free electrons. The by-product of the dopant incorporation is hydrogen molecules, which are pumped away from the

MOCVD chamber. This description is simplistic; the actual reaction is considerably more complex, and we do not get into the details. We state that the complex nature is responsible for the observed difference in the doping efficiency between growth with AsH_3 and with *t*-BAs.

The usual *p*-type dopant is carbon tetrachloride (CCl_4). Unlike the *n*-type dopant, the by-product (Cl_2) during the incorporation of carbon into GaAs is rather reactive. Chlorine etches GaAs, forming mostly GaCl. It also reacts with hydrogen, producing HCl, which etches GaAs. In addition, chlorine reacts with the hydrocarbon in the growth chamber and forms $CH_{\xi}Cl_{4-\xi}$ (where $\xi = 1, 2,$ or 3) compounds, grabbing the carbon atoms that would otherwise be incorporated into the lattice and act as doping atoms. Therefore, a MOCVD growth using CCl_4 can be tricky.

Other *p*-type dopants include diethylzinc (DEZn) and trimethylaresine (TMAs). The zinc in DEZn is the element that actually dopes the semiconductor. Zinc, unlike carbon, has a relatively high diffusion coefficient, similar to Be used in MBE. Device reliability becomes an issue when DEZn is used as the dopant. Often the device characteristics can drift with time, as a Zn-doped HBT is biased continuously. Prior to the use of carbon doping, such a dopant was the available choice. At the present, it is not much used. TMAs is an interesting carbon-dopant alternative to CCl_4. In Table 1-1 it was listed along with arsine, TEAs, and *t*-BAs because it is also known as a source of arsenic. The chemical makeup of TMAs is such that it releases carbon during the MOCVD reaction, acting as a source of carbon besides giving off As. Therefore, it is possible that without any other dopant in the chamber, the use of TMAs will result in a *p*-type GaAs layer. Although CCl_4 is a popular carbon dopant, TMAs remains in use. Trimethylarsine is sometimes called trimethylarsenic. Its abbreviation can be confusing at times. It is often written as TMAs, but it has also been written as TMA, which could be confused with trimethylaluminum.

Just as in MBE, the preferred wafer orientation for MOCVD growth is along the [100] direction. In-situ growth monitors such as RHEED are not available in MOCVD. Therefore, calibration of the epitaxy growth rate in a MOCVD run typically requires wet etching of the calibration samples. The growth rate is inferred from the measured layer thickness divided by the growth time.

§ 1-4 LATTICE-MISMATCHED LAYERS

Even though the primary epitaxial layers of the GaAs-based HBTs are either GaAs or AlGaAs, it is sometimes advantageous to insert thin layers of InGaAs in a device structure. In growing AlGaAs layers, we do not worry about the possible generation of misfit dislocations, since the lattice constants for AlAs and for GaAs differ by only 0.14%. In contrast, the lattice constant of InAs is 6.0584 Å, which is 62% larger than that of GaAs. Growing a thick layer of InAs directly on a GaAs substrate inevitably generates a high density of misfit dislocations (and the material eventually becomes amorphous). This was the same problem that prevented the early heterostructures based on other material systems from becoming functional.

If the thickness of an InGaAs layer grown on GaAs is kept below a certain critical thickness, the strain generated due to lattice mismatch can be absorbed entirely by the lattice. The resulting growth is then free of dislocations, with the lattice constant of the grown InGaAs layer being the same as that of the GaAs layer underneath. Since lattice mismatch does not occur in strained layers, subsequent growth of GaAs or AlGaAs layers on top does not suffer from degraded qualities. The exact critical thickness of an $In_{\xi}Ga_{1-\xi}As$ layer depends on the indium mole fraction, ξ, as well as the growth conditions. Many theoretical and experimental studies have been directed toward finding the critical thickness. Generally, 300 Å is well below the critical thickness for an indium composition fraction of 15%. This strained layer does not degrade upon thermal cycling up to 300°C. It is also believed that the critical thickness is two monolayers for InAs and $\sim$40 Å for $In_{0.5}Ga_{0.5}As$.

Strained InGaAs layers are useful for applications that do not require thick layers. One example is the base layer in a *Pnp* HBT. Since this layer is designed to be only $\sim$300 Å, it can be made of strained InGaAs as long as the indium composition is below 15%. The strained InGaAs base layer has two distinct advantages over the conventional GaAs layer. One relates to the fact that InGaAs layer allows a heavier doping level, allowing the base resistance of the *Pnp* HBT to be reduced. Another advantage relates to the availability of a selective etching solution that etches AlGaAs and GaAs but stops at InGaAs, simplifying the fabrication of the HBTs with an InGaAs base. Strained InGaAs layer also finds application in field-effect transistors, for use as the electron channel layer. This is discussed in § 5-6.

Unstrained InGaAs layers with thicknesses exceeding the critical thicknesses are also used in devices. As the misfit dislocations are generated, the lattice constants of the unstrained layers relax to their natural bulk values. Subsequent GaAs or AlGaAs layers grown on top will be amorphous. Therefore, unstrained layers are not suitable as part of the active device layers. If they are used, unstrained InGaAs layers are found exclusively as the cap layer that appears at the top of a HBT or a field-effect transistor. Because it is the last layer to be grown, concerns about subsequent amorphous layer growth on top are immaterial. More important, InGaAs layers allow a heavier doping level than conventional GaAs. The unstrained InGaAs cap layers therefore reduce the emitter or the drain/source contact resistance dramatically.

§ 1-5 BASIC DEVICE PHYSICS

To predict quantitatively the electrical behaviour of a semiconductor device, we need to know the carrier distributions and express them in appropriate equations. Generally, in this book we attempt to describe the derivation or physical origin of every equation. The derivation can be based on experimental data, in which case the equation is phenomenological. The derivation can also be based on some equations, which in turn are derived from other, more fundamental equations. However, we need to end the loop of derivation somewhere, simply taking some basic equations as axioms. In this section, many equations are simply given without derivation. The derivations of these fundamental device physics equations are found readily in elementary textbooks.

The first two fundamental equations relate to electrostatics:

$$\frac{d\varepsilon}{dx} = \frac{q}{\epsilon_s}(p - n + N_d - N_a) \tag{1-8}$$

$$\varepsilon = -\frac{dV}{dx} \tag{1-9}$$

Equation (1-8) states that the derivative of the electric field (ε) with respect to the distance (x) is equal to the net charge concentration at that location divided by the dielectric constant of the semiconductor material (ϵ_s). For GaAs, the relative dielectric constant is 13.1. Since the free-space dielectric constant (ϵ_0) is 8.85×10^{-14} F/cm, ϵ_s; which is the product of the relative dielectric constant and the free-space dielectric constant, is equal to 1.159×10^{-12} F/cm. The net charge is the electron charge (q) times the carrier concentration, which is equal to the sum of the free hole concentration (p) and the donor doping density (N_d) minus the sum of the free electron (n) and the acceptor doping density (N_a). Like ϵ_0, q is a universal constant; it is equal to 1.6×10^{-19} C.

Equation (1-9) defines the electric potential (V) as a scalar quantity such that the negative of its gradient is equal to the electric field. Its physical significance relates to the work done in carrying a charge from one point to another. The Poisson equation is obtained when the electric field ε in Eq. (1-8) is replaced by $-dV/dx$ given in Eq. (1-9). The poisson equation relates the electric potential directly to the net charge concentration. In this book, however, we refer to both Eqs. (1-8) and (1-9) as the Poisson equations.

When working with a device composed of several distinct semiconductor materials, there is a possibility that Eq. (1-9) becomes incorrect. As is, the electric field is due purely to the space charges. It does not include a "quasi-electric field" component ($\varepsilon_{\text{quasi}}$), which accounts for the contribution from the gradient of the electron affinity (χ_e). The electron affinity of a material is the energy required to bring a free electron at the conduction band edge to the outside of the semiconductor (vacuum level). It is a constant in a given material but varies when different materials are joined together. We shall refer to the field component due to the space charges as the Poisson electric field, $\varepsilon_{\text{Poisson}}$. Therefore, we should have written Eq. (1-9) as

$$\frac{d\varepsilon_{\text{Poisson}}}{dx} = \frac{q}{\epsilon_s}(p - n + N_d - N_a) \tag{1-10}$$

However, we wrote Eq. (1-9) the way we did because that is the form with which most readers are familiar. We just need to be a little bit careful in applying Eq. (1-8) when the problem under study involves with a nonzero quasi-electric field. Expressed mathematically, the quasi-electric field is

$$\varepsilon_{\text{quasi}} = -\frac{d\chi_e}{dx} \tag{1-11}$$

The total electric field is the sum of the two components:

$$\varepsilon = \varepsilon_{\text{Poisson}} + \varepsilon_{\text{quasi}} \tag{1-12}$$

We cite an example illustrating the importance of the quasi-electric field in device applications. Consider an AlGaAs base layer in which the aluminium mole fraction is varied spatially. Because the electron affinity is not constant with position, a quasi-electric field is established. The overall electric field is nonzero, even though the entire base region remains charge neutral so that the electrostatic potential V is zero. As far as the current conduction is concerned, a quasi-electric field is every bit as good as a Poisson electric field. It moves electrons in the opposite direction while pushing holes along its direction. Therefore, if the electron affinity variation is designed correctly, the quasi-electric field can be used to fasten the minority carrier movement in the base and thereby reduce the base transit time.

A basic equation governing the numbers of electron and hole carriers is given by

$$p \cdot n = n_i^2 = N_c N_v \exp\left(-\frac{E_g}{kT}\right) \tag{1-13}$$

This equation holds for either extrinsic or intrinsic materials, although it may fail in heavily doped samples (see problem 12). For convenience, we will neglect such a subtlety and apply the relationship in heavily doped samples. Equation (1-13) states that under thermal equilibrium, the product of the free-electron carrier concentration (n) and the free-hole concentration (p) is a constant for a given material, equal to the square of the intrinsic carrier concentration (n_i). It turns out that n_i^2 for a given semiconductor has an inversely exponential dependence on its energy gap, with the proportional constant being the product of the conduction band density of states (N_c) and the valence-band density of states (N_v). Generally, N_c and N_v do not vary significantly from one semiconductor to another. Therefore, the difference in the intrinsic carrier concentrations of various semiconductors results primarily from the difference in the energy gap.

Example 1-2:

Compare the intrinsic carrier concentrations of GaAs and Si at room temperature. For GaAs, $N_c = 4.7 \times 10^{17}$ cm^{-3}, $N_v = 7.0 \times 10^{18}$ cm^{-3}, and $E_g = 1.424$ eV. For Si, $N_c = 2.8 \times 10^{19}$ cm^{-3}, $N_v = 1.04 \times 10^{19}$ cm^{-3}, and $E_g = 1.12$ eV.

When the lattice temperature (T) is at room temperature of 298 K, n_i^2 according to Eq. (1.13) is $(4.7 \times 10^{17})(7.0 \times 10^{18})\exp[-1.424/(298 \cdot 8.629 \times 10^{-5})]$. Equivalently, $n_i = 1.79 \times 10^6\ \text{cm}^{-3}$. For silicon, for which $E_g = 1.12\ \text{eV}$, the intrinsic carrier concentration at room temperature is $1.45 \times 10^{10} \text{cm}^{-3}$, which is roughly four orders of magnitude greater than that of GaAs.

The intrinsic carrier concentration depends strongly on temperature. If the temperature dependencies of N_c and N_v are neglected, n_i in GaAs increases by six orders of magnitude of about $10^{12}\ \text{cm}^{-3}$ at $T = 600$ K. For Si under such an operating temperature, n_i has a value on the order of $10^{14}\ \text{cm}^{-3}$, which approaches the practical doping levels used in silicon transistors.

Most often a semiconductor is intentionally doped. If it is doped as n-type, donor impurities introduce an energy level E_d in the forbidden gap. For most practical donors, E_d is within only few kT of the conduction band edge such that all of the extra electrons acquire enough thermal energy to move to the conduction band at room temperature. Therefore, the equilibrium free electron concentration (n_0) is taken to be equal to the doping density N_d. Since the $n \cdot p$ product is n_i^2 (Eq. 1-13), the equilibrium free-hole electron concentration (p_0) is equal to n_i^2/N_d:

$$n_0 = N_d \quad p_0 = \frac{n_i^2}{N_d} \qquad (n\text{-type material}) \tag{1-14}$$

Conversely, for p-type material, the hole concentration can be assumed to equal to the p-type doping density N_a. We therefore have

$$p_0 = N_a \quad n_0 = \frac{n_i^2}{N_a} \qquad (p\text{-type material}) \tag{1-15}$$

Although seemingly obvious, these statements about Eq. (1-14) are correct only if $N_d \gg n_i$. If the doping concentration for some reason does not deviate strongly from the intrinsic carrier concentration (such as when the device operates under high operating temperatures), then a more accurate expression should be used. In the absence of an external bias, the electric field is zero. According to Eq. (1-8), we have in a n-type material in which $N_a = 0$,

$$n = p + N_d \qquad (n\text{-type material}) \tag{1-16}$$

There are two unknowns, n and p. The second equation that enables us to establish the two unknowns is Eq. (1-13), $p \cdot n = n_i^2$. Solving these two equations together, we find that under thermal equilibrium n and p are

$$n = n_0 = \frac{\sqrt{N_d^2 + 4n_i^2} + N_d}{2} \quad (n\text{-type material}) \tag{1-17}$$

$$p = p_0 = \frac{\sqrt{N_d^2 + 4n_i^2} - N_d}{2} \quad (n\text{-type material}) \tag{1-18}$$

It is easily verified that these two equations reduce to Eq. (1-14) when $N_d \gg 4n_i^2$. A similar derivation obtains for p-type material:

$$p = p_0 = \frac{\sqrt{N_a^2 + 4n_i^2} + N_a}{2} \quad (p\text{-type material}) \tag{1-19}$$

$$n = n_0 = \frac{\sqrt{N_a^2 + 4n_i^2} - N_a}{2} \quad (p\text{-type material}) \tag{1-20}$$

Once the electron and hole concentrations are ascertained from Eqs. (1-14) and (1-15), we can determine the Fermi level (E_f) of the semiconductor during thermal equilibrium. The Fermi level is derived from statistical mechanics, and represents the chemical potential of the electrons in the semiconductor system. The chemical potential governs the flow of particles between systems, just as the temperature governs the flow of energy. This concept is elaborated on in Chapter 2, where we analyze a p-n junction consisted of two dissimilar systems. For the present discussion, we treat the Fermi level to be an energy level that is constant in the entire semiconductor. The Fermi level in an n-type layer relates to the electron concentration according to

$$n = n_i \exp\left(\frac{E_f - E_i}{kT}\right) = N_c \exp\left(\frac{E_f - E_c}{kT}\right) \tag{1-21}$$

Similarly, the Fermi level in a p-type layer relates to the hole concentration according to

$$p = n_i \exp\left(\frac{E_i - E_f}{kT}\right) = N_v \exp\left(\frac{E_v - F_f}{kT}\right) \tag{1-22}$$

These two equations allow us to locate the Fermi relative to either the conduction band edge (E_c), the valence band edge (E_v), or the intrinsic energy level (E_i). E_i is approximately but not exactly equal to $(E_c + E_v)/2$, due to the fact that N_c is not exactly equal to N_v. In this book, we minimize the use of E_i and usually make the energy reference with respect to either E_c or E_v.

The energy of an electron is measured by the energy difference with respect to the Fermi level, as shown in Fig.1-19. The separation of the Fermi level from the

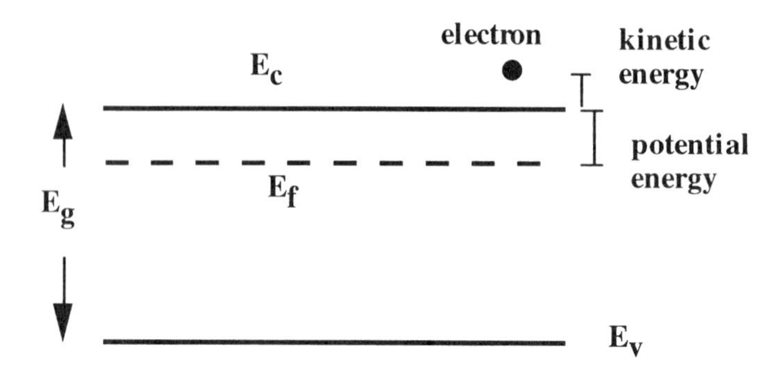

FIGURE 1-19. Band diagram showing the electron potential energy and kinetic energy.

conduction band edge is taken as the potential energy of an electron, while the amount of energy above the conduction band edge represents the kinetic energy.

N_c and N_v for major semiconductors are well known and are easily found in tables. For $Al_\xi Ga_{1-\xi}As$ material, however, N_c and N_v are generally calculated indirectly from some more fundamental formula, because listing values for all possible variations of ξ between 0 and 1 is impractical. The conduction band density of states is expressed in terms of the density of state effective electron mass, m^*_{de}, which is generally different from the conduction effective mass discussed later. The relationship, based on a derivation for ideal gas particles, is given by

$$N_c = 2\left(\frac{2\pi kT}{h^2}\right)^{3/2} m^{*3/2}_{de} \tag{1-23}$$

where h is the Planck constant, equal to 6.626×10^{-34} J-s. The effective mass in $Al_\xi Ga_{1-\xi}As$ was determined to have the following dependence:

$$m^*_{de}(\xi) = 0.067 + 0.083\xi \quad 0 \leq \xi \leq 0.45 \tag{1-24}$$

This equation applies to $\xi < 0.45$, representing the density of states effective mass for Γ band electrons. When $\xi > 0.45$, AlGaAs is an indirect energy gap material, and the minimum conduction band edge is located at the X valley. Because the constant-energy surface for the X valley is an ellipsoid rather than a sphere as in the Γ valley, the effective mass equation assumes a different form. We discuss the details in § 1-7. Once m^*_{de} is determined from Eq. (1-24), N_c as a function of the aluminum mole fraction can be calculated.

The valence band, for the purpose of estimating N_v, can be viewed as consisting of two degenerate bands. One band is characterized by a light hole effective mass (m^*_{lh}) and another is characterized by a heavy hole effective mass (m^*_{hh}). Their dependence on the aluminum mole fraction (for $0 \le \xi \le 1$) is given by

$$m^*_{lh}(\xi) = 0.087 + 0.063\xi \tag{1-25}$$

$$m^*_{hh}(\xi) = 0.62 + 0.14\xi \tag{1-26}$$

Once the density state effective hole masses are known, the valence band density of states is calculated from

$$N_v = 2\left(\frac{2\pi kT}{h^2}\right)^{3/2}\left(m^{*3/2}_{lh} + m^{*3/2}_{hh}\right) \tag{1-27}$$

The equations for the effective masses of GaAs-based materials, InP-based materials, as well as Si, are found in § 1-7.

Example 1-3:

Consider an n-type $Al_{0.35}Ga_{0.65}As$ layer in which the free electron concentration is $1 \times 10^{16}\ \text{cm}^{-3}$. Find the energy difference between the Fermi level and the conduction band edge. Draw a band diagram. The energy gap of $Al_{0.35}Ga_{0.65}As$ is 1.86 eV.

The aluminum mole fraction is $\xi = 0.35$. According to Eq. (1-24), the density of states effective electron mass is 0.09605 times the free electron mass. From Eq. (1-23), the density of states at $T = 300$ K is calculated as

$$N_c = 2\left[\frac{2\pi(1.38 \times 10^{-23})300}{(6.624 \times 10^{-34})^2}\right]^{3/2}\left[0.09605(9.1 \times 10^{-31})\right]^{3/2} = 3.72 \times 10^{23}\ \text{m}^{-3}$$

Note that the above calculation is carried out in MKS units, since it involves mass unit conversion. Generally, the cgs system is the system of choice for semiconductor physics calculation, and it is used throughout the rest of the examples in this book, unless otherwise noted. Before proceeding further, we convert the previous result to cgs units:

$$N_c = 3.72 \times 10^{17}\text{cm}^{-3}$$

According to Eq. (1-21), $E_c - E_f = kT\ln(N_c/n) = kT\ln[3.72 \times 10^{17})/(1 \times 10^{16})] = 0.093$ eV at $T = 300$ K. Figure 1-20 shows the band diagram of this n-type material, indicating the relative position of the Fermi level with respect to the conduction band edge.

In the band diagram we also drew the vacuum energy level (E_0), which represents the energy that an electron would have if it were just free of the

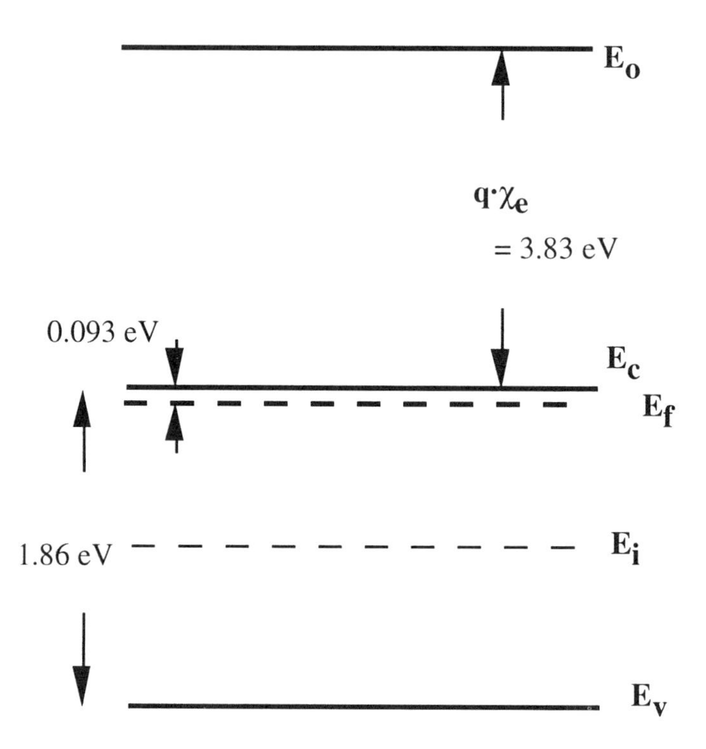

FIGURE 1-20. Band diagram of the *n*-type $Al_{0.35}Ga_{0.65}As$ layer discussed in Example 1-3.

influence of the given material. When two distinct materials are joined together, the conduction band and valence band edges may be discontinuous due to the band discontinuities between the two materials. However, the vacuum energy level is always continuous, even in a heteromaterial system. If it were not a single-valued function in space, we could conceive means to extract work from an equilibrium situation by emitting electrons and then reabsorbing them at an infinitesimal distance where the vacuum level had changed value. The electron affinity when multiplied by an electron charge, $q \cdot \chi_e$, specifies the energy required to release an electron from the bottom of the conduction band to the vacuum level. The value of $q \cdot \chi_e$ varies from 4.07 eV in GaAs to 3.83 eV in $Al_{0.35}Ga_{0.65}As$.

Two crucial assumptions leading to Eqs. (1-21)–(1-22) are that there exists a negligible amount of energy states inside the energy gap, and the probability of the electron occupancy in an energy level E_n above E_c (or the hole occupancy in an energy level below E_v) is determined by the so-called Maxwell–Boltzmann statistics. The first assumption is generally correct; the number of states at E_d or E_a introduced by dopants is negligible compared to the number of states in the conduction and valence bands. The second assumption, in contrast, can be questionable at times. The Maxwell–Boltzmann statistics treat the electrons as distinguishable particles and allow any number of electrons to occupy the same energy state. Given such an assumption, the probability that an energy level E_n is occupied by an electron can be shown to be $\exp[(E_f - E_n)/kT]$ [1]. Realistically, it is impossible to distinguish one electron from another. Furthermore, the electrons obey the Pauli exclusion principle, which states that no two electrons may occupy the same energy state simultaneously. The statistics accounting for these two restrictions are called the Fermi–Dirac statistics, from which an accurate expression for the probability of the electron occupancy of an energy level E_n is derived. It is found to be $1/\{1 + \exp[(E_n - E_f)]/kT]\}$. Comparing this expression to the Maxwell–Boltzmann expression, we find that the latter is a good approximation of the Fermi–Dirac expression only when $\exp[(E_f - E_n)]/kT] \ll 1$. Since the lowest available energy state in the conduction band is the conduction band edge, we replace E_n by E_c, so the condition for the Maxwell–Boltzmann statistics to work then becomes $\exp[(E_f - E_c)]/kT] \ll 1$ or, equivalently, $\exp[(E_c - E_f)]/kT] \gg 1$. This condition requires that $E_c - E_f > 3kT$. In Example 1-3, we found that the Fermi level of n-type $Al_{0.35}Ga_{0.65}As$ doped at 1×10^{16} cm^{-3} is such that $E_c - E_f = 0.093$ eV. This amounts to about $3.6kT$, which is larger than $3kT$. Therefore, the Maxwell–Boltzmann approximation embodied in Eq. (1-21), which was used to calculate the position of E_f, is fairly accurate. If $E_c - E_f$ becomes $< 3kT$ due to heavy doping, more accurate expressions than those of Eqs. (1-21)–(1-22) need to be used, albeit these equations are more complicated. The degree of complexity depends on the required accuracy to approximate the following integral:

$$\frac{n}{N_c} = \frac{2}{\sqrt{\pi}} \int_0^{\infty} \frac{\sqrt{x}}{1 + \exp[x - (E_f - E_c)/kT]} dx \tag{1-28}$$

where $x = (E_n - E_c)/kT$. Since the equation is obtained from the Fermi–Dirac statistics, the integral is called the Fermi–Dirac integral. Equation (1-28) is the ultimate equation relating the Fermi level to the free electron concentration. The integral is, nonetheless, not easy to deal with because the integration cannot be carried out analytically. With a numerical method, this integral is evaluated as shown in Fig. 1-21. If $E_f - E_c \ll 3kT$, then the integral can be approximated as an integration of $x^{1/2}\exp(-x)$, which leads to the simplified equation given in Eq. (1-21). Figure 1-21 demonstrates that the Maxwell–Boltzmann approximation is indeed accurate as long as $E_f - E_c \ll 3kT$. For $E_f - E_c > 3kT$, Eq. (1-21) becomes erroneous. Under this condition, an expression based on the polynomial expansion can be used to approximate the integral. It is known as the Joyce–Dixon approximation:

$$\frac{E_f - E_c}{kT} = \ln\left(\frac{n}{N_c}\right) + A_1\left(\frac{n}{N_c}\right) + A_2\left(\frac{n}{N_c}\right)^2 + A_3\left(\frac{n}{N_c}\right)^3 + A_4\left(\frac{n}{N_c}\right)^4 \tag{1-29}$$

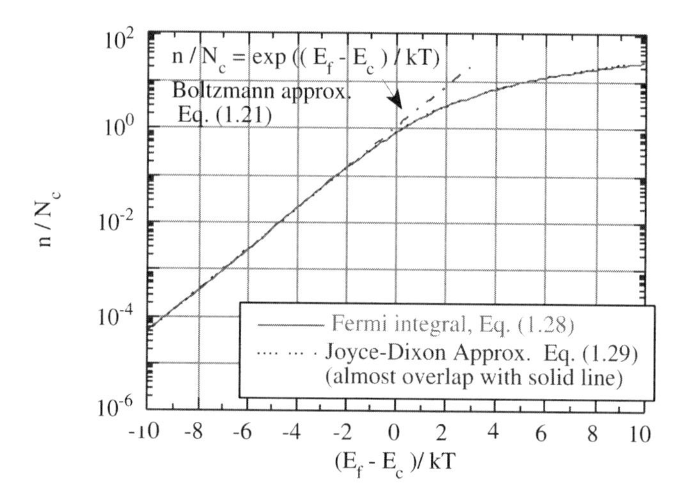

FIGURE 1-21. Fermi integral of Eq. (1-28) and its approximations by Eqs. (1-21) and (1-29).

where $A_1 = 3.53553 \times 10^{-1}$ (or $A_1 = 1/\sqrt{8}$); $A_2 = -4.95009 \times 10^{-3}$; $A_3 = 1.48386 \times 10^{-4}$, and $A_4 = -4.42563 \times 10^{-6}$. The validity of the equation is seen in Fig. 1-21. A drawback of using this equation is that it is difficult to determine n when E_f is given, although it is straightforward to calculate E_f once n is known.

Similar statements can be made for calculating the hole occupancy probability. For a hole energy state E_p, the probability is $1/\{1 + \exp[(E_f - E_p/kT]\}$ according to the Fermi–Dirac statistics, and is approximately equal to $\exp[(E_p - E_f)/kT]$ according to the Maxwell–Boltzmann statistics. The Maxwell-Boltzmann expression represents a good approximation only when $\exp[(E_p - E_f)/kT] \ll 1$. Since the highest available energy state in the valence band is the valence band edge, we replace E_p by E_v and the condition for the Maxwell–Boltzmann statistics to work then becomes $\exp[(E_v - E_f)\ /kT] \ll 1$ or, equivalently, $E_f - E_v \gg 3kT$. Under such circumstances, Eq. (1-22) is applicable. If the Maxwell–Boltzmann statistics fail, then the Joyce–Dixon approximation of the holes can be used:

$$\frac{E_v - E_f}{kT} = \ln\left(\frac{p}{N_v}\right) + A_1\left(\frac{p}{N_v}\right) + A_2\left(\frac{p}{N_v}\right)^2 + A_3\left(\frac{p}{N_v}\right)^3 + A_4\left(\frac{p}{N_v}\right)^4 \quad (1\text{-}30)$$

where the A_1, A_2, A_3, and A_4 are identical to those of Eq. (1-29).

When $E_c - E_f < 3kT$ in an n-type material or $E_f - E_v < 3kT$ in a p-type material, the semiconductor is said to be *degenerate*. According to Fig. 1-21, these conditions correspond to $n/N_c > 0.05$ or $p/N_v > 0.05$, respectively. This means that the assumptions embodied in Eqs. (1-21)–(1-22) are no longer valid and the Fermi level cannot be determined from the Maxwell–Boltzmann approximation.

Example 1-4:

Draw the band diagram of the base layer of a HBT with the p-type GaAs base layer doped at $N_a = 3 \times 10^{19}$ cm^{-3}. N_v for GaAs is 4.7×10^{18} cm^{-3}. The energy gap of GaAs is 1.424 eV.

Since N_v is smaller than N_a, this sample is degenerate. Equation (1-22) is therefore inadequate and the more exact Eq. (1-30) needs to be used. We find that, since $p \approx N_a = 3 \times 10^{19}$ cm^{-3},

$$\frac{E_v - F_f}{kT} = \ln\left(\frac{3 \times 10^{19}}{4.7 \times 10^{18}}\right) + A_1\left(\frac{3 \times 10^{19}}{4.7 \times 10^{18}}\right) + A_2\left(\frac{3 \times 10^{19}}{4.7 \times 10^{18}}\right)^2 + A_3\left(\frac{3 \times 10^{19}}{4.7 \times 10^{18}}\right)^3 + A_4\left(\frac{3 \times 10^{19}}{4.7 \times 10^{18}}\right)^4$$

Alternatively, we can use Fig. 1-21. First, we locate $p/N_v = 6.383$ on the y axis. Correspondingly, $(E_v - E_f)/kT = 4.01$ is located from the x axis. Hence, $E_v - E_f = 0.103$ eV at room temperature. The band diagram is shown in Fig. 1-22.

The following approximation of the Fermi level in degenerate semiconductors is sometimes used in this book. Rather than identifying the Fermi level using Fig. 1-21, we simply draw the Fermi level on top of the valence band edge for p-type material, and on top of the conduction band edge for n-type material.

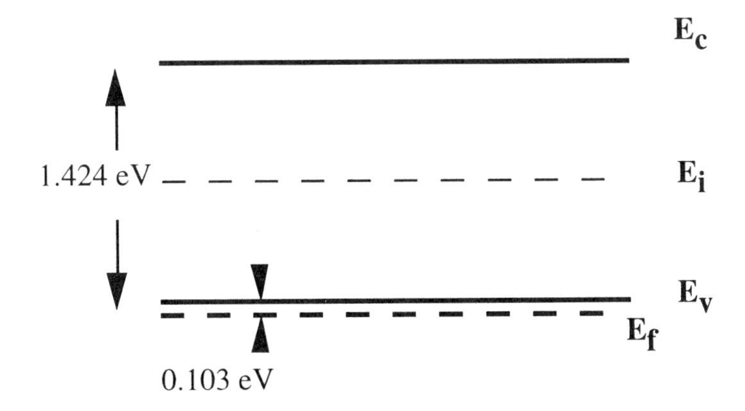

FIGURE 1-22. Band diagram of the p-type GaAs discussed in Example 1-4.

Now that the band diagram concept has been introduced, we are ready to discuss another equation relating to electrostatics. Suppose that for some reason a band diagram is already known; then the electric field may be inferred directly from the band diagram rather than by invoking the Poisson equations (Eqs. 1-8 and 1-9). Since the potential energy of an electron is $E_c - E_f$ and the potential is related to the potential energy by the charge $(-q)$, we can define the electron potential as

$$\psi = -\frac{1}{q}(E_c - E_f) \tag{1-31}$$

The electric field is the negative of the spatial gradient of the potential ($\varepsilon = -d\psi/dx$). Hence,

$$\varepsilon = \frac{1}{q}\frac{dE_c}{dx} = \frac{1}{q}\frac{dE_v}{dx} \tag{1-32}$$

In Eq. (1-9), the electrostatic potential V is the potential due solely to the space charge. The electric field calculated by taking the derivative of V does not include the quasi-electric field. In contrast, the electron potential defined in Eq. (1-31) accounts for both the electrostatic potential as well as the electron affinity. Consequently, the electric field expressed in Eq. (1-32) includes both the $\varepsilon_{\text{Poisson}}$ and $\varepsilon_{\text{quasi}}$ components.

If only the Poisson type of field is desired, without considering the quasi-electric field component, then the field is obtained by noting that $E_0 = E_c + q\chi_e$ (E_0 is the vacuum level). Taking the derivative, we find that

$$\frac{1}{q}\frac{dE_0}{dx} = \varepsilon + \frac{d\chi_e}{dx} \tag{1-33}$$

Combining the above equation with Eqs. (1-11) and (1-12), we derive

$$\varepsilon_{\text{Poisson}} = \frac{1}{q}\frac{dE_0}{dx} \tag{1-34}$$

Example 1-5:

Sometimes the band diagram of an abrupt heterojunction (no alloy grading) is drawn as shown in Fig. 1-23. The conduction band discontinuity is due to the fact that the materials at the two sides of the junction have different electron affinities. Plot the electrostatic field profile due to space charges and describe what is wrong with the diagram. Assume that the dielectric constant for the two materials is the same and that there is no interface charge. (The

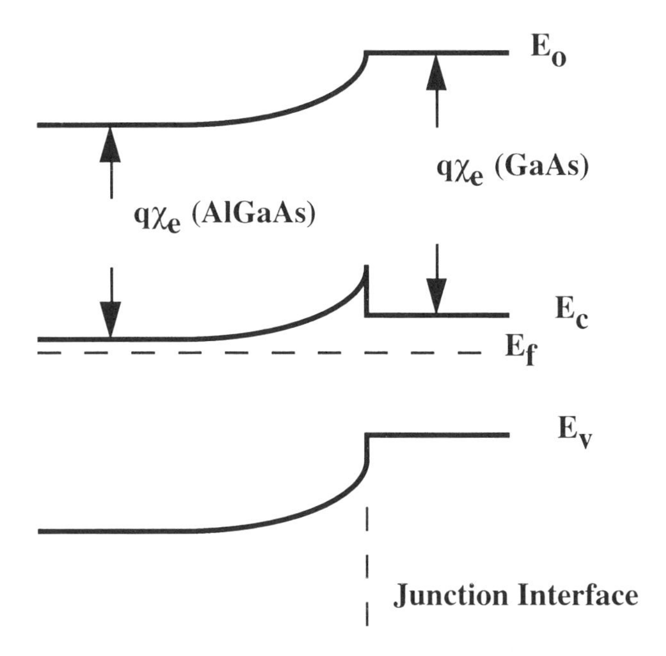

FIGURE 1-23. Band diagram of an abrupt heterojunction discussed in Example 1-5.

construction of the band diagram for heterojunctions will be discussed in Chapter 2.)

The electric field component originating from the space charges is equal to the derivative of the vacuum level with respect to the position. Figure 1-23 shows that E_0 remains constant with a position at the right side of the junction, suggesting that $\varepsilon_{\text{Poisson}}$ is zero there. To the left of the junction, $\varepsilon_{\text{Poisson}}$ has a large magnitude near the junction interface, although it decreases toward zero at distances far from the junction. Therefore, right on the junction interface, there is a discontinuity in $\varepsilon_{\text{Poisson}}$. However, according to the Maxwell equation, the normal components of $\varepsilon_{\text{Poisson}}$ at the two sides of a junction should be the same if the junction does not have any interfacial charge. The discontinuity in the electric field suggested by Fig. 1-23 is thus incorrect.

Many of the band diagrams to be constructed in Chapter 2, especially those involved with a p^+-N heterojunction, do in fact appear to be like Fig. 1-23. The reason for the seemingly abrupt change of slope in the conduction band profile there is as follows. When one side of a junction is heavily doped, the depletion thickness is so small that it is barely discernable in the normal scale. The band profile appears to be flat on that side of the junction, when in reality the band does bend in a fashion to preserve the continuity of the electric field.

Previous equations govern the carrier distribution, the electric field, and the potential profile. The next two equations go one step further, relating the carrier concentrations to the current density:

$$J_n = q\mu_n n\varepsilon + qD_n \frac{dn}{dx} \tag{1-35}$$

$$J_p = q\mu_p p\varepsilon - qD_p \frac{dp}{dx} \tag{1-36}$$

where $\varepsilon = \varepsilon_{\text{Poisson}} + \varepsilon_{\text{quasi}}$. Equation (1-35) states that the electron current density (J_n) consists of two components: a drift current and a diffusion current. The drift component, appearing as the first term, is proportional to the product of the electric field and the electron concentration, with the proportionality constant denoted as the electron mobility (μ_n). The second term, the diffusion component, is proportional to the gradient of the charge density, with the proportionality constant denoted as the electron diffusion coefficient (D_n). It turns out that there exists a particular relationship between the two material parameters, μ_n and D_n. If the exact Fermi–Dirac statistics are used, a variational form of the Fermi–Dirac integral given in Eq. (1-28) needs to be evaluated. We give the end results as

$$\frac{D_n}{\mu_n} = \frac{kT}{q}\left[1 + A_1\left(\frac{n}{N_c}\right) + 2A_2\left(\frac{n}{N_c}\right)^2 + 3A_3\left(\frac{n}{N_c}\right)^3 + \cdots\right] \tag{1-37}$$

where A_1, A_2, and A_3 are identical to those used in Eq. (1-29). When the ratio of n to N_c is small, then the above equation simplifies to the so-called Einstein relationship:

$$D_n = \frac{kT}{q}\mu_n \tag{1-38}$$

In this book, we exclusively use the approximate equation given by the Einstein relationship. The preceding discussion applies equally to the holes whose current density is given in Eq. (1-36). The minor difference between Eqs. (1-35) and (1-36) lies in the signs in front of the diffusion terms. This is because the hole has a positive charge while the electron charge is negative. When the gradient is positive ($dp/dx > 0$), the number of holes decreases as x increases. The holes therefore move in the $-x$ direction, leading to a negative diffusive current. The equations relating the hole diffusion coefficient D_p and the hole mobility (μ_p) are analogous to those given in Eqs. (1-37) and (1-38).

Example 1-6:

An n-type GaAs bar is in equilibrium. Its doping concentration as a function of the distance x is given as $N_d(x) = N_{d0}\exp(-x/\lambda)$, where both the highest doping concentration N_{do} and the characteristic length of the doping decay λ are known. Find the electric field profile.

Since this is an n-type material, we approximate $n(x)$ as $N_d(x)$ in accordance with Eq. (1-14). $N_d(x)$ is plotted in Fig. 1-24. Because there are more electrons on the left side of the bar, electrons tend to diffuse from the left to the right, establishing a negative diffusive current. However, since the overall current is zero at any position of the bar, there must exist positive drift current at any given position x that exactly cancels out the diffusive current. A positive drift current implies the presence of a positive electric field that points from the left toward the right. In terms of equations, we use Eq. (1-35) and write

$$0 = q\mu_n n\varepsilon + qD_n\frac{dn}{dx}$$

$$\varepsilon(x) = -\left(\frac{D_n}{\mu_n}\right)\left[\frac{1}{n(x)}\right]\left[\frac{dn(x)}{dx}\right] = -\left(\frac{kT}{q}\right)\left[\frac{1}{N_{do}\exp(-x/\lambda)}\right] \times N_{do}\exp\left(-\frac{x}{\lambda}\right)\left(-\frac{1}{\lambda}\right)$$

Therefore, $\varepsilon(x) = +kT/q(1/\lambda)$. It turns out that the electric field in this case is constant with position. In general, when the doping profile is not exponential-like, the electric field established by the doping density gradient is a function of position.

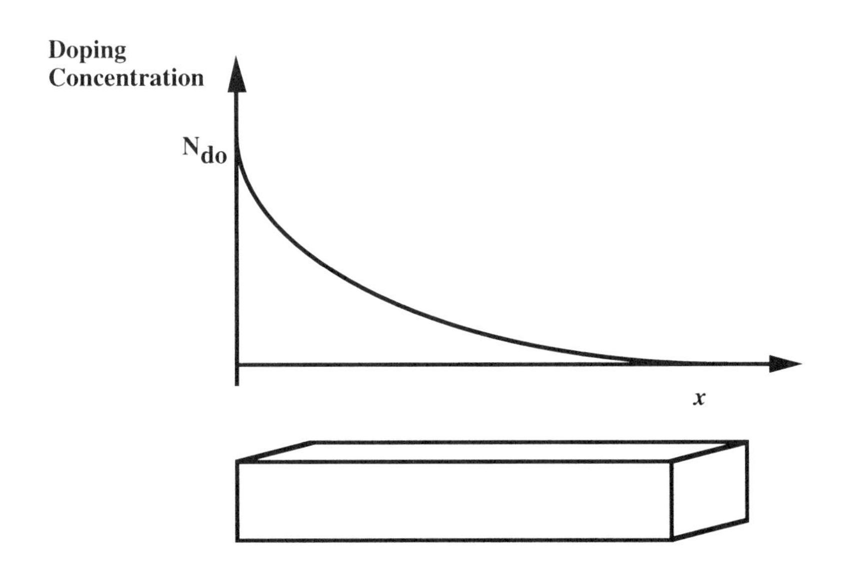

FIGURE 1-24. The n-type GaAs bar discussed in Example 1-6.

The product $\mu_n\varepsilon$ in the drift component of the electron current density is basically the velocity of the electron. The larger the electric field, the faster the electron travels, and thus the larger is the current density. The velocity of an electron, however, cannot increase indefinitely. When the electric field increases to an extreme, the velocity saturates at a value known as the electron saturation velocity (v_{sat}), which has some dependence on the lattice temperature, the doping density, and other device parameters. A generally accepted value is 1×10^7 cm/s for GaAs and Si. When the electron velocity saturates, we replace Eq. (1-35) by the following equation:

$$J_n = qnv_{\text{sat}} + qD_n \frac{dn}{dx} \tag{1-39}$$

Similarly for holes, if the hole velocity saturates, then J_p is written as

$$J_p = qpv_{\text{sat}} - qD_p \frac{dp}{dx} \tag{1.40}$$

§ 1-6 CONTINUITY EQUATIONS AND QUASI-NEUTRALITY ASSUMPTION

We develop the continuity equations that are used extensively to study the carrier transport in a bipolar transistor. Consider a differential length Δx of a semiconductor sample whose area is A in the yz plane, as shown in Fig. 1-25. The electron current entering the shaded incremental volume of A is $AJ_n(x)$. The electron current leaving the volume is $AJ_n(x + \Delta x)$. In an incremental time Δt,

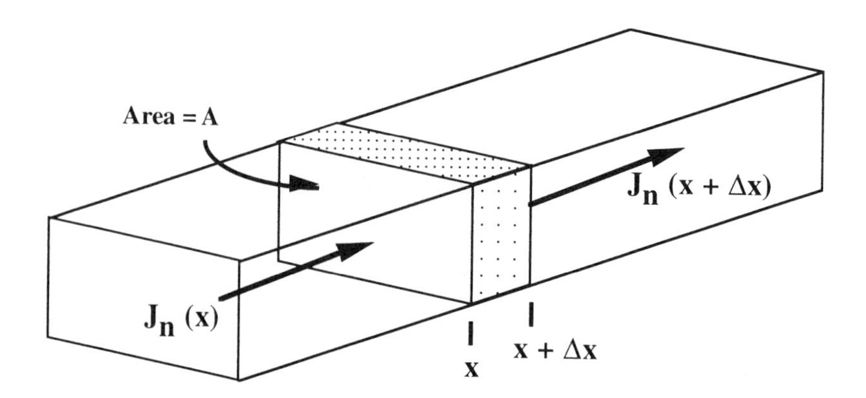

FIGURE 1-25. The region of interest in a semiconductor bar where the continuity equation is developed.

the amount of electrons accumulated in the incremental volume due to the difference in the incoming and outgoing current flow is given by

$$\frac{J_n(x) - J_n(x + \Delta x)}{-q} A\, \Delta t \tag{1-41}$$

The denominator is $-q$ rather than q because the electron charge is negative. Two other factors affect the total number of electrons accumulating in the volume in an incremental time. It is possible that during this time, some amount of the accumulated electrons recombine with the holes. It is also possible that, due to external means, some additional electrons are generated. The net number of accumulated electrons decreases in the former case and increases even more in the latter case. The rate of carrier recombination is proportional to the excess electron concentration (δn) and is typically given as

$$U_e = \frac{\delta n}{\tau_n} = \frac{n - n_0}{\tau_n} \tag{1-42}$$

where τ_n, the electron recombination lifetime, is a material parameter that depends on the doping levels, among other variables. The unit for the recombination rate is $1/\text{cm}^3$-s. There is not a fixed equation for the generation rate like the recombination rate given in Eq. (1-42). It is simply expressed as G_n. After the electron loss and increase due to the recombination and generation is accounted for, the number of electrons accumulated in the incremental volume in an amount Δt is

$$\frac{J_n(x) - J_n(x + \Delta x)}{-q} A\, \Delta t - \frac{n - n_0}{\tau_n} A\, \Delta x \Delta t + G_n A\, \Delta x \Delta t \tag{1-43}$$

Dividing this quantity by the incremental volume $A\,\Delta x$ and by the incremental time Δt, we obtain an electron continuity equation:

$$\frac{\Delta n}{\Delta t} = \frac{-1}{q} \frac{J_n(x) - J_n(x + \Delta x)}{\Delta x} - \frac{n - n_0}{\tau_n} + G_n \tag{1-44}$$

As Δx approaches zero, $J_n(x + \Delta x)$ is expanded in a Taylor series as $J_n(x) + \partial J_n(x)/\partial x \Delta x$. Replacing Δt by ∂t and Δx by ∂x, we then obtain the differential form of the continuity equation:

$$\frac{\partial n}{\partial t} = \frac{1}{q} \frac{\partial J_n(x)}{\partial x} - \frac{n - n_0}{\tau_n} + G_n \tag{1-45}$$

Substituting J_n from Eq. (1-35) into the above equation, we obtain the continuity equation for the electrons:

$$D_n \frac{\partial^2 n}{\partial x^2} + \mu_n \frac{\partial}{\partial x}(n\varepsilon) + G_n - \frac{n - n_0}{\tau_n} = \frac{\partial n}{\partial t} \tag{1-46}$$

Following a similar procedure for the hole carriers, we write the hole generation rate as G_p and the hole recombination rate as

$$U_h = \frac{\delta p}{\tau_p} = \frac{p - p_0}{\tau_p} \tag{1-47}$$

where δp is the excess hole carrier concentration. Because the recombination lifetimes are properties of the carriers themselves, τ_n and τ_p generally have different (although similar) values. The conservation of charge in the incremental volume requires that

$$\frac{\partial p}{\partial t} = -\frac{1}{q}\frac{\partial J_p(x)}{\partial x} - \frac{p - p_0}{\tau_p} + G_p \tag{1-48}$$

Substituting J_p from Eq. (1-36), we obtain the continuity equation for the holes:

$$D_p \frac{\partial^2 p}{\partial x^2} - \mu_p \frac{\partial}{\partial x}(p\varepsilon) + G_p - \frac{p - p_0}{\tau_p} = \frac{\partial p}{\partial t} \tag{1-49}$$

Whenever an electron is recombined, a hole disappears in the process. Therefore, the hole recombination rate is identical to the electron recombination rate, even during non-steady-state situations. That is,

$$U_e = \frac{n - n_0}{\tau_n} = U_h = \frac{p - p_0}{\tau_p} \tag{1-50}$$

Similarly, whenever an electron is generated, a hole is also generated. Therefore, the carrier generation rates are equal:

$$G_n = G_p \tag{1-51}$$

When excess electrons are present in a semiconductor, nature tends to restore the thermal equilibrium by reducing the excess electrons through recombination. The reverse of the recombination, carrier generation, is also possible. But since the generation is opposite to nature's way to restore equilibrium when excess carriers are present, some type of energy must be given to the semiconductor to generate these carriers. A simplistic example is to shine light onto GaAs. The electrons in the valence band of the direct-energy-gap semiconductor, upon absorbing the light energy, which is higher than the energy gap energy, are excited to the conduction band. The generation of the electrons in the conduction band leaves holes behind in the valence band. Consequently, the generation is often called *electron-hole pair generation*, to emphasize the fact that an equal amount of electrons and holes is generated in this process. In situations where there is a deficiency of carriers (n and p are smaller than their thermal equilibrium values), carrier generation occurs spontaneously to restore equilibrium, without the need to supply external energy.

Both electron and the hole continuity equations apply to n-type and p-type samples. They describe the electron and hole profiles under arbitrary bias conditions. It is rather cumbersome to work with two equations, and there is a desire to combine them into a single equation. Toward this goal, we perform a thought experiment first. Suppose that the GaAs material is biased with an externally applied electric field (ε_{app}) as shown in Fig. 1-26. A beam of light shines on the center portion of the material, generating electron-hole pairs locally in the center. A simplistic reasoning of what would happen next is the following. Since the electric field points from left to right, the excess holes, being positive charges, move along the field direction toward the right terminal. The amount of time it takes for the hole pulse to arrive at the right terminal is equal to $d/(\mu_p \varepsilon_{app})$, where d is the distance marked on the figure. After such time, there

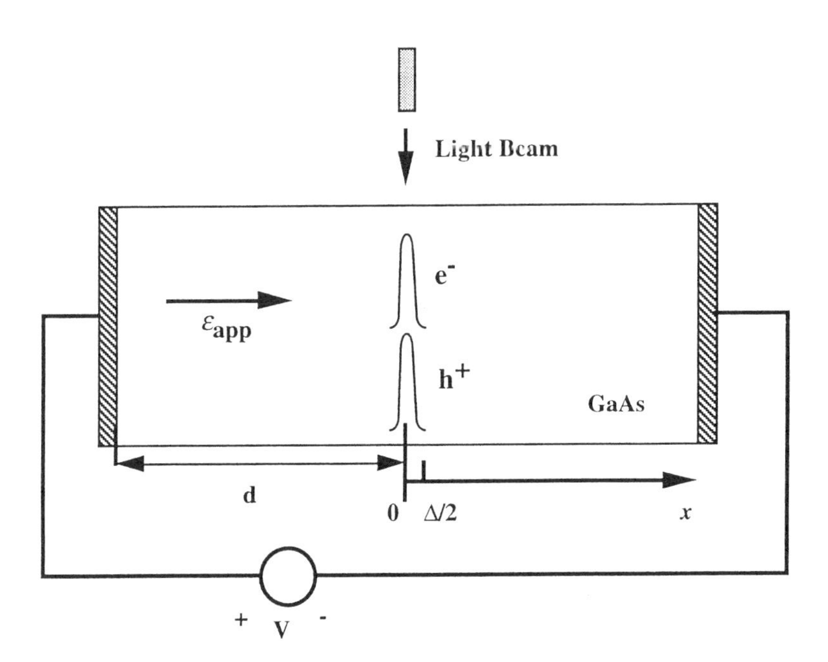

FIGURE 1-26. Pulses of electron-hole pairs are generated with a light source. An external electric field is applied to the semiconductor bar.

is no more hole current. Likewise, the electrons that are negatively charged travel in the opposite direction, toward the left terminal. The amount of time this takes is $d/(\mu_n \varepsilon_{app})$, which is shorter than that for the holes because the electron mobility is larger than the hole mobility.

The preceding description, though logical superficially, does not capture what actually happens. Before the electron pulse and the hole pulse separate completely from one another, an internal electric field unaccounted for in the previous picture is established. We illustrate the sequence of events in Fig. 1-27. At $t = 0$, the two pulses overlap completely. At $t > 0$, the faster-diffusing species, the electron, tends to move ahead of the slower-diffusing species, the hole. The externally applied field causes the two sides of the distribution about their peak to be somewhat asymmetrical. This effect, however, is neglected for now, since the concentration gradient is so large and the current is primarily diffusive. As time elapses, the electrons continue to diffuse faster than the holes. However, they are not free to move independently of the holes. As shown in Fig. 1-27*b*, the separation in the electron and hole pulses creates an internal electric field (ε_{int}) that tends to retard the faster-diffusing electrons and pull the slower holes. Because of the microscopic distance involved with this charge separation, the internal field is large enough to bind the electron and the hole pulses to travel together. Figure 1-27*c* illustrates this concept, that although both carriers continuously diffuse toward the outer region, the two pulses are separated only by a finite distance over which a strong internal electric field exists. The effective diffusion coefficient of the overall pulse is expected to be smaller than that of the faster-diffusing species, but is larger than that of the slower-diffusing species. The externally applied electric field skews the carrier profile, without changing the fact that both carriers travel together as a lumped pulse.

The electric field ε given in Eqs. (1-46) and (1-49) consists of both the externally applied electric field and the internal field. That is,

$$\varepsilon = \varepsilon_{app} + \varepsilon_{int} \tag{1-52}$$

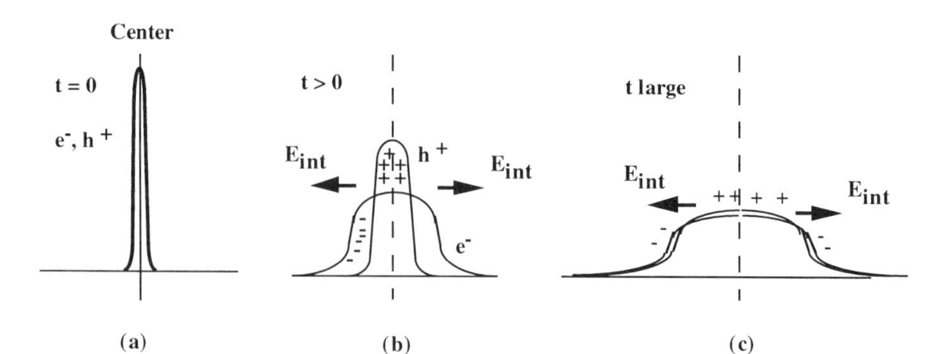

FIGURE 1-27. Time evolution of the electron-hole pairs of Fig. 1-26: (*a*) time = 0; (*b*) time slightly greater than 0; (*c*) after a long period of time.

For simplicity, we consider a uniform sample so that ε_{app} is constant. Although ε_{app} is known, ε_{int} is not. Along with n and p, there are a total of three unknowns in these two continuity equations. A third equation is required to determine uniquely the free electron concentration, the free hole concentration, and the internal electric field in the semiconductor. The third equation comes from Eq. (1-8), which is rewritten here:

$$\frac{d\varepsilon}{dx} = \frac{d\varepsilon_{int}}{dx} = \frac{q}{\epsilon_s}(p - n + N_d - N_a) \tag{1-53}$$

Unfortunately, there are no analytical solutions to the system of three equations (Eqs. 1-46, 1-49, and 1-53). Some approximations must be made to obtain tractable results for the three unknowns n, p, and ε_{int}. The approximation typically invoked at this point is the *quasi-neutrality approximation*. This approximation assumes that the excess electron density $\delta n = n - n_0$ is exactly balanced by an excess hole concentration $\delta p = p - p_0$ at any given location (n_0 and p_0 denote the electron and hole concentrations at thermal equilibrium, respectively). To a certain degree of accuracy, the charge balance agrees with intuition because just a small imbalance in the charges establishes a large enough field to prevent the electrons and holes from separating from each other. Certainly this assumption cannot be exactly correct. If the electrical neutrality truly existed, then there would not be any internal field. We therefore cannot use the quasi-neutrality assumption in Eq. (1-53) to determine the magnitude of the internal electric field. Instead, we determine ε_{int} from the following procedure. According to Eqs. (1-35) and (1-36), the total electrical current density J is

$$J = J_p + J_n = \sigma\varepsilon + q(D_n - D_p)\frac{\partial n}{\partial x} \tag{1-54}$$

where σ is the *conductivity* of the material and is equal to

$$\sigma = q\mu_n n + q\mu_p p \tag{1-55}$$

Equation (1-54) can be rewritten as

$$\varepsilon = \frac{J}{\sigma} - \frac{kT}{q}\frac{(\mu_n/\mu_p) - 1}{n(\mu_n/\mu_p) + p}\frac{\partial n}{\partial x} \tag{1-56}$$

The first-term is the externally applied electric field, leaving the second term to represent the internal field:

$$\varepsilon_{int} = -\frac{kT}{q}\left[\left(\frac{\mu_n}{\mu_p}\right) - 1\right]\left[n\left(\frac{\mu_n}{\mu_p}\right) + p\right]^{-1}\frac{\partial n}{\partial x} \tag{1-57}$$

We now invoke the quasi-neutrality assumption to solve the equations governing the excess carrier concentrations. Mathematically, this assumption is expressed as

$$\delta n(x) = n(x) - n_0 = \delta p(x) = p(x) - p_0 \tag{1-58}$$

Implicit in this assumption is the following useful equation:

$$n(x) - p(x) = n_0 - p_0 \tag{1-59}$$

We consider only the samples with uniform doping levels. Therefore, n_0 and p_0 appearing in the previous equation are constants. In addition, $d(\delta n)/dx = dn/dx$ and $d(\delta p)/dx = dp/dx$ in these samples. Equations (1-46) and (1-49) can be rewritten as

$$D_n \frac{\partial^2}{\partial x^2}(\delta n) + \mu_n \varepsilon \frac{\partial}{\partial x}(\delta n) + \mu_n n \frac{\partial}{\partial x} \varepsilon + G_n - \frac{\delta n}{\tau_n} = \frac{\partial}{\partial t}(\delta n) \tag{1-60}$$

$$D_p \frac{\partial^2}{\partial x^2}(\delta n) - \mu_p \varepsilon \frac{\partial}{\partial x}(\delta n) - \mu_p p \frac{\partial}{\partial x} \varepsilon + G_n - \frac{\delta n}{\tau_n} = \frac{\partial}{\partial t}(\delta n) \tag{1-61}$$

The recombination term in the second equation is written as $\delta n/\tau_n$, not $\delta n/\tau_p$. This result is consistent with Eq. (1-50). Further, we replace G_p in the second equation by its equivalent, G_n. The two simultaneous equations with two unknowns can be solved algebraically. By first multiplying Eq. (1-60) by $p\mu_p$ and Eq. (1-61) by $n\mu_n$, and then adding the two resultant equations together, we cancel the terms involving $\partial\varepsilon/\partial x$. With the identity of Eq. (1-58), we obtain

$$\frac{n\mu_n D_p + p\mu_p D_n}{n\mu_n + p\mu_p} \frac{\partial^2}{\partial x^2}(\delta n) - \frac{\mu_n \mu_p (n_0 - p_0)}{n\mu_n + p\mu_p} \varepsilon \frac{\partial}{\partial x}(\delta n) + G_n - \frac{\delta n}{\tau_n} = \frac{\partial}{\partial t}(\delta n) \tag{1-62}$$

This equation is called the *ambipolar equation*, which governs the behavior of excess electrons and holes in a semiconductor given the quasi-neutrality assumption. For convenience, we define the ambipolar diffusion coefficient and the ambipolar mobility as follows:

$$D^\wedge = \frac{n\mu_n D_p + p\mu_p D_n}{n\mu_n + p\mu_p} = \frac{(n+p) D_n D_p}{n D_n + p D_p} \tag{1-63}$$

$$\mu^\wedge = \frac{\mu_n \mu_p (n_0 - p_0)}{n\mu_n + p\mu_p} \tag{1-64}$$

These definitions allow Eq. (1-62) to take on a more succinct form:

$$D^{\wedge}\frac{\partial^2}{\partial x^2}(\delta n) - \mu^{\wedge}\varepsilon\frac{\partial}{\partial x}(\delta n) - \frac{\delta n}{\tau_n} = \frac{\partial}{\partial t}(\delta n) \tag{1-65}$$

The ambipolar equation put in this form resembles the continuity equation given in Eq. (1-46). A close examination of the two equations reveals a subtle difference between them. The term involving the electric field has the form $d(n\varepsilon)/dx$ in the continuity equation, but has the form of $\varepsilon dn/dx$ in the ambipolar equation. The two equations are identical in form only if the applied electric field is constant; otherwise, the two equations are different.

$D^{\wedge}$ and $\mu^{\wedge}$ both depend on n, and therefore on δn. Consequently, an analytical solution of the ambipolar equation is difficult. The determination of an analytical solution becomes simplified when $D^{\wedge}$ and $\mu^{\wedge}$ are relatively independent of the carrier concentration. Such is the case in extrinsic materials in which either the n-type or the p-type dopant concentration greatly exceeds the intrinsic carrier concentration. From inspection of Eqs. (1-63) and (1-64), we find that $D^{\wedge} \approx D_p$ and $\mu^{\wedge} \approx \mu_p$ in an n-type material where $n \gg p$, and that $D^{\wedge} \approx D_n$ and $\mu^{\wedge} \approx \mu_n$ in a p-type material where $p \gg n$.

While $D^{\wedge}$ and $\mu^{\wedge}$ are fairly constant in extrinsic materials, the recombination rate that appears as the third term of the ambipolar equation can depend on the carrier concentration. As is made clear in Eq. (1-50), this term is written either as $\delta n/\tau_n$, as in Eq. (1-41), or, equivalently, as $\delta p/\tau_p$ according to Eq. (1-47). In general, both τ_n and τ_p are functions of excess carrier concentrations. The question is whether we can safely assume either τ_n or τ_p to be concentration independent in certain situations, just as $D^{\wedge}$ and $\mu^{\wedge}$ are approximately constant in extrinsic materials. Consider an n-type semiconductor whose doping concentration is $N_d = 3 \times 10^{16}\text{cm}^{-3}$. According to Eq. (1-14), $n_0 = 3 \times 10^{16}\text{cm}^{-3}$, while $p_0 = 1.1 \times 10^{-4}\text{cm}^{-3}$. Suppose that an excess density of electron-hole pairs is created such that $\delta n = \delta p = 1 \times 10^{14}$ cm^{-3}. The net concentration of the majority electron concentration becomes $n = n_0 + \delta n = 3.01 \times 10^{16}$ cm^{-3}, which is about the same at the thermal equilibrium value of n_0. In contrast, the minority hole concentration becomes $p = p_0 + \delta p = 1 \times 10^{14}$ cm^{-3}, which greatly exceeds the thermal equilibrium value of p_0. As far as a hole is concerned, its chance of meeting an electron and subsequently recombining with it does not change much, since the electron concentrations before and after the introduction of the excess carriers are relatively the same. We therefore expect τ_p to remain constant in this n-type material. The opposite scenario is encountered for an electron. Its chance of meeting a hole changes significantly after the introduction of the excess carriers, because the hole concentration increases by ~ 18 orders of magnitude after thermal equilibrium is perturbed. We then expect τ_n in the n-type material to be strongly concentration dependent. Conversely, τ_n for the minority electron in a p-type material is relatively constant, but τ_p for the majority hole is not.

In summary, for an n-type material, we rewrite the ambipolar equation as

$$D_p \frac{\partial^2}{\partial x^2}(\delta p) - \mu_p \varepsilon \frac{\partial}{\partial x}(\delta p) + G_p - \frac{\delta p}{\tau_p} = \frac{\partial}{\partial t}(\delta p) \qquad (n\text{-type material}) \quad (1\text{-}66)$$

where D_p, μ_p, and τ_p can all be treated as constants. Likewise, for a p-type material, the ambipolar equation is written as

$$D_n \frac{\partial^2}{\partial x^2}(\delta n) - \mu_n \varepsilon \frac{\partial}{\partial x}(\delta n) + G_n - \frac{\delta n}{\tau_n} = \frac{\partial}{\partial t}(\delta n) \qquad (p\text{-type material}) \quad (1\text{-}67)$$

where D_n, μ_n, and τ_n can all be treated as constants. Solving both the electron and the hole continuity equations together in extrinsic materials is tantamount to solving a minority continuity equation only, provided that the externally applied electric field is constant. Although it seems that only the minority carrier behavior is critical to a solution, we implicitly assume certain behavior about the majority carrier as well, that the majority carriers quickly redistribute themselves such that quasi-neutrality is maintained at all time and at every location.

Example 1-7:

In the Haynes–Shockley experiment shown in Fig. 1-26, a light pulse generates electron-hole pairs in a localized region with a pulse width that is infinitely thin (from $x = -\Delta/2$ to $+\Delta/2$). To simplify the problem, we consider the n-type semiconductor bar to be infinitely long ($d = \infty$), extending in the directions to both sides of the localized light pulse. An external electric field of magnitude ε is applied. Assume that N is the number of electron-hole pairs generated per unit area. Find the evolution of the hole pulse and the electron pulse as a function of position and time, given that the carrier recombination life times and mobilities are known.

From the quasi-neutrality assumption, both the electron and hole pulses stay together as they move. Since this is an n-type semiconductor, the minority hole continuity equation should be used as the governing differential equation. The parameter D_n or τ_n, which applies to the majority electrons, is irrelevant. The pulses of both the excess electrons and the excess holes, which are assumed to be the same shape anywhere and anytime, are determined purely by the minority properties such as D_p and τ_p. We caution that it is their excess concentrations that have the same shape ($\delta n = \delta p$), not the overall carrier profiles ($n \neq p$). If the electric field points in the $+x$ direction, then the minority hole pulse moves toward the right. The majority electron pulse, which stays with the hole pulse, also moves to the right. Immediately after the pulse generation, $G_p = 0$. Therefore, the current continuity equation governing the hole pulse is

$$D_p \frac{\partial^2}{\partial x^2}(\delta p) - \mu_p \varepsilon \frac{\partial}{\partial x}(\delta p) - \frac{\partial p}{\tau_p} = \frac{\partial}{\partial t}(\delta p)$$

Since the bar is considered infinitely long, the boundary conditions are $(+\infty) = (-\infty) = 0$. The initial condition is

$$\lim_{t \to 0, \Delta \to 0} \int_{-\Delta/2}^{\Delta/2} \delta p(x, t)\, dx = N$$

Because this type of problem is not encountered in the rest of the book, we do not discuss the solution technique, but instead state the solution that satisfies the differential equation as well as both the boundary conditions and the initial condition:

$$\delta p(x, t) = \frac{N}{\sqrt{4\pi D_p t}} \exp\left[-\frac{(x - \mu_p \varepsilon t)^2}{4\pi D_p t} \right]$$

The evolution of the excess carrier profile as a function of time is illustrated in Fig. 1-28, for a case where $\mu_p = 100\ \text{cm}^2/\text{V-s}$, $N = 10^{12}\ \text{cm}^{-2}$, and $\varepsilon = 1 \times 10^4$ V/cm. It is seen from the above equation that the peak of the pulse travels at a speed equal to $\mu_p \varepsilon$. As the pulse travels, the carriers simultaneously diffuse, making the peak concentration of the pulse decrease with time.

Once the minority carrier distribution is determined from the minority carrier continuity equations (Eqs. 1-66 and 1-67), the minority carrier current is

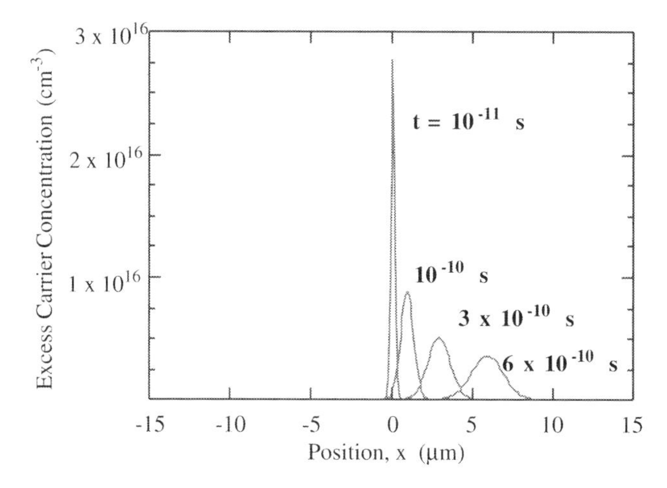

FIGURE 1-28. Calculated time evolution of the excess carrier concentration discussed in Example 1-7.

obtained from Eq. (1-35) for the minority electrons and Eq. (1-36) for the minority holes. Determination of the majority carrier current is trickier, however. We might be tempted to determine the majority carrier distribution as the sum of the doping concentration and the minority distribution established by solving the minority carrier continuity equation. We could then plug the result into Eq. (1-35) or (1-36) to find the majority carrier current. This straightforward procedure, unfortunately, does not result in an accurate solution, because the electric field appearing in these equations includes the internal field contribution (which is assumed to be zero in the quasi-neutrality approximation) in addition to the externally applied field. Since the majority carrier concentration is large, any slight error in the approximation of the actual electric field leads ultimately to a large error in the current calculation. We do not encounter a similar difficulty in the minority carrier current calculation because the minority carrier concentration is small and the internal field does not contribute significantly to the minority current. Instead of calculating the current directly from Eqs. (1-35) and (1-36), the majority carrier current is often obtained by noting that the overall current flow in a semiconductor sample is constant. If the overall current is somehow known, then the majority carrier current is obtained by subtracting the minority carrier current from the total current.

Once the majority current is established, the electric field in the quasi-neutral region is known. In many minority current calculations, the field is so small that the drift component of the minority current can be considered to be zero. However, the field in the quasi-neutral region, though small, is finite, and the drift component of the majority current can be appreciable simply because of the large number of majority carriers. The voltage drops across the neutral regions are obtained by integrating the electric fields there. This procedure is illustrated in a *p-n* junction problem discussed in Chapter 2 (see Example 2-8).

The ambipolar equation governing the carrier profiles is a differential equation. To determine the carrier profiles uniquely, we need to know the boundary conditions of the carriers. They are given as

$$D_n \frac{\partial n}{\partial x} + \mu_n n \varepsilon = -s(n - n_0) \qquad \text{(surface normal points in } +x \text{ direction)} \tag{1-68a}$$

$$D_n \frac{\partial n}{\partial x} + \mu_n n \varepsilon = s(n - n_0) \qquad \text{(surface normal points in } -x \text{ direction)} \tag{1-68b}$$

$$-D_p \frac{\partial p}{\partial x} + \mu_p p \varepsilon = s(p - p_0) \qquad \text{(surface normal points in } +x \text{ direction)} \tag{1-69a}$$

$$D_p \frac{\partial p}{\partial x} - \mu_p p \varepsilon = s(p - p_0) \qquad \text{(surface normal points in } -x \text{ direction)} \tag{1-69b}$$

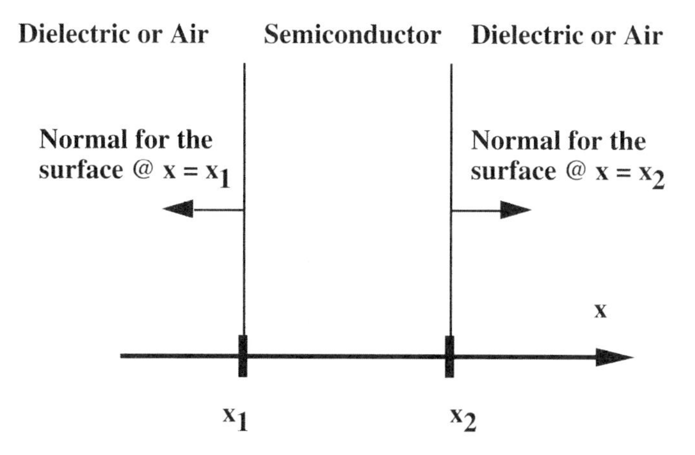

FIGURE 1-29. The directions of the surface normals at the two sides of a semiconductor sample.

The surface recombination velocity (s, in cm/s) is the only parameter required to characterize the surface. It is generally considered to be independent of the excess carrier concentration. Equations (1-68) and (1-69) differ by a sign change, depending on the direction of the surface normal, which is clarified in Fig. 1-29. A highly reactive surface is characterized by a surface recombination velocity exceeding 1×10^6 cm/s, whereas a passivated surface has a recombination velocity of the order of 1×10^3 cm/s. A derivation of these boundary conditions is found in Ref. [1].

Example 1-8:

The semiinfinite, long, n-type bar shown in Fig. 1-30 is exposed to a constant light source. The electron-hole pairs are generated uniformly in the sample, at a rate of G_p. If the surface recombination velocity is s, find the carrier concentrations (both electrons and holes) as a function of position, x.

We invoke the quasi-neutrality assumption. The excess electron and hole carrier profiles are assumed to be indentical ($\delta n = \delta p$). Because the bar is doped n-type, the continuity equation for the hole is used. We are interested in the steady-state solution, and the external electric field is 0. Simplifying the terms of Eq. (1-66), we write

$$D_p \frac{\partial^2}{\partial x^2}(\delta p) + G_p - \frac{\delta p}{\tau_p} = 0$$

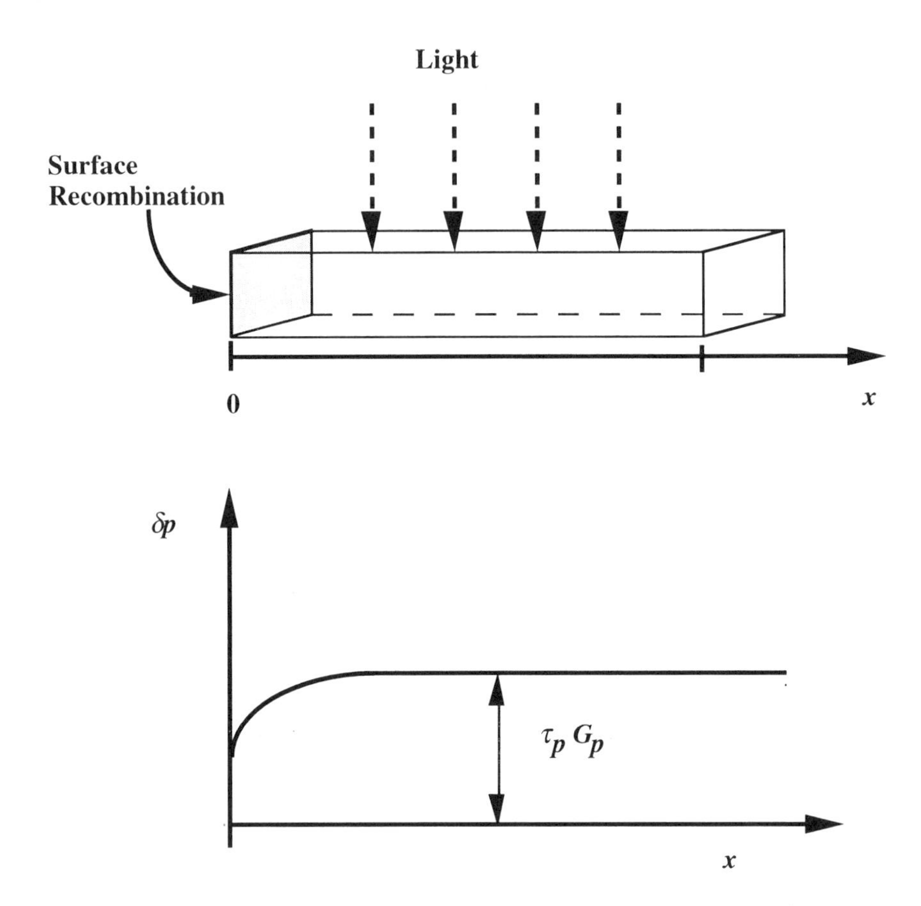

FIGURE 1-30. The semiinfinite, long, *n*-type bar discussed in Example 1-8.

The appropriate boundary condition at $x = 0$, according to Eq. (1-20), is

$$D_p \left. \frac{\partial \delta p}{\partial x} \right|_{x=0} = s\delta p$$

Since there is really not a surface at $x = \infty$, we can apply Eq. (1-20) and assume the surface recombination velocity there to be zero. Therefore,

$$D_p \left. \frac{\partial \delta p}{\partial x} \right|_{x=\infty} = 0$$

From elementary differential equation analysis, we find the solution that satisfies both boundary conditions to be

$$\delta p(x) = \tau_p G_p \left[1 - \frac{s\tau_p \exp(-x/\sqrt{\tau_p D_p})}{s\tau_p + \sqrt{\tau_p D_p}} \right] \tag{1-70}$$

The excess carrier concentration is plotted in Fig. 1-30. At large values of x, the excess hole concentration is equal to $\tau_p G_p$.

Example 1-8 is seemingly limited in application because it considers a semi-infinite bar, but it can actually be used to perform a quick approximation for more complicated problems. The following example illustrates this concept.

Example 1-9:

A light that shines on the MMIC (monolithic micorwave IC) structure shown in Fig. 1-31 has an intensity such that $G_p = 10^{18}$ carriers/cm^3-s. The generation rate is uniform throughout the sample. The surface recombination velocities at the metal interfaces is 10^7 cm/s. The interface between the depleted AlGaAs passivation ledge and the GaAs bulk on the top surface has a surface recombination velocity of 10^3 cm/s. Treat this MMIC structure to be infinitely wide in the horizontal directions such that the two sides of the MMIC structure can be considered to have a zero surface recombination velocity. Sketch the excess carrier concentrations in the structure. Assume that the wafer was ground to a thickness of 100 μm before the back-side metal was applied. The 100-μm wafer thickness is that typically used in a MMIC circuit design. $N_d = 10^{16}$ cm^{-3}, $\tau_p = 1 \times 10^{-8}$ s, and $D_p = 9.1$ cm^2/V-s.

Figure 1-30 and Eq. (1-70) show that the number of the generated carriers in a semiinfinitely long bar saturates at a constant value at a distance $\sim 3\sqrt{D_p \tau_p}$ away from the surface. This is the distance at which the exponential term in Eq. (1-70) drops to a near-zero value. We find the characteristic length $\sqrt{D_p \tau_p}$ to be

$$\sqrt{D^p \tau_p} = \sqrt{9.1(1 \times 10^{-8})} = 3 \times 10^{-4} \text{ cm}$$

Three times this distance is therefore 9 μm. This means that, from the top surface, it takes 9 μm of depth before the excess hole concentration saturates at

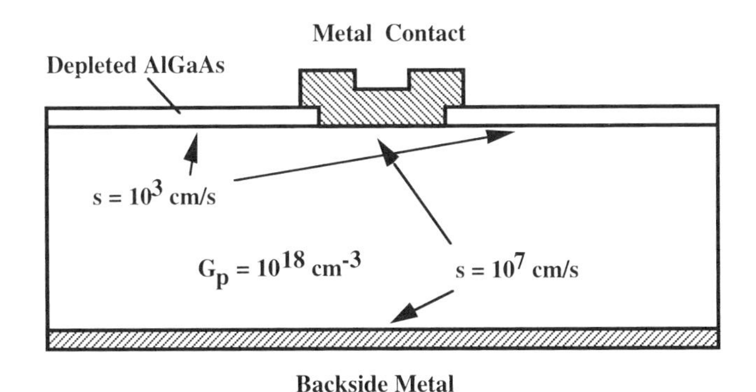

FIGURE 1-31. The semiconductor structure discussed in Example 1-9.

a constant value equal to $G_p\tau_p$. From the bottom surface, it also takes 9 μm upward before the excess hole concentration saturates at the constant value of $G_p\tau_p$. Any point in the bulk of the wafer at 9 μm away from any surface therefore has an excess hole concentration of

$$G_p\tau_p = 1 \times 10^{18}(1 \times 10^{-8}) = 1 \times 10^{10}\ \text{cm}^{-3}$$

At either the top or the bottom, the excess carrier concentration is given by

$$\delta p(0) = \tau_p G_p \left[1 - \frac{s\tau_p \exp(-x/\sqrt{\tau_p D_p})}{s\tau_p + \sqrt{\tau_p D_p}} \right]\Bigg|_{x=0} = \tau_p G_p \left(1 - \frac{s\tau_p}{s\tau_p + \sqrt{\tau_p D_p}} \right)$$

For the metal/semiconductor interface (either the top contact or the entire bottom), $s = 10^7$ cm/s, so the excess carrier concentration there is

$$\delta p = 1 \times 10^{-8}(1 \times 10^{18}) \left[1 - \frac{1 \times 10^7(1 \times 10^{-8})}{1 \times 10^7(1 \times 10^{-8}) + \sqrt{(1 \times 10^{-8} \cdot 9)1}} \right]$$
$$= 3 \times 10^7\ \text{cm}^{-3}$$

At the A1GaAs/GaAs interface, $s = 10^3$ cm/s, and the excess carrier concentration there is

$$\delta p = 1 \times 10^{-8}(1 \times 10^{18}) \left[1 - \frac{1 \times 10^3(1 \times 10^{-8})}{1 \times 10^3(1 \times 10^{-8}) + \sqrt{(1 \times 10^{-8}) \cdot 9.1}} \right]$$
$$= 9.7 \times 10^9\ \text{cm}^{-3}$$

A schematic contour of the excess carrier concentration is shown in Fig. 1-32.

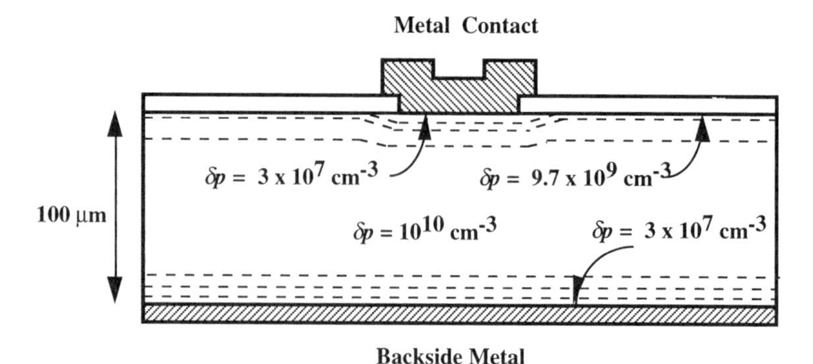

FIGURE 1-32. The minority carrier concentration contours of the structure shown in Fig. 1-31.

There are other basic device equations that have not been discussed in this section. Some that relate to the *p-n* junction are described in Chapter 2. Other equations, such as the impact ionization equations, are used rarely and are limited to certain applications. We therefore describe these equations in the appropriate sections.

§ 1-7 MATERIAL PARAMETERS

Several material parameters are used extensively in this book. We tabulate these parameters in various semiconductor materials in Appendix E. The parameters related to the heterojunction characteristics of two semiconductor materials are given in Appendix F. In the following, we give some description to each of the parameters, not necessarily in order of their appearance in Appendix E.

The relative dielectric constant, ϵ_r, when multiplied by the dielectric permittivity of the free space ($\epsilon_0 = 8.85 \times 10^{-14}$ F/cm), is equal to the dielectric permittivity of the semiconductor, ϵ_s. A high value of ϵ_r (or ϵ_s) gives rise to a high value of capacitance per unit area when the dielectric is sandwiched between two metal plates.

N_c is the conduction band density of states. (In this book, N_C, with capital C in the subscript, refers to collector doping.) According to Eq. (1-21), the value of N_c relative to the free electron concentration dictates the location of the electron Fermi level. N_v is the valence band density of states. According to Eq. (1-22), its value relative to the free hole concentration dictates the location of the hole Fermi level. n_i is the intrinsic carrier concentration, the concentration of thermally generated carriers at a given temperature. The values listed in the appendix are those at room temperature. The values increase exponentially with the energy gap E_g according to Eq. (1-13).

k_{th} is the thermal conductivity. The higher the value, the more efficient is the ability of the material to conduct heat. It is desirable to have a high k_{th} value, so that the power dissipation in a device is channeled away efficiently, without heating up the device considerably above the ambient temperature.

The surface recombination velocity, s, relates the excess carrier concentration at the surface to the fluxes of current coming toward the surface. The relationship, described in Eqs. (1-68) and (1-69), often form the boundary conditions of a given problem. The value listed in Appendix E is for a semiconductor exposed to air (or other dielectric). When the surface of interest is an interface between two semi-conductors, the values of s can be dramatically different, as shown in Appendix F.

τ_n is the recombination lifetime of the minority electron carriers. As described in the previous section, it is relatively independent of the excess electron carrier concentration in a *p*-type material. Since the majority of HBTs are of the *Npn* type, the electron recombination lifetime in the heavily doped *p*-type base is a well-studied device parameter. It is found to depend on the acceptor doping level in the base [2]:

$$\tau_n = \left(\frac{N_a}{1 \times 10^{10}} + \frac{N_a^2}{1.6 \times 10^{29}} \right)^{-1} \qquad (N_a \text{ in cm}^{-3} \text{ and } \tau_n \text{ in s}) \qquad (1\text{-}71)$$

τ_p is the recombination lifetime of the minority hole carriers. It is relatively independent of the excess hole carrier concentration in an *n*-type material. There has not been a detailed experimental study of τ_p, mainly because *Pnp* HBTs are not the predominant type of HBTs. From the compiled data of Ref. [3], we formulate the minority hole recombination lifetime as

$$\tau_p = \left(\frac{N_d^{0.693}}{5.4 \times 10^4} + \frac{N_d^{2.54}}{1 \times 10^{40}} \right)^{-1} \qquad (N_d \text{ in cm}^{-3} \text{ and } \tau_p \text{ in s}) \qquad (1\text{-}72)$$

The recombination lifetimes expressed in Eqs. (1-71) and (1-72) are plotted as a function of the doping level. The results are shown in Fig. 1-33.

The carrier mobility is a parameter that characterizes the drift motion of the carrier in the presence of an electric field. It generally decreases with temperature because of increased scattering between the carrier and the vibrating lattice, which is termed *phonon scattering*. It also decreases with the amount of doping impurity in the lattice due to increased scattering with the impurity atoms,

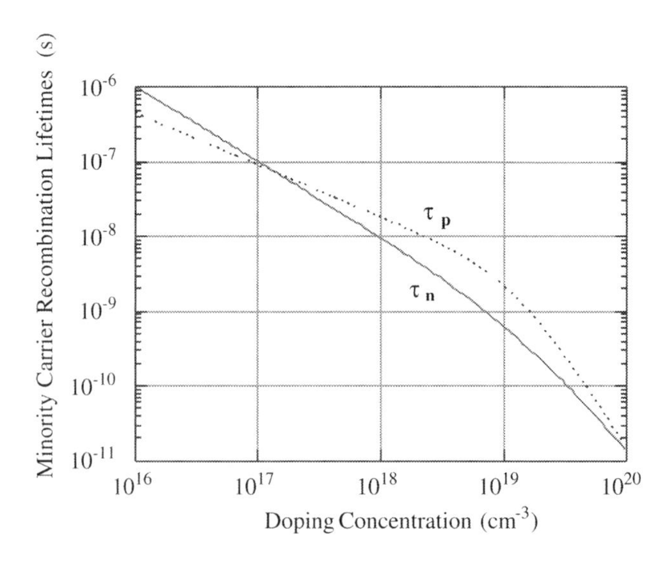

FIGURE 1-33. Minority carrier recombination lifetimes as a function of doping concentration in GaAs.

termed *impurity scattering*. Depending on the relative magnitude of either scattering mechanism, the mobility exhibits different dependencies on the lattice temperature and the doping impurity. In a heavily doped material such as the base layer of a HBT, impurity scattering typically dominates and the mobility shows a weak dependence on the temperature but is critically dependent on the doping level. In this book, we neglect the temperature dependence of the mobility, and consider mobility to be solely a function of the doping level.

We distinguish between the majority carrier mobility and the minority carrier mobility. They differ because a minority electron in a p-type material and a majority electron in an n-type material experience somewhat different amounts of scatterings. For the majority electron mobility and the majority hole mobility, we adopt the results of Ref. [4]:

$$\mu_n = \frac{7200}{[1 + (5.51 \times 10^{-17}) N_d]^{0.233}} \quad \text{(majority carrier mobility)} \quad (1\text{-}73)$$

$$\mu_p = \frac{380}{[1 + (3.17 \times 10^{-17}) N_a]^{0.266}} \quad \text{(majority carrier mobility)} \quad (1\text{-}74)$$

In these expressions, N_d and N_a are in cm^{-3} and μ_n and μ_p are in cm^2/V-s.

The majority carrier mobilities in GaAs according to Eqs. (1-73) and (1-74) are shown in Fig. 1-34. Some experimental data for the majority electron mobility in GaInP are shown in Fig. 1-35.

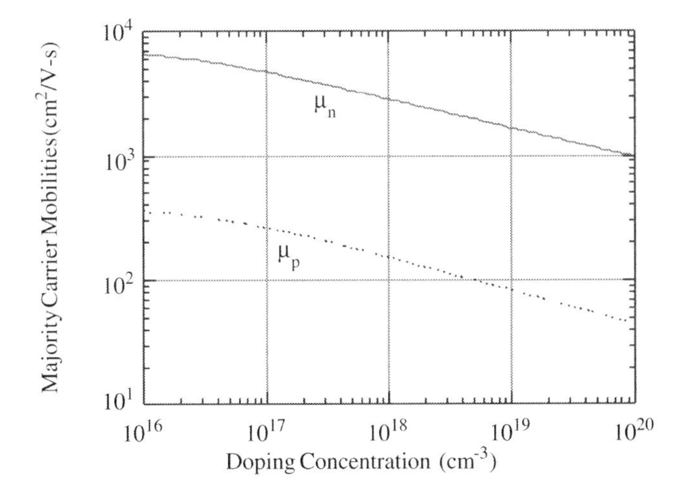

FIGURE 1-34. Majority carrier mobilities as a function of doping concentration in GaAs.

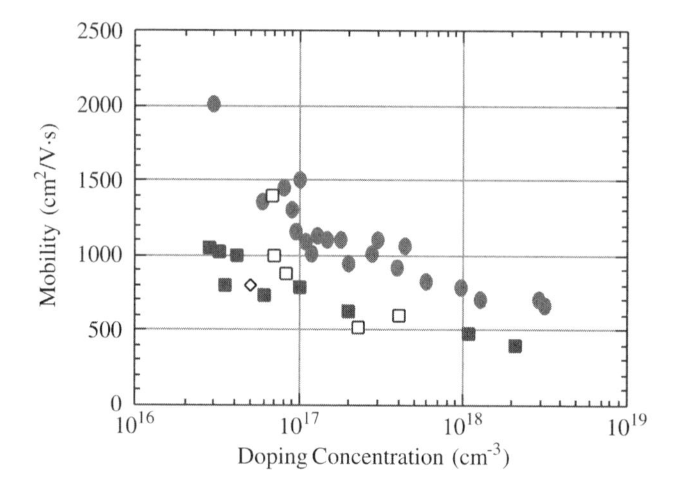

FIGURE 1-35. Electron majority carrier mobility measured from various GaInP layers. (From W. Liu, E. Beam, T. Kim, and A. Khatibzadeh. Recent developments in GaInP/GaAs heterojunction bipolar transistors, in M. F. Chang, ed., *Current Trends in Heterojunction Bipolar Transistors*, Singapore: World Scientific, 1996. © World Scientific, reprinted with permission.)

For minority electron mobility, we adopt the result of Ref. [2]:

$$\mu_n = 8300\left(1 + \frac{N_a}{(3.98 \times 10^{15}) + N_a/641}\right)^{-1/3} \quad \text{(minority carrier mobility)} \tag{1-75}$$

Unfortunately, there has not been a systematic study on the minority hole mobility (at least not known to the author). There are data on the minority diffusion coefficients, which may be used to infer the minority hole mobility. For simplicity, we will just assume that the minority hole mobility is the same as the majority hole mobility:

$$\mu_p = \frac{380}{[1 + (3.17 \times 10^{-17})N_d]^{0.266}} \quad \text{(minority carrier mobility)} \tag{1-76}$$

In Eqs. (1-75) and (1-76), N_d and N_a are in cm^{-3} and μ_n and μ_p are in $cm^2/V\text{-}s$. The minority carrier mobilities in GaAs according to Eqs. (1-75) and (1-76) are shown in Fig. 1-36.

Figure 1-37 plots the electron carrier velocity in an n-type semiconductor as a function of electric field. Two curves are shown, one for a lightly doped sample ($N_d = 1 \times 10^{13}$ cm^{-3}) and one for a heavily doped sample ($N_d = 5 \times 10^{18}$ cm^{-3}). When the electric field is small, the drift velocities of both samples increase

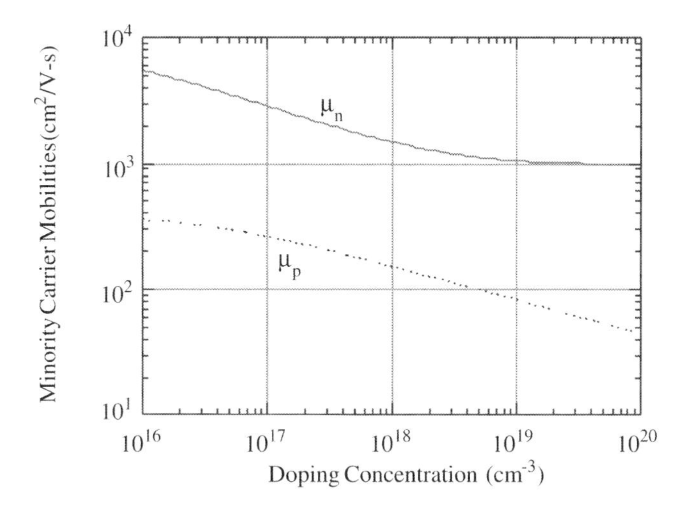

FIGURE 1-36. Minority carrier mobilities as a function of doping concentration in GaAs.

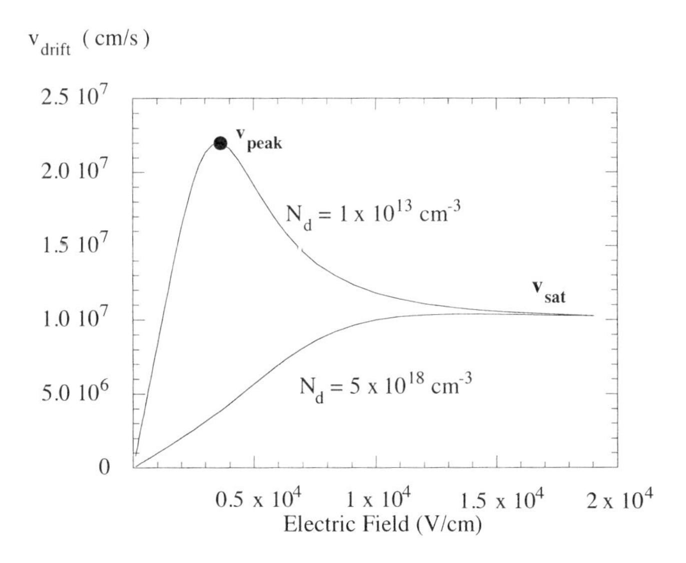

FIGURE 1-37. Electron drift velocity as a function of externally applied electric field in GaAs. (From W. Liu, Microwave and d.c. characterizations of *Npn* and *Pnp* HBTs, Ph.D. dissertation, Stanford University, Stanford CA, 1991.)

linearly with the field, with slopes equal to their respective mobilities. As the field increases to an extreme, the carrier velocities of both samples saturate at a value known as the *saturation velocity* v_{sat}. A physical reason for this phenomenon is as follows. Even at thermal equilibrium, the electron carriers are not static. They move randomly at a thermal velocity determined by the lattice temperature, which is about 1×10^7 cm/s for all semiconductors at room temperature. Although all carriers move rapidly, the net current is zero because their random movement on average yields a net zero velocity. When an electric field is applied to the sample, the net velocity becomes nonzero. The linear relationship between the drift velocity and the applied field occurs only when the resulting drift velocity is considerably smaller than the random thermal velocity, such that the thermal and drift motions can be considered independently. However, as the drift velocity approaches the thermal velocity, both motions need to be considered simultaneously. Under such circumstances, there exist highly efficient scattering processes that cause the carriers to quickly lose energy to the lattice, thus preventing the carriers from picking up additional kinetic energy. The drift velocity then approaches a limiting value and displays no dependence on the electric field.

Because we are concerned primarily with *Npn* HBTs and *n*-type field-effect transistors, the more relevant saturation velocity is the electron saturation velocity, which we denote as v_{sat}. The values listed in Appendix E are those at the room temperature. They are a function of the lattice temperature generally. For GaAs, an accepted equation describing the temperature dependence is

$$v_{sat} = (1.28 - 0.0015 \cdot T) \times 10^7 \text{ cm/s} \tag{1-77}$$

T, the lattice temperature, is in kelvins. For simplicity, the saturation velocity for the holes is taken to be equal to that expressed above.

At the medium range of electric field, the electron velocity neither increases linearly with field nor saturates at a specific value. According to Fig. 1-37, the exact dependence of the electron velocity depends on the doping level of the sample. In a heavily doped sample, the electron drift velocity increases monotonically with decreasing slope, reaching the value v_{sat} asymptotically. In a lightly doped sample, however, the electron drift velocity attains a value higher than v_{sat} before decreasing back toward the saturation value. This is explained with the help of Fig. 1-38, which shows the energy band structure of GaAs. The figure is a plot of energy of the electron as a function of the wave vector **k** of the electron wave function. (Note that the band diagrams presented in § 1-1 plot the electron energy as a function of position while the wave vector is [000]. Here, we are plotting the band diagram at a particular position, showing the electron energy as a function of the wave vector). For the present purpose of gathering the most critical information from the band structure, we do not need to delve into a detailed discussion of the wave vector. We need to be concerned with only three important directions. When **k** is in the [000] direction, the band structure surrounding it is called the Γ valley. When the wave vector is in the [100] direction, the band structure surrounding it is called the X valley. When **k** is in the [111] direction, the band structure surrounding it is called the L valley. These three directions are particularly important because the energy band structures at such directions in

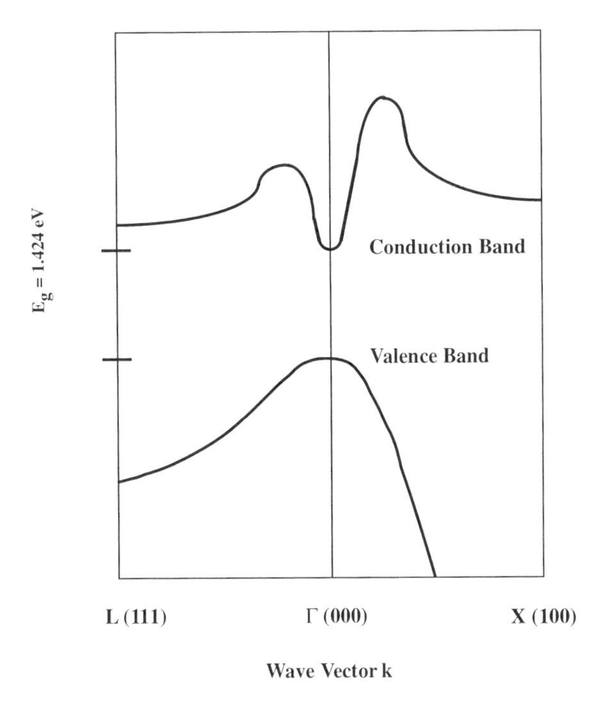

FIGURE 1-38. Schematic energy diagram of a direct-energy-gap material.

most semiconductors have either the minimum or a local minimum in the conduction band.

The effective mass at the valleys can be obtained from the following equation:

$$m^* = \left(\frac{h}{2\pi}\right)^2 \left(\frac{d^2E}{dk^2}\right)^{-1} \tag{1-78}$$

It is inversely proportional to the second derivative of the electron energy with respect to the wave vector. For the Γ valley, the electron energy increases rapidly with **k**. Hence the effective mass associated with the Γ valley is small. The L valley and X valley effective masses, in contrast, are significantly larger, as evidenced by their relatively flat curvatures in their E with respect to k. In GaAs, the Γ valley portion of the conduction band is closest to the valence band, and the energy gap of GaAs is taken to be their energy separation. When an electron gains enough energy and leaves the valence band for the conduction band, it most likely lands on a state in the Γ valley, since this contains the lower energy states. The electrical properties of the electron are therefore determined by the properties in the Γ valley. Naturally, the electron effective mass in GaAs is that associated with the Γ valley, rather than any other valley minima.

As shown in Fig. 1-38, the energy minimum in the X valley is higher than the minimum in the L valley, which in turn is higher than the minimum in the Γ valley. As aluminum is incorporated in the GaAs to form $Al_{\xi}Ga_{1-\xi}As$, these relative valley positions in GaAs do not remain unchanged. The values of the various valleys as a function of composition are shown in Fig. 1-39. As shown, for $\xi < 0.45$, the energy gap remains direct gap, determined by the Γ valley, whereas for $\xi > 0.45$, the energy gap is indirect and is determined by the X valleys. It is also interesting to observe that the L valley energy increases with aluminum composition at a faster rate than the X valley energy. The L valley energy eventually surpasses the X valley energy, also at ξ near 0.45.

At low electric fields, the electrons acquire relatively small amounts of kinetic energy and stay at the bottom of the Γ conduction band. The Γ valley is characterized by a small electron effective mass (m_e^*), manifested by the sharp parabola shape there. For GaAs, m_e^* is about 0.067 times the free electron mass, m_0, which is equal to 9.1×10^{-31} kg. This small effective mass is responsible for the high electron mobility in GaAs. As the field is increased, the electrons gain enough energy, occupying energy levels with higher energy values than those of the L-valley energy states. In their continual quest for states with low energy, they are easily scattered (i.e., with a change of the **k** vector by interactions with phonons) from the Γ valley to the L valley. The L valley has a notably larger effective mass than the Γ valley, as evidenced by the rather flat parabola shape in the L valley. As the electron transfers take place, the electron mobility decreases

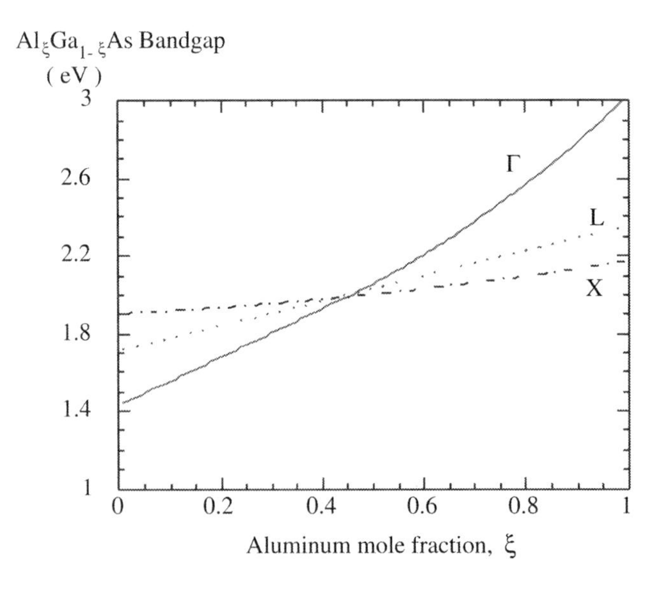

FIGURE 1-39. Values of the Γ, L, and X valleys as a function of the aluminum composition in an AlGaAs material. (From W. Liu, Microwave and d.c. Characterizations of *Npn* and *Pnp* HBTs, Ph.D. dissertation, Stanford University, Stanford CA, 1991.)

in value, and a negative differential mobility is in effect. This chain of reactions translates into a reduced electron drift velocity. This phenomenon of decreasing velocity with increasing field is referred to as the *Gunn effect*. It is a physical basis for the velocity overshoot phenomenon to be discussed in § 4-5. As the field is continuously increased, the drift velocity finally saturates. The field at which the electron drift velocity reaches its highest possible value is called the *critical electric field* (ε_{crit}), above which the velocity decreases monotonically toward v_{sat}.

The Gunn effect is observed for relatively low background doping levels. If the semiconductor is heavily doped, such as 5×10^{18} cm^{-3}, there exists significant impurity scattering, in addition to the phonon scattering. Consequently, the mobility of the electrons in the Γ valley is lowered to such an extent that the negative differential mobility is not observed at all.

The parameter $E_{\Gamma-L}$ is the energy separating the minimums of the Γ and L valleys in the conduction band. The higher the value of $E_{\Gamma-L}$, the higher will be the critical electric field.

The valence band structure of GaAs is completely different from that of the conduction band. For the valence band, there exists only one band extremum, in the Γ direction. This is to be contrasted with the several conduction band extrema of the conduction band, such as in the L and X directions, in addition to Γ. Therefore, the Gunn effect is not observed for holes, for any background doping. The hole drift velocity simply increases monotonically to v_{sat}, similar to the behavior of the electron drift velocity in the heavily doped sample shown in Fig. 1-37. The hole effective mass (m_h^*) is generally larger than m_e^*, as suggested by the flatter parabola shape of the valence band in comparison to the conduction band.

One observation from Fig. 1-38 is that the band structure of GaAs has a minimum in the conduction band and a maximum in the valence band at the same **k** value at the Γ valley. An electron making a smallest-energy transition from the valence band to the conduction band does so without a change in the **k** value. Therefore, GaAs is said to be a direct-energy-gap material. Such property allows laser to be made in the GaAs material system. In contrast, silicon is an indirect-energy-gap material. Figure 1-40 illustrates the band structure of silicon, showing that although the valence band has a maximum in the Γ direction, the minimum of the conduction band resides in the X valley. Therefore, an electron making a smallest-energy transition necessarily involves a change in the **k** vector (a change in the momentum of the electron). Also shown in Fig. 1-40 is that there is a significant energy separation between the X-valley minimum and the next local energy minimum. Therefore, Gunn effects that depends on electron transfer between energy valleys are not observed in silicon material.

Based on Fig. 1-38, we take the effective mass of an $Al_\xi Ga_{1-\xi}As$ material to be that of the Γ valley when $\xi < 0.45$, and that of the X valley when $\xi > 0.45$. The constant energy surfaces with respect to the energy minimum in the Γ valley are spherical. In this case, the effective mass defined in Eq. (1-78) is the same in all directions. It turns out that a single effective mass is not enough to

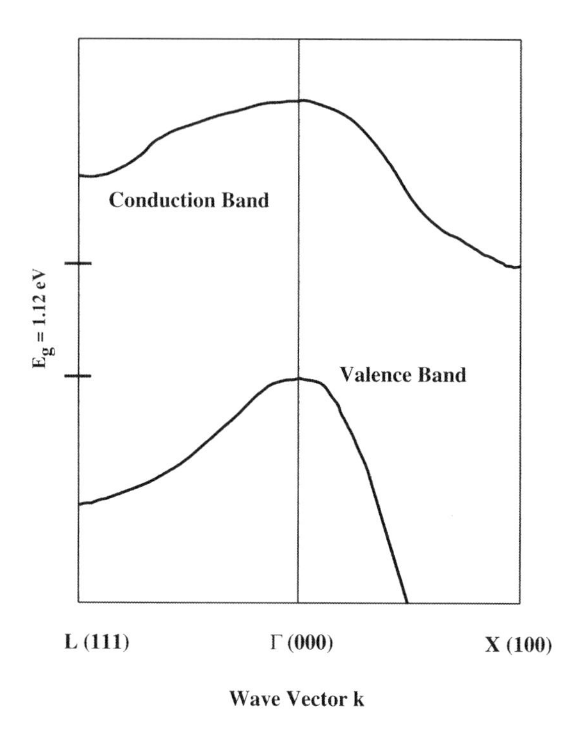

FIGURE 1-40. Schematic energy diagram of an indirect-energy-gap material.

characterize the X-valley electrons completely. The constant energy surfaces with respect to the energy minimum in the X valley are ellipsoidal. One longitudinal effective mass (m_{lX}^*) and one transverse effective mass (m_{tX}^*) are required. The electron effective mass in $Al_\xi Ga_{1-\xi}As$ is given by [5]

$$m_\Gamma^* = \begin{cases} (0.067 + 0.083\xi)m_0 & \text{when } \xi \leq 0.45 \\ \text{not used} & \text{when } \xi > 0.45 \end{cases} \tag{1-79a}$$

$$m_{tX}^* = \begin{cases} \text{not used} & \text{when } \xi < 0.45 \\ (0.23 - 0.04\xi)m_0 & \text{when } \xi \geq 0.45 \end{cases} \tag{1-79b}$$

$$m_{lX}^* = \begin{cases} \text{not used} & \text{when } \xi < 0.45 \\ (1.3 - 0.2\xi)m_0 & \text{when } \xi \geq 0.45 \end{cases} \tag{1-79c}$$

where m_0 is the free electron mass. Equation (1-79) is of course an oversimplification. Although such approximation is fairly accurate at the extreme values of ξ, it is problematic at ξ values close to 0.45. The simplification assumes that the all of the electrons populate the Γ valley when ξ is right below 0.45 and suddenly all switch to the X valley the moment ξ increases above 0.45. In reality, there is a gradual transition from a purely Γ-valley occupation at $\xi < 0.45$ to a purely X-valley occupation at $\xi > 0.45$.

The masses are still not those to be used in typical equations to calculate the device current or the density of states. When calculating the conduction properties of the semiconductor, the effective electron mass (m_e^*) to be used in the current calculation is given by

$$m_e^* = \begin{cases} m_\Gamma^* & \text{when } \xi < 0.45 \\ \frac{1}{6}(2 \cdot m_{tX}^* + 4\sqrt{m_{lX}^* \cdot m_{tX}^*}) & \text{when } \xi \geq 0.45 \end{cases} \tag{1-80}$$

For the calculation of the density of states, the density-of-state effective mass of electrons is given by

$$m_{de}^* = \begin{cases} m_\Gamma^* & \text{when } \xi < 0.45 \\ 6^{2/3}(m_{lX}^* \cdot m_{tX}^* \cdot m_{tX}^*)^{2/3} & \text{when } \xi \geq 0.45 \end{cases} \tag{1-81}$$

The conduction band density of states relates to m_{de}^* in the following manner:

$$N_c = 2\left(\frac{2\pi kT}{h^2}\right)^{3/2} m_{de}^{*3/2} \tag{1-82}$$

The hole effective mass in $Al_\xi Ga_{1-\xi}As$ is less complicated, since the valence band maximum is always along the Γ [000] direction, independent of ξ. Though it is not shown explicitly in Fig. 1-38, a valence band actually consists of two degenerate bands. Since the curvatures of the bands are different, one is called the heavy hole band, characterized by a heavy hole effective mass (m_{hh}^*) , and another, the light hole band, characterized by a light hole effective mass (m_{lh}^*). Their dependence on the aluminum mole fraction is given by

$$m_{lh}^*(\xi) = 0.087 + 0.063\xi \tag{1-83}$$

$$m_{hh}^*(\xi) = 0.62 + 0.14\xi \tag{1-84}$$

The hole effective mass, for the purpose of calculating transport characteristics, is given by:

$$m_h^* = \frac{m_{hh}^{*3/2} + m_{lh}^{*3/2}}{m_{hh}^{*1/2} + m_{lh}^{*1/2}} \qquad (0 \leq \xi \leq 1) \tag{1-85}$$

For the calculation of the density of states, the density-of-state effective mass of holes is given by

$$m_{dh}^{*} = (m_{lh}^{*3/2} + m_{hh}^{*3/2})^{2/3} \qquad (0 \leq \xi \leq 1) \tag{1-86}$$

The valence band density of states related to m_{dh}^{*} in the following manner:

$$N_v = 2\left(\frac{2\pi kT}{h^2}\right)^{3/2} m_{dh}^{*3/2} \tag{1-87}$$

We proceed to describe the heterojunction parameters listed in Appendix F. We mentioned previously that the energy gap of the $Al_\xi Ga_{1-\xi}As$ material is determined by the Γ valley at $\xi < 0.45$, but is determined by the X valley at $\xi > 0.45$. The energy gap (in eV) is given as

$$E_g(Al_\xi Ga_{1-\xi}As) = \begin{cases} 1.424 + 1.247\xi & 0 \leq \xi \leq 0.45 \\ 1.900 + 0.125\xi + 0.143\xi^2 & \xi > 0.45 \end{cases} \tag{1-88}$$

The energy gap difference (ΔE_g) in an $Al_\xi Ga_{1-\xi}As/GaAs$ heterojunction is shared between the conduction band and the valence band. That is, $\Delta E_g = \Delta E_c + \Delta E_v$, where ΔE_c is the conduction band discontinuity and ΔE_v is the valence band discontinuity. According to Ref. [6], the amount of ΔE_v is linearly dependent on the aluminum mole fraction (ξ) for all values of ξ. ΔE_v and ΔE_c in an $Al_\xi Ga_{1-\xi}As/GaAs$ heterojunction are therefore given by

$$\Delta E_v(\xi) = 0.55\xi \qquad \text{(in eV)} \tag{1-89}$$

$$\Delta E_c(\xi) = \begin{cases} (1.247 - 0.55)\xi & 0 \leq \xi \leq 0.45 \\ 0.476 + (0.125 - 0.55)\xi + 0.143\xi^2 & \xi > 0.45 \quad \text{(in eV)} \end{cases} \tag{1-90}$$

The electron affinity variation with ξ is just the complementary of the variation in ΔE_c. Knowing that $q\chi_e$ for GaAs is 4.07 eV, we write find $q\chi_e(\xi)$ from the condition that $\Delta E_c(\xi) + [q\chi_e(\xi) - 4.07] = 0$:

$$q\chi_e(\xi) = \begin{cases} 4.07 - 0.697\xi & 0 \leq \xi \leq 0.45 \\ 3.594 + 0.425\xi - 0.143\xi^2 & \xi > 0.45 \quad \text{(in eV)} \end{cases} \tag{1-91}$$

The preceding equations are for an $Al_\xi Ga_{1-\xi}As/GaAs$ heterojunction for all values of aluminum composition ξ. ξ can have any value between 0 and 1, since AlAs and GaAs have nearly the same lattice constants. In other material systems, such as $Ga_{0.51}In_{0.49}P/GaAs$, $InP/In_{0.51}Ga_{0.47}As$, or $In_{0.52}Al_{0.48}As/In_{0.53}Ga_{0.47}As$, we need only be concerned with certain fixed gallium, indium, or aluminum compositions. This is because only at such fractions can the two materials be lattice-matched.

Another material constant of interest for $Al_\xi Ga_{1-\xi}As$ with varying ξ is the dielectric constant. It is given as

$$\epsilon_s(\xi) = (13.18 - 3.12\xi)(8.85 \times 10^{-14})\ \text{F/cm} \tag{1-92}$$

We conclude with a discussion of InGaAs. The energy gap (in eV) of $In_\xi Ga_{1-\xi}As$ material without strain is [7]

$$E_g(In_\xi Ga_{1-\xi}As) = 1.42 - 1.615\xi + 0.555\xi^2 \tag{1-93}$$

A less accurate but commonly used expression is obtained by linear extrapolation between the InAs energy gap of 0.35 eV and the GaAs energy gap of 1.42 eV. This expression is

$$E_g(In_\xi Ga_{1-\xi}As) = 1.42 - 1.07\xi \tag{1-94}$$

REFERENCES

1. McKelvey, J. P. (1996). *Solid State and Semiconductor Physics*. New York: Harper & Row.
2. Tiwari, S., and Wright, S. L. (1990). "Material properties of p-type GaAs at large dopings." *Appl. Phys. Lett.* **56,** 563–565.
3. (1990). *Properties of Gallium Arsenide*, 2nd ed. London: INSPEC, The Institution of Electrical Engineers.
4. Sunderland, D., and Dapkus, P. D. (1987). "Optimizing N-p-n and P-n-p heterojunction bipolar transistors for speed." *IEEE Trans. Electr. Dev.* **34,** 367–377.
5. Adachi, S. (1985, Aug. 1). "GaAs, AlAs, and $Al_xGa_{1-x}As$: Material parameters for use in research and device applications." *J. Appl. Phys.* **58,** R1–R29.
6. Batey, J., and Wright, S. L. (1986). "Energy band alignment in GaAs:(Al,Ga) As heterostructures: The dependence on alloy composition." *J. Appl. Phys.* **59,** 200–209.
7. Adachi, S. (1982). "Material parameters of $In_{1-x}Ga_xAs_yP_{1-y}$ and related binaries." *J. Appl. Phys.* **53,** 8775–8792.

PROBLEMS

1. Draw the band diagrams of: (**a**) a p-type GaAs doped at $1 \times 10^{20}\ \text{cm}^{-3}$; (**b**) an n-type $Al_{0.35}Ga_{0.65}As$ doped at $5 \times 10^{17}\ \text{cm}^{-3}$; (**c**) an n-type GaAs whose hole concentration is $10^{-6}\ \text{cm}^{-3}$.

2. A uniformly doped semiconductor is heated unevenly across two regions, with region A being hotter than region B. A probe is placed in region A and another in region B. The voltmeter connecting across the two probes (with zero current flowing through the voltmeter) indicates that region A has a higher potential. Is the material n-type or p-type?

3. An n-type GaAs is doped at $1 \times 10^{16}\ \mathrm{cm}^{-3}$. Find $E_f - E_c$ at a substrate temperature of 600 K. Repeat the exercise for an n-type silicon with the same doping concentration.

4. Consider an n-type GaAs doped at $1 \times 10^{16}\ \mathrm{cm}^{-3}$. (**a**) Find the minority carrier lifetime of this material, using the equations given in § 1-7. (**b**) The material is exposed to a constant light source, and 10^{17} electron-hole pairs are generated per cubic centimeter per second. What are the steady-state electron and hole concentrations with the light turned on? (**c**) At time $t = 0$ the light is switched off. Find the carrier concentrations as a function of time. What is the time constant of the decrease of the carrier concentrations?

5. An n-type GaAs bar (uniformly doped at $5 \times 10^{18}\ \mathrm{cm}^{-3}$) with a 100-$\mu\mathrm{m}^2$ cross section has a length of 1 cm. A voltage source of 10 V is connected between its two ends. (**a**) If the GaAs bar is maintained at 300 K, What is the current flow? (**b**) Suppose that the GaAs bar heats up by the power dissipation in accordance with this relationship: $T = T_0 + R_{\mathrm{th}} \cdot V \cdot I$, where $T_0 = 300$ K, $R_{\mathrm{th}} = 200$ K/W, V is the applied bias, and I is the current flow. Determine the current flow in this case assuming that the mobility decreases by 30% for every 30 K increase in temperature.

6. Consider a Haynes–Shockley experiment on an n-type silicon bar. If a pulse of holes is injected at $x = 0$, $t = 0$, and if the maximum of the hole pulse reaches a probe at $x = 100\ \mu\mathrm{m}$ at $t = 4$ ns, determine the hole mobility. Assume that a voltage of 5 V is maintained between the two ends of the 1-cm-long bar.

7. An n-type GaAs bar is infinitely long in both the $+x$ and the $-x$ directions. At the $x = 0$ cross section, excess carriers are created at a rate of $g_s\ \mathrm{cm}^{-2}\text{-}\mathrm{s}^{-1}$. There is no externally applied electric field.

a. Write out the continuity equation at $x \neq 0$, and show that the solution for $\delta p(x)$ is $A \exp(-x/L_p)$ for $x > 0$ and $A \exp(x/L_p)$ for $x < 0$, where A is an unknown integration constant.

b. Write the continuity equation at $x = 0$ and integrate the equation from $x = -\Delta$ to $+\Delta$. Take the limit of Δ to 0 and show that

$$D_p \left. \frac{d\,\delta p}{dx} \right|_{-\Delta}^{+\Delta} + g_s = 0$$

c. Show that the integration constant $A = 2L_p g_s / D_p$.

d. From now on, consider only regions at $x > 0$. Show that the minority hole diffusion current $J_p^{\mathrm{diff}} = kT\mu_p \delta p(x)/L_p$. Similarly, show that the diffusive component of the majority electron current $J_n^{\text{diff}} = -kT\mu_n\delta p(x)/L_p$.

e. Show that the hole drift current due to the internal electric field is equal to

$$J_p^{\text{drift}} = -kT\frac{p\mu_p(\mu_n - \mu_p)}{p\mu_p + n\mu_n}\frac{\delta p}{L_p}$$

f. Show that the electron drift current due to the internal electric field is equal to

$$J_n^{\text{drift}} = kT\frac{n\mu_n(\mu_n - \mu_p)}{p\mu_p + n\mu_n}\frac{\delta p}{L_p}$$

g. Show that indeed the magnitude of $J_{pT} = J_p^{\text{diff}} + J_p^{\text{drift}}$ is equal to $J_{nT} = J_n^{\text{diff}} + J_n^{\text{drift}}$.

h. Determine the following ratios: J_p^{diff}/J_{pT}; $J_p^{\text{drift}}/J_{pT}$; J_p^{diff}/J_{nT}; and $J_n^{\text{drift}}/J_{nT}$.

i. Show that the ratios in (**h**) degenerate to 1, 0, μ_n/μ_p, $(\mu_p - \mu_n)/\mu_p$, respectively, in heavily doped n-type material. This result demonstrates that the minority drift current due to the internal electric field is negligible and the minority current is due mainly to diffusion. In addition, the majority carrier current due to the internal electric field is significant. This current component needs to be large enough to assure quasi-neutrality throughout the entire sample. It compensates the difference between the diffusion currents, which is a result of the difference between μ_n and μ_p.

8. Consider again the n-type GaAs bar of Problem 7. Again, excess carriers are created at a rate of g_s cm^{-2}-s^{-1} at the $x = 0$ cross section. This time, however, an externally applied electric field ε is present.

a. Write out the continuity equation at $x \neq 0$, and show the following with A being an integration constant:

$$\delta p(x) = A\exp\left[\frac{x}{L_p}\left(\frac{\mu_p\varepsilon L_p}{2D_p} - \sqrt{\left(\frac{\mu_p\varepsilon L_p}{2D_p}\right)^2 + 1}\right)\right] \qquad \text{for } x > 0$$

$$\delta p(x) = A\exp\left[\frac{x}{L_p}\left(\frac{\mu_p\varepsilon L_p}{2D_p} + \sqrt{\left(\frac{\mu\varepsilon L_p}{2D_p}\right)^2 + 1}\right)\right] \qquad \text{for } x > 0$$

b. Assume that the applied electric field is large. Show that the integration constant A is

$$A = g_s\left(\mu_p\varepsilon + \frac{D_p}{\mu_p\varepsilon\tau_p}\right)^{-1}$$

c. What is the total current density flowing in the bar? The sample's doping level is N_d.

9. For a particular sample, the probability of finding electrons in states an energy kT above E_c is e^{-10} at room temperature. Determine the location of the Fermi level with respect to E_c in terms of kT.

10. An n-GaAs bar of length L contains a linearly graded doping profile that varies from $N_d = N_d(0) = 10^{16}\ \text{cm}^{-3}$ at $x = 0$ to $N_d = N_d(L) = 2 \times 10^{18}\ \text{cm}^{-3}$ at $x = L$. The bar is being maintained under thermal equilibrium at room temperature.

a. Assume the charge neutrality approximation. Derive a formula for the electric field $\varepsilon(x)$ in terms of $N_d(0)$ and $N_d(L)$.

b. Draw the energy diagram for the bar, showing qualitatively the behavior of E_f, E_c, and E_v as a function of x.

c. An acceptor impurity concentration of $9 \times 10^{15}\ \text{cm}^{-3}$ is added uniformly to the semiconductor. What effect will the added impurity concentration have on the electron concentration within the bar?

11. In Example 1-6, we determined the electric field of an n-type semiconductor bar. We assumed that charge neutrality exists, so the electron concentration is equal to the dopant concentration, at any location. However, Eq. (1-8) states that the electric field is zero if charge neutrality truly exists everywhere in the semiconductor. Clarify this paradox with the analysis developed in § 1-6.

12. Equating n_i^2 to $N_c N_v \exp(-E_g/kT)$ as shown in Eq. (1-13) implicitly assumes the Maxwell-Boltzmann statistics.

a. Show that if Eqs. (1-21) and (1-22) are used, then the np product is equal to $N_c N_v \exp(-E_g/kT)$.

b. Show that such equivalence fails if Eqs. (1-29) and (1-30) are used.

c. Does $np = n_i^2$ remain valid as we dope a nondegenerate semiconductor into being degenerate?

CHAPTER 2

TWO-TERMINAL HETEROJUNCTION DEVICES

§ 2-1 p^+-N HETEROJUNCTION UNDER THERMAL EQUILIBRIUM

A *p-n* junction is formed when a *p*-type doped portion of the semiconductor is joined with an *n*-type doped portion. As a fundamental component for functions such as rectification, the *p-n* junction forms the basic unit of a bipolar transistor. If both the *p*-type and the *n*-type regions are of the same semiconductor material, the junction is called a *homojunction*. If the junction layers are made of different semiconductor materials, it is a *heterojunction*. A heterojunction bipolar transistor (HBT) consists of a heterojunction in the base-emitter junction and a homojunction in the base-collector junction. A *double* heterojunction bipolar transistor (DHBT) is made of two heterojunctions, both in the base-emitter and the base-collector junctions.

As a matter of convention, if the *n*-type doped semiconductor material has a higher energy gap than the *p*-type doped material, it is denoted a *p-N heterojunction*. The use of capital and lowercase letters connotes the relative size of the energy gap. Conversely, if the *p*-type doped material has a larger energy gap than the *n*-type material, the junction is referred to as a *P-n* heterojunction. In this section as well as the following sections, we study the electrostatic and current–voltage relations of a heterojunction, but the results are readily applicable to homojunctions. We limit our scope to a *step junction*, in which the doping levels change abruptly from *n*-type of one particular concentration to *p*-type of another concentration. Besides its simplicity in analysis, a step junction is the most prevalent form of junction in III-V devices. With the control of epitaxial growth techniques such as MBE and MOCVD, the abrupt change in the doping type across a heterojunction takes place in a monolayer or so. Other types of junctions, such as graded junctions in which the doping density varies linearly across the junction, are seldom used in III-V applications. [Note: In III-V technologies, grading can mean either the grading (or varying) of the dopant concentration or the alloy composition. As has been noted, generally, the doping levels in III-V device layers are constant. Therefore, unless otherwise noted, the word *grading* shall refer specifically to the grading of alloy composition.]

Consider, for example, a *p-N* heterojunction formed with a *p*-type GaAs layer doped at 3×10^{19} cm^{-3} and an *N*-type AlGaAs layer doped at 1×10^{16} cm^{-3}. The band diagrams of individual layers were shown in Figs. 1-20 and 1-22. For convenience, these two band diagrams are drawn side by side in Fig. 2-1, with their vacuum levels aligned to illustrate the difference in the Fermi levels in the charge-neutral condition. From statistical physics, the Fermi level is the chemical potential of the electrons in a semiconductor system. It governs the flow of particles between systems, just as the temperature governs the flow of energy. When two systems with a single chemical species are brought into contact, there will be no net particle flow if they have the same chemical potential. Conversely, if their chemical potentials differ, then particles will flow from the system at the higher chemical potential to the system at the lower chemical potential. The flow persists until their chemical potentials become equal.

We return to the discussion of Fig. 2-1. As calculated in Examples 1-3 and 1-4, $E_c - E_f = 0.093$ eV for the *N* side and $E_f - E_v = -0.103$ V for the *p* side. When the two regions are brought into contact, the Fermi level on the *N* side is initially higher than that on the *p*-side. Based on the above description about the chemical potential, the electrons in a *p-N* junction tend to flow from regions of higher Fermi level (*N* regions) to regions of lower Fermi levels (*p* regions). This agrees intuitively with the fact that the electrons tend to diffuse from the region

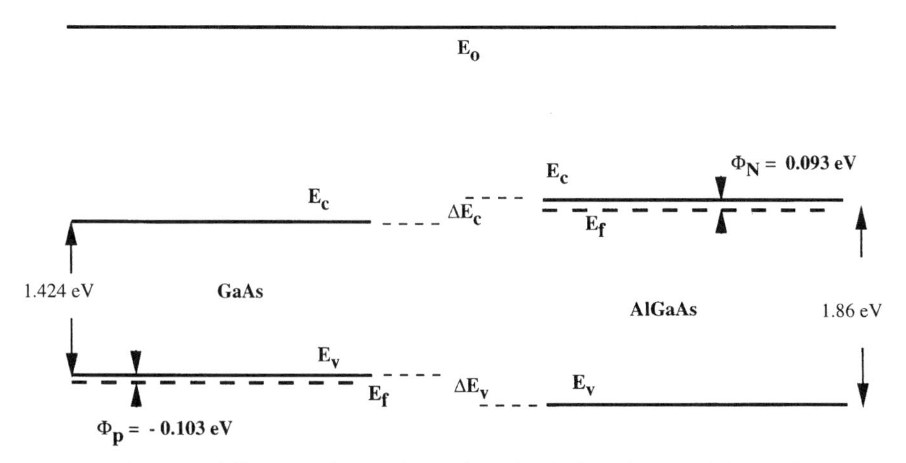

FIGURE 2-1. Band diagram of a *p-N* heterojunction before the two sides reach thermal equilibrium.

concentrated with more electrons to the region with less electrons. On the other hand, the holes flow from the region with a lower Fermi level toward the region with a higher Fermi level. Therefore, the holes flow from the p-type layer toward the N-type layer consistent with the fact that there are more holes in the p-type region and they tend to diffuse to the region with less holes. As these mobile carriers move toward the other sides, they leave behind the uncompensated dopant atoms near the junction. In the p-type region, the uncompensated acceptors are ions with negative charges. In the N-type region, the uncompensated donors are positive ions. Therefore, the carrier diffusion results in a electric field pointing from the N-type region to the p-type region. This electric field retards further electron diffusion from the N-type toward the p-type as well as hole diffusion in the opposite direction. Hence, there is a natural negative feedback mechanism such that the more carrier movement takes place, the larger the electric field becomes and the tendency of the carrier movement is reduced. Eventually, an equilibrium condition sets in and the tendency of the carrier diffusion is exactly counterbalanced by the electric field that impedes the carrier movement. At this precise movement, the Fermi levels at the two regions are aligned. The above exchange of carriers occurs on a very short time scale, so the whole process can be thought of as instantaneous. The two sides immediately adjacent to the junction where the dopants become uncompensated are called the *depletion regions*, meaning that they are depleted of mobile carriers. (Another name for the depletion region is the *space-charge region*). At the two extreme ends away from the actual junction, carrier movement has never occurred. Their dopant atoms are still compensated by their respective electrons or holes. These are called the *neutral regions* because the net charge concentrations there are zero.

Given that the Fermi level during thermal equilibrium lines up throughout the entire semiconductor, we draw a horizontal line in the band diagram to represent the constant Fermi level. At regions far away from the junction where the semiconductor remains neutral, the relative positions of E_f with respect to E_c and to E_v are unmodified from those prior to the joining of the two sides. This is shown in Fig. 2-2. The conduction band edges at the two sides are not at the same level, since it is the Fermi level that must be aligned. The same can be said about the valence band edges. The conduction and valence band edges across the depletion region must be connected somehow to form continuous curves. The exact manner at which they are connected is discussed shortly. For now, we simply connect them qualitatively, in a way that makes some sense. There are discontinuities in the conduction and the valence bands right at the junction interface, denoted by ΔE_c and ΔE_v, respectively. Their values are given by Eqs. (1-90) and (1-89), respectively.

To ascertain the manner in which the conduction and valence band edges are connected across the depletion region, we solve the Poisson equation of Eq. (1-8). Let $x = 0$ correspond to the junction boundary, with the p side in the $-x$ direction and the N side in the $+x$ direction. We denote $x = -X_{p0}$ as the boundary separating the neutral and the depletion regions on the p side, and

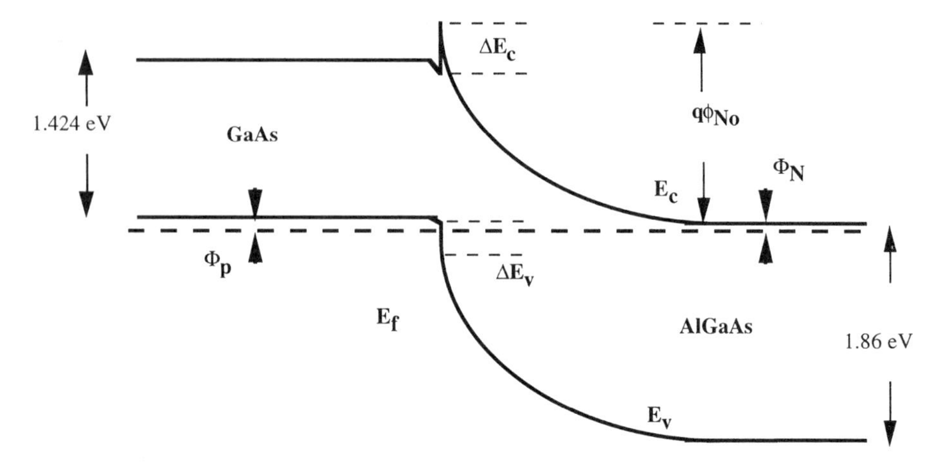

FIGURE 2-2. Band diagram of the *p-N* heterojunction of Fig. 2-1 after the two sides reach thermal equilibrium.

X_{N0} the boundary on the N side. The subscript 0 emphasizes that we are considering the thermal equilibrium condition. Generally, X_p and X_N denote the depletion boundaries in the presence of an external bias, a condition that is examined in the subsequent section. In the p-side depletion region, where $p \approx n \approx N_d \approx 0$, the net charge concentration is the negative acceptor density. In the N-side depletion region, where $n \approx p \approx N_a \approx 0$, the net charge concentration is the donor density. Therefore, Eq. (1-8) simplifies to

$$\frac{d\varepsilon}{dx} = -\frac{q}{\epsilon_p} N_a \qquad \text{for} - X_{p0} < x < 0 \tag{2-1a}$$

$$\frac{d\varepsilon}{dx} = \frac{q}{\epsilon_N} N_d \qquad \text{for } 0 < x < X_{N0} \tag{2-1b}$$

where ϵ_p and ϵ_N are the dielectric constants of the p-type and the N-type semiconductor materials, respectively. The neglect of the mobile carriers constitutes the *depletion approximation*, which is often assumed in an analysis of a *p-n* homojunction as well. We critique this assumption at the end of this section.

After integrating Eqs. (2-1a) and (2-1b) and applying the boundary conditions that the electric fields at $-X_{p0}$ and at X_{N0} are zero, we find

$$\varepsilon(x) = -\frac{q}{\epsilon_p} N_a(x + X_{p0}) \qquad \text{for } -X_{p0} < x < 0 \tag{2-2a}$$

$$\varepsilon(x) = -\frac{q}{\epsilon_N} N_d(X_{N0} - x) \qquad \text{for } 0 < x < X_{N0} \tag{2-2b}$$

We concentrate on analyzing the p side temporarily. The potential profile is obtained by integrating $\varepsilon(x)$ in accordance with Eq. (1-9):

$$V(X) = -\int \varepsilon(x)\,dx = \frac{q}{\epsilon_p} N_a\left(\frac{x^2}{2} + X_{p0}x\right) + C \qquad \text{for } -X_{p0} < x < 0 \tag{2-3}$$

The integration constant C is evaluated once a boundary condition is known. Since it is the relative value of potentials rather than their absolute values that is of importance, we arbitrarily define the zero potential at a convenient location, namely, at $x = -X_{p0}$. With the boundary condition that $V(-X_{p0}) = 0$, $V(x)$ is written as

$$V(X) = \frac{q}{\epsilon_p} N_a\left(\frac{x^2}{2} + X_{p0}\,x + \frac{X_{p0}^2}{2}\right) \qquad \text{for } -X_{p0} < x < 0 \tag{2-4}$$

The built-in potential on the p side (ϕ_{p0}) is the difference in the potentials at $x = 0$ and $x = -X_{p0}$. It is readily verified from Eq. (2-4) that

$$\phi_{p0} = \frac{q}{2\epsilon_p} N_a X_{p0}^2, \tag{2-5}$$

where ϕ_{p0} is a positive number. The band diagram, being an energy diagram for the negatively charged electron, shows that the electron energy decreases with the associated increase in $V(x)$.

We now work on the potential profile on the N side. The potential is again obtained by integrating the appropriate electric profile. Taking the boundary condition that $V(0) = \phi_{p0}$, we obtain

$$V(x) = \frac{q}{\epsilon_N} N_d\left(X_{N0}\,x - \frac{x^2}{2}\right) + \phi_{p0} \qquad \text{for } 0 < x < X_{N0} \tag{2-6}$$

The built-in potential on the N side (ϕ_{N0}) is the difference in the potentials at $x = 0$ and $x = X_{N0}$. $V(X_{N0}) - V(0)$, according to Eq. (2-6), is equal to

$$\phi_{N0} = \frac{q}{2\epsilon_N} N_d X_{N0}^2 \tag{2-7}$$

The overall built-in junction potential (ϕ_{bi}) is the potential difference from the neutral region on one side the neutral region on the other. It is equal to

$$\phi_{\text{bi}} = \phi_{N0} + \phi_{p0} = \frac{q}{2\epsilon_N} N_d X_{N0}^2 + \frac{q}{2\epsilon_p} N_a X_{p0}^2 \tag{2-8}$$

Generally, the phrase "built-in potential" without a qualifier refers to the junction built-in potential given by Eq. (2-8). Thus far, both X_{p0} and X_{N0} are unknown quantities. These two unknowns can be determined by solving two linearly independent equations relating these two variables. One of the required equations is obtained by enforcing the continuity of the electric flux density $D = \epsilon\varepsilon$ in the absence of an interfacial charge density at the junction:

$$\epsilon_p \varepsilon(0^-) = \epsilon_N \varepsilon(0^+) \tag{2-9}$$

Substituting $\varepsilon(0^-)$ from Eq. (2-2a) and $\varepsilon(0^+)$ from Eq. (2-2b), we obtain the following condition:

$$N_a X_{p0} = N_d X_{N0} \tag{2-10}$$

This equation is sometimes called the charge conservation relationship. The equality ensures that charge neutrality exists in the overall p-N junction.

The second equation relating X_{N0} and X_{p0} is found by taking the ratio of the built-in potentials on the N side to those on the p side:

$$\frac{\phi_{N0}}{\phi_{p0}} = \frac{\epsilon_p N_d X_{N0}^2}{\epsilon_N N_a X_{p0}^2} \tag{2-11}$$

Substituting the charge conservation relationship into the above equation, we rewrite the ratio as

$$\frac{\phi_{N0}}{\phi_{p0}} = \frac{\epsilon_p N_a}{\epsilon_N N_d} \tag{2-12}$$

From this equation, the built-in potential of the junction is

$$\phi_{\text{bi}} = \left(1 + \frac{\phi_{N0}}{\phi_{p0}}\right)\phi_{p0} = \left(1 + \frac{\epsilon_p N_a}{\epsilon_N N_d}\right)\phi_{p0} \tag{2-13}$$

Alternatively, from an examination of the band diagram of Fig. 2-2, ϕ_{bi} is equal to

$$\phi_{bi} = \frac{E_{gp} + \Delta E_c - \Phi_p - \Phi_N}{q} \tag{2-14}$$

where E_{gp} is the energy gap of the p-side material (which has the smaller energy gap). Φ_p and Φ_N represent the differences of the Fermi levels with respect to the valence band edge and the conduction band edge, respectively, in the neutral regions. That is,

$$\begin{aligned} \Phi_p &= E_f - E_v|_{x=-\infty} \\ \Phi_N &= E_c - E_f|_{x=\infty} \end{aligned} \tag{2-15}$$

Both Φ_p and Φ_N are calculated from the equilibrium statistics equations given in Eqs. (1-21) and (1-22). Once the doping levels are specified, ϕ_{bi} for the heterojunction is readily calculated from Eq. (2-14). In turn, ϕ_{p0} can be obtained from Eq. (2-13). Other quantities such as ϕ_{N0} and $V(x)$ are then obtained from the aforementioned equations.

Example 2-1:

Draw a band diagram for a (p) GaAs/(N) $Al_{0.35}Ga_{0.65}As$ heterojunction at thermal equilibrium. The doping level and relevant material parameters are identical to those of Examples 1-3 and 1-4. That is, for the p-side GaAs, $N_a = 3 \times 10^{19}$ cm^{-3}; and for the N-side AlGaAs, $N_d = 1 \times 10^{16}$ cm^{-3}.

From the cited examples, we found that $\Phi_p = -0.103$ eV and $\Phi_N = 0.093$ eV. ΔE_0, according to Eq. (1-90), is 0.244 eV. E_{gp}, the energy gap of the narrower gap material (GaAs), is 1.424 eV. Therefore, the built-in voltage is found from Eq. (2-14):

$$\phi_{bi} = 1.424 + 0.244 - (-0.103) - 0.093 = 1.678 \text{ V}$$

From Eq. (1-92), the relative dielectric constants are 13.18 for GaAs and 12.09 for $Al_{0.35}Ga_{0.65}As$. Equation (2-13) is used to find ϕ_{p0}:

$$\phi_{p0} = 1.678 \div \left[1 + \frac{13.18(3 \times 10^{19})}{12.09(1 \times 10^{16})}\right] = 5.13 \times 10^{-4} \text{ V}$$

ϕ_{N0}, which is equal to $\phi_{bi} - \phi_{p0}$, is $1.678 - 5.13 \times 10^{-4} \approx 1.6775$ V. It is clear from this calculation that if one side is significantly more heavily doped than the other, practically all of the built-in potential drops across the depletion region of the lightly doped side. Once ϕ_{N0} and ϕ_{p0} are determined, X_{p0} and X_{N0} are calculated from Eqs. (2-5) and (2-7):

$$X_{N0} = \sqrt{\frac{2\epsilon_N}{qN_d}\phi_{N0}} = \sqrt{\frac{2(12.09)(8.85 \times 10^{-14})}{(1.6 \times 10^{19})(1 \times 10^{16})}1.677} = 4.74 \times 10^{-5} \text{ cm}$$

$$X_{p0} = \sqrt{\frac{2\epsilon_p}{qN_a}\phi_{p0}} = \sqrt{\frac{2(13.18)(8.85 \times 10^{-14})}{(1.6 \times 10^{19})(3 \times 10^{19})}5.13 \times 10^{-4}} = 1.58 \times 10^{-8} \text{ cm}$$

The depletion thickness of the heavily doped side is merely 1.58 Å, which is practically zero. With the calculated parameters, the band diagram is drawn as shown in Fig. 2-3. The band profiles vary parabolically with position as given by Eqs. (2-4) and (2-6).

To facilitate comparison among material systems, we also draw the band diagrams of the following heterojunctions: AlAs/GaAs (Fig. 2-4),

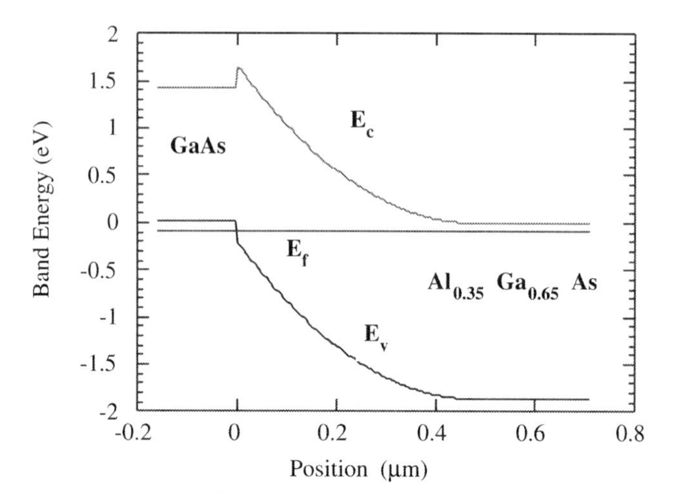

FIGURE 2-3. Band diagram of the (p) GaAs/(N)$Al_{0.35}Ga_{0.65}As$ heterojunction discussed in Example 2-1.

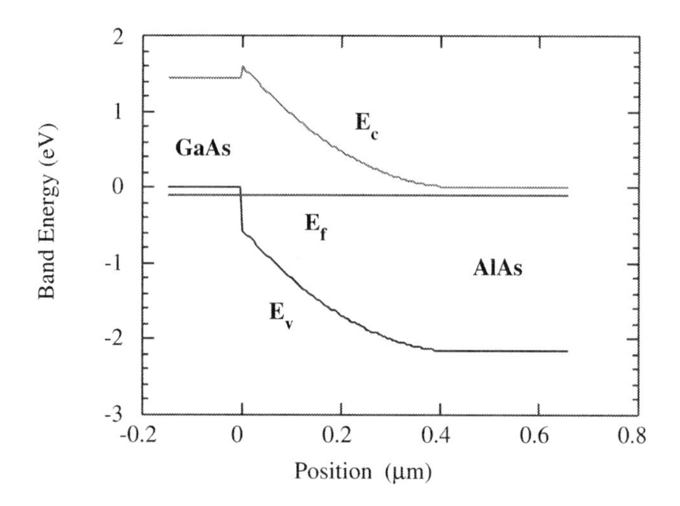

FIGURE 2-4. Band diagram of (p) GaAs/(N)AlAs heterojunction.

$InP/In_{0.53}Ga_{0.47}As$ (Fig. 2-5), $In_{0.52}Al_{0.48}As/In_{0.53}Ga_{0.47}As$ (Fig. 2-6), ordered $Ga_{0.51}In_{0.49}P$/GaAs (Fig. 2-7), and disordered $Ga_{0.51}In_{0.49}P$/GaAs (Fig. 2-8). In each case, the smaller-energy-gap material is doped p-type at a concentration of 3×10^{19} cm^{-3} and the larger-energy-gap material is doped N-type at a concentration of 1×10^{16} cm^{-3}. The relevant material parameters such as N_c, N_v, and ΔE_c are found in Appendix E and F.

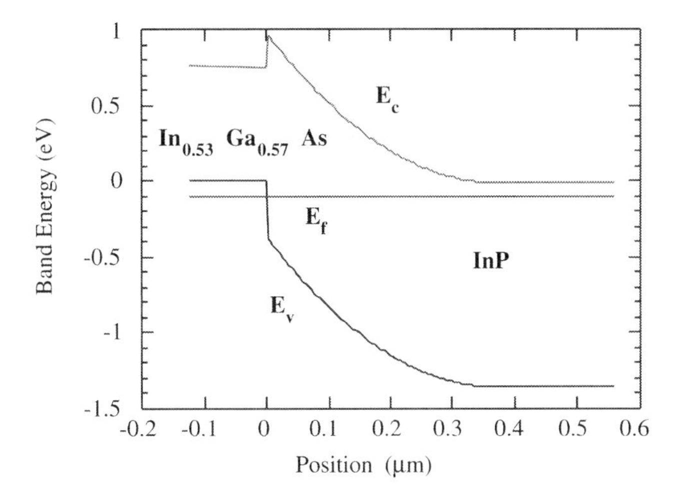

FIGURE 2-5. Band diagram of an InP/$In_{0.53}Ga_{0.47}As$ heterojunction.

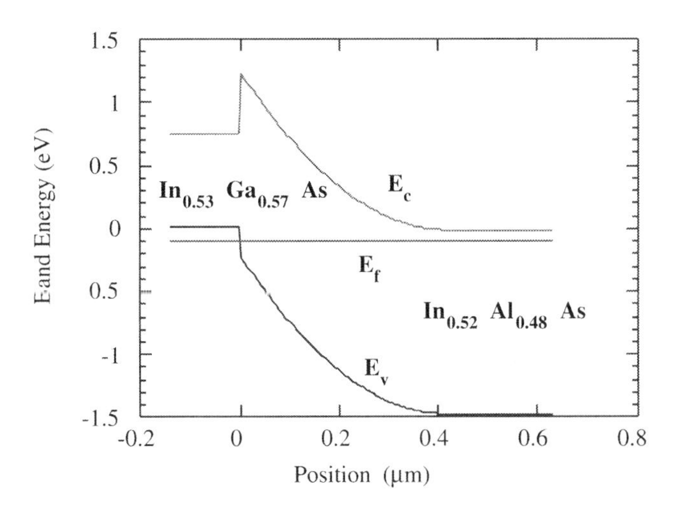

FIGURE 2-6. Band diagram of an $In_{0.52}Al_{0.48}As/In_{0.53}Ga_{0.47}As$ heterojunction.

The equations we have developed are applicable to a homojunction as well. Because $\epsilon_p = \epsilon_N$ and $\Delta E_c = 0$ in a homojunction, some simplification in the numerical calculation is expected.

Although they are not often considered in device design, the carrier concentrations inside the p-N junction are somewhat interesting. Because of the discontinuities in the conduction and the valence bands at the heterojunction, the carrier concentrations are discontinuous at $x = 0$. This fact is observed in the next example.

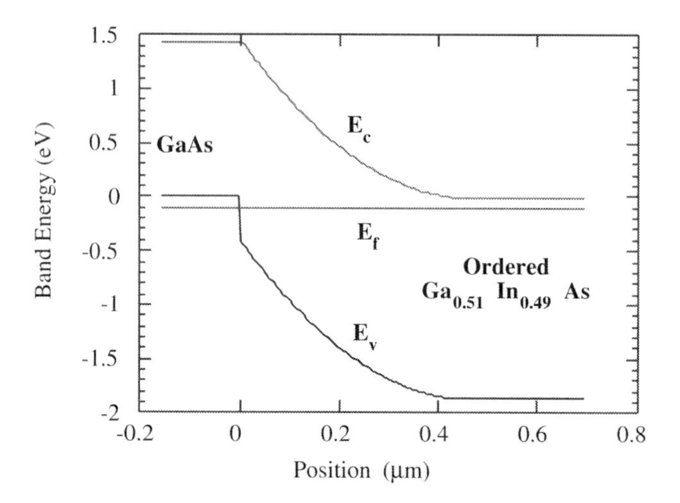

FIGURE 2-7. Band diagram of an ordered $Ga_{0.51}In_{0.49}P$/GaAs heterojunction.

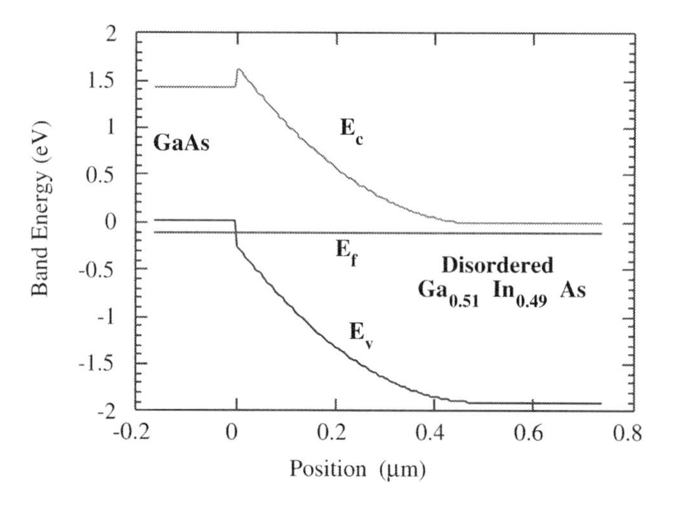

FIGURE 2-8. Band diagram of a disordered $Ga_{0.51}In_{0.49}P$/GaAs heterojunction.

Example 2-2:

Plot roughly the carrier concentrations as a function of distance for the *p-N* junction of Example 2-1.

The first step towards plotting the carrier concentrations is to establish the intrinsic carrier concentrations on both sides of the junction. n_i for GaAs, found from Appendix E, is 1.79×10^6 cm^{-3}. The intrinsic carrier concentration for $Al_{0.35}Ga_{0.65}As$, in contrast, requires some calculation. We determined in Example 1-2 that the conduction band density of states for AlGaAs is $N_c = 3.72 \times 10^{17}\ \text{cm}^{-3}$. We now determine N_v, which is obtained indirectly through the use of the heavy-hole and light-hole effective masses described in § 1-7. From Eqs. (1-83) and (1-84), with ξ being 0.35, $m^*_{\text{lh}} = 0.087 + 0.063(0.35) = 0.109$ and $m^*_{\text{hh}} = 0.62 + 0.14(0.35) = 0.669$. The valence band density of states is calculated from Eq. (1-27).

$$N_v = 2\left[\frac{2\pi(1.38\times10^{-23})300}{(6.624\times10^{-34})^2}\right]^{3/2}\{[0.109(9.1\times10^{-31})]^{3/2} + [0.669(9.1\times10^{-31})]^{3/2}\}$$
$$= 7.3\times10^{24}\ \text{m}^{-3} \quad \text{or} \quad 7.3\times10^{18}\ \text{cm}^{-3}$$

The energy gap for $Al_{0.35}Ga_{0.65}As$ is obtained from Eq. (1-88) to be $1.424 + 1.247(0.35) = 1.86$ eV. With $N_c = 3.72\times10^{17}\ \text{cm}^{-3}$, $N_v = 7.3\times10^{18}\ \text{cm}^{-3}$, and $E_g = 1.86$ eV, we find n_i at $T = 300$ K according to Eq. (1-13):

$$n_i = \sqrt{N_c N_v}\exp\left(-\frac{E_g}{2kT}\right) = 3.6\times10^2\ \text{cm}^{-3}$$

In the following, we assume the Maxwell–Boltzmann approximation of the Fermi integral for simplicity, using the relationships given in Eqs. (1-21) and (1-22). For the N side, we see from the band diagram that $(E_c - E_f)|_{x=0} = (E_c - E_f)|_{x=\infty} + q\phi_{NO}$. Dividing both sides by $-kT$ and then taking the exponential, we find that

$$N_c\exp\left(-\frac{E_c - E_f}{kT}\bigg|_{x=0}\right) - N_c\exp\left(-\frac{E_c - E_f}{kT}\bigg|_{x=\infty}\right)\exp\left(-\frac{q\phi_{NO}}{kT}\right) \tag{2-16}$$

Simplifying the terms, we arrive at a relationship governing the surface electron concentration to the built-in potential across the N-type depletion region:

$$n|_{x=0+} = N_d\exp\left(-\frac{q\phi_{NO}}{kT}\right) \tag{2-17}$$

Substituting the values $N_d = 1\times10^{16}\ \text{cm}^{-3}$ and $\phi_{NO} = 1.6775$ V, we find $n|_{x=0+} = 5.9\times10^{-13}\ \text{cm}^{-3}$. Using $n_i = 3.6\times10^2\ \text{cm}^{-3}$ for AlGaAs, we find $p|_{x=0+} = n_i^2/n = 2.2\times10^{17}\ \text{cm}^{-3}$. At the depletion edge, $n|_{x=X_{NO}} = N_d = 1\times10^{16}\ \text{cm}^{-3}$ and $p|_{x=X_{NO}} = n_i^2/n = 1.3\times10^{-11}\ \text{cm}^{-3}$. A relationship similar to Eq. (2-17) exists for holes on the N side:

$$p|_{x=0+} = \frac{n_i^2}{N_d}\exp\left(\frac{q\phi_{NO}}{kT}\right) = p|_{x=X_{NO}}\exp\left(\frac{q\phi_{NO}}{kT}\right) \tag{2-18}$$

We now calculate the carrier concentrations on the p side. Similar to the process outlined for Eq. (2-16) that leads to Eq. (2-17), the hole concentration on the p side is given by

$$p|_{x=0-} = N_a \exp\left(\frac{q\phi_{p0}}{kT}\right) \tag{2-19}$$

Substituting the values $N_a = 3 \times 10^{19}$ cm^{-3} and $\phi_{p0} = 5.13 \times 10^{-4}$ V, we find $p|_{x=0-} = 2.94 \times 10^{19}$ cm^{-3}. Using $n_i = 1.79 \times 10^6$ cm^{-3} for GaAs, $n|_{x=0-} = n_i^2/p = 1.09 \times 10^{-7}$ cm^{-3}. At the depletion edge, $p|_{x=X_{p0}} = N_a = 3 \times 10^{19}$ cm^{-3} and $n|_{x=-X_{p0}} = n_i^2/p = 1.07 \times 10^{-7}$ cm^{-3}.

We tabulate these critical concentrations on a $\log_{10}$ scale:

	$n\|_{x=0+}$	$p\|_{x=0+}$	$n\|_{XN0}$	$p\|_{x=XN0}$	n_i(AlGaAs)	$p\|_{x=0-}$	$n\|_{x=0-}$	$p\|_{x=Xp0}$	$n\|_{x=-Xp0}$	n_i(GaAs)
(cm^{-3})	5.9E-13	2.2E17	1E16	1.3E-11	3.6E2	2.94E19	1.09E-7	3E19	1.07E-7	1.79E6
$\log_{10}$	−12.2	17.34	16	−10.9	2.56	19.47	−6.96	19.48	−6.97	6.25

We plot the carrier concentration as a function of distance, as shown in Fig. 2-9. The intrinsic carrier concentration on each side is constant, and the

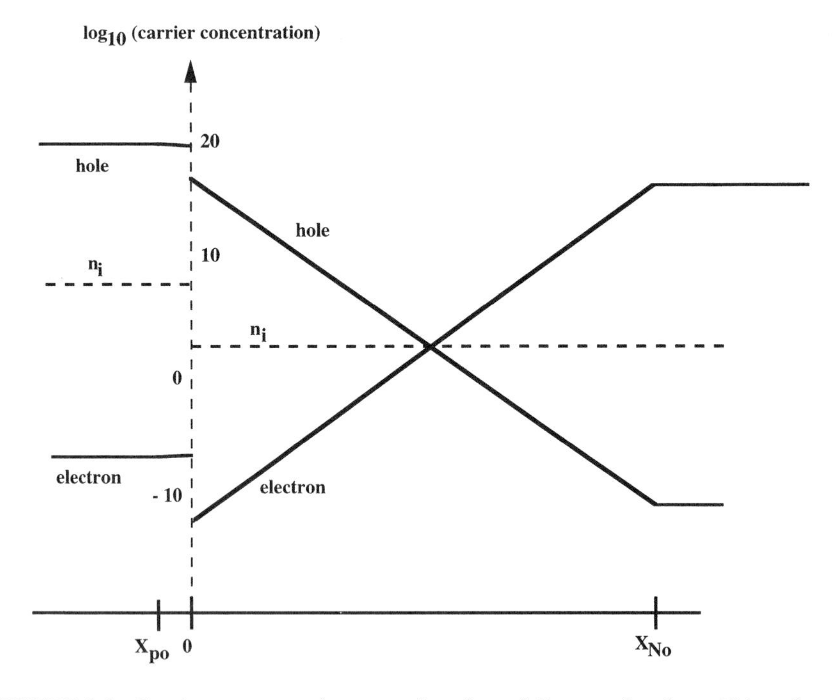

FIGURE 2-9. Carrier concentrations as a function of distance for the p-N junction of Example 2-2.

electron and hole profiles are symmetrical with respect to it. The carrier concentrations are discontinuous at the junction interface, due to the band discontinuities between the two different materials.

There is one subtle error associated with the above calculation. At the N-AlGaAs side, we calculated the minority hole concentration at the junction interface to be 2.2×10^{17} cm^{-3}. This value is larger than N_d, which is only 1×10^{16} cm^{-3}. In the depletion approximation used to derive all previous equations, we assumed that the net charge at $X_{NO} > x > 0$ is exactly equal to the uncompensated donor density N_d. With the newly realized fact that $p > N_d$, the whole analysis is therefore incorrect. The finding that depletion approximation can fail in a p-n junction with asymmetrical doping levels was published [1]. Fortunately, as is demonstrated in the following, the depletion approximation is adequate for practical consideration in which the potential profile is the main concern.

We revisit the Poisson equation and examine how we can incorporate the mobile carrier into our analysis. Combining Eqs. (1-8) and (1-9), we write

$$\frac{d^2V(x)}{dx^2} = -\frac{q}{\epsilon_s}(p - n + N_d - N_a) \tag{2-20}$$

On the p side, the depletion approximation holds up, and Eqs. (2-2a) and (2-4) remain applicable. On the N side, where the depletion approximation breaks down, $N_a = 0$ and n can still be approximated at zero. The hole concentration, p, however, can no longer be neglected in this more accurate analysis. The hole concentration as a function of position is inferred from Eq. (2-18) and is given by

$$p(x) = \frac{n_i^2}{N_d} \exp\left[\frac{q\phi_{NO} - q[V(x) - \phi_{pO}]}{kT}\right] \tag{2-21}$$

We quickly check the validity of this equation. At $x = 0^+$, $V(x) = \phi_{pO}$, so the above equation becomes identical to Eq. (2-18). At $x = X_{NO}$, $V(x) = \phi_{pO} + \phi_{NO}$. Therefore, $p(x)$ becomes n_i^2/N_d, as expected. Together with Eq. (2-20), we arrive at the following differential equation, which does not employ the depletion approximation:

$$\frac{d^2V(x)}{dx^2} = -\frac{q}{\epsilon_s}\left(N_d + \frac{n_i^2}{N_d} \exp\left[\frac{q\phi_{NO} - q[V(x) - \phi_{pO}]}{kT}\right]\right) \tag{2-22}$$

Unfortunately, there is no analytical solution for this nonlinear ordinary differential equation. Even though the potential as a function of the distance cannot be ascertained, we can still proceed with the problem somewhat. We integrate both sides of the above equation with respect to $dV(x)$. The following is a useful mathematical identity:

$$\int \frac{d^2V}{dx^2}\, dV = \int \frac{d^2V}{dx^2}\frac{dV}{dx}\, dx = \int \frac{dV}{dx}\left(\frac{d^2V}{dx^2}\, dx\right) = \int \frac{dV}{dx}\, d\left(\frac{dV}{dx}\right) \tag{2-23}$$

Integrating Eq. (2-22) from $x = X_{NO}$ (where $dV/dx = \varepsilon = 0$ and $V = \phi_{po} + \phi_{NO}$) to any arbitrary x, we obtain

$$\int_0^{dV/dx} \frac{dV}{dx}\, d\left(\frac{dV}{dx}\right) = -\frac{q}{\epsilon_N}\int_{\phi_{NO}+\phi_{po}}^{V(x)} N_d + \frac{n_i^2}{N_d}\exp\left[\frac{q\phi_{NO} - q[V(X) - \phi_{po}]}{kT}\right] dV \tag{2-24}$$

Despite the fact that we cannot solve for $V(x)$ analytically, the above equation renders an expression for the electric field profile, which is $-dV/dx$. Carrying out the integration, we find:

$$\frac{1}{2}\varepsilon(x)^2 = \frac{q}{\epsilon_N}\left(N_d[\phi_{po} + \phi_{NO} - V(x)] + \frac{n_i^2}{N_d}\frac{kT}{q} \times \left\{\exp\left[\frac{q\phi_{po} + q\phi_{NO} - qV(x)}{kT}\right] - 1\right\}\right) \tag{2-25}$$

It is easily verified that the above equation satisfies the boundary conditions that $\varepsilon = 0$ and $V = \phi_{po} + \phi_{NO}$ at $x = X_{NO}$.

We invoke the continuity relationship for the electric flux because we assume that there is no interface charge at the junction. So, $\varepsilon(0^+) = \epsilon_p\varepsilon(0^-)/\epsilon_N$. Since the depletion approximation remains valid on the p side, Eq. (2-2a), which describes the electric field at $x = 0^-$, is still applicable. Substituting $x = 0$ into the equation, we obtain

$$\varepsilon(0^+) = -\frac{q}{\epsilon_N} N_a X_{po} \tag{2-26}$$

With this information about $\varepsilon(0^+)$ and that $V(0^+) = \phi_{po}$, Eq. (2-25) gives the following condition:

$$\frac{q^2}{2\epsilon_N^2} N_a^2 X_{po}^2 = \frac{q}{\epsilon_N}\left\{N_d\phi_{NO} + \frac{n_i^2}{N_d}\frac{kT}{q}\left[\exp\left(\frac{q\phi_{NO}}{kT}\right) - 1\right]\right\} \tag{2-27}$$

With the identity that $\phi_0 = qN_aX_0^2/(2\epsilon_p)$, the above equation is rewritten as

$$\frac{\epsilon_p}{\epsilon_N} N_a\phi_{p0} = N_d\phi_{N0} + \frac{kT}{q}\frac{n_i^2}{N_d}\left[\exp\left(\frac{q\phi_{N0}}{kT}\right) - 1\right] \tag{2-28}$$

Previously, when the hole was neglected in the N-side depletion region, the second term of the right-hand side was zero. We then easily obtained the ratio of ϕ_{N0}/ϕ_{p0}, as given in Eq. (2-12), and subsequently, ϕ_{N0} and ϕ_{p0} could be obtained individually as demonstrated in Example 2-1. At present, when the hole concentration is being considered, the determination of the ϕ_{N0}/ϕ_{p0} ratio becomes difficult because Eq. (2-28) in its entirety is a transcendental equation. To facilitate the solution, we note that $\phi_{p0} = \phi_{bi} - \phi_{N0}$ and substitute it into the above equation. An equation with a single unknown ϕ_{N0} thus results:

$$\frac{\epsilon_p}{\epsilon_N} N_a(\phi_{bi} - \phi_{N0}) - N_d\phi_{N0} = \frac{kT}{q}\frac{n_i^2}{N_d}\left[\exp\left(\frac{q\phi_{N0}}{kT}\right) - 1\right] \tag{2-29}$$

In Example 2-1, for the (p) GaAs/(N) $Al_{0.35}Ga_{0.65}As$ heterojunction with $N_d = 1 \times 10^{16}$ cm^{-3} and $N_a = 3 \times 10^{19}$ cm^{-3}, we find ϕ_{bi} to be 1.678 V. Substituting the value $\epsilon_p/\epsilon_N = 13.18/12.09$, $kT/q = 0.0258$ V, and $n_i = 3.6 \times 10^2$ cm^{-3} for AlGaAs, we determine ϕ_{N0} numerically from Eq. (2-29) to be 1.6773 V. This value does not differ much at all from the value based on depletion approximation (1.6775 V)! Therefore, we continue to use the depletion approximation to draw the band diagram. However, we acknowledge a limitation of the depletion approximation. Since the minority hole concentration adds constructively to the net positive charge concentration on the N side, the actual electric field is higher than that obtained with the depletion approximation. Consequently, we can infer that the actual breakdown voltage of a p-n junction is lower than that predicted by the depletion approximation.

§ 2-2 p^+-N HETEROJUNCTION UNDER EXTERNAL BIAS

Figure 2-10 illustrates the situation when an external bias is applied to the p-N junction, with the p side chosen as the positive terminal. When V_a is positive, the p-N junction is said to be *forward-biased*. Conversely, when V_a is negative, the p-N junction is said to be *reverse-biased*. We again use depletion approximation to find the qualitative features of the band diagram under nonequilibrium.

Under an external bias, the actual numerical values of the depletion thicknesses and built-in potentials of the two neutral regions differ from those under equilibrium. To distinguish these values, we use X_N here to denote the depletion thickness on the N side under an external bias, as opposed X_{N0}, which applies to the thermal equilibrium condition. Similarly, we use X_p rather than X_{p0} for the p side. The built-in potentials under the bias condition are written as ϕ_p and ϕ_N,

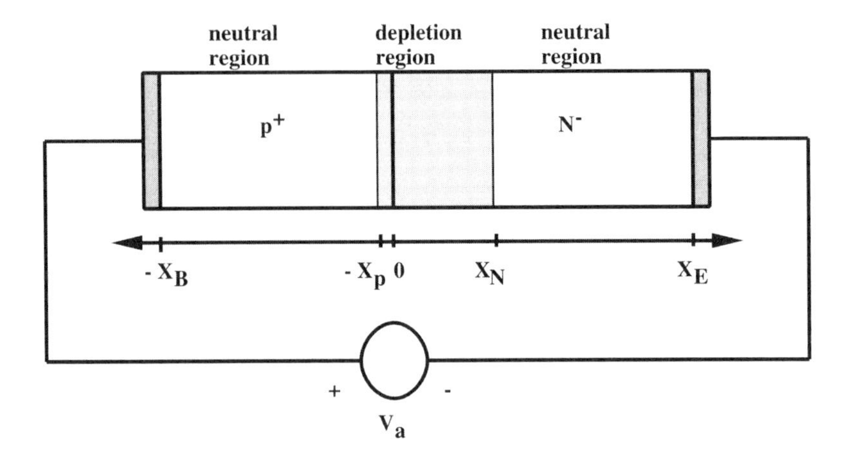

FIGURE 2-10. A *p-N* junction under an external bias of V_a. The positive terminal of the bias voltage is assigned to the *p* side. The depletion region is shaded.

rather than ϕ_{p0} and ϕ_{N0}. We further break down V_a into two components, with V_{ap} being the portion of the applied bias that drops on the *p* side of the junction and V_{aN}, the voltage drop on the *N* side. Hence,

$$V_a = V_{ap} + V_{aN} \tag{2-30}$$

Solving the differential equations posed in Eq. (2-1) and applying the boundary condition that $\varepsilon(X_p) = \varepsilon(X_N) = 0$, we obtain

$$\varepsilon(x) = -\frac{q}{\epsilon_p} N_a(x + X_p) \qquad \text{for} -X_p < x < 0 \tag{2-31a}$$

$$\varepsilon(x) = -\frac{q}{\epsilon_N} N_d(X_N - x) \qquad \text{for } 0 < x < X_N \tag{2-31b}$$

Let us focus on the *p* side briefly. The potential profile is obtained by integrating $\varepsilon(x)$ in accordance with Eq. (1-9). Similar to the previous section, we arbitrarily choose the potential to be zero at $x = -X_p$. With the boundary condition that $V(X_p) = 0$, $V(X)$ is then written as

$$V(x) = \frac{q}{\epsilon_p} N_a\left(\frac{x^2}{2} + X_p x + \frac{X_p^2}{2}\right) \qquad \text{for} -X_p < x < 0 \tag{2-32}$$

The built-in potential on the *p* side (ϕ_p) is the difference in the potentials at $x = 0$ and $x = -X_p$. It is by our definition equal to $\phi_{p0} - V_{ap}$. Therefore,

$$\phi_{p0} - V_{ap} = \phi_p = \frac{q}{2\epsilon_p} N_a X_p^2 \tag{2-33}$$

We now work on the potential profile on the N side, which is obtained by integrating the electric field profile of Eq. (2-31b). Taking the boundary condition that $V(0) = \phi_p$, we find

$$V(x) = \frac{q}{\epsilon_N} N_d \left(X_N x - \frac{x^2}{2} \right) + \phi_p \qquad \text{for } 0 < x < X_N \tag{2-34}$$

The built-potential on the N side (ϕ_N) is the difference in the potentials at $x = 0$ and $x = X_N$. It is equal to

$$\phi_{N0} - V_{aN} = \phi_N = \frac{q}{2\epsilon_N} N_d X_N^2 \tag{2-35}$$

We now seek the values of X_p and X_N, which are unknown thus far. We again assume that there is no interface charge at the junction surface, so the electric flux density $D = \epsilon\varepsilon$ is continuous:

$$\epsilon_p \varepsilon(0^-) = \epsilon_N \varepsilon(0^+) \tag{2-36}$$

Substituting $\varepsilon(0^-)$ from Eq. (2-31a) and $\varepsilon(0^+)$ from Eq. (2-31b), we obtain the charge conservation relationship:

$$N_a X_p = N_d X_N \tag{2-37}$$

Another equation relating X_N and X_p is found by taking the ratio of the built-in potentials on the N side to the p side, which, with the help of the above equation, is simplified to the following:

$$\frac{\phi_N}{\phi_p} = \frac{\phi_{N0} - V_{aN}}{\phi_{p0} - V_{ap}} = \frac{\epsilon_p N_d X_N^2}{\epsilon_N N_a X_p^2} = \frac{\epsilon_p N_a}{\epsilon_N N_d} \tag{2-38}$$

The following is an identity based on our definitions of V_{aN} and V_{ap}:

$$\phi_{\text{bi}} - V_a = (\phi_{N0} - V_{aN}) + (\phi_{p0} - V_{ap}) \tag{2-39}$$

An algebraic manipulation of the above two equations leads to

$$\phi_{p0} - V_{ap} = \frac{1}{1 + \epsilon_p N_a / \epsilon_N N_d} (\phi_{\text{bi}} - V_a) \tag{2-40}$$

Since ϕ_{bi}, V_a, and ϕ_{p0} are known quantities, V_{ap} is obtained from the above equation. Once V_{ap} is known, V_{aN} is determined from Eq. (2-39). The depletion thicknesses are then calculated according to

$$X_p = \sqrt{\frac{2\epsilon_p(\phi_{p0} - V_{ap})}{qN_a}} \tag{2-41}$$

$$X_N = \sqrt{\frac{2\epsilon_N(\phi_{N0} - V_{aN})}{qN_d}} \tag{2-42}$$

Example 2-3:

Draw the band diagram when the (p) GaAs/(N) $Al_{0.35}Ga_{0.65}As$ heterojunction of Example 2-1 is forward-biased with a $V_a = 0.5$ V. The relevant parameters given and calculated in the previous examples include: $N_a = 3 \times 10^{19}\,\text{cm}^{-3}$; $N_d = 1 \times 10^{16}\,\text{cm}^{-3}$; $\phi_{\text{bi}} = 1.678$ V; $\phi_{N0} = 1.6775$ V; $\phi_{p0} = 5.13 \times 10^{-4}$ V; $\epsilon_p/\epsilon_N = 13.18/12.09$; $kT/q = 0.0258$ V; and $n_i = 3.6 \times 10^2\,\text{cm}^{-3}$ for AlGaAs and $1.79 \times 10^6\,\text{cm}^{-3}$ for GaAs.

According to Eq. (2-40),

$$5.13 \times 10^{-4} - V_{ap} = \frac{1}{1 + 13.18(3 \times 10^{19})/12.09(1 \times 10^{16})}(1.678 - 0.5)$$

V_{ap} is calculated to be 1.53×10^{-4} V. $V_{aN} = V_a - V_{ap} = 0.5 - 1.53 \times 10^{-4} = 0.49985$ V. The built-in voltages on the two sides of the junction, which are critical to the construction of the band diagram, are the following: $\phi_p = \phi_{p0} - V_{ap} = (5.13 \times 10^{-4}) - (1.53 \times 10^{-4}) = 3.6 \times 10^{-4}$; $\phi_N = \phi_{N0} - V_{aN} = 1.6775 - 0.49985 = 1.17765$ V. The depletion thicknesses, according to Eqs. (2-41) and (2-42), are $X_P = 1.32$ Å and $X_N = 0.397$ µm. Compared to the values $X_{p0} = 1.58$ Å and $X_{N0} = 0.474$ µm at thermal equilibrium, we see that the depletion thicknesses shrink under forward bias. The band diagram is constructed as shown in Fig. 2-11.

The analysis developed applies equally well to the forward- and the reverse-biased conditions. The following example demonstrates how the band diagram is constructed under a reverse bias.

Example 2-4:

Solve Example 2-3, except that a reverse bias of -5 V is applied to the p-N junction.

We again begin with Eq. (2-40), with $V_a = -5$ V,

$$5.13 \times 10^{-4} - V_{ap} = \frac{1}{1 + 13.18(3 \times 10^{19})/12.09(1 \times 10^{16})}(1.678 + 5)$$

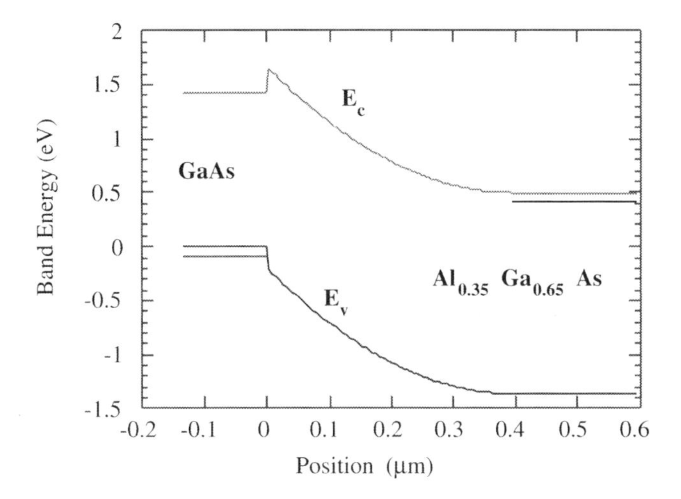

FIGURE 2-11. The band diagram of the GaAs/$Al_{0.35}Ga_{0.65}As$ heterojunction of Example 2-1 when it is forward-biased with $V_a = 0.5$ V.

V_{ap} is calculated to be -2.554×10^{-3} V. $V_{aN} = V_a - V_{ap} = -5 + (2.554 \times 10^{-3}) = -4.9974$ V. The built-in voltages in each side of the junction, which are critical to the construction of the band diagram, are the following: $\phi_p = \phi_{p0} - V_{ap} = (5.13 \times 10^{-4}) + (2.554 \times 10^{-3}) = 3.07 \times 10^{-3}$ V; $\phi_N = \phi_{N0} - V_{aN} = 1.6775 + 4.9974 = 6.675$ V. The depletion thicknesses, according to Eqs. (2-41) and (2-42), are $X_p = 2.73$ Å and $X_N = 0.945$ μm. Comparing these values to $X_{p0} = 1.58$ Å and $X_{N0} = 0.474$ μm at the thermal equilibrium, we find that the depletion thicknesses expand in a reverse bias. The band diagram is constructed as shown in Fig. 2-12.

Only the Fermi levels in the neutral regions are drawn on the band diagrams of Figs. 2-11 and 2-12. They are at different levels, separated by an energy amount equal to the product of the electron charge and the applied voltage ($q \cdot V_a$). One natural question is, what is the Fermi level inside the depletion region? This question is, in reality, somewhat contradictory. Fermi level, in a strict definition, is the chemical potential of the electrons when the semiconductor is in thermal equilibrium. Any external bias, whether a forward or a reverse bias, upsets the thermal equilibrium and the Fermi level is undefinable. However, we are accustomed to calculating carrier densities from Fermi levels, such as Eqs. (1-21) and (1-22). Under external biases, we introduce the so-called *quasi-Fermi levels*, which are defined as the levels such that the electrons and holes can still be calculated from equations similar to those of Eqs. (1-21) and (1-22). The reason that the prefix "quasi" is used is the following. In a nonequilibrium situation such as in a forward bias, the electrons are injected into the p region and the holes are injected into the N region. The excess carriers

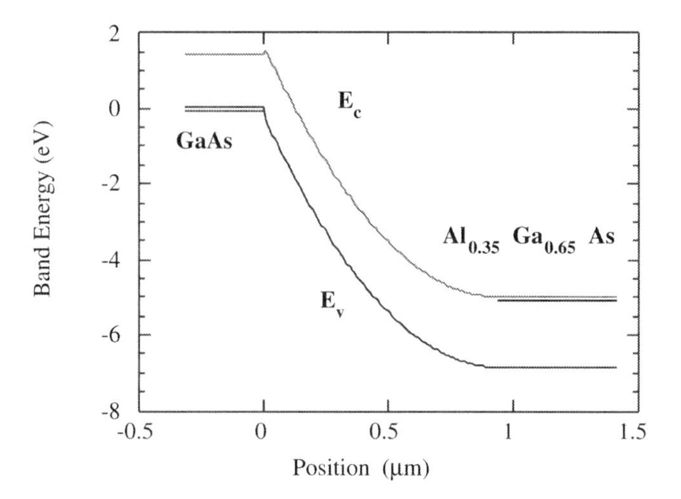

FIGURE 2-12. The band diagram of the GaAs/$Al_{0.325}Ga_{0.65}As$ heterojunction of Example 2-1 when it is reverse-biased with $V_a = -5$ V.

eventually recombine. The typical recombination times in GaAs are on the order of 1 ns. This is much longer than the times ($\sim$1 ps) required at the room temperature for the conduction electrons to reach thermal equilibrium with each other in the conduction band, and for holes to reach thermal equilibrium with each other in the valence band. The distributions of electrons and holes are very close to the thermal equilibrium distribution, although the total number of holes is not in equilibrium with the total number of electrons. Because the distributions are not far from those at equilibrium, the situation is called "quasi-equilibrium" and the characteristic levels used to find the electrons and holes are called "quasi-Fermi levels."

We denote the quasi-Fermi level for the electrons as E_{fn} and the quasi-Fermi level for the holes as E_{fp}. By definition, the electrons and holes are governed by the following equations:

$$n = n_i \exp\left(\frac{E_{fn} - E_i}{kT}\right) = N_c \exp\left(\frac{E_{fn} - E_c}{kT}\right) \tag{2-43}$$

$$p = n_i \exp\left(\frac{E_i - E_{fp}}{kT}\right) = N_v \exp\left(\frac{E_v - E_{fp}}{kT}\right) \tag{2-44}$$

The quasi-Fermi levels have been determined for an idealized *p-n* homojunction [2]. We extend the analysis for *p-N* heterojunctions in § 2-5, where we construct the quasi-Fermi levels in the depletion region. Although the levels we drew in the neutral regions should be referred to as the quasi-Fermi levels, we occasionally call them Fermi levels, since their relative position with the band edges does not change from those under the thermal equilibrium condition.

There is an interesting property about the definition of the quasi-Fermi levels. For the electrons, we derive another expression for the electron current based on Eq. (1-35):

$$J_n = q\mu_n\left(n\varepsilon + \frac{kT}{q}\frac{dn}{dx}\right) = q\mu_n\left\{\frac{n}{q}\frac{dE_c}{dx} + \frac{kT}{q}\left[n\frac{1}{kT}\left(\frac{dE_{fn}}{dx} - \frac{dE_c}{dx}\right)\right]\right\}$$

$$= q\mu_n n\frac{1}{q}\left(\frac{dE_{fn}}{dx}\right) \tag{2-45}$$

The electron conduction current is directly proportional to the spatial gradient of the electron quasi-Fermi level. A hole current expression is obtained analogously:

$$J_p = q\mu_p p\frac{1}{q}\left(\frac{dE_{fp}}{dx}\right) \tag{2-46}$$

§ 2-3 *p-N⁺*, *P⁺-n*, AND *P-n⁺* HETEROJUNCTIONS

As is discussed in Chapter 3, the base layer of a HBT has a narrower energy gap and a heavier doping concentration than the emitter material. The p^+-N heterojunction described in the last section is suitable for an *Npn* HBT that consists of an N-type, wide-energy-gap, lightly doped emitter and a p-type, narrow-energy-gap, heavily doped base. A p-N^+ heterojunction, in contrast, is not particularly useful for HBT because the N-type emitter would be more heavily doped than the p-type base. However, studying this heterojunction in some detail clarifies several concepts as well as completes our analysis of the p-N heterojunction in general.

When a p-N^+ heterojunction is formed, the Fermi levels of the two sides must line up during thermal equilibrium, as shown in Fig. 2-13. $\Phi_N = E_c - E_f$ is drawn to be small to reflect the fact that the N side is heavily doped. Likewise, $\Phi_p = E_f - E_v$ has a larger magnitude because the p-side doping is light. Based on the experience gained in previous sections, ϕ_N (the built-in potential on the N side) is practically zero, whereas ϕ_p (the built-in potential on the p side) is relatively large. Since both Φ_N and ϕ_N are small, the conduction band energy on the N^+ side is barely above the Fermi level. Even at the junction interface, where we normally expect a depletion region, the band tilts only slightly. In contrast, the conduction band discontinuity ΔE_c for a GaAs/$Al_{0.35}Ga_{0.65}As$ heterojunction is fairly large (0.244 eV). With ΔE_c greatly exceeding the sum of Φ_N and ϕ_N, the conduction band on the p side necessarily dips below the Fermi level. This fact immediately warns us that the depletion approximation and the Boltzmann approximation cannot be correct on the p side near the junction interface. We encountered a similar situation in § 2-1, where the depletion approximation also failed. There are two differences, however. First, in § 2-1, we sensed no immediate

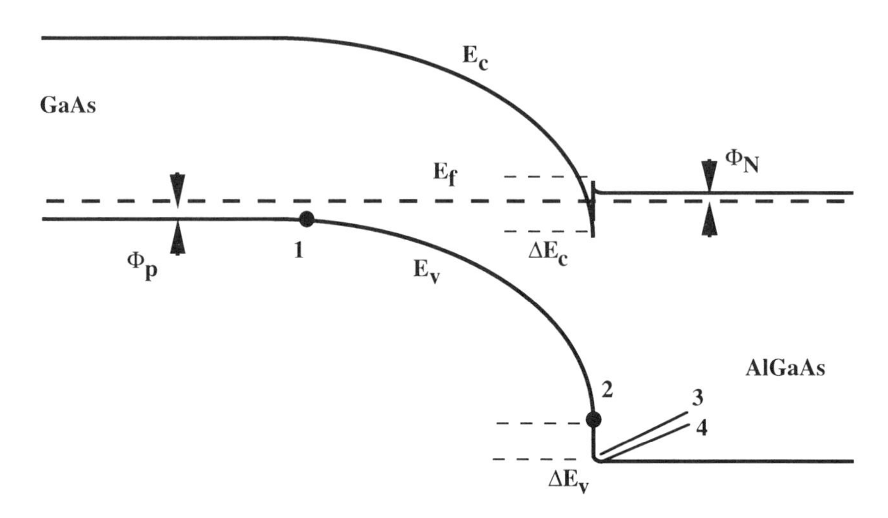

FIGURE 2-13. Band diagram of a p-N^+ heterojunction under thermal equilibrium.

warning of the problem of depletion approximation until we started plotting the actual carrier concentration. Here, just sketching the band diagram alerts us to the importance of mobile carriers. Second, in §2-1, the minority hole carrier was incorporated into the Poisson equation using an expression based on the Boltzmann approximation. In the present case, the conduction band edge on the p side of the p-N^+ heterojunction is below the Fermi level, so the Fermi–Dirac statistics must be used to calculate the minority electron concentration.

We lay out the framework to determine the band diagram of a p-N^+ heterojunction, taking into account the nonnegligible electron concentration in the p-GaAs layer. For simplicity (and without losing generality), we choose to use only the first term of the Joyce–Dixon approximation polynomial of Eq. (1-29). That is, we express the relationship between the free electron concentration and the Fermi level as

$$\frac{E_f - E_c}{kT} = \ln\left(\frac{n}{N_c}\right) + \frac{1}{\sqrt{8}}\left(\frac{n}{N_c}\right) \tag{2-47}$$

The Poisson equation for the p side, instead of Eq. (2-31a), is now written in accordance with Eq. (1-8). For completeness, we include both the electron and the hole carriers (not just the electron, which must be included):

$$\frac{d\varepsilon}{dx} = -\frac{q}{\epsilon_p}(N_a + n - p) = -\frac{q}{\epsilon_p}\left(N_a + n - \frac{n_i^2}{n}\right) \qquad (-X_p < x < 0) \tag{2-48}$$

Since E_c relates to ε through the fact that $1/q(dE_c/dx) = \varepsilon$, Eqs. (2-47) and (2-48) form a system of two simultaneous equations with two unknowns, $\varepsilon(x)$ and $n(x)$. Differentiating Eq. (2-47) with respect to x and noting that E_f is invariant with x, we find

$$-\frac{q}{kT}\varepsilon = \frac{N_c}{n}\frac{d}{dx}\left(\frac{n}{N_c}\right) + \frac{1}{\sqrt{8}}\frac{d}{dx}\left(\frac{n}{N_c}\right) = \frac{1}{n}\frac{dn}{dx} + \frac{1}{\sqrt{8}N_c}\frac{dn}{dx} \qquad (-X_p < x < 0) \tag{2-49}$$

Multiplying Eq. (2-48) by Eq. (2-49), we obtain

$$-\frac{q}{kT}\varepsilon\frac{d\varepsilon}{dx} = -\frac{q}{\epsilon_p}\left(N_a + n - \frac{n_i^2}{n}\right)\left[\frac{1}{n}\frac{dn}{dx} + \frac{1}{\sqrt{8}N_c}\frac{dn}{dx}\right] \qquad (-X_p < x < 0) \tag{2-50}$$

Some algebraic manipulation simplifies this equation to

$$\frac{\epsilon_p}{2kT}\frac{d\varepsilon^2}{dx} = \frac{n}{\sqrt{8}N_c}\frac{dn}{dx} + \left(1 + \frac{N_a}{\sqrt{8}N_c}\right)\frac{dn}{dx} + \left(N_a - \frac{n_i^2}{\sqrt{8}N_c}\right)\frac{1}{n}\frac{dn}{dx} - \frac{n_i^2}{n^2}\frac{dn}{dx} \tag{2-51}$$

We integrate both sides with respect to dx and get

$$\frac{\epsilon_p}{2kT}\varepsilon^2 = \frac{n^2}{2\sqrt{8}N_c} + \left(1 + \frac{N_a}{\sqrt{8}N_c}\right)n + \left(N_a - \frac{n_i^2}{\sqrt{8}N_c}\right)\ln(n) + \frac{n_i^2}{n} + C \tag{2-52}$$

where C is an integration constant, readily determined from the condition that $\varepsilon(-X_{p0}) = 0$ and $n(-X_{p0}) = n_i^2/N_a$. Once C is evaluated, we find $\varepsilon(0^-)$ expressed in terms of n_0, which is defined as $n(0^-)$:

$$\begin{aligned}&\frac{N_c}{N_a}\frac{1}{2\sqrt{8}N_c^2}\left(n_0^2 - \frac{n_i^4}{N_a^2}\right) + \left(\frac{1}{N_c} + \frac{N_a}{\sqrt{8}N_c^2}\right)\left(n_0 - \frac{n_i^2}{N_a}\right)\\ &+ \left(\frac{N_a}{N_c} - \frac{n_i^2}{\sqrt{8}N_c^2}\right)\ln\left(\frac{n_0 N_a}{n_i^2}\right) + \frac{n_i^2}{N_c}\left(\frac{1}{n_0} - \frac{N_a}{n_i^2}\right) = \frac{\epsilon_p}{2kTN_a}\varepsilon(0^-)^2\end{aligned} \tag{2-53}$$

Although the depletion approximation fails on the p side, it remains valid on the N side. (The expression would become more unnecessarily complicated though still manageable if we were to include the mobile carriers in the N-side depletion region as well.) Therefore, the equations developed in §2-2 for $0 < x < X_N$ are still applicable. In particular, we find that Eqs. (2-31b) and (2-35) can be combined to yield

$$\epsilon_N \varepsilon^2(0^+) = 2qN_d\phi_N = 2qN_d(\phi_{\text{bi}} - \phi_p) \tag{2-54}$$

ϕ_{bi} was given in Eq. (2-14) and is repeated here to stress that both the N-p^+ and N^+-p heterojunctions have the same built-in voltage (more discussion later):

$$\phi_{\text{bi}} = \frac{E_{gp} + \Delta E_c - \Phi_p - \Phi_N}{q} \quad \text{(abrupt } N\text{-}p \text{ heterojunction)} \tag{2-55}$$

The built-in voltage of the junction is known once the doping levels are specified. The second variable appearing in Eq. (2-54), the built-in potential on the p side (ϕ_p), is not yet known. Fortunately, we can express ϕ_p in terms of $n(0^-)$ by

$$\begin{aligned}\phi_p &= \frac{E_{gp}}{q} + \frac{kT}{q}\ln\left(\frac{N_a}{N_v}\right) + \left.\frac{E_f - E_c}{q}\right|_{x=0^+} \\ &= \frac{E_{gp}}{q} + \frac{kT}{q}\ln\left(\frac{N_a}{N_v}\right) + \frac{kT}{q}\left[\ln\left(\frac{n_0}{N_c}\right) + \frac{n_0}{\sqrt{8}N_c}\right]\end{aligned} \tag{2-56}$$

Plugging ϕ_p into Eq. (2-54), we end up with an expression for $\varepsilon(0^+)$ that depends solely on the unknown n_0. $\varepsilon(0^-)$ in Eq. (2-53) is also purely a function of n_0. Because Eqs. (2-54) and (2-53) are related through the fact that $\epsilon_p\varepsilon(0^-) = \epsilon_N\varepsilon(0^+)$, they are combined to form a transcendental equation about n_0, whose value can be determined through some numerical techniques. Once n_0 is ascertained, the values of all other variables are found. The exact band diagram of a p-N^+ heterojunction can then be constructed. We do not attempt to carry out the calculation because the p-N^+ heterojunction is not a practical element in HBTs. However, this analysis makes clear how Fermi–Dirac statistics are incorporated in a situation where the depletion approximation fails.

We elaborate on the built-in voltage. For many cases, writing ϕ_{bi} for a p^+-N heterojunction like that of Eq. (2-14) is merely a matter of intuition. However, the correct expression for ϕ_{bi} in a p-N^+ heterojunction is not entirely obvious. Sometimes it is mistakenly thought that, among other misconceptions, ϕ_{bi} is the potential difference between the conduction band edges of the N layer at $x = -\infty$ and of the p layer at $x = +\infty$. It is helpful to remember that ϕ_{bi} is the sum of the electrostatic potentials on both sides. If we focus on the conduction band, it is equal to the sum of $E_c(x = -\infty) - E_c(0^-)$ on the p side and $E_c(0^+) - E_c(\infty)$ on the N side. In other words, ϕ_{bi} is the total amount of bending in the conduction band (or the valence band). We carry out one exercise, using Fig. 2-13 as the band diagram. The built-in voltage is equal to the sum of

$E_v(x = -\infty) - E_v(0^-)$ on the p side and $E_v(0^+) - E_v(\infty)$ on the N side. It can be seen that $\phi_{bi} = V_{12} + V_{34}$, and

$$\phi_{bi} = \frac{E_{gN} - \Delta E_v - \Phi_p - \Phi_N}{q} \tag{2-57}$$

This is equivalent to the built-in potential given in Eq. (2-55).

Perhaps an easier method to identify the built-in voltage is to draw the band diagram in a space-charge-neutral situation (which is not the thermal equilibrium condition.) This is shown in Fig. 2-14. The built-in potential is the potential separation between the Fermi levels on the two sides. It is obvious from this figure that, whether we have a p^+-N or a p-N^+ heterojunction, the built-in potential is given in Eq. (2-55).

We comment on the sign of the built-in voltage. For any p-n junction, it is customary to take the p side as the positive terminal and the n side as the negative terminal. Under this convention, it is seen from Fig. 2-14 that with respect to the negative terminal (n side), the Fermi level at the positive terminal (p side) has a lower energy. This means that, in terms of electron potential, the p side is positive with respect to the n side. Hence, the built-in potential is positive. From this description, the built-in potential is always positive for a p-n homojunction and is generally positive for a heterojunction. However, as shown in § 2-7, it is possible to have a negative built-in potential in an n-N or p-P heterojunction.

We had examined the p^+-N heterojunction in § 2-1 and the p-N^+ heterojunction just now. We also want to study the n-P heterojunction. For a Pnp HBT, an n^+-P heterojunction is what is usually designed. However, the following

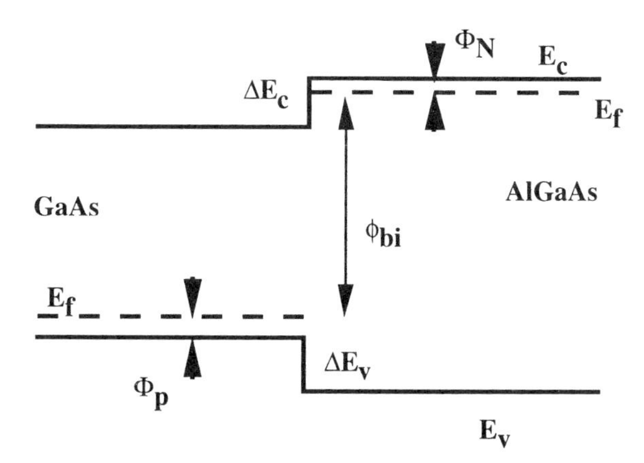

FIGURE 2-14. Band diagram of the p-N^+ heterojunction of Fig. 2-13 under the space-charge-neutral situation. This diagram facilitates the determination of the built-in voltage.

description fits equally well for an n-P^+ heterojunction. The band diagrams are shown in Figs. 2-15 (n^+-P) and 2-16 (n-P^+).

The built-in potential is the energy difference between the Fermi levels under the space-charge-neutral condition. Unlike the p-N heterojunction studies before, ϕ_{bi} for the P-n heterojunction takes on a slightly different form:

$$\phi_{\text{bi}} = \frac{E_{gP} - \Delta E_c - \Phi_P - \Phi_n}{q} \qquad \text{(abrupt } P\text{-}n \text{ heterojunction)} \qquad (2\text{-}58)$$

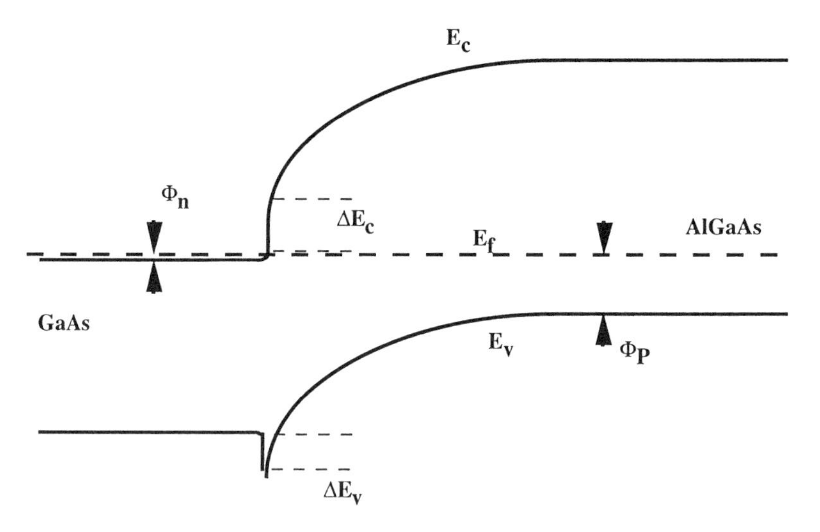

FIGURE 2-15. Band diagram of an n^+-P heterojunction under thermal equilibrium.

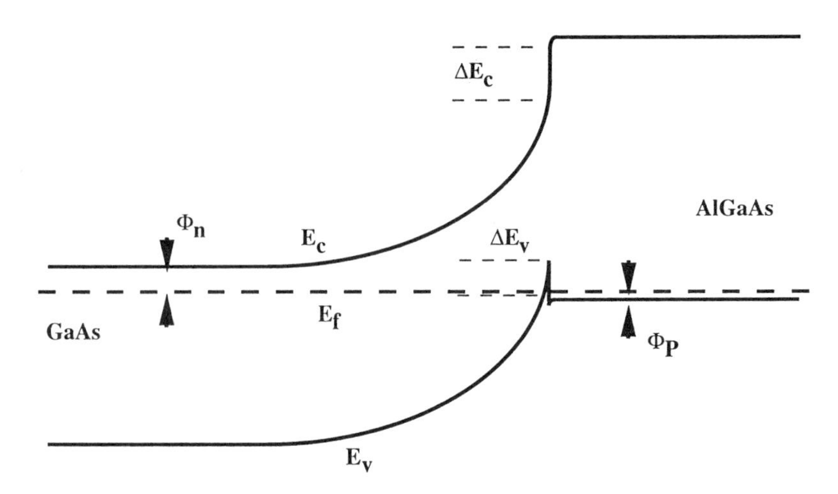

FIGURE 2-16. Band diagram of an n-P^+ heterojunction under thermal equilibrium.

where E_{gP} is the energy gap of the larger-energy-gap P-type material. Φ_P and Φ_n have similar definitions in Eq. (2-15):

$$\Phi_P = E_f - E_v|_{x=+\infty} \tag{2-59a}$$

$$\Phi_n = E_c - E_f|_{x=-\infty} \tag{2-59b}$$

The whole derivation of the electric field and potential profiles under thermal equilibrium is similar to that of the p-N heterojunction. We list the important relationships in the following:

$$\varepsilon(x) = -\frac{q}{\epsilon_n} N_d(x + X_{n0}) \qquad \text{for } -X_{n0} < x < 0 \tag{2-60a}$$

$$\varepsilon(x) = -\frac{q}{\epsilon_P} N_a(X_{P0} - x) \qquad \text{for } 0 < x < X_{P0} \tag{2-60b}$$

$$V(x) = \frac{q}{\epsilon_n} N_d\left(\frac{x^2}{2} + X_{n0}x + \frac{X_{n0}^2}{2}\right) \qquad \text{for } -X_{n0} < x < 0 \tag{2-61a}$$

$$V(x) = \frac{q}{\epsilon_P} N_d\left(X_{P0}x - \frac{x^2}{2}\right) + \phi_{n0} \qquad \text{for } 0 < x < X_{P0} \tag{2-61b}$$

$$\phi_{n0} = \frac{q}{2\epsilon_n} N_d X_{n0}^2 \tag{2-62}$$

$$\phi_{P0} = \frac{q}{2\epsilon_P} N_a X_{P0}^2 \tag{2-63}$$

$$\epsilon_n \varepsilon(0^-) = \epsilon_P \varepsilon(0^+) \tag{2-64}$$

$$N_d X_{n0} = N_a X_{P0} \tag{2-65}$$

$$\phi_{\text{bi}} = \left(1 + \frac{\phi_{P0}}{\phi_{n0}}\right)\phi_{n0} = \left(1 + \frac{\epsilon_n N_d}{\epsilon_P N_a}\right)\phi_{n0} \tag{2-66}$$

$$\begin{aligned} \phi_{\text{bi}} &= \phi_{P0} + \phi_{n0} = \frac{q}{2\epsilon_P} N_a X_{P0}^2 + \frac{q}{2\epsilon_n} N_d X_{n0}^2 \\ &= \frac{E_{gP} - \Delta E_c - \Phi_P - \Phi_n}{q} \end{aligned} \tag{2-67}$$

The procedure to calculate the band diagram is as follows. First, E_{gP}, ΔE_c, Φ_P, and Φ_n are calculated from the specified doping levels and material parameters. Second, ϕ_{bi} is determined from Eq. (2-67). Once ϕ_{bi} is found, ϕ_{n0} is calculated from Eq. (2-66) and consequently ϕ_{P0} is calculated from $\phi_{\text{bi}} - \phi_{n0}$. These pieces of information are then used to calculate the depletion thickness X_{n0} and X_{P0} from Eqs. (2-62) and (2-63).

In the presence of an external bias, we divide the externally applied voltage V_a into two components: one dropping across the P side (V_{aP}) and another across the n side (V_{an}) of the junction:

$$V_a = V_{aP} + V_{an} \tag{2-68}$$

The development of the equations follows closely to that in § 2-2.

$$\varepsilon(x) = -\frac{q}{\epsilon_n} N_d(x + X_n) \qquad \text{for } -X_n < x < 0 \tag{2-69a}$$

$$\varepsilon(x) = -\frac{q}{\epsilon_P} N_a(X_P - x) \qquad \text{for } 0 < x < X_P \tag{2.69b}$$

$$V(x) = \frac{q}{\epsilon_n} N_d\left(\frac{x^2}{2} + X_n x + \frac{X_n^2}{2}\right) \qquad \text{for } -X_n < x < 0 \tag{2-70a}$$

$$V(x) = \frac{q}{\epsilon_P} N_a\left(X_P x - \frac{x^2}{2}\right) + \phi_n \qquad \text{for } 0 < x < X_P \tag{2-71b}$$

$$\phi_{n0} - V_{an} = \phi_n = \frac{q}{2\epsilon_n} N_d X_n^2 \tag{2-71}$$

$$\phi_{P0} - V_{aP} = \phi_P = \frac{q}{2\epsilon_P} N_a X_P^2 \tag{2-72}$$

$$\epsilon_n \varepsilon(0^-) = \epsilon_P \varepsilon(0^+) \tag{2-73}$$

$$N_d X_n = N_a X_P \tag{2-74}$$

$$\frac{\phi_P}{\phi_n} = \frac{\phi_{P0} - V_{aP}}{\phi_{n0} - V_{an}} = \frac{\epsilon_n N_d X_P^2}{\epsilon_P N_a X_n^2} = \frac{\epsilon_n N_d}{\epsilon_P N_a} \tag{2-75}$$

$$\phi_{\text{bi}} - V_a = (\phi_{P0} - V_{aP}) + (\phi_{n0} - V_{an}) \tag{2-76}$$

$$\phi_{n0} - V_{an} = \frac{1}{1 + \epsilon_n N_d / \epsilon_P N_a}(\phi_{\text{bi}} - V_a) \tag{2-77}$$

Since ϕ_{bi}, V_a, and ϕ_{P0} are known from the thermal equilibrium analysis that always precedes a nonequilibrium analysis, V_{an} is obtainable from Eq. (2-77). Once V_{an} is known, V_{aP} is determined from Eq. (2-76). The depletion thicknesses are then calculated according to

$$X_n = \sqrt{\frac{2\epsilon_n(\phi_{n0} - V_{an})}{qN_d}} \tag{2-78}$$

$$X_P = \sqrt{\frac{2\epsilon_P(\phi_{P0} - V_{aP})}{qN_a}} \tag{2-79}$$

§ 2-4 GRADED HETEROJUNCTIONS

An abrupt heterojunction has discontinuities in both the conduction and the valence bands. The discontinuities result mainly because the heterojunction is formed by juxtaposing two dissimilar materials, without an intermediate region in which the semiconductor properties of one material vary gradually to those of another. Taking the band diagram of Fig. 2-2 for instance, we see that the discontinuity creates a spike in the potential profile that obtrudes out the otherwise smooth conduction band level. When the electron carriers leave the N side for the p^+ side, they must first climb over the barrier before descending in the potential drop and become thermalized at the conduction band edge of the p^+ side. This barrier spike impedes the current flow and is sometimes undesirable. But a heterojunction needs not to be abrupt. With conventional epitaxial growth techniques such as MBE and MOCVD, it is not difficult to grade the alloy content (such as aluminium in AlGaAs) gradually from zero to a certain value over a distance. The resulting heterojunction is called a *graded heterojunction*. The distance through which the alloy mole fraction is varied is called the *grading distance*. If the graded heterojunction is designed properly, the potential spike is removed.

We describe the procedure to construct the band diagram of a graded heterojunction. The procedure applies to N-p and p-N heterojunctions in general, but in the following we specifically describe this procedure in the context of a p^+-N heterojunction (which is particularly important in graded Npn HBTs). In order to reduce the mathematical complexities, we approximate ϕ_{p0} (the built-in potential on the p^+ side) as zero. The value was found to be on the order of 10^{-4} V in a comparable p^+-N abrupt heterojunction and we do not expect its value to change much in a graded heterojunction. We also use the depletion approximation to solve the Poisson equation on the N side. Although the previous two approximations are reasonable, there is one more assumption that requires more caution. We assume that the band diagram can be drawn by summing the Poisson potential and the *energy gap potential* (to be described shortly). This summation constitutes the so-called *superposition method*, whose practicality is addressed later in this section. (The Poisson potential is the name used in this book; sometimes it is referred to as the semiconductor potential).

Our first example concerns an abrupt p^+-N heterojunction, which was the subject of discussion in § 2-1. Once the reader becomes familiar with the superposition method, we then reapply it to construct the band diagram of a graded p^+-N heterojunction.

We focus on the drawing of the conduction band edge. Since $\phi_{p0} = 0$, ϕ_{N0} is equal to ϕ_{bi}, which in turn is given by

$$\phi_{\mathrm{bi}} = \phi_{N0} = \frac{E_{gp} + \Delta E_c - \Phi_p - \Phi_N}{q} \tag{2-80}$$

If the doping levels are anywhere close to either N_c or N_v for their respective type, then the Joyce–Dixon approximation must be used. We take the first term of Eqs. (1-29) and (1-30), and express Φ_p and Φ_N as

$$\Phi_N = E_c - E_f = kT\left(\ln\frac{N_d}{N_c} + \frac{1}{\sqrt{8}}\frac{N_d}{N_c}\right) \tag{2-81}$$

$$\Phi_p = E_f - E_v = kT\left(\ln\frac{N_a}{N_v} + \frac{1}{\sqrt{8}}\frac{N_a}{N_v}\right) \tag{2-82}$$

The electric field is zero everywhere outside the N-side depletion region. Within the depletion region, it is expressed, similar to Eq. (2-2b), as

$$\varepsilon_{\text{Poisson}}(x) = -\frac{q}{\epsilon_N} N_d(X_{NO} - x) \qquad \text{for } 0 < x < X_{NO} \tag{2-83}$$

Here we replace our usual symbol for the electric field by $\varepsilon_{\text{Poisson}}$ to emphasize its origin in the space charges. We use ε to denote the actual electric field experienced by the mobile carrier, which includes another component resulting from the spatial variation of the electron affinity. Right now we are still working on an abrupt heterojunction in which the aluminum content remains fixed throughout the N layer; hence, $\varepsilon_{\text{Poisson}} = \varepsilon$. However, in the subsequent consideration of the graded heterojunction, the equality will not hold.

The equation relating the Poisson electric field and the voltage was given in Eq. (1-9). For the discussion of this section, we replace our usual potential symbol V by φ_{Poisson} to emphasize that φ_{Poisson} is related directly to $\varepsilon_{\text{Poisson}}$ but not the quasi-electric field component. Therefore, we rewrite Eq. (1-9) as

$$\varepsilon_{\text{Poisson}}(x) = -\frac{d\varphi_{\text{Poisson}}}{dx} \tag{2-84}$$

The Poisson potential satisfies the following boundary conditions:

$$\varphi_{\text{Poisson}}|_{x \le 0} = 0 \qquad \varphi_{\text{Poisson}}|_{x \ge X_{NO}} = -\phi_{NO} \tag{2-85}$$

Integrating Eq. (2-84) using the Poisson electric field from Eq. (2-83) as well as the above boundary conditions, we obtain

$$\varphi_{\text{Poisson}}(x) = \begin{cases} 0 & \text{for } x \le 0 \\ -\dfrac{q}{2\epsilon_N} N_d(X_{NO} - x)^2 + \phi_{NO} & \text{for } 0 < x < X_{NO} \\ \phi_{NO} & \text{for } x \ge X_{NO} \end{cases} \tag{2-86}$$

with the condition that

$$\phi_{NO} = \frac{q}{2\epsilon_N} N_d X_{NO}^2 \tag{2-87}$$

Since ϕ_{NO} is already known from Eq. (2-80), this equation allows us to calculate the N-side depletion thickness X_{NO}.

The band potential (φ_g) relates the relative amount of the electron affinity of the various materials. Using the valence band edge of the p^+ side as the reference, we equate $(-q\varphi_g)$ to the conduction band edge in the absence of any electrostatic potential. Note that $-q$ is multiplied to the potential φ_g because the band diagram plots the electron energy. It is given by

$$-q\varphi_g(x) = \begin{cases} E_{gp} & \text{for } x < 0 \\ E_{gp} + \Delta E_c & \text{for } x > 0 \end{cases} \tag{2-88}$$

where E_{gp} is the energy gap on the p^+ side. For the GaAs/AlGaAs heterojunction under consideration, it is the energy gap of GaAs.

The conduction band edge $E_c(x)$ is given as the sum of the two potentials (the superposition method):

$$E_c(x) = -q\varphi_{\text{Poisson}}(x) + [-q\varphi_g(x)] \tag{2-89}$$

Carrying out the addition, we write $E_c(x)$ with reference to the valence band edge:

$$E_c(x) = \begin{cases} E_{gp} & \text{for } x \le 0 \\ E_{gp} + \Delta E_c + \dfrac{q^2}{2\epsilon_N} N_d(X_{NO} - x)^2 - q\phi_{NO} & \text{for } 0 < x < X_{NO} \\ E_{gp} + \Delta E_c - q\phi_{NO} & \text{for } x \ge X_{NO} \end{cases} \tag{2-90}$$

The conduction band profile depicted by the above equation is identical to that of the band diagram shown in Fig. 2-3. The valence band edge is easily obtained by spacing appropriately the energy gap from $E_c(x)$.

We now apply the same procedure to a graded (p^+) GaAs/(N) AlGaAs heterojunction. As a first exercise, we examine a linearly graded junction whose aluminum mole fraction in the N-$Al_\xi Ga_{1-\xi}As$ layer increases linearly from 0 to the maximum value ξ_{max} (ξ_{max} is 0.35 for typical HBT applications). Despite the grading of the AlGaAs layer, its doping level remains constant at N_d, and the doping level on the p^+ side is still N_a. To simplify the analysis, we take the dielectric constant to be the same for GaAs and AlGaAs of varying the aluminum mole fraction.

Some physical pictures of the grading distance are in order before we proceed. Figure 2-17 illustrates the qualitative band diagrams of p^+-N heterojunctions under the space-charge-neutral condition and under thermal equilibrium. We assume that the grading of the material is not too abrupt, so that the Fermi level at any position within the grading layer can still be determined from the bulk equations given in § 1-5. This assumption is always made implicitly. It ensures that the semiconductor properties at any position depends solely on the aluminum mole fraction at that position. This assumption would fail if we vary ξ from 0 to ξ_{max} in, say, only 1 Å. Nonetheless, in practical designs for which the grading distances are on the order of 100 Å, this assumption is valid.

N_c is relatively the same for $Al_{\xi}Ga_{1-\xi}As$ with varying aluminum mole fraction. Because the N-type doping level is constant, the Fermi level with respect to the conduction band edge shown in Fig. 2-17 is relatively the same. When the grading distance ($X_{grading}$) is short, Fig. 2-17*a* shows that the conduction band has a barrier, similar to that encountered in an abrupt heterojunction. When $X_{grading}$ lengthens, the conduction band profile looks considerably smoother. In these two cases, we have assumed the grading distance to be shorter than the depletion thickness. As a result, the built-in potential is known precisely, as indicated in Fig. 2-17 and given in Eq. (2-55). In Fig. 2-18, which is applicable to a situation in which $X_{grading}$ exceeds the depletion thickness, the

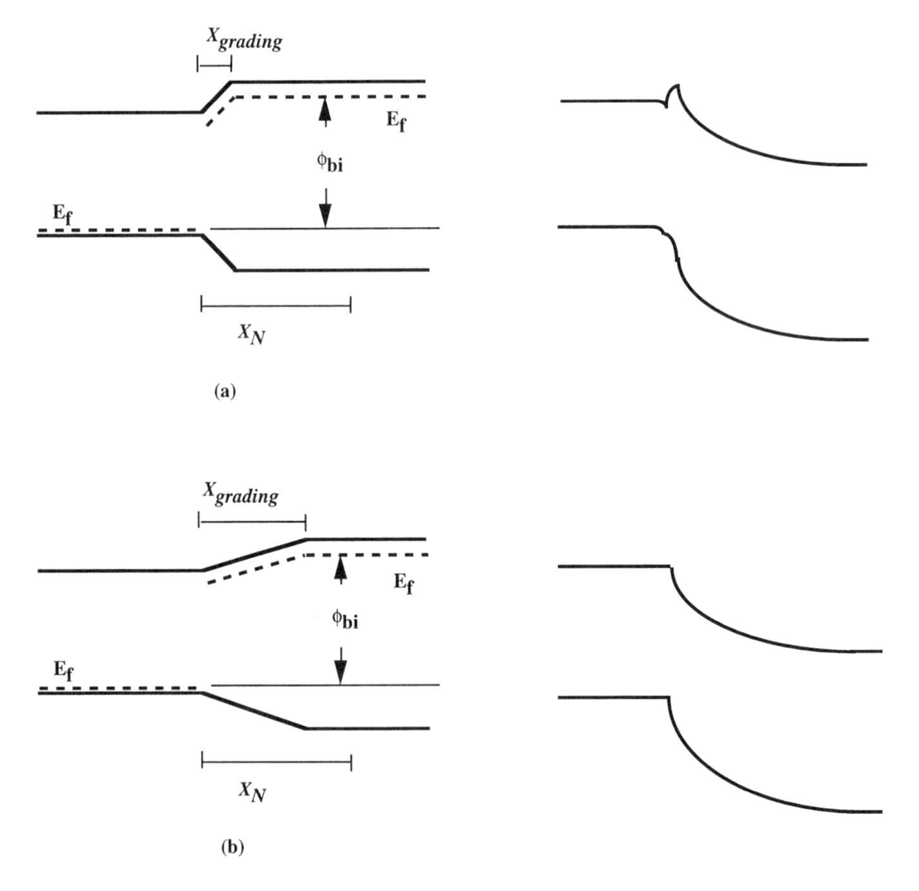

FIGURE 2-17. Band diagrams of p^+-N heterojunctions with a short and a long grading distance. The diagrams under the space-charge-neutral condition are shown on the left, and the diagrams under thermal equilibrium are shown on the right.

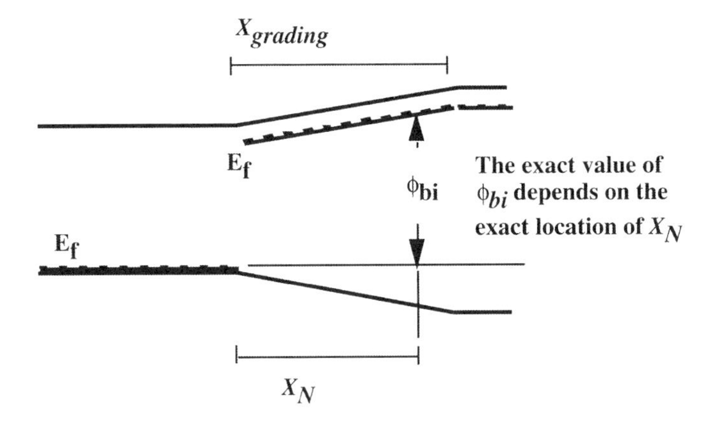

FIGURE 2-18. Space-charge-neutral band diagram of a p^+-N heterojunction whose grading distance exceeds the depletion thickness. The determination of the built-in voltage is not straightforward.

built-in potential depends on the actual value of the depletion thickness. Without the actual value of ϕ_{NO} known a priori, we cannot solve φ_{Poisson} explicity [i.e., Eq. (2-86) will contain an unknown, ϕ_{NO}]. In this case, $\varphi_{\text{Poisson}}(x)$ depends on $\phi_g(x)$ and they cannot be determined independently. The superposition method of calculating individually φ_{Poisson} and ϕ_g and then summing them at the end thus breaks down. A more general analysis to be discussed in § 2-7 is required. Nonetheless, the superposition method is useful in most device applications, since the grading distance is usually less than or equal to the depletion thickness.

Because we are considering situations in which $X_{\text{grading}} < X_{NO}$, $\phi_{\text{bi}} = \phi_{NO}$ is still given by Eq. (2-14) with a minor modification of the $\Delta E_{c,\,\text{max}}$ term:

$$\phi_{NO} = \frac{E_{gp} + \Delta E_{c,\text{max}} - \Phi_p - \Phi_N}{q} \tag{2-91}$$

$\Delta E_{c,\,\text{max}}$ for the graded junction is simply the discontinuity between GaAs and $\text{Al}_\xi\text{Ga}_{1-\xi}\text{As}$ when $\xi = \xi_{\text{max}}$. Specifically, from Eq. (1-90),

$$\Delta E_{c,\,\text{max}} = (1.247 - 0.55)\,\xi_{\text{max}} \qquad \text{(in eV)} \tag{2-92}$$

Because the doping levels are constant in the graded junction, $\varphi_{\text{Poisson}}(x)$ is identical to that given in Eq. (2-86). $\varphi_g(x)$ in the case of linearly graded $\text{GaAs/Al}_\xi\text{Ga}_{1-\xi}\text{As}$ heterojunction is given by

$$-q\varphi_g(x) = \begin{cases} E_{gp} & \text{for } x \le 0 \\ E_{gp} + \Delta E_{c,\,\text{max}} \dfrac{x}{X_{\text{grading}}} & \text{for } 0 < x < X_{\text{grading}} \\ E_{gp} + \Delta E_{c,\text{max}} & \text{for } x \ge X_{\text{grading}} \end{cases} \tag{2-93}$$

Applying Eq. (2-89), we find:

$$E_c(x) =$$

$$\begin{cases} E_{gp} & \text{for } x \leq 0 \\ E_{gp} + \Delta E_{c,\max} \dfrac{x}{X_{\text{grading}}} + \dfrac{q^2}{2\epsilon_s} N_d(X_{NO} - x)^2 - q\phi_{NO} & \text{for } 0 < x < X_{\text{grading}} \\ E_{gp} + \Delta E_{c,\max} + \dfrac{q^2}{2\epsilon_s} N_d(X_{NO} - x)^2 - q\phi_{NO} & \text{for } X_{\text{grading}} < x < X_{NO} \\ E_{gp} + \Delta E_{c,\max} - q\phi_{NO} & \text{for } x \geq X_{NO} \end{cases} \quad (2\text{-}94)$$

The discussion thus far has been for the thermal equilibrium condition. When $V_a \neq 0$, we can still use the above equations, except for replacing X_{NO} and ϕ_{NO} in the proceeding equations by X_N and ϕ_N, respectively. We keep in mind our assumption that X_{grading} should be smaller than X_N in order for the superposition technique to work properly.

Example 2-5:

Draw the conduction band profile for a graded (p^+) GaAs/(N) $Al_\xi Ga_{1-\xi}As$ heterojunction with ξ varying linearly from 0 to 0.35 in a grading distance $X_{\text{grading}} = 300$ Å. $N_a = 3 \times 10^{19}$ cm^{-3}; $N_d = 5 \times 10^{17}$ cm^{-3}. The dielectric constant is assumed to be constant at $\epsilon_s = 13.18(8.85 \times 10^{-14})$ F/cm for $Al_\xi Ga_{1-\xi}As$ with varying ξ. N_c for $Al_{0.35}Ga_{0.65}As$ is 3.72×10^{17} cm^{-3} and N_v for GaAs is 4.7×10^{18} cm^{-3}. Repeat the drawings for $V_a = 1.0$, 1.2, and 1.4 V.

$\Delta E_{c,\max} = 1.247\xi_{\max} = 0.697(0.35) = 0.244$ eV. From Fig. 1-21, we find that for $N_d/N_c = 1.344$, $E_f - E_c = 0.8\,kT = 0.021$ eV and $N_a/N_v = 6.383$, $E_v - E_f = 4.01\,kT = 0.103$ eV. Hence, $\Phi_N = -0.021$ eV and $\Phi_p = -0.103$ eV. ϕ_{NO} (taken to be ϕ_{bi}), according to Eq. (2-14), is $1.424 + 0.244 + 0.021 + 0.103 = 1.792$ V. X_{NO}, according to Eq. (2-7), is thus 723 Å. At $V_a = 1.4$ V, X_N which is calculated from Eq. (2-42) to be 338 Å, is still greater than X_{grading}. Therefore, at all three biases, the depletion thickness is larger than X_{grading} and the superposition technique is valid.

From these numbers, we calculate specifically for $V_a = 1.2$ V, so $\phi_N = 1.792 - 1.2 = 0.592$ V and $X_N = 4.15 \times 10^{-6}$ cm. According to Eq. (2-94), we have $E_c(x) = 1.424$ eV for $x < 0$ and $1.424 + 0.244 - 1.792 = -0.124$ eV for $x > X_N = 4.15 \times 10^{-6}$ cm. At intermediate x values between $x = 0$ and $x = X_{\text{grading}} = 3 \times 10^{-6}$ cm,

$$E_c(x) = 1.424 + 0.244 \frac{x}{300 \times 10^{-5}} + 3.43 \times 10^{10}(4.15 \times 10^{-5} - x)^2 - 0.592 \text{ eV}$$

Between $x = X_{\text{grading}}$ and $x = X_N$, $E_c(x)$ is given by

$$E_c(x) = 1.424 + 0.244 + 3.43 \times 10^{10}(4.15 \times 10^{-5} - x)^2 - 0.592 \text{ eV}$$

$E_c(x)$ at all three biases is calculated with this procedure, and the results are shown in Fig. 2-19.

Example 2-5 demonstrates that when linear grading is used, a dip in the conduction band profile may occur at high forward biases. This is not desirable, as it may trap electron carriers and affect the carrier transport. The fundamental cause of the dip is that the Poisson potential varies parabolically, whereas the energy gap potential varies linearly. A popular grading scheme is *parabolic grading*. It will be shown that this particular grading scheme cancels the parabolic variation of the Poisson potential, resulting in a linear variation of the conduction band edge without the dip.

We are again considering a (p^+) GaAs/(N) $Al_\xi Ga_{1-\xi}As$ heterojunction with N_a and N_d as the doping levels. Since we have not altered the doping level, φ_{Poisson} remains unchanged despite the change in the aluminum grading scheme. We modify Eq. (2-86) slightly, however, to reflect the fact that we now allow V_a to be nonzero:

$$\varphi_{\text{Poisson}}(x) = \begin{cases} 0 & \text{for } x \le 0 \\ \dfrac{q}{2\epsilon_s} N_d(X_N - x)^2 - \phi_N & \text{for } 0 < x < X_N \\ -\phi_N & \text{for } x \ge X_N \end{cases} \qquad (2\text{-}95)$$

where ϕ_N is $\phi_{N0} - V_a$ and X_N was given in Eq. (2-42).

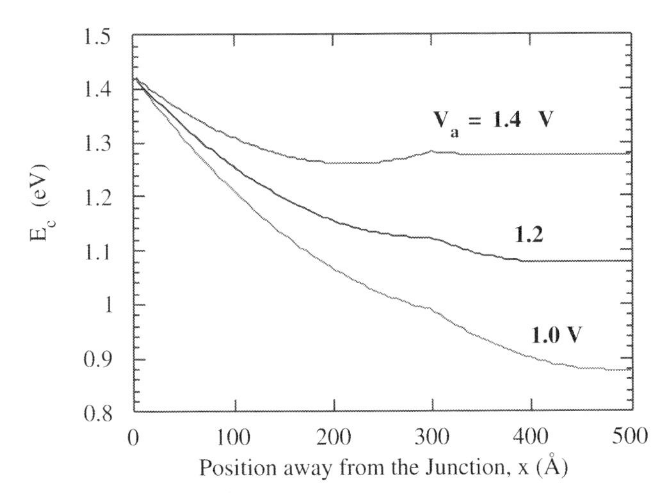

FIGURE 2-19. Conduction band profile for the *linearly* graded GaAs/$Al_\xi Ga_{1-\xi}As$ heterojunction discussed in Example 2-5.

The parabolic grading scheme requires the grading thickness to be the depletion thickness when the applied bias is equal to $(E_{gp} - \Phi_N - \Phi_p)/q$. In this manner, $q\phi_{N0} - qV_a$ has precisely the value of $\Delta E_{c,\,\max}$, and

$$X_{\text{grading}} = \sqrt{\frac{2\epsilon_s}{qN_d}(\phi_{N0} - V_a)}\,\Bigg|_{v_a=(E_{gp}-\Phi_N-\Phi_p)/q} = \sqrt{\frac{2\epsilon_s}{qN_d}\frac{\Delta E_{c,\,\max}}{q}} \tag{2-96}$$

The parabolic grading is such that the aluminum mole fraction ξ varies from 0 at $x = 0$ to $\xi_{\max}$ at $x = X_{\text{grading}}$ in the following manner:

$$\xi(x) = \begin{cases} 0 & \text{for } 0 \geq x \\ \xi_{\max} - \xi_{\max}\left(1 - \dfrac{x}{X_{\text{grading}}}\right)^2 & \text{for } 0 \leq x \leq X_{\text{grading}} \\ \xi_{\max} & \text{for } X_{\text{grading}} \leq x \end{cases} \tag{2-97}$$

With this particular choice of grading, the energy gap potential at $0 \leq x \leq X_{\text{grading}}$ is

$$\begin{aligned} -q\varphi_g &= E_{gp} + \Delta E_{c,\max} - \Delta E_{c,\max}\frac{1}{X^2_{\text{grading}}}(X_{\text{grading}} - x)^2 \\ &= E_{gp} + \Delta E_{c,\max} - \frac{q^2 N_d}{2\epsilon_s}(X_{\text{grading}} - x)^2 \end{aligned} \tag{2-98}$$

We have used Eq. (2-96) in the derivation of Eq. (2-98). Note that at $x = 0$, $\varphi_g = \Delta E_{c,\max} - 2\epsilon_s X^2_{\text{grading}}/(qN_d)$. Again applying the definition of X_{grading}, we find that $-q\varphi_g$ is indeed equal to 0 at $x = 0$. At the other extreme, when $x = X_{\text{grading}}$, $-q\varphi_g$ becomes $\Delta E_{c,\max}$, also as expected. To summarize, we write $-q\varphi_g(x)$ for any position x as

$$-q\varphi_g(x) = \begin{cases} E_{gp} & \text{for } 0 \geq x \\ E_{gp} + \Delta E_{c,\max} - \dfrac{q^2 N_d}{2\epsilon_s}(X_{\text{grading}} - x)^2 & \text{for } 0 \leq x \leq X_{\text{grading}} \\ E_{gp} + \Delta E_{c,\max} & \text{for } X_{\text{grading}} \leq x \end{cases} \tag{2-99}$$

$E_c(x) = -q\varphi_{\text{Poisson}}(x) + [-q\varphi_g(x)]$. A quick examination of Eqs. (2-95) and (2-99) reveals that the terms involving x^2 cancel. The resulting conduction band varies linearly with x, a goal achievable only with a parabolic grading. After some algebraic manipulation and noting that $q^2 N_d(X_N^2 - X^2_{\text{grading}})/$

$(2\epsilon_s) = q\phi_N - \Delta E_{c,\max}$, we write $E_c(x)$ as

$$E_c(x) = \begin{cases} E_{gp} & \text{for } x \leq 0 \\ E_{gp} - \dfrac{q^2 N_d}{\epsilon_s}(X_N - X_{\text{grading}})x & \text{for } 0 \leq x \leq X_{\text{grading}} \\ E_{gp} + \Delta E_{c,\max} + \dfrac{q^2}{2\epsilon_s} N_d (X_N - x)^2 - q\phi_N & \text{for } X_{\text{grading}} < x < X_N \\ E_{gp} + \Delta E_{c,\max} - q\phi_N & \text{for } x \geq X_N \end{cases} \quad (2\text{-}100)$$

Example 2-6:

Design a (p^+) GaAs/(N) $Al_\xi Ga_{1-\xi}As$ graded heterojunction such that the conduction band varies linearly with position. ξ is varied from 0 to 0.35. Just as the diode in Example 2-5, $N_a = 10^{19}$ cm^{-3}, $N_d = 5 \times 10^{17}$ cm^{-3}, and $\epsilon_s = 13.18(8.85 \times 10^{-14})$ F/cm.

From Example 2-5, $\Delta E_{c,\max} = 1.247\xi_{\max} = 0.244$ eV, $\Phi_N = -0.021$ eV, $\Phi_p = -0.103$ eV, $\phi_{NO} = 1.792$ V. For the parabolic grading scheme to work, the grading distance needs to be that given in Eq. (2-96):

$$X_{\text{grading}} = \sqrt{\frac{2\epsilon_s}{qN_d}\frac{\Delta E_{c,\max}}{q}} = \sqrt{\frac{2(13.18)(8.85 \times 10^{-14})}{(1.6 \times 10^{-19})(5 \times 0^{17})}0.244} = 2.67 \times 10^{-6} \text{ cm}$$

The aluminum mole fraction should be varied from 0 at the junction interface at 0.35 at $X_{\text{grading}} = 267$ Å in the manner given by Eq. (2.97):

$$\xi(x) = 0.35 - 0.35\left(1 - \frac{x}{267\ \text{Å}}\right)^2$$

This way, the conduction band profile is given by Eq. (2-100). We plot the results for $V_a = 1.0$, 1.2, and 1.4 V in Fig. 2-20. The figure demonstrates that there is no dip feature in the parabolic grading scheme.

Although parabolic grading produces a linear E_c profile as calculated from the superposition method, we need to be careful about one implicit assumption of the method that has not been mentioned. It concerns the very assumption of the depletion approximation used to write the Poisson potential. When V_a is large, there is a large amount of mobile carriers due to current conduction. These carriers alter the Poisson potential from having a simple parabolic dependence on the distance. Reference [3] gives a detailed account of how the band profile is affected by the carriers. Because of the complexity of the problem when the carriers are taken into consideration, parabolic grafting is still a popular grading scheme (if the heterojunction is to be graded). However, we should also be aware of the limitation of the scheme, especially when the emitter doping level is light.

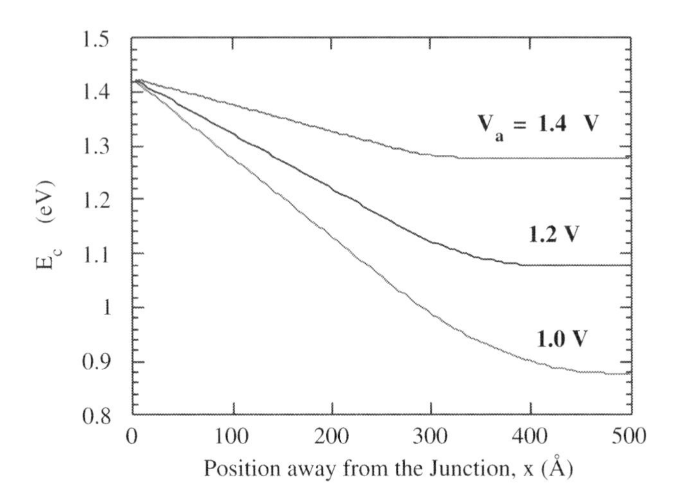

FIGURE 2-20. Conduction band profile for the *parabolically graded* $GaAs/Al_\xi Ga_{1-\xi}As$ heterojunction discussed in Example 2-6.

The junction built-in potential for an abrupt *p-N* heterojunction was given in Eq. (2-55). For a graded *p-N* heterojunction, the built-in potential is determined from the band diagram under the charge-neutral condition, as shown in Fig. 2-17*b*. ϕ_{bi} is equal to the difference in the Fermi levels at the two sides of the junction:

$$\phi_{bi} = \frac{E_{gp} - \Phi_p - \Phi_N}{q} \qquad \text{(graded } p\text{-}N \text{ heterojunction)} \qquad (2\text{-}101)$$

where Φ_p and Φ_N were given in Eq. (2-15).

We summarize some differences between the abrupt and graded heterojunctions, using the *p-N* heterojunction as an example. First (and obviously), the band discontinuity is removed in the graded junction. Second, the holes from the *p* side experience an additional ΔE_v energy barrier in the abrupt heterojunction to that in a homojunction. In the graded junction, they experience an incremental energy barrier of $\Delta E_c + \Delta E_v$ compared to the homojunction. The hole barrier is higher for the graded heterojunction than for the abrupt heterojunction. Finally, the built-in potential is smaller in the graded junction than in the abrupt junction, by an amount ΔE_c. The property of smaller built-in potential leads to a smaller turn-on voltage in the graded heterojunction.

§ 2-5 DIODE CURRENT–VOLTAGE CHARACTERISTICS

The presence of a potential spike in an abrupt heterojunction and its absence in a graded heterojunction suggest that the current-conduction mechanisms differ in these two types of heterojunctions. It is widely accepted that the potential spike in the depletion region limits the carrier conduction in the abrupt heterojunction. For the graded heterojunction, it is the rate of carrier diffusion outside the depletion region that limits the current conduction. We first develop the relationship between the applied bias V_a and the diode current I_D in the abrupt heterojunction. The derivation is based on the theories of the semiconductor-metal junction, whose band diagrams under various bias conditions are shown in Fig. 2-21.

The semiconductor shown in the figure is doped n-type. As discussed in § 1-1, a metal can be considered as a group of materials whose conduction band overlaps its valence band or whose valence band is partially filled. Therefore, the electrons all reside in the energy states right near its Fermi energy level. During thermal equilibrium, the Fermi level in the whole system is flat and coincides with the Fermi level inside the semiconductor. The critical parameters in this band diagram are the energy barrier ϕ_B and the applied bias V. (Note:

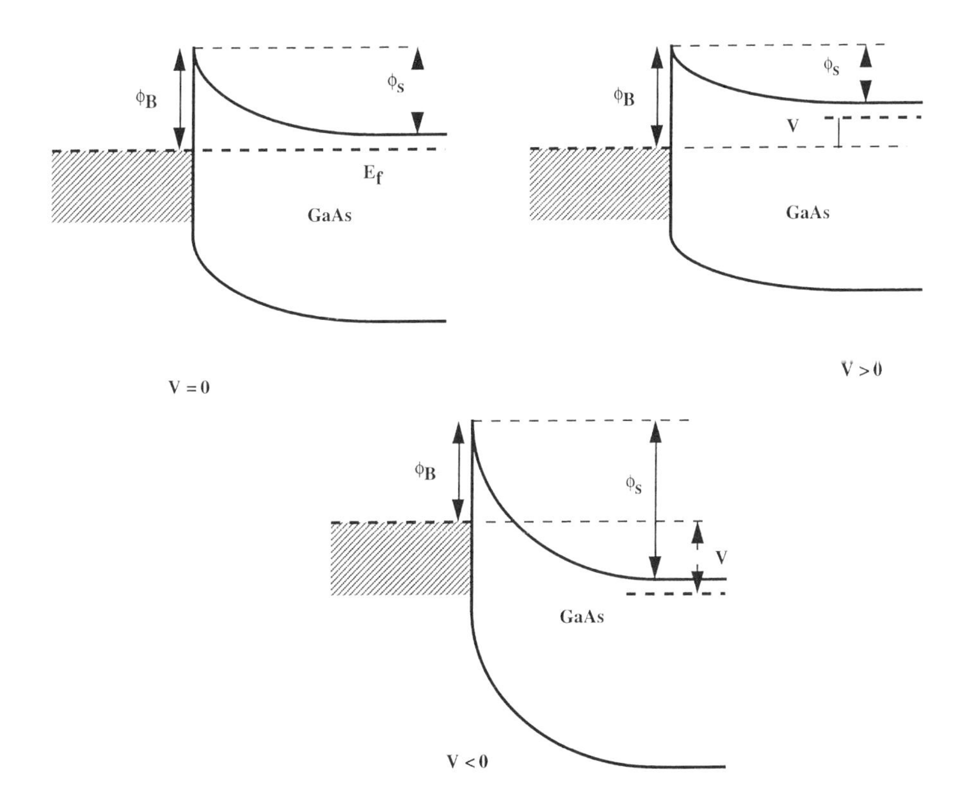

FIGURE 2-21. Band diagrams of metal-semiconductor junctions at thermal equilibrium, forward bias, and reverse bias. ϕ_s, the semiconductor potential, is the potential difference between that at the bulk and the surface of the semiconductor. ϕ_B is considered to be a material constant, unaffected by the applied bias.

V denotes the bias of a metal-semiconductor junction while V_a is the bias of a semiconductor heterojunction, such as a p-N heterojunction.) ϕ_B is assumed to be a fixed potential barrier independent of the bias. Because the metal is taken to be positive terminal, Fig. 2-21b shows that in a forward-biased condition, the quasi-Fermi level of the electrons is raised by an amount qV in comparison to the metal Fermi level.

The dominant current-conduction mechanism in a metal-semiconductor junction is thermionic emission, that the electrons from the semiconductor can enter the metal only if they acquire enough thermal energy to overcome the potential barrier $\phi_s = \phi_B - V$. When V increases, the barrier ϕ_s becomes smaller and more electrons flow into the metal. Since the electrons move from the negative terminal (semiconductor) to the positive terminal (the metal), the current is positive, consistent with the fact that it is forward-biased. From a study of the free electrons, the current density due to thermionic emission is given by

$$J_{\text{thermionic}} = A^* T^2 \exp\left(\frac{-q\phi_B}{kT}\right)\left[\exp\left(\frac{qV}{kT}\right) - 1\right] \tag{2-102a}$$

A^*, the effective Richardson's constant, is

$$A^* = \frac{4\pi q m_e^* k^2}{h^3} \tag{2-102b}$$

where h is Plank's constant and k is the Boltzmann's constant; m_e^* is the effective conduction electron mass; and A^* has the value 120 A/cm^2-K^2 if the free electron mass is used in place of m_e^*.

Based on the thermionic current density expression for the metal-semiconductor system, we derive a current–voltage (I–V) relationship for the abrupt p-N heterojunction. Two important parameters in Eq. (2-102) are ϕ_B and V. We examine ϕ_B first. ϕ_B is the potential barrier separating the electron Fermi level in the metal and the tip of the conduction band in the semiconductor. If we denote the analog of ϕ_B in the metal-semiconductor junction as ϕ_B' in the p-N heterojunction, then ϕ_B' represents the potential barrier between the Fermi level on the p side and the tip of the conduction band on the N side. Let us return to the band diagram of a p^+-N heterojunction shown in Fig. 2-2. It is easier to visualize the magnitude of ϕ_B' by blocking out the p^+ portion of the band diagram. ϕ_B' is found to be

$$\phi_B' = \phi_{NO} + \frac{\Phi_N}{q} \tag{2-103}$$

V in the metal-semiconductor junction is the applied voltage, which drops completely in the semiconductor (metal does not sustain a voltage drop). The analog of V in the abrupt heterojunction, defined as V', is the voltage drop across the N-type semiconductor. In term of our previous terminology, V' is simply V_{aN}, which is smaller than the actual applied bias (V_a):

$$V' = V_{aN} \tag{2-104}$$

We work out these two parameters. According to Eqs. (2-12) and (2-13),

$$\phi'_B = \phi_{N0} + \frac{\Phi_N}{q} = \frac{\epsilon_p N_a}{\epsilon_N N_d}\phi_{p0} + \frac{\Phi_N}{q} = \frac{\epsilon_p N_a}{\epsilon_N N_d}\phi_{\text{bi}}\left(1 + \frac{\epsilon_p N_a}{\epsilon_N N_d}\right)^{-1} + \frac{\Phi_N}{q}$$

$$= \left(\frac{E_{gp}}{q} + \frac{\Delta E_c}{q} + \frac{\epsilon_N N_d}{\epsilon_p N_a}\frac{\Phi_N}{q} - \frac{\Phi_p}{q}\right)\left(1 + \frac{\epsilon_N N_d}{\epsilon_p N_a}\right)^{-1} \tag{2-105}$$

As for V', we start from an identity of Eq. (2-38):

$$\frac{\phi_{N0} - V_{aN}}{\phi_{p0} - V_{ap}} = \frac{\epsilon_p N_a}{\epsilon_N N_d} \tag{2-106}$$

Cross-multiplying the terms, substituting $V_a - V_{aN}$ for V_{ap}, and finally grouping the terms with V_{aN}, we obtain an equivalent identity.

$$\left(1 + \frac{\epsilon_p N_a}{\epsilon_N N_d}\right)V_{aN} = \frac{\epsilon_p N_a}{\epsilon_N N_d}V_a + \left(\phi_{N0} - \frac{\epsilon_p N_a}{\epsilon_N N_d}\phi_{p0}\right) \tag{2-107}$$

According to Eq. (2-12), the term enclosed in parentheses on the right-hand side degenerates to 0. Therefore, the above identity yields

$$V' = V_{aN} = \frac{\epsilon_p N_a}{\epsilon_N N_d}\left(1 + \frac{\epsilon_p N_a}{\epsilon_N N_d}\right)^{-1} V_a \tag{2-108}$$

Now that we have obtained both V' and ϕ'_B, we substitute their values into Eq. (2-102) and finalize the expression for the diode current as

$$I_D = A_D A^* T^2 \exp\left(-\frac{1}{kT}\frac{E_{gp} + \Delta E_c + K\Phi_N - \Phi_p}{1 + K}\right)\left[\exp\left(\frac{q}{kT}\frac{V_a}{1 + K}\right) - 1\right] \tag{2-109}$$

where A_D is the diode area and K is defined to be

$$K = \frac{\epsilon_N N_d}{\epsilon_p N_a} \tag{2-110}$$

Example 2-7:

A graduate student measures the I–V characteristics of an abrupt p^+-N heterojunction, as shown in Fig. 2-22. Estimate the ratio of the doping levels N_d/N_a in this diode.

The dotted line is the numerical best fit to the diode I–V characteristics. We restrict the fitted current range in the medium to avoid nonideal features, such as flattening of current due to resistance at high current levels and leakage current at low current levels. The equation for the fit is shown in the same figure, indicating that the current increases exponentially with $25.526V_a$. We equate 25.526 to $q/(\eta kT)$, where η is the ideality factor. Since $kT/q = 0.0258$, $\eta = 1.52$. According to Eq. (2-109), this ideality factor is simply $1 + K$. K is thus equal to 0.52. If we neglect the differences in the dielectric constants of the two heteromaterials, we establish from Eq. (2-110) that

$$K = \frac{\epsilon_N N_d}{\epsilon_p N_a} \approx \frac{N_d}{N_a} = 0.52$$

This K value suggests that the doping levels are not too different; one is about half of the other. The graduate student later checked his log book and found that the N_d/N_a ratio used in the epitaxial growth was much smaller than that

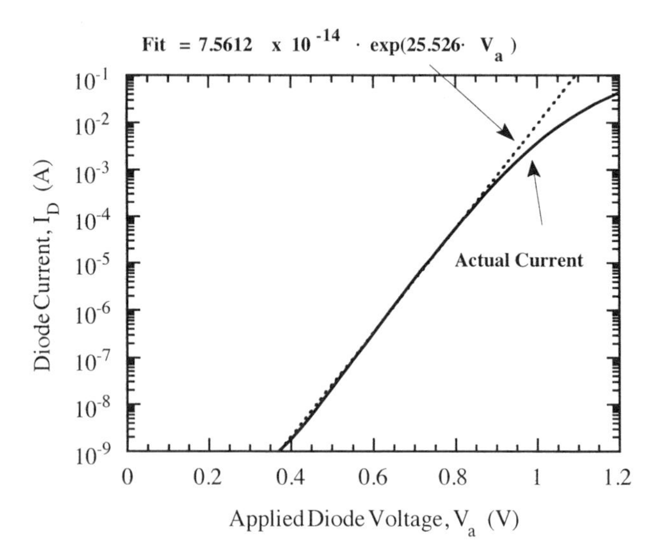

FIGURE 2-22. I–V characteristics of the abrupt p^+-N heterojunction discussed in Example 2-7. The dotted line is a least-square fit at the low to medium current levels where the contact resistances do not affect the exponential dependence of the diode current on the applied bias.

determined in the example. In fact, it was the range of 10^{-2}. At this point, there are two possibilities. One, the graduate students's controller for the growth system malfunctioned during the wafer growth and the doping levels of the two sides of the heterojunctions were not too different. Second, and more plausibly, the implicit assumption of Eqs. (2-109) that the current is limited purely by the thermionic emission is actually incorrect. It is found that when tunneling is a significant part of the overall current conduction, the ideality factor can approach values like 1.52 although the K ratio is only on the order of 0.01 [4]. Nonetheless, Eq. (2-109) remains to be a popular equation describing the I–V of a heterojunction.

We now consider the current transport in a graded heterojunction. Because the conduction band spike of an abrupt heterojunction is removed by the grading, we do not expect the thermionic emission current expression of Eq. (2-109) to apply to the graded heterojunction. Instead, the similarity between the band diagrams of a graded heterojunction and a homojunction suggests that the current conduction mechanism is identical to that found in a p-n homojunction. Without a band discontinuity, the current flows in the depletion region of a graded heterojunction without impediment. The amount of current flow is limited by the amount of carriers that can diffuse in a unit time in the quasi-neutral regions. Figure 2-23 illustrates the band diagrams of a forward- and a reverse-biased graded p^+-N heterojunction. We explain the drawing of the quasi-Fermi levels shortly, after we determine the current–voltage relationship for the junction.

An important assumption used in homojunction theory is adopted here to derive I_D as a function of V_a in the graded heterojunction. We assume that the carriers of the same species on opposite edges of the depletion region have the same distribution in energy, by virtue of the efficiency of the collision processes that occur in that region. As a consequence, the minority electron quasi-Fermi level at the edge of the p side has reached the equilibrium value as the majority electron Fermi level of the N side, and vice versa for the holes. In terms of the band diagram of Fig. 2-23, our assumption states that the value of E_{fp} at point 1 is equal to the equilibrium value at the far left, and the value of E_{fn} at point 2 is the same as that at the far right. We focus temporarily on the electron distribution. According to the band diagram under an external bias, we find the electron concentration ratio at the junction edges to be

$$\frac{n(X_N)}{n(-X_p)} = \frac{N_c \exp[(E_{fn} - E_c)/kT]_{x=X_N}}{N_c \exp[(E_{fn} - E_c)/kT]_{x=-X_p}} = \exp\left(\frac{q\phi_{\text{bi}} - qV_a}{kT}\right) \tag{2-111}$$

In the absence of the bias, we denote the electron concentrations at the edges as $n_0(X_N)$ and $n_0(-X_p)$ to emphasize these are equilibrium concentrations. From the band diagram under thermal equilibrium, we find:

$$\frac{n_0(X_N)}{n_0(-X_p)} = \frac{N_c \exp\{[E_f - E_c(x = +\infty)]/kT\}}{N_c \exp\{[E_f - E_c(x = -\infty)]/kT\}} = \exp\left(\frac{q\phi_{\text{bi}}}{dT}\right) \tag{2-112}$$

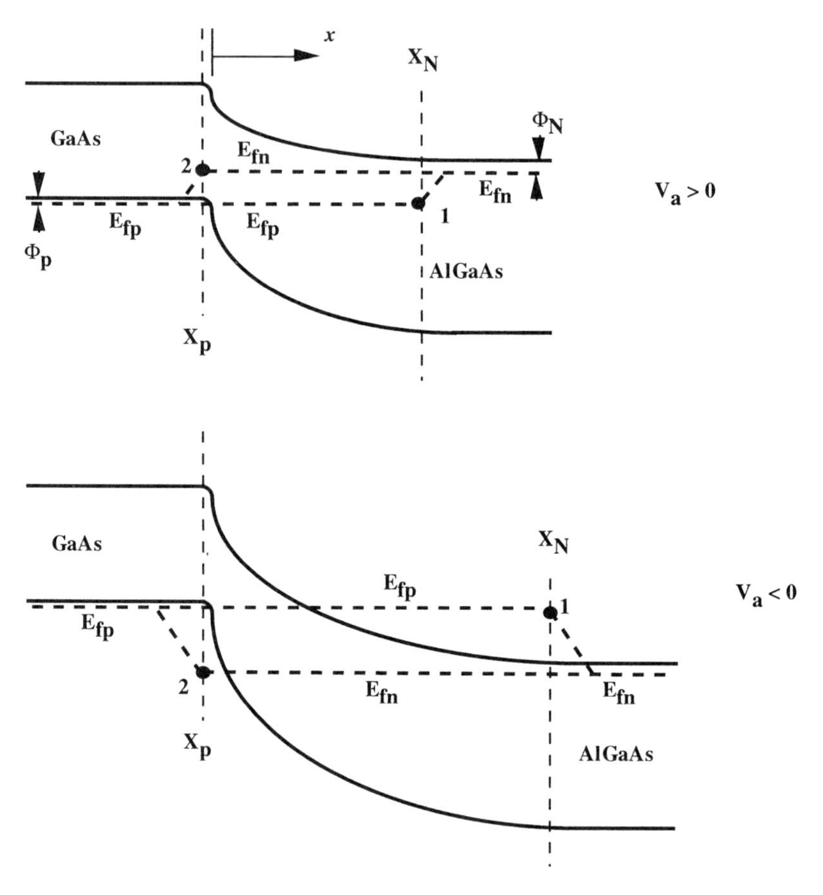

FIGURE 2-23. Band diagrams of a forward- and a reverse-biased graded p^+-N heterojunction.

Combining these two equations, we obtain an important boundary condition:

$$n(-X_p) = n_0(-X_p) \exp\left(\frac{qV_a}{kT}\right) \tag{2-113}$$

The derivation is likewise repeated for holes, leading to another boundary condition:

$$p(X_N) = p_0(X_N) \exp\left(\frac{qV_a}{kT}\right) \tag{2-114}$$

Equations (2-113) and (2-114) are sometimes referred to as the laws of the junction. The first equation states that, under an external bias V_a, the minority electron concentration at the edge of the p-side depletion region is amplified by a factor $\exp(qV_a/kT)$. The same amplification factor also occurs for the minority hole concentration at the edge of the N-side depletion region. These two junction laws form the boundary conditions necessary for the solution of the carrier diffusion equation in the neutral regions of the diode, which is the subject of the following.

Since we stated that the diffusion of carriers in the neutral regions limit the overall current flow, we seek an equation governing the carrier concentration in the neutral regions. In § 1-6, we developed the ambipolar equation, which was shown to reduce to the continuity equation when the external field is constant. For the problem under consideration, we can assume that the entirety of the applied bias drops across the depletion region. The electric field in the neutral regions is therefore zero, and the equation governing the electron carrier on the p-side of the graded junction is given by Eq. (1-67):

$$D_n \frac{\partial^2 n}{\partial x^2} - \frac{n - n_0}{\tau_n} = 0 \tag{2-115}$$

We have replaced $\partial n/\partial t$ in the original equation by 0 because we are interested in the steady state. D_n and τ_n are the diffusion coefficient and recombination lifetime, respectively, for the minority electrons in the p-type material. $n(x)$ is the electron concentration to be determined, and n_0 is the electron concentration under thermal equilibrium, equal to n_i^2/N_a on the p side. Equation (2-11) is rewritten as

$$\frac{d^2 \delta n(x)}{dx^2} - \frac{\delta n(x)}{L_n^2} = 0 \tag{2-116}$$

where δn, the excess electron concentration, is equal to $n(x) - n_0$. L_n, called the *minority electron diffusion length*, is given by

$$L_n = \sqrt{\tau_n D_n} \tag{2-117}$$

With τ_n typically being 1 ns and D_n being 25 cm^2/s, L_n is on the order of 1.6 μm.

A similar second-order linear ordinary differential equation is developed for the minority holes on the N side:

$$\frac{d^2 \delta p(x)}{dx^2} - \frac{\delta p(x)}{L_p^2} = 0 \tag{2-118}$$

where δp is defined as $p(x) - p_0$, with p_0 being the thermal equilibrium hole concentration on the N side, equal to n_i^2/N_d. L_p, the *minority hole diffusion length*, is given by

$$L_p = \sqrt{\tau_p D_p} \tag{2-119}$$

With τ_p typically being 1 ns and D_p being 1 cm^2/s, L_p is on the order of 0.3 μm.

The similarity in the differential equations for the electrons and holes allows us to solve for the hole concentration only. The electron profile is easily reproduced once the hole profile is known. At present we concentrate on only Eq. (2-118). To facilitate solving the equation, we assume that the metal contact to the p side is located at $x = +\infty$, where the hole concentration is at its thermal equilibrium value. In this coordinate system, the boundary conditions for Eq. (2-118) are obtained from Eq. (2-114) for $x = X_N$ and from the thermal equilibrium value for $x = \infty$:

$$\delta p(X_N) = p_0\left[\exp\left(\frac{qV_a}{kT}\right) - 1\right] \qquad \delta p(\infty) = 0 \tag{2-120}$$

To avoid possible confusion in the future, we denote n_i^{WG} as the intrinsic carrier concentration in the bulk of the N-type material. It will be distinguished from n_i^{NG}, which is the intrinsic carrier concentration in the bulk of the p-type material. (WG and NG stand for wide gap and narrow gap, respectively.) The differential equation of Eq. (2-116) has the following general solution:

$$\delta p(x) = A\exp\left(-\frac{x}{L_p}\right) + B\exp\left(\frac{x}{L_p}\right) \tag{2-121}$$

A and B are the unknown constants to be evaluated from the boundary conditions of Eq. (2-120). Since $\delta p = 0$ at $x = \infty$, B must be zero. The constant A is determined from the remaining boundary condition, and the solution of $\delta p(x)$ is given by

$$\delta p(x) = \frac{n_i^{\mathrm{WG}^2}}{N_d}\left[\exp\left(\frac{qV_a}{kT}\right) - 1\right]\exp\left(-\frac{x - X_N}{L_p}\right) \tag{2-122}$$

Similarly, the excess electron concentration is

$$\delta n(x) = \frac{n_i^{\mathrm{NG}^2}}{N_a}\left[\exp\left(\frac{qV_a}{kT}\right) - 1\right]\exp\left(\frac{x + X_p}{L_n}\right) \tag{2-123}$$

Conceptually, the diode current could be computed in several ways. We could, for example, examine one particular position x on the N side. The diode current would then be

$$I_D = I_n(x) + I_p(x) \tag{2-124}$$

Equation (2-124) states that the diode current is the sum of the electron current and the hole current. Because of the continuity of the drain current, this equation is valid for all x. Although $I_n(x)$ varies with x, $I_p(x)$ also varies with x in a complementary fashion such that the total current remains constant. There is a problem in evaluating I_D this way, however. At a particular position x on the N side, $I_p(x)$ is a minority carrier current. It is evaluated from $\delta p(x)$ in accordance with Eq. (1-36):

$$I_p(x) = -qA_D D_p \frac{d\delta p(x)}{dx} + qA_D \mu_p p \varepsilon(x) \approx -qA_D D_p \frac{d\delta p(x)}{dx} \qquad (2\text{-}125)$$

We have substituted $\varepsilon(x) = 0$, since x is in the neutral region. Despite that $I_p(x)$ is known, $I_n(x)$, the majority electron current, cannot be ascertained from a similar current expression:

$$I_n(x) = qA_D D_n \frac{d\delta n(x)}{dx} + qA_D \mu_n n \varepsilon(x) \qquad (2\text{-}126)$$

As discussed in § 1-6 in relation to the quasi-neutrality assumption, the electric field $\varepsilon(x)$ for the majority carrier cannot be approximated as 0, as was done for the minority holes. $\varepsilon(x)$ for the majority carrier current must be determined exactly. Otherwise, even a small field, when multiplied by the large amount of the majority electron concentration in the N-type region, leads to the nonnegligible current. Therefore, we are not able to evaluate the majority electron current from Eq. (2-126), although we could for the minority hole current.

Conversely, we could consider a particular position x on the p side instead. This time, we could determine the minority electron current $I_n(x)$, but not the majority hole current $I_p(x)$. The overall I_D, hence, is still indeterminate.

A clever way to resolve this difficulty is to consider the following. Suppose that there is no carrier generation or recombination in the depletion region. The minority electron current at $-X_p$ of the p side will be equal to the majority electron current at X_N of the N side. Similarly, the minority hole current at X_N of the N side will be equal to the majority hole current at $-X_p$ of the p side. The diode current is then evaluated according to the following manipulation:

$$I_D = I_n(x = X_N) + I_p(x = X_N) = I_n(-X_p) + I_p(X_N) \qquad (2\text{-}127)$$

Since $I_n(-X_p)$ and $I_p(X_N)$ are both minority carrier currents, they can be evaluated with the approximation that $\varepsilon \approx 0$ in the quasi-neutral regions. Using Eqs. (1-35), (1-36), and (2-127), we find

$$I_D = qA_D D_n \frac{d\delta n(x)}{dx}\bigg|_{x=-X_p} + qA_D D_p \frac{d\delta p(x)}{dx}\bigg|_{x=X_n}$$

$$= qA_D \left(\frac{D_p}{L_p} \frac{n_i^{\mathrm{WG}^2}}{N_d} + \frac{D_n}{L_n} \frac{n_i^{\mathrm{NG}^2}}{N_a} \right) \left[\exp\left(\frac{qV_a}{kT} \right) - 1 \right] \qquad (2\text{-}128)$$

Equation (2-128) is the so-called *long-diode current-voltage expression.* It was derived based on the assumption that the semiconductor bar is infinitely long.

In fact, we used a boundary condition in Eq. (2-120) that $\delta p(\infty) = 0$ to yield the hole profile in Eq. (2-122). Likewise, we used the boundary condition that $\delta n(-\infty) = 0$ to yield the electron profile in Eq. (2-123). Let us consider a practical diode with finite lengths as shown in Fig. 2-10. We denote the p^+ side of the diode as the base and its length as X_B. We call the N side the emitter and its length X_E. The proper boundary conditions are therefore $\delta p(X_E) = 0$ and $\delta n(X_B) = 0$. We see from Eqs. (2-122) and (2-123) that the new boundary conditions are fulfilled as long as $X_B \gg 3L_n$ and $X_E \gg 3L_p$. Therefore, under this long-diode situation, Eq. (2-128) remains valid.

However, if $X_B \ll 3L_n$ and $X_E \ll 3L_p$, then Eq. (2-128) needs to be modified. The derivation of this short-diode case is as follows. We first rewrite the boundary condition for the emitter side as

$$\delta p(X_N) = p_0\left[\exp\left(\frac{qV_a}{kT}\right) - 1\right] \qquad \delta p(X_E) = 0 \tag{2-129}$$

The solution to the governing differential equation of Eq. (2-118) satisfying above boundary conditions is

$$\delta p(x) = p_0\left[\exp\left(\frac{qV_a}{kT}\right) - 1\right]\frac{\exp[(x - X_E)/L_p] - \exp[-(x - X_E)/L_p]}{\exp[(X_N - X_E)/L_p] - \exp[-(X_N - X_E)/L_p]} \tag{2-130}$$

Because we are concerned with the short diode, we know that $X_N - X_E \ll L_p$ (and therefore $x - X_E \ll L_p$). Using the approximation that $\exp(x) \sim 1 + x$ when x is small, we simplify the above equation to

$$\delta p(x) = \frac{n_i^{\mathrm{WG}^2}}{N_d}\left[\exp\left(\frac{qV_a}{kT}\right) - 1\right]\left(\frac{x - X_E}{X_N - X_E}\right) \tag{2-131}$$

The minority carrier concentration in a short diode therefore decreases linearly with the distance measured from the depletion edge, X_N. A similar expression for the base side is given by (X_B and X_p are both positive numbers)

$$\delta n(x) = \frac{n_i^{\mathrm{NG}^2}}{N_a}\left[\exp\left(\frac{qV_a}{kT}\right) - 1\right]\left(\frac{x + X_B}{X_B - X_p}\right) \tag{2-132}$$

The concentration profiles of both the long and the short diodes are illustrated qualitatively in Fig. 2-24. Note that the excess carrier concentration is higher on the p^+ side. Although N_a is larger than N_d, $(n_i^{\mathrm{NG}})^2$ on the p^+ side is much larger than $(n_i^{\mathrm{WG}})^2$ on the N^- side.

Once the excess minority carrier concentration profiles are determined, the diode current is obtained in a similar manner as that described for the long

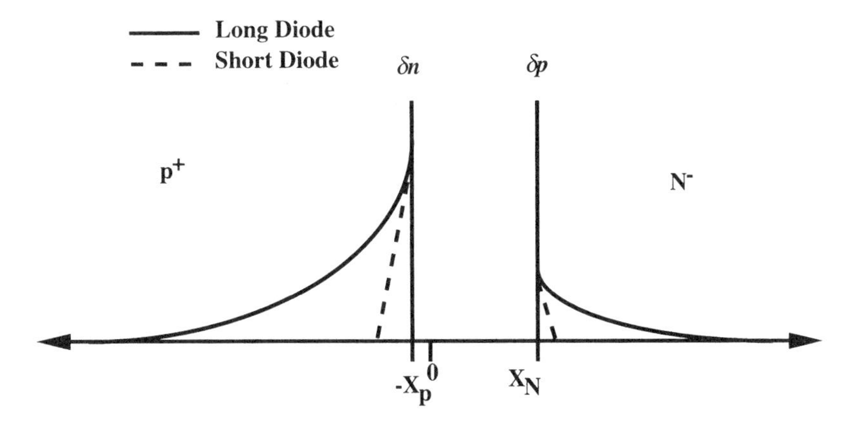

FIGURE 2-24. Minority carrier concentrations as a function of position. The dependence is exponential for the long diode and linear for the short diode.

diode. The diode current is given by

$$I_D = qA_D\left(\frac{D_p}{X_E - X_N}\frac{n_i^{\mathrm{WG}^2}}{N_d} + \frac{D_n}{X_B - X_p}\frac{n_i^{\mathrm{NG}^2}}{N_a}\right)\left[\exp\left(\frac{qV_a}{kT}\right) - 1\right] \tag{2-133}$$

Comparing Eq. (2-133) to Eq. (2-128), we find that they are equivalent in form. The only modification is that L_n and L_p in Eq. (2-128) are replaced by the distances of the neutral regions of the short diodes in Eq. (2-133).

This diode current expression for the abrupt heterojunction has a similar form as the expression for the graded heterojunction given in Eq. (2-133). Although there is a T^2 dependence in the abrupt heterojunction expression, it is insignificant in comparison to the exponential dependence on the temperature. Therefore, to a first-order approximation, we generalize the diode current to be

$$I_D = I_0\left[\exp\left(\frac{qV_a}{\eta kT}\right) - 1\right] \tag{2-134}$$

where I_0 is a constant. η, the ideality factor, is different for the two types of heterojunctions:

$$\eta = \begin{cases} 1 + \dfrac{\epsilon_N N_d}{\epsilon_p N_a} & \text{for abrupt heterojunctions} \\ 1 & \text{for graded heterojunctions} \end{cases} \tag{2-135}$$

(Note: The equation for the abrupt heterojunctions assumes that the thermionic emission is the primary current conduction mechanism. If tunneling is also important in the abrupt heterojunction, η will be larger than that given above. See Ref. [4] for more detail.)

In the reverse-biased condition, the factor $\exp(qV_a/\eta kT) \ll 1$, and the diode current becomes independent of the bias at a value equal to $-I_0$. The fact that diode current in a reverse bias is supposed to be saturated at a constant value gives rise to the name *(reverse) saturation current* for I_0. In the next section, we shall see that whereas I_D in the forward-biased condition remains accurate, I_D as given by Eq. (2-134) requires some modification in the reverse-biased condition.

Example 2-8:

An ideal *p-n* homojunction is shown in Fig. 2-25, with a forward bias applied to result a total current density J_T flowing through the junction. This *p-n* junction is a short diode; hence, the carrier concentrations vary linearly with distance as shown in the figure. Let the minority carrier mobilities for the holes and electrons be μ_p and μ_n, respectively, and the *p*-type and *n*-type doping be N_a and N_d, respectively. Assume that these four parameters are such that the total hole current (J_p) is identical to the total electron current (J_n). That is, $J_p = J_n = J_T/2$. Find the voltage drop across the quasi-neutral regions, V_1 and V_2, assuming the thickness of neutral region on either side is W.

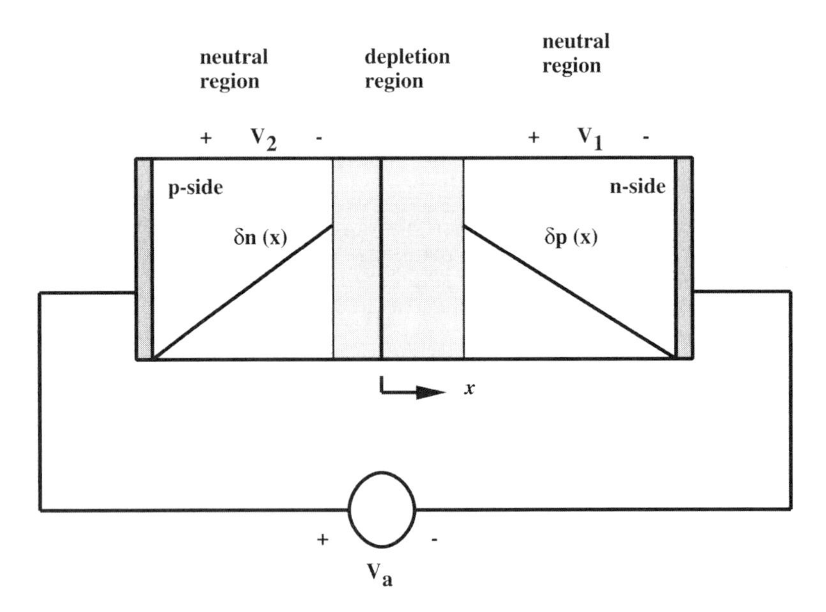

FIGURE 2-25. The *p-n* diode discussed in Example 2-8. The neutral layer thicknesses are short so that the minority carrier concentrations on both sides vary linearly with position.

Both J_n and J_p contain a diffusion and a drift component. We express J_T as

$$J_T = J_p + J_n = (J_{p\text{-diff}} + J_{p\text{-drift}}) + (J_{n\text{-diff}} + J_{n\text{-drift}})$$

On the N side, the hole carrier is the minority carrier. In the quasi-neutral region, where the electric field is small, we approximate $J_{p\text{-drift}}$ as zero (although $J_{n\text{-drift}}$ is significant due to the large amount of electrons there.) The diffusion components are given by

$$J_{p\text{-diff}} = -qD_p \frac{dp}{dx}$$

$$J_{n\text{-diff}} = qD_n \frac{dn}{dx}$$

According to the quasi-neutrality assumption discussed in § 1-6, the excess hole and excess electron concentrations are equal in the quasi-neutral region. We write the total electron concentration as $n = n_0 + \delta n$ and the total hole concentration as $p = p_0 + \delta p$. Since both $n_0 = N_d$ and $p_0 = n_i^2/N_d$ are constant with space, we equate dp/dx with dn/dx. This equality, when substituted into the preceding expressions for the diffusion currents, leads to $J_{n\text{-diff}} = -(D_n/D_p)J_{p\text{-diff}}$. Therefore, J_T is equal to

$$J_T = J_{n\text{-drift}} + J_{p\text{-diff}}\left(1 - \frac{D_n}{D_p}\right)$$

It is given in the problem that $J_T = 2J_{p\text{-diff}}$. Thus, $J_{n\text{-drift}}$ on the n side is

$$J_{n\text{-drift}} = J_{p\text{-diff}}\left(1 + \frac{D_n}{D_p}\right) = \frac{J_T}{2}\left(1 + \frac{\mu_n}{\mu_p}\right)$$

Since $J_{n\text{-drift}} = q\mu_n N_d \varepsilon_1$, the electric field on the n side is found to be

$$\varepsilon_1 = \frac{J_T}{2q\mu_n N_d}\left(1 + \frac{\mu_n}{\mu_p}\right)$$

Analogously, the electric field on the p side is found to be

$$\varepsilon_2 = \frac{J_T}{2q\mu_p N_a}\left(1 + \frac{\mu_p}{\mu_n}\right)$$

The total voltage drop $V_1 + V_2$ is given by

$$V_1 + V_2 = W(\varepsilon_1 + \varepsilon_2) = \frac{WJ_T}{2q\mu_n N_d}\left(1 + \frac{\mu_n}{\mu_p}\right) + \frac{WJ_T}{2q\mu_p N_a}\left(1 + \frac{\mu_p}{\mu_n}\right)$$

$$= \frac{WJ_T}{2q}\left(\frac{1}{\mu_n} + \frac{1}{\mu_p}\right)\left(\frac{1}{N_d} + \frac{1}{N_a}\right)$$

We construct the quasi-Fermi levels for the entire heterojunction. (In § 2-4, we drew only the Fermi levels in the neutral regions.) We consider the location $x = X_N$ first. From the laws of the junction (Eq. 2-114), the minority hole concentration is greater than its equilibrium concentration by a factor of $\exp(qV_a/kT)$. Therefore, the quasi-Fermi level for the minority hole is spaced by a magnitude V_a away from the Fermi level of the thermal equilibrium. In the region $x > X_N$, Eq. (2-122) states that the minority hole carrier concentration decreases exponentially with a characteristic length L_p as x increases. Therefore, on the energy band diagram whose vertical scale is exponential with respect to the carrier density, the minority hole quasi-Fermi level decreases linearly with x. Eventually, the quasi-Fermi level reaches the Fermi level of the thermal equilibrium at $(x - X_n) \approx 3L_p$. Above such a distance, the quasi-Fermi level overlaps with the Fermi level of the thermal equilibrium because the excess hole density there is zero.

On the p side, the hole is the majority carrier. The quasi-Fermi level coincides with the p-side Fermi level under thermal equilibrium, which is spaced at a distance qV_a from the N-side Fermi level under thermal equilibrium. Therefore, the hole quasi-Fermi levels at X_N and $-X_p$ are the same level. This equality tempts us to draw the quasi-Fermi level in the depletion region as a horizontal line, connecting the quasi-Fermi level at the two sides of the junction. This is indeed a widely accepted practice. A discussion similar to that given for the hole can be repeated for the electrons. The band diagram with the quasi-Fermi levels is hence as shown in Fig. 2-23. We emphasize that such a construction of quasi-Fermi levels is only approximate, with increasing error as the position is moved farther into the junction, away from the depletion edge. (Although the absolute magnitudes of the quasi-Fermi levels do not deviate much from those drawn, the carrier concentrations can be off by orders of magnitude because of their exponential dependence on the quasi-Fermi levels.)

Regardless of the approximation involved, if we accept the Fermi levels in the depletion region to be as shown in Fig. 2-23, then $E_{fn} - E_{fp}$ is a constant, equal to qV_a. If we evaluate $n(x)$ and $p(x)$ inside the depletion region using Eqs. (2-43) and (2-44), we find that the $p \cdot n$ product satisfies the following relationship:

$$n \cdot p = n_i^2 \exp\left(\frac{qV_a}{kT}\right) \tag{2-136}$$

In a forward bias, the $n \cdot p$ product greatly exceeds its thermal equilibrium value of n_i^2. This is expected because large amounts of carriers are injected from both sides of the junction into the depletion regions. In a reversed-biased condition, where V_a is negative, the exponential factor is practically zero and the $n \cdot p$ product is much less that its thermal equilibrium value.

§ 2-6 SPACE-CHARGE RECOMBINATION AND GENERATION CURRENTS

The diode current expression for the graded heterojunction was obtained by assuming the continuity of the electron and hole currents through the depletion region. Equivalently, we assumed that there is no carrier generation or recombination inside the depletion region. In reality, carrier recombination does occur in a forward-biased junction, and carrier generation takes place in a reverse-biased junction. The overall diode current needs to include a component to account for either the recombination or the generation. Because the diode current in a forward-biased junction is large, the additional recombination current component is comparatively small and can be safely neglected. (But in a transistor in which the space-charge recombination current is not negligible compared to the base current, the space-charge current can still have a profound impact on the transistor's performance.) The generation current in a reverse-biased junction, in contrast, is often larger than I_0 predicted in the previous section. In fact, the current of practical p-n junction in reverse bias exhibits strong voltage dependence, rather than having a constant value equal to $-I_0$. In the following, we first study the recombination current in the abrupt heterojunction and then in the graded heterojunction. We study the generation current later.

The most important physical process resulting in the generation and recombination of carriers in the depletion region is the capture and emission of carriers through localized states in the energy gap. These localized states act as stepping stones for an electron from the valence band first to reside in these states and then to be transferred to the conduction band (or vice versa). The energy states are always present because of lattice imperfections as well as impurity atoms. A model that formulates this process adequately is the Shockley–Read–Hall theory. The most important result of this theory is its derivation of the recombination rate (U, in cm^{-3}-s^{-1}), which is given approximately by

$$U = \frac{np - n_i^2}{(n + p + 2n_i)\tau} \tag{2-137}$$

where τ is the recombination rate of the carriers in the region, assumed here to be equal for the electron and for the hole. In a forward-biased junction, $np \gg n_i^2$ (Eq. 2-136). Therefore, U is a positive number and it is correctly referred to as the recombination rate. In a reverse-biased junction, $np \ll n_i^2$. In this case, U is negative, and we often refer to the same expression as the generation rate.

If the recombination rate is known everywhere in the depletion region, then the space-charge recombination current is proportional to the integral of U inside the whole depletion region:

$$I_{\text{rec}} = qA_D \int_{-X_p}^{X_N} U(x)\,dx = qA_D \int_{-X_p}^{0} U^{\text{NG}}(x)\,dx + qA_D \int_{0}^{X_N} U^{\text{WG}}(x)\,dx \tag{2-138}$$

where U^{NG} and U^{WG} are the recombination rates in the narrow-energy-gap and wide-energy-gap materials, respectively. The carrier concentrations of an abrupt heterojunction were plotted in Fig. 2-9. It was shown that on the p^+ side, the hole concentration greatly exceeds the electron concentration. The recombination rate in the region $-X_p < x < 0$ is thus given by

$$U^{\text{NG}} = \frac{np - n_i^2}{(n + p + 2n_i)\tau} \approx \frac{np}{p\tau} = \frac{n_i^2 \exp(qV_a/kT)}{p\tau} \tag{2-139}$$

We now substitute $p = n_i \exp(E_i - E_{fp})$, where E_{fp}, a constant, is at the same level as the Fermi level under thermal equilibrium. Therefore, we obtain

$$U^{\text{NG}}(x) = \frac{n_i}{\tau} \exp\left(-\frac{E_i(x) - E_{fp}}{kT}\right) \exp\left(\frac{qV_a}{kT}\right) \tag{2-140}$$

We developed in §2-1 that $E_c(x)$, and hence $E_i(x)$, vary parabolically with distance. However, the amount of band bending is minute due to its heavy doping in the narrow-energy-gap material. From the detailed analysis (§2-1) that accounts for the mobile holes in the N-side depletion region, we find ϕ_p to be less than kT/q. For the analysis here, we shall assume it to be equal to kT/q. Furthermore, the depletion thickness X_p is small in comparison to the overall depletion region. Therefore, it is fair to approximate the conduction band as decreasing linearly (rather than parabolically) by an amount ϕ_p in a distance X_p. That is, $E_i(x)$ is approximated by [5]

$$E_i(x) = E_i(-\infty) - q\phi_p\left(1 + \frac{x}{X_p}\right) \tag{2-141}$$

where $E_i(-\infty)$ is the equilibrium value at $x \le -X_p$. The expression correctly shows that $E_i(x)$ decreases to $E_i(-\infty) - q\phi_p$ at the junction interface ($x = 0$). Taking the derivative of Eq. (2-141) leads to the following identity:

$$dE_i(x) = -\frac{q\phi_p}{X_p}\,dx \tag{2-142}$$

The recombination current in the narrow-energy-gap material (I_{rec}^{NG}) is evaluated as

$$
\begin{aligned}
I_{rec}^{NG} &= \frac{qA_D n_i^{NG}}{\tau} \exp\left(\frac{qV_a}{kT}\right) \int_{-X_p}^{0} \exp\left(-\frac{E_i(x) - E_{fp}}{kT}\right) dx \\
&= \frac{qA_D n_i^{NG}}{\tau\phi_p} \exp\left(\frac{qV_a}{kT}\right) \int_{E_i(-\infty)}^{E_i(\infty)-q\phi_p} \exp\left(-\frac{E_i - E_{fp}}{kT}\right) dE_i \\
&= \frac{A_D X_p kT}{\tau\phi_p} n_i^{NG} \exp\left(-\frac{E_i - E_{fp}}{kT}\right) \exp\left(\frac{qV_a}{kT}\right)\left[\exp\left(\frac{q\phi_p}{kT}\right) - 1\right] \\
&= \frac{A_D X_p kT}{\tau\phi_p} \frac{(n_i^{NG})^2}{N_a} \exp\left(\frac{qV_a}{kT}\right)\left[\exp\left(\frac{q\phi_p}{kT}\right) - 1\right]
\end{aligned}
\tag{2-143}
$$

As mentioned, we assume ϕ_p to have a value of kT/q in a heavily doped p-type region. Therefore, we can express I_{rec}^{NG} as

$$
I_{rec}^{NG} = 1.7 \frac{qA_D X_p}{\tau} \frac{(n_i^{NG})^2}{N_a} \exp\left(\frac{qV_a}{kT}\right) \tag{2-144}
$$

Note that I_{rec}^{NG} increases with V_a with an ideality factor of unity rather than 2.

We now determine the recombination current in the wider-energy-gap region on the lightly doped N side. The following derivation is typically employed to study the recombination current in a homojunction diode. Instead of evaluating $U^{WG}(x)$ exactly in the region $0 < x < X_N$, we identify the maximum recombination rate. Neglecting n_i in the expression for U in Eq. (2-137), we find the U reaches its maximum value when $n = p = n_i \exp(qV_a/2kT)$. U_{max} is given by

$$
U_{max} = \frac{n_i}{2\tau} \exp\left(\frac{qV_a}{2kT}\right) \tag{2-145}
$$

We assume that $U = U_{max} =$ constant for the entire depletion region on the N side. This assumption leads to

$$
I_{rec}^{WG} = qA_D \int_0^{X_N} \frac{n_i^{WG}}{2\tau} \exp\left(\frac{qV_a}{2kT}\right) dx = \frac{qA_D n_i^{WG} X_N}{2\tau} \exp\left(\frac{qV_a}{2kT}\right) \tag{2-146}
$$

The recombination current in the wide-energy-gap material increases with V_a with an ideality factor of 2. This differs from the recombination current in the narrow-energy-gap material. The overall recombination current is $I_{rec}^{NG} + I_{rec}^{WG}$, equal to

$$
I_{rec} = 1.7 \frac{qA_D X_p}{\tau} \frac{(n_i^{NG})^2}{N_a} \exp\left(\frac{qV_a}{kT}\right) + \frac{qA_D n_i^{WG} X_N}{2\tau} \exp\left(\frac{qV_a}{2kT}\right) \tag{2-147}
$$

I_{rec} can have an ideality factor approaching either 1 or 2, depending on which component dominates. The ratio of these two components is

$$\frac{I_{\text{rec}}^{\text{NG}}}{I_{\text{rec}}^{\text{WG}}} = \frac{n_i^{\text{WG}} N_a}{3.4(n_i^{\text{NG}})^2} \frac{X_N}{X_p} \exp\left(\frac{qV_a}{2kT}\right) = \frac{n_i^{\text{WG}} N_a^2}{3.4(n_i^{\text{NG}})^2 N_d} \exp\left(\frac{qV_a}{2kT}\right) \quad (2\text{-}148)$$

We give a first-order estimate of V_a required to make the ratio equal to unity. Since the p side is heavily doped, we approximate E_{fp} there to lie on the valence band edge. From Eq. (2-44), we then write the following equation:

$$N_a = n_i^{\text{NG}} \exp\left(\frac{E_i - E_{fp}}{kT}\right) \approx n_i^{\text{NG}} \exp\left(\frac{E_{gp}}{2kT}\right) \quad (2\text{-}149)$$

When $I_{\text{rec}}^{\text{NG}}$ is equal to $I_{\text{rec}}^{\text{WG}}$, V_a is determined from Eq. (2-148) using the above relationship:

$$V_a = \frac{2kT}{q}\left[\ln\left(\frac{n_i^{\text{WG}} N_a}{3.4 n_i^{\text{NG}} N_d}\right) + \ln\left(\frac{N_a}{n_i^{\text{NG}}}\right)\right] = \frac{E_{gp}}{q} + \frac{2kT}{q}\ln\left(\frac{n_i^{\text{WG}} N_a}{3.4 n_i^{\text{NG}} N_d}\right) \quad (2\text{-}150)$$

For a (p^+) GaAs/(N) $Al_{0.35}Ga_{0.65}As$ heterojunction, we find $n_i^{\text{WG}} = 3.6 \times 10^2\,\text{cm}^{-3}$ and $n_i^{\text{NG}} = 1.79 \times 10^6\,\text{cm}^{-3}$. If $N_a = 3 \times 10^{19}\,\text{cm}^{-3}$ and $N_d = 5 \times 10^{17}\,\text{cm}^{-3}$, the second component is roughly equal to 0.325 V. In other words, V_a needs to be equal to $1.424 + 0.325\,\text{V} = 1.75\,\text{V}$ before the first component exceeds the second component. For most practical bias ranges, the space charge recombination current is therefore dominated by the second component, that which takes place in the wide-energy-gap material. Furthermore, the ideality factor of the recombination current exhibits an ideality factor of 2.

We now determine the recombination current in a graded p^+-N heterojunction. In particular, we are interested in a parabolically graded heterojunction, since it presumably results in a linear conduction band profile, which is desirable from the viewpoint of current conduction. From the previous analysis of the abrupt heterojunction, we do not expect the recombination current on the p^+ side to be appreciable. Consequently, we consider only recombination current that takes place in the N-side depletion region. We could carry out the same reasoning employed for the abrupt heterojunction, approximating U by $U_{\max}$ given in Eq. (2-145). The advantage of this simplification is that an analytical expression of I_{rec} is obtainable, leading to the conclusion that the ideality factor for I_{rec} is 2. In the following, however, we do not repeat the same analysis. Instead, we determine both the recombination rate and the recombination current exactly, using a simple computer program. In this manner, we verify numerically our claim that I_{rec} has a value close to 2.

To simplify the writing of the following equations, we assume that N_c and N_v for $Al_{\xi}Ga_{1-\xi}As$ are the same as those for GaAs, independent of ξ. Further, we list only the equation for the recombination current at $x < X_{\text{grading}}$. The recombination current elsewhere can be easily reproduced by the reader. (Actually, the recombination current elsewhere can be neglected. This is because, although some recombination current occurs at $X_N > x > X_{\text{grading}}$, the region is of higher energy gap and the carrier concentrations there are significantly less than those in the narrower-energy-gap region within $X_{\text{grading}} > x > 0$.)

Figure 2-23*a* illustrates the band diagram of a forward-biased junction, showing the relative positions of the quasi-Fermi levels with respect to the conduction and valence band edges. We have obtained $E_c(x)$ for the parabolically graded junction in § 2-4. It is repeated here for convenience (for $X_{\text{grading}} > x > 0$):

$$E_c(x) = E_{gp} - \frac{qN_d}{\epsilon_s}(X_N - X_{\text{grading}})x \tag{2-151}$$

X_{grading} was given by Eq. (2-96). The energy gap of the graded material is a function of x because the aluminum mole fraction varies parabolically in the region. $E_g(x)$ is simply $E_{gp} + 1.247\xi(x)$, which is simplified to the following in accordance with Eq. (2-97)

$$E_g(x) = E_{gp} + 2\Delta E_{c,\max}\frac{x}{X_{\text{grading}}} - \Delta E_{c,\max}\left(\frac{x}{X_{\text{grading}}}\right)^2 \tag{2-152}$$

The valence band edge is obtained from $E_c(x)$ by subtracting $E_g(x)$. The relationship between $\Delta E_{c,\max}$ and X_{grading} of Eq. (2-96) was used in the derivation:

$$\begin{aligned} E_v(x) &= \Delta E_{c,\max}\left(\frac{x}{X_{\text{grading}}}\right)^2 - \left[\frac{qN_d}{\epsilon_s}(X_N - X_{\text{grading}}) + \frac{2\Delta E_{c,\max}}{X_{\text{grading}}}\right]x \\ &= \frac{qN_d}{\epsilon_s}\left(\frac{x^2}{2} - X_N x\right) \end{aligned} \tag{2-153}$$

The coordinate system of the graded heterojunction is such that $E_c(0) = E_{gp}$ and $E_v(0) = 0$. With $E_{fp} - E_v = \Phi_p$, E_{fp} in this coordinate system is simply equal to Φ_p. Since the spacing between E_{fn} and E_{fp} is equal to qV_a, we write

$$E_{fp} = \Phi_p \quad E_{fn} = \Phi_p + qV_a \tag{2-154}$$

The recombination current, according to the definition of Eq. (2-138), is given by

$$I_{\text{rec}} = \frac{qA_D}{\tau}\int_0^{X_{\text{grading}}} n_i(x)^2 \exp\left(\frac{qV_a}{kT}\right)\frac{1}{[n(x) + p(x)]}\,dx$$

$$= \frac{qA_D}{\tau} \int_0^{X_{\text{grading}}} N_c N_v \exp\left[\frac{qV_a - E_g(x)}{kT}\right]$$

$$\times \left\{ N_c \exp\left[\frac{\Phi_p + qV_a - E_c(x)}{kT}\right] + N_v \exp\left[\frac{E_v(x) - \Phi_p}{kT}\right] \right\}^{-1} dx \tag{2-155}$$

where $E_g(x)$, $E_c(x)$, and $E_v(x)$ were given in Eqs. (2-151)–(2-153). We perform the numerical calculation for a p-N heterojunction with $N_a = 5 \times 10^{18}$ cm^{-3} and $N_d = 5 \times 10^{17}$ cm^{-3}. Figure 2-26 shows the calculated recombination rate as a function of x. The base doping and emitter doping used in the calculation are 5×10^{18} cm^{-3} and 5×10^{17} cm^{-3}, respectively. The applied bias is 1.35 V. The junction is taken to have $\xi_{\max} = 0.3$. In the calculation, $\tau(x)$ is taken to be a constant, equal to 10^{-9} s. Knowing that $x = 0$ corresponds to the metallurgical junction boundary, we find from Fig. 2-26 that most of recombination occurs with 60 Å of the junction. This result is expected from Fig. 2-27, which shows the carrier concentrations as a function of distance.

The space-charge recombination current is plotted as a function of applied bias and shown in Fig. 2-28. The conditions used for the calculations are the same as used for Fig. 2-26. The solid line in the figure is a fitted curve, indicating that the ideality factor η is about 1.8 if I_{rec} is written in the form $I_0 \exp(qV_a/\eta kT)$. This value is close to 2, as predicted in our previous analysis, which assumed that $U = U_{\max}$ throughout the entire depletion region. For simplicity, the

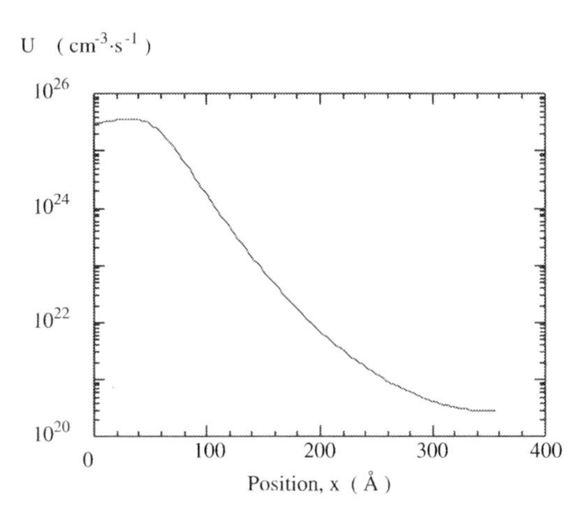

FIGURE 2-26. Recombination rate as a function of position for a forward-biased p^+-N heterojunction. The base doping and emitter doping used in the calculation are 5×10^{18} cm^{-3} and 5×10^{17} cm^{-3}, respectively. The applied bias is 1.35 V. The junction is taken to have $\xi_{\max} = 0.3$. (From W. Liu, Microwave and d.c. characterizations of *Npn* and *Pnp* HBTs, Ph.D. dissertation, Stanford University, Stanford, CA, 1991.)

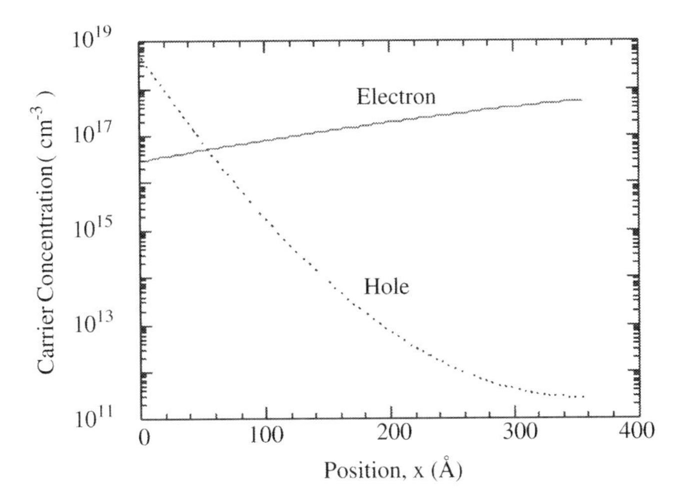

FIGURE 2-27. Calculated carrier concentrations as a function of position. (From W. Liu, Microwave and d.c. characterizations of *Npn* and *Pnp* HBTs, Ph.D. dissertation, Stanford University, Stanford, CA, 1991.)

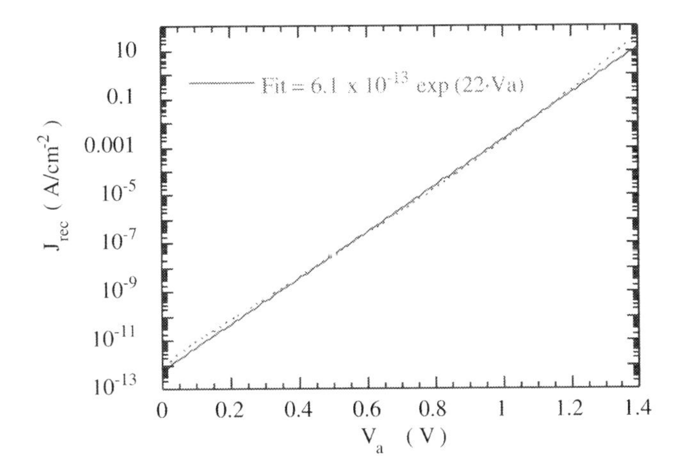

FIGURE 2-28. Calculated space-charge recombination current of the diode. (From W. Liu, Microwave and d.c. characterizations of *Npn* and *Pnp* HBTs, Ph.D. dissertation, Stanford University, Stanford, CA, 1991.)

ideality factor of the recombination current is often said to be 2, although it is actually less than 2.

The space-charge generation current occurs in a reverse-biased junction. In normal transistor operation, only the base-collector junction is reverse-biased. Since the base-collector junction in HBTs is a homojunction, we consider the generation current only in a homojunction. As mentioned previously, the generation rate is given by the same U expression of Eq. (2-137). With a reverse bias, $pn = n_i^2 \exp(qV_a/kT)$ is approximately zero. We therefore treat both n and p to be much smaller than n_i, and the generation rate simplifies to

$$U = \frac{pn - n_i^2}{(p + n + 2n_i)\tau} = -\frac{n_i}{2\tau} \tag{2-156}$$

A minus sign indicates that it is opposite of the carrier recombination; the magnitude of the generation rates is $n_i/(2\tau)$. We note that, for a homojunction, n_i is constant throughout the depletion region. The generation current is given by

$$I_{\text{gen}} = -qA_D \int_{-X_p}^{X_N} U(x)\,dx = \frac{qA_D n_i}{2\tau}(X_N + X_p) \tag{2-157}$$

Although I_{gen} appears to be insensitive to voltage, it varies with V_a in a square-root fashion due to the dependence of X_N and X_p on V_a. According to Eqs. (2-41) and (2-42), we write

$$I_{\text{gen}} = \frac{qA_D n_i}{2\tau}\left[\sqrt{\frac{2\epsilon_s}{qN_d}(\phi_{NO} - V_{aN})} + \sqrt{\frac{2\epsilon_s}{qN_a}(\phi_{pO} - V_{ap})}\right] \tag{2-158}$$

where ϕ_{NO} and ϕ_{pO} are the built-in potentials on the N side and p side, respectively. They can be obtained from the equations of § 2-1.

§ 2-7 ISOTYPE HETEROJUNCTIONS

Unlike a p-n junction, an *isotype* (n-N or p-P) heterojunction is not critical to the operation of HBTs or FETs. It is, however, a widely used element to improve the device performance in bipolar and field-effect transistors alike. An example of an isotype heterojunction in a HBT is that formed with the emitter cap layer (n-GaAs) and the active emitter layer (N-AlGaAs). The wide-energy-gap AlGaAs is the material responsible for the transistor properties of HBTs. If the emitter contact were placed directly on AlGaAs, the emitter contact resistance would be undesirably high. Generally, a metal contact on GaAs yields much less contact resistance, and GaAs is the preferred material on which the emitter contact is formed. A simple solution to retain the AlGaAs emitter with a minimized emitter resistance is to include both the AlGaAs and GaAs layers as an integral part of the emitter.

If the isotype heterojunction is not graded (i.e., juxtaposing the n-GaAs and N-AlGaAs layers abruptly), then the band diagram look schematically like Fig. 2-29. The method to compute the band diagram is similar to that for a p-N

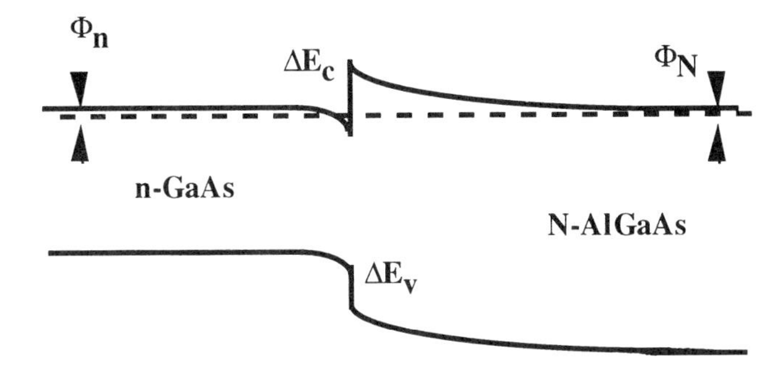

FIGURE 2-29. Band diagram of an *n*-GaAS/*N*-AlGaAs heterojunction under thermal equilibrium.

heterojunction. On the *n* side of the *n*-*N* heterojunction, the conduction band bends down. This means that there is an appreciable amount of electrons nearby the junction and the depletion approximation may be incorrect in the *n* layer. Two actions are required to account accurately for these mobile electrons. First, obviously, we need to amend the depletion approximation by including the mobile electrons in the Poisson equation. Second, we need to employ the Joyce–Dixon approximation rather than the Boltzmann approximation, since the Fermi level on the *n* side is close to the conduction band edge. As far as the *N* side is concerned, no departure from the depletion approximation is necessary, since the conduction band bends upward, away from the Fermi level.

The built-in potential, a critical parameter to the solution of the band profile, is obtained using the method outlined in § 2-3. It is equal to the separation between the Fermi levels of the two sides under the charge-neutral condition. As will be discussed shortly, the *n* side of an *n*-*N* heterojunction is taken to be the positive terminal. Therefore,

$$\phi_{\mathrm{bi}} = (\Phi_n + \Delta E_c - \Phi_N)/q \qquad (n\text{-}N \text{ heterojunction}) \qquad (2\text{-}159)$$

It is possible to have a negative built-in potential! If somehow the *N* side is so lightly doped that Φ_N exceeds ΔE_c, then ϕ_{bi} can easily be negative. In a *p*-*n* heterojunction, the built-in potential is always positive if ΔE_c and ΔE_v of the two material systems are positive.

As for a *p*-*P* heterojunction, it is customary to define the *P* side as the positive terminal. Therefore,

$$\phi_{\mathrm{bi}} = (\Phi_p + \Delta E_v - \Phi_P)/q \qquad (p\text{-}P \text{ heterojunction}) \qquad (2\text{-}160)$$

We have no desire to carry out a detailed analysis for the abrupt heterojunction, since an isotype heterojunction in III–V applications is generally graded. We mention only some qualitive characteristics about an abrupt *n*-*N* heterojunction. For the purpose of obtaining a simplified picture, we assume that if the electron accumulation takes place in the n layer, then the conduction band bends down and remains pinned at the Fermi level. It is customary to define the n side of an n-N heterojunction as the positive terminal. With this reference choice, electrons are injected from the N side into the other side in a forward bias but are blocked by a potential barrier in a reverse bias, just as in a p-N heterojunction.

The band diagrams under the external bias conditions are illustrated in Fig. 2-30. The current flowing through the n-N junction can be approximated with the thermionic expression, with the barrier energy ϕ_B being the potential barrier between the Fermi level on the n side and the tip of the conduction band on the N side. It is equal to

$$\phi_B = \frac{1}{q}(\Delta E_c - \Phi_n) \tag{2-161}$$

If we further assume that all voltage drops on the depleted N side such that V in Eq. (2-102) is equal to V_a, the applied voltage, then we write

$$J_{n\text{-}N} \sim A^* T^2 \exp\left(\frac{\Phi_n - \Delta E_c}{kT}\right)\left[\exp\left(\frac{qV_a}{kT}\right) - 1\right] \tag{2-162}$$

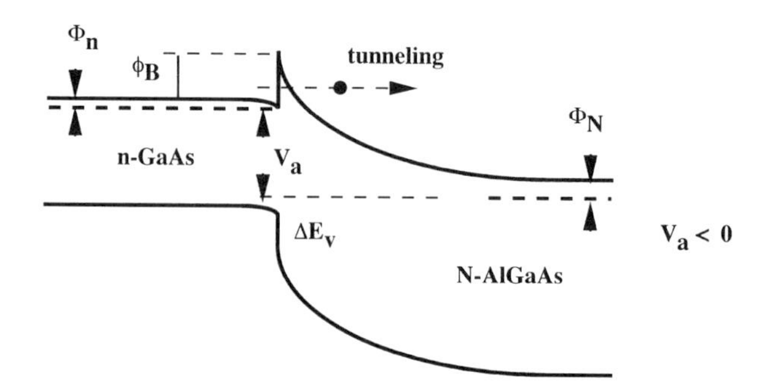

FIGURE 2-30. Band diagram of an n-N heterojunction under the forward- and reverse-bias conditions.

According to this equation, $J_{n\text{-}N}$ at negative applied voltages is small, but becomes large at positive applied voltages. This rectifying behavior, however, is not likely to be observed in practice. In the reverse-biased situation, as shown in Fig. 2-30, there will be an appreciable amounts of tunneling such that the reverse leakage current is significant.

As mentioned in the introduction, the n-N heterojunction in a III-V device is not used as a rectifier. It serves as a transition between the metal contact material and a transistor layer underneath the metal. The transition from the wide-energy-gap to the narrow-energy-gap material should be smooth, minimizing the electron reflection by a potential barrier. An abrupt heterojunction is therefore undersirable and an isotype heterojunction in a HBT or FET is always graded. The question is, what should be the grading distance to ensure a smooth transition from one material to the other?

To determine the desired grading distance, we start with the Poisson equation, Eq. (1-8). We replace V by φ_{Poisson} to stress that the potential is due to the space charges. Later we will add the energy gap potential φ_g to account for the grading of the alloy composition, in a manner similar to that performed in § 2-4. For simplicity, we make the doping level on both the n and the N sides the same, at N_d. We later comment on situations when the doping levels differ. The Poisson equation is given by

$$\frac{d^2\varphi_{\text{Poisson}}}{dx^2} = -\frac{q}{\epsilon_s}(N_d - n) \tag{2-163}$$

Equation (2-163) applies to $-\infty < x < \infty$. For convenience, we place the N region at $x < 0$ and the n region at $x > 0$. Since we are considering an n-N heterojunction, the hole and the acceptor concentrations are approximately zero. The material parameters, ϵ_s in the preceding equations and N_c in the following equations, are assumed to be constant, independent of the alloy composition. For simplicity, we shall use Boltzmann statistics, equating N_d and n to the following:

$$N_d = N_c \exp\left[\frac{E_f - E_c(-\infty)}{kT}\right] \tag{2-164}$$

$$n(x) = N_c \exp\left[\frac{E_f - E_c(x)}{kT}\right] \tag{2-165}$$

where $E_c(-\infty)$ is the conduction band energy at $x = -\infty$. Substituting these two equations into the Poisson equation, we obtain

$$\frac{d^2\varphi_{\text{Poisson}}}{dx^2} = -\frac{qN_d}{\epsilon_s}\left\{1 - \exp\left[\frac{E_c(-\infty) - E_c(x)}{kT}\right]\right\} \tag{2-166}$$

There are two variables in the above equation, φ_{Poisson} and $E_c(x)$. They are related by the superposition method employed in § 2-4, namely,

$$E_c(x) = -q\varphi_{\text{Poisson}} + (-q\varphi_g) \tag{2-167}$$

We write down the expression for $(-q\varphi_g)$ in the case of a linear grading of the aluminum composition from $\xi = \xi_{\max}$ at $x = -X_{\text{grading}}/2$ to $\xi = 0$ at $x = +X_{\text{grading}}/2$. An intuitive expression for such a grading scheme is simply $\xi(x) = \xi_{\max} - x/X_{\text{grading}}$ for $-X_{\text{grading}}/2 < x < +X_{\text{grading}}/2$; $\xi(x) = \xi_{\max}$ for $-X_{\text{grading}}/2 < x$; and $\xi(x) = 0$ for $x > +X_{\text{grading}}/2$. However, this expression is problematic for our formulation here. As is shown shortly, the differential equation we are about to solve contains the second spatial derivative of φ_g. The intuitive expression, though simple, contains discontinuities in the first, and therefore the second derivatives. These discontinuities would cause unnecessary numerical nuances. We therefore approximate the linear grading of the aluminium composition by a hyperbolic tangent function that has continuous second derivatives:

$$\xi(x) = \frac{\xi_{\max}}{2}\left[1 - \tanh\left(\frac{x}{X_{\text{grading}}}\right)\right] \tag{2-168}$$

Again, this expression is applicable to the entire range of x between $-\infty$ and ∞. At $x = -\infty$, the hyperbolic tangent is equal to -1, so $\xi(x) = \xi_{\max}$. Conversely, at $x = +\infty$, the hyperbolic tangent is equal to 1, hence $\xi(x) = 0$. We therefore write the energy gap potential as

$$-q\varphi_g = E_{gp} + \frac{\Delta q\chi_e}{2}\left[1 - \tanh\left(\frac{x}{X_{\text{grading}}}\right)\right] \tag{2-169}$$

where $\Delta q\chi_e$ is the difference between the electron affinity energies between the two dissimilar materials. In the case of an $Al_{0.35}Ga_{0.65}As/GaAs$ heterojunction, $\Delta q\chi_e = 0.244$ eV. From Eq. (2-167), the Poisson potential is found to be

$$\varphi_{\text{Poisson}} = \frac{E_c(x) - (-q\varphi_g)}{-q} = \frac{E_c(x)}{-q} - \frac{E_{gp}}{-q} + \frac{\Delta\chi_e}{2} - \frac{\Delta\chi_e}{2}\tanh\left(\frac{x}{X_{\text{grading}}}\right) \tag{2-170}$$

We now define a normalized variable to simplify the construction of the numerical program:

$$\Psi \triangleq \frac{E_c(x) - E_c(-\infty)}{kT} \tag{2-171}$$

Ψ measures the amount of deviation of the conduction band energy from its value at $x = -\infty$. With this definition, φ_{Poisson} is rewritten as

$$\varphi_{\text{Poisson}} = -\frac{kT}{q}\Psi - \frac{kT}{q}E_c(-\infty) + \frac{E_{gp}}{q} + \frac{\Delta\chi_e}{2} - \frac{\Delta\chi_e}{2}\tanh\left(\frac{x}{X_{\text{grading}}}\right) \tag{2-172}$$

Substituting this equation back into the Poisson equation (Eq. 2-166), we find a differential equation in terms of the variable Ψ:

$$\frac{kT}{q}\frac{d^2\Psi}{dx^2} + \frac{\Delta\chi_e}{2}\frac{d^2}{dx^2}\tanh\left(\frac{x}{X_{\text{grading}}}\right) = \frac{qN_d}{\epsilon_s}[1 - \exp(-\Psi)] \tag{2-173}$$

This equation is further simplified by defining a so-called *Debye length* (L_D) and normalized variable z:

$$L_D = \sqrt{\frac{kT\epsilon_s}{q^2N_d}} \tag{2-174}$$

$$z = \frac{x}{L_D} \tag{2-175}$$

After some algebraic manipulation, we arrive at the final equation:

$$\frac{d^2\Psi}{dz^2} = 1 - \exp(-\Psi) - \frac{\Delta\chi_e}{2kT}\frac{d^2}{dz^2}\tanh\left(\frac{L_D}{X_{\text{grading}}}z\right) \tag{2-176}$$

The appropriate boundary conditions are

$$\Psi(z = -\infty) = \Psi(z = \infty) = 0 \tag{2-177}$$

This nonlinear ordinary differential equation does not have an analytical solution. Numerical techniques are required for its solution. We present the results in Fig. 2-31, with various values of L_D/X_{grading}.

Although the above analysis assumes that the doping level is the same in the two materials, this assumption can be relaxed without significant error. If the doping levels are constant but not equal, N_D may be taken as the doping level on the low-electron-affinity side (N region). The exact doping level in the n region is less important because only a small fraction of the built-in potential is dropped across the accumulation region.

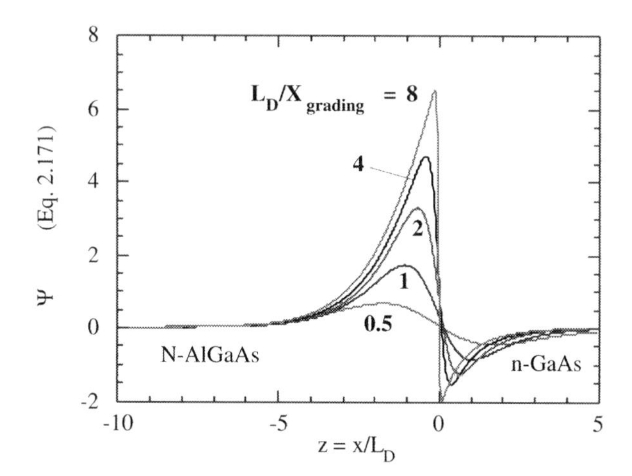

FIGURE 2-31. Calculated band energy (expressed in the normalized variable defined in Eq. (2-171)) as a function of position. The abruptness of the heterojunction is reduced as the grading distance increases.

Example 2-9:

What is a good grading distance for a (n)GaAs/(N)Al$_{0.35}$Ga$_{0.65}$As heterojunction if the potential barrier is to be minimized? The doping level is the same in both materials, equal to $2 \times 10^{17}\,\text{cm}^{-3}$.

The Debye length is, according to Eq. (2-174),

$$L_D = \sqrt{\frac{0.0258(13.18)(8.85 \times 10^{-14})}{(1.6 \times 10^{-19})(2 \times 10^{17})}} = 9.7 \times 10^{-7}\ \text{cm}$$

From Fig. 2-31, we see that when the L_D/X_{grading} ratio is smaller than $\frac{1}{2}$, the amount of potential barrier is only on the order of $1kT$. We therefore design X_{grading} to exceed $2L_D \approx 200$ Å.

REFERENCES

1. Kennedy, D. P. (1975). "The potential and electric field at the metallurgical boundary of an abrupt *p-n* semicondutor junction." *IEEE Trans. Electr. Dev.* **22,** 988–994.
2. Shockley, W. (1949). "The theory of *p-n* junctions in semiconductors and *p-n* junction transistors." *Bell Syst. Tech. J.* **28,** 435.
3. Tiwari, S., and Frank, D. J. (1988). "Barrier and recombination effects in the base-emitter junction of heterostructure bipolar transistors." *Appl. Phys. Lett.* **52,** 993–995.
4. Liu, W. (1998). *Handbook of III–V Heterojunction Bipolar Transistors.* New York: Wiley & Sons. Chapter 3.

5. The development of the recombination current in the narrow-bandgap side of the junction follows the lecture notes of Harris, J. (1988). "Physics of advanced electronic devices." EE graduate course, Stanford University. See also Ref. [1].

PROBLEMS

1. Construct the band diagrams of the following abrupt junctions at thermal equilibrium: (N) $Al_{0.35}Ga_{0.65}As$/(p) GaAs; (N) InP/(p) $In_{0.53}Ga_{0.47}As$. Consider two doping designs for each of the four junctions: **(a)** $N_d = 2 \times 10^{17}$ cm^{-3} and $N_a = 5 \times 10^{18}$ cm^{-3}; **(b)** $N_d = 1 \times 10^{18}$ cm^{-3} and $N_a = 1 \times 10^{20}$ cm^{-3}. For GaAs, use $N_v = 4.7 \times 10^{18}$ cm^{-3}.

2. Plot the minority carrier profiles of the two junctions in Problem 1. Assume Eq. (1-13) work for degenerate semiconductors as well as non-degenerate semiconductors.

3. Redraw the band diagrams of Problem 1 for a forward bias of 0.5 V, and for a reverse bias of -5 V.

4. Not all parabolic grading schemes are created equal. It is known that, as demonstrated in §2-4, a parabolic grading results in a linear conduction band profile in the conduction band. However, there are several kinds of "parabolic" grading schemes. It is the purpose of this problem to show that some parabolic grading schemes do not yield the desired conduction band profile,

Consider a p^+-N heterojunction identical to that of Example 2-6. However, suppose that instead we grade the aluminium content in the following fashion:

$$\xi(x) = \begin{cases} 0 & \text{for } 0 \geq x \\ \xi_{\max} - \xi_{\max}\left(1 - \dfrac{x^2}{X_{\text{grading}}^2}\right) & \text{for } 0 \leq x \leq X_{\text{grading}} \\ \xi_{\max} & \text{for } X_{\text{grading}} \leq x \end{cases}$$

This grading is parabolic. It resembles the correct parabolic grading of Eq. (2-97), but a closer examination shows that they differ. Derive the $\varphi_g(x)$ corresponding to this $\xi(x)$. The Poisson potential for this p^+-N heterojunction is taken to be the same as that of Eq. (2-86). Show that the resulting conduction band energy does not vary linearly with position. Assume the thermal equilibrium condition.

5. Design a (p^+) GaAs/(N) $Al_\xi Ga_{1-\xi}As$ graded heterojunction such that the conduction band varies linearly with position. ξ is varied from 0 to 0.35. $N_a = 1 \times 10^{20}$ cm^{-3}; $N_d = 1 \times 10^{17}$ cm^{-3}. Draw the band diagrams for $V_a = 1.0$ and 1.2 V.

6. The ideality factor of a (p^+) GaAS/(N) $Al_{0.35}Ga_{0.65}As$ heterojunction is measured to be 1.2. Is it a graded or an abrupt heterojunction? Estimate the ratio of the doping concentrations of the two sides.

7. Consider a long GaAs homojunction diode under the reverse bias. Assume that the carrier lifetime is 1ns for holes or electrons, independent of the doping levels. One side of the junction is to be doped with 10^{17} cm^{-3} and the other, 10^{16} cm^{-3}.

a. Which side should be n-type to minimize the magnitude of the current?

b. What is the value of the reverse saturation current density?

c. What is the generation current density with a -2 V bias? Use $N_v = 4.7 \times 10^{18}$ cm^{-3} for GaAs.

8. Calculate the recombination currents on both the N and the p^+ sides of an abrupt (N) $Al_{0.35}Ga_{0.65}As/(p^+)$ GaAs heterojunction. $N_a = 3 \times 10^{19}$ cm^{-3}; $N_d = 2 \times 10^{17}$ cm^{-3}; $V_a = 1.2$ V; $\tau = 10^{-9}$ s; and $A_D = 100$ μm^2.

9. Assume Boltzmann statistics in the following analysis of a homojunction. The n-type doping is N_d; the p-type doping concentration is N_a; and the dielectric constant is ϵ_s.

a. Show that the mobile hole and electron concentrations obey the following relationships in the depletion regions of both sides:

$$n(x) = N_d \exp\left[-\frac{q}{kT}(\phi_{bi} - V)\right]$$

$$p(x) = N_a \exp\left(-\frac{qV}{kT}\right)$$

where V satisfies the boundary condition that $V(-\infty) = 0$ and $V(\infty) = \phi_{bi}$.

b. Write out the Poisson equation for the p side from $x = -\infty$ to 0. Multiply both sides of the Poisson equation by $2\,dV/dx$. Show that the following equality results:

$$\frac{d}{dx}\left(\frac{dV}{dx}\right)^2 = \frac{2q}{\epsilon_s} N_a \frac{d}{dx}\left\{V + \frac{kT}{q}\exp\left(-\frac{qV}{kT}\right) + \frac{N_d}{N_a}\frac{kT}{q}\exp\left[-\frac{q}{kT}(\phi_{bi} - V)\right]\right\}$$

c. Integrate the differential equation in **(b)** from $x = -\infty$ to 0. Apply the boundary conditions that $V(-\infty) = 0$ and $V(0) = \phi_{p0}$. Show that the electric field at $x = 0^-$, equal to $-dV/dx$ at $x = 0^-$, is given by

$$\varepsilon(0^-)^2 = \frac{2q}{\epsilon_s} N_a \left\{ \phi_p - \frac{kT}{q}\left[1 - \exp\left(-\frac{q\phi_{p0}}{kT}\right)\right] + \frac{N_d}{N_a}\frac{kT}{q}\exp\left(-\frac{q\phi_{\text{bi}}}{kT}\right)\left[\exp\left(\frac{q\phi_{p0}}{kT}\right) - 1\right]\right\}$$

d. Similarly for the n side, write out the Poisson equation and multiply both sides by $2\,dV/dx$. Show that the following equality results:

$$\frac{d}{dx}\left(\frac{dV}{dx}\right)^2 = -\frac{2q}{\epsilon_s} N_d \frac{d}{dx}\left\{ V - \frac{N_a}{N_d}\frac{kT}{q}\exp\left(-\frac{qV}{kT}\right) - \frac{kT}{q}\exp\left[-\frac{q}{kT}(\phi_{\text{bi}} - V)\right]\right\}$$

e. Integrate differential equation in **(d)** from $x = 0$ to ∞. Apply the boundary conditions that $V(0) = \phi_{p0}$ and $V(\infty) = \phi_{\text{bi}}$. Show that the electric field at $x = 0^+$ equal to $-dV/dx$ at $x = 0^+$, is given by

$$\varepsilon(0^+)^2 = \frac{2q}{\epsilon_s} N_d \left(\phi_{\text{bi}} - \phi_{p0} - \frac{kT}{q}\left\{1 - \exp\left[-\frac{q}{kT}(\phi_{\text{bi}} - \phi_{p0})\right]\right\} + \frac{N_d}{N_a}\frac{kT}{q}\left[\exp\left(-\frac{q\phi_{p0}}{kT}\right) - \exp\left(-\frac{q\phi_{\text{bi}}}{kT}\right)\right]\right)$$

f. By equating $\varepsilon(0^+)$ with $\varepsilon(0^-)$, show that

$$\phi_{p0} = \frac{N_d}{N_a + N_d}\phi_{\text{bi}} + \frac{N_a - N_d}{N_a + N_d}\frac{kT}{q}\left[1 - \exp\left(-\frac{q\phi_{\text{bi}}}{kT}\right)\right]$$

g. Show that ϕ_{n0}, equal to $\phi_{bi} - \phi_{p0}$, is

$$\phi_{n0} = \frac{N_a}{N_a + N_d}\phi_{\text{bi}} + \frac{N_a - N_d}{N_a + N_d}\frac{kT}{q}\left[1 - \exp\left(-\frac{q\phi_{\text{bi}}}{kT}\right)\right]$$

h. Calculate the numerical values of ϕ_{p0} and ϕ_{n0}. The diode has $N_d = 1 \times 10^{16}\ \text{cm}^{-3}$ and $N_a = 1 \times 10^{17}\ \text{cm}^{-3}$. This analysis accounts for both the majority and minority mobile carriers in the "depletion" regions. Is there significant voltage drop in the heavily doped side?

10. Calculate the built-in potential of an n-i-n GaAs homojunction with N_d being $10^{17}\ \text{cm}^{-3}$ on one side and $10^{15}\ \text{cm}^{-3}$ on the other side.

11. A metal-semiconductor junction is formed by evaporating tungsten in contact with silicon. The barrier height is 0.45 eV and the n-type silicon is doped at $5 \times 10^{16}\ \text{cm}^{-3}$.

a. Draw the band diagram at the thermal equilibrium.

b. A light beam shines on this junction, creating electron-hole pairs. Which way would the current flow within the device when the junction is connected into a circuit? We denote the amount of current flow as I_{light}.

c. Under the bias setup of **(b)**, what is the maximum voltage that can be measured across the junction (at zero output current)?

12. Which is the more accurate expression for the total diode current, Eq. (2-133) or the following?

$$I_D = qA_D\left(\frac{D_p}{X_E - X_N}\frac{n_i^{\text{WG}^2}}{N_d} + \frac{D_n}{X_B - X_P}\frac{n_i^{\text{NG}^2}}{N_p}\right)\left[\exp\left(\frac{qV_a}{kT}\right) - 1\right] + I_{\text{rec}}$$

$$= qA_D\left(\frac{D_p}{X_E - X_N}\frac{n_i^{\text{WG}^2}}{N_d} + \frac{D_n}{X_B - X_P}\frac{n_i^{\text{NG}^2}}{N_a}\right)\left[\exp\left(\frac{qV_a}{kT}\right) - 1\right]$$

$$+ 1.7\frac{qA_D X_p}{\tau}\frac{(n_i^{\text{NG}})^2}{N_a}\exp\left(\frac{qV_a}{kT}\right) + \frac{qA_D n_i^{\text{WG}} X_N}{2\tau}\exp\left(\frac{qV_a}{2kT}\right)$$

13. Equation (2-158) suggests that there is a finite generation current even as the magnitude of the p-n junction reverse bias approaches zero. Is this physically possible? Explain.

CHAPTER 3

HBT D.C. CHARACTERISTICS

§ 3-1 BASIC TRANSISTOR OPERATION

A bipolar junction transistor (BJT) consists of three main layers, the emitter, the base, and the collector. The emitter and the collector are doped to the same type, whereas the base layer sandwiched in between has the opposite type of doping. These three layers form two *p-n* junctions, connected back to back. For convenience, we will discuss the operation of an *npn* transistor in which the electron is the conducting carrier between the emitter that emits the electron and the collector that collects the electron. Its dual, the *pnp* transistor, operates in the same way in principle, except that the hole is the primary conducting carrier.

In the normal operating condition, the base-emitter junction is forward-biased at V_{BE} ($V_{BE} > 0$) and the base-collector junction is reverse-biased at V_{BC}($V_{BC} < 0$). This is called the *forward-active* mode of operation. Suppose for a moment that the base layer is thick, as illustrated in Fig. 3-1*a*. The minority carrier concentration in the entire transistor is shown in the figure. According to the junction law, the minority carrier concentration at the edge of a depletion region is amplified (if $V_a > 0$) or reduced (if $V_a < 0$) by a factor of $\exp(qV_a/kT)$ in comparison to its thermal equilibrium concentration, where V_a is the applied voltage to a *p-n* junction. At the emitter side of the base layer, the minority carrier concentration is high because the base-emitter junction is forward-biased. Conversely at the collector side of the base, the carrier concentration approaches nil because the base collector junction is reverse-biased. If the base layer is thick enough (long-diode case of § 2-5), the electron concentration starts out with a high concentration at the emitter edge and then decreases

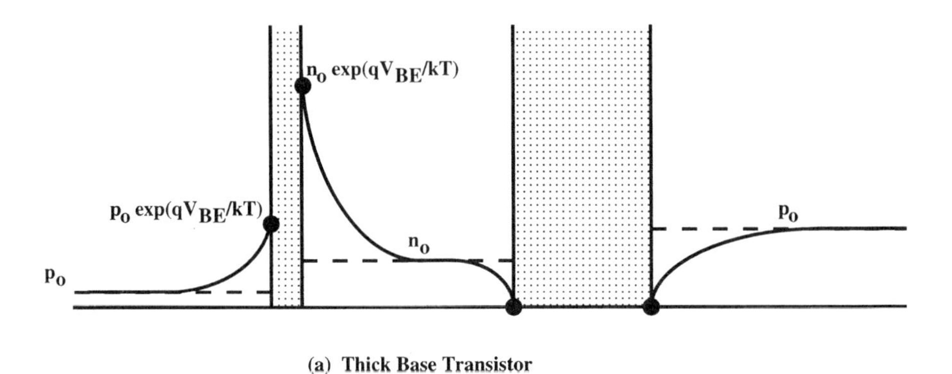

(a) Thick Base Transistor

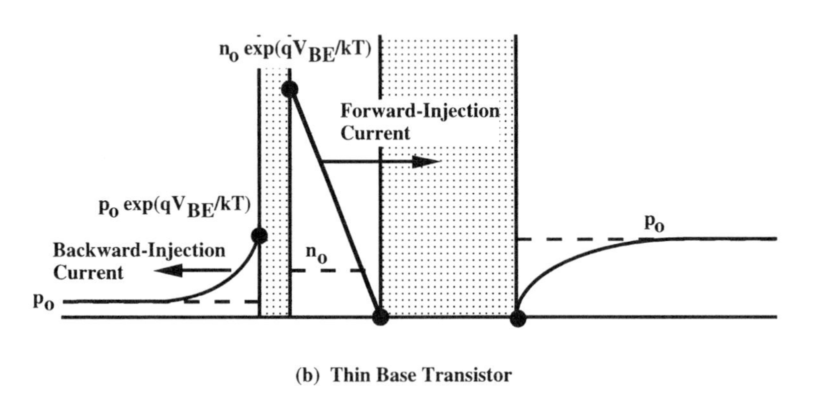

(b) Thin Base Transistor

FIGURE 3-1. (*a*) The electron minority carrier concentration is an *npn* homojunction bipolor transistor with an excessively thick base layer. The shaded regions are the depletion regions of the base-emitter and the base-collector junctions. (*b*) The electron minority carrier concentration in an *npn* homojunction bipolor transistor with a properly designed (thin) base layer. The minority carrier concentration in the base varies linearly with position.

exponentially to n_0 in the middle (Eq. 2-123). n_0, representing the equilibrium electron concentration in the base, has a small value equal to n_i^2/N_B. The electron concentration stays at n_0 for some distance before it finally decreases to 0 at the collector edge. Because we are concerned with a uniformly doped base, the base electric field is zero. Without a drift component, the electron current is due entirely to carrier diffusion, whose magnitude is proportional to the slope of the carrier concentration.

Near the emitter side, the electron is high; but after three diffusion lengths away from the emitter junction, the electron concentration decreases to and remains flat at n_0 in the middle of the base layer. All of the excess electrons are recombined. Without any spatial gradient in the electron concentration, the electron current is zero. It remains 0 until near the collector edge of the base layer, where a concentration gradient exists. However, the magnitude of the current is insignificant, since n_0 is a small value.

Normally the transistor is designed so that the hole injection from the base to both the emitter and the collector is negligible. The current flowing across the two junctions is made up mostly of electrons. With a large electron current injecting into the base from the emitter but only a small electron current leaving the base for the collector, a large portion of the injected electrons must exist the base layer through the base contact. Because the base layer in an *npn* transistor is *p*-type, it is the hole rather than the electron that moves through the base contact. The prior suggestion of electrons exiting the base layer means essentially that a large number of holes enter the base layer to recombine with the excess electrons inside. The magnitude of this base current is large, reflecting the difference between the emitter and the collector currents. This scenario is undesirable, as the electron carriers are not collected at the collector, but instead are recombined in the base.

Consider now a much thinner base layer than the previous case. We redraw qualitatively the minority carrier concentration in the three layers of the transistor, shown in Fig. 3-1*b*. As before, the carrier concentration is high at the emitter side and nearly zero at the collector side. Because the base layer is thin, the excess electron concentration at the emitter side decreases rapidly toward zero at the collector side, without ever saturating at a constant value of n_0. Although the mathematical form governing such a carrier profile is exponential as before, an exponential decrease across a short distance is approximately linear. Therefore, Fig. 3-1*b* shows a linear decrease of electron concentration from the emitter to the collector side, similar to that observed for a short diode as discussed in § 2-5.

The electron diffusion current as a function of the position in the base is obtained by taking the spatial derivative of the minority electron profile. Because the electron concentration varies nearly linearly, the electron current at the emitter side is essentially the same as that at the collector side. With these two currents roughly equal, the base current is small, as desired. Physically, the base current is small in a thin-base transistor because a thin base allows the injected electrons from the emitter to diffuse quickly through the base without being recombined. Upon reaching the base-collector junction, the electrons are quickly swept across the collector by the high electric field in the reverse-biased junction.

The collected electrons constitute the collector current of the transistor. By controlling the amount of the base-emitter bias, we vary the amount of the electron injection from the emitter to the base, and consequently, the amount of electrons collected at the collector. This current flow, however, is independent of the amount of the collector-base bias, as long as the base-collector junction is reverse-biased so that the minority electron concentration at the collector side of the base layer is maintained zero. The property that the signal current (collector current) is modified by the input bias (base-emitter bias) but unaffected by the output bias (collector-base bias) fulfills the requirement of a sound three-terminal device. As the input port and output port are isolated from one another, complicated circuits based on the transistors can be easily designed.

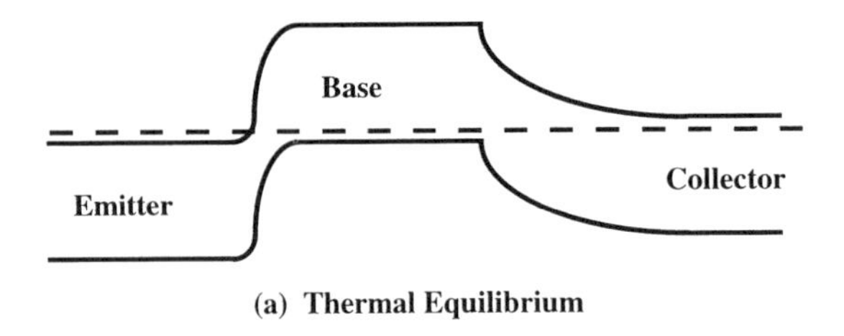

(a) Thermal Equilibrium

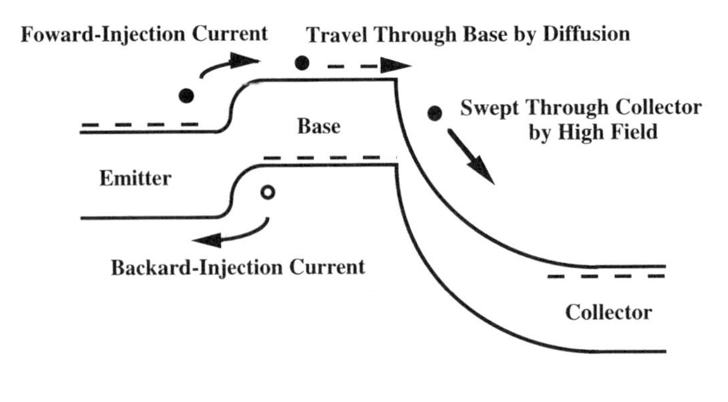

(b) Normal Operation

FIGURE 3-2. Band diagrams of an *npn* homojunction bipolar transistor under (*a*) thermal equilibrium (*b*) and in the normal, forward-active mode.

Figure 3-2 illustrates the band diagram of an *npn* BJT in thermal equilibrium and in normal operation. The energy gaps are the same throughout the entire transistor. The Fermi levels as well as the depletion regions of the band diagram reflect the fact that the doping level in a homojunction bipolar transistor is highest for the emitter and lowest for the collector.

We have not mentioned another current component that always exists, besides the desired collector current flow. As shown in the Fig. 3-1*b*, this current is composed of holes from the base that are back-injected into the emitter. These holes cannot originate from the *n*-type collector or the *n*-type emitter. They must be supplied externally through the base contact. The current flow due to the hole movement is thus called the *base current*. The base current does not contribute to the signal current flowing from the input to the output port. Instead, it represents the amount of current that must be supplied externally to the transistor before some output signal is obtained. The bipolar transistor is so named to emphasize that both the electron and the hole play significant roles in the device operation. The goal of an *npn* transistor design is to maximize the electron current flow and simultaneously minimize the hole current.

After the minority carrier continuity equation is solved in the transistor, the base current and collector current are given by

$$I_{Bp} = \frac{qA_E D_{pE}}{X_E} \frac{n_{iE}^2}{N_E} \exp\left(\frac{qV_{BE}}{kT}\right) \tag{3-1}$$

$$I_C = \frac{qA_E D_{nB}}{X_B} \frac{n_{iB}^2}{N_B} \exp\left(\frac{qV_{BE}}{kT}\right) \tag{3-2}$$

We denote the base current as I_{Bp} rather than I_B because the back-injection current is just one component of the overall base current. We will consider other components in the next section. For now, we are concerned exclusively with I_{Bp} because it is the dominant base current component in a silicon BJT. In the above expressions, A_E is the emitter area, N_E is the emitter doping level, X_E is the emitter thickness, D_{pE} is the minority hole diffusion coefficient in the emitter, n_{iE} is the intrinsic carrier concentration in the emitter, N_B is the base doping level, X_B is the base thickness, D_{nB} is the minority electron diffusion coefficient in the base, and n_{iB} is the intrinsic carrier concentration in the base. For transistors whose $X_E \gg L_{pE}$ (the diffusion length in the emitter), X_E needs to be replaced by L_{pE}. The replacement follows from a similar logic as discussed in §2-5. For homojunction BJTs, $n_{iB} = n_{iE}$. Further, on a first-order analysis, $D_{pE} \approx D_{nB}$ and $X_E \approx X_B$. Therefore, the Eqs. (3-1) and (3-2) demonstrate that the emitter doping level in a homojunction BJT needs to exceed the base doping level in order for I_C to exceed I_{Bp}.

In a sense we have traded off an important transistor design freedom to make the collector current larger than the base current. For many applications, we would actually prefer the base doping to be high and the emitter doping to be below. A high base doping allows a transistor to have a low base resistance and thus higher power gain. A low emitter doping reduces the base-emitter junction capacitance, leading to improved high-frequency performance. However, because of the nature of the homojunction transistor operation, these advantages must be sacrificed in the interest of minimizing the base current.

The homojunction BJT is the type of transistor that was first reduced to practice. Through decades of improvement, it continues to be dominant in applications requiring high speeds. Among all the developments through which BJTs have evolved, perhaps the most innovating change is the replacement of the homojunction emitter material by a larger-energy-gap material, thus forming a heterojunction bipolar transistor (HBT). Figures 3-3 and 3-4 illustrate the band diagrams of abrupt and graded *Npn* HBTs, respectively. The capital *N* in *Npn* rather than a lowercase letter *n* is used to reflect the characteristic that the emitter is made of a larger-energy-gap material. The qualifier "abrupt" or "graded" describes the nature of the base-emitter heterojunction. It is implicitly understood that the base-collector junction of a HBT is a homojunction. The figures reflect the fact that, despite their close resemblance with the BJT band diagram, the emitter has a larger energy gap than the rest of the transistor.

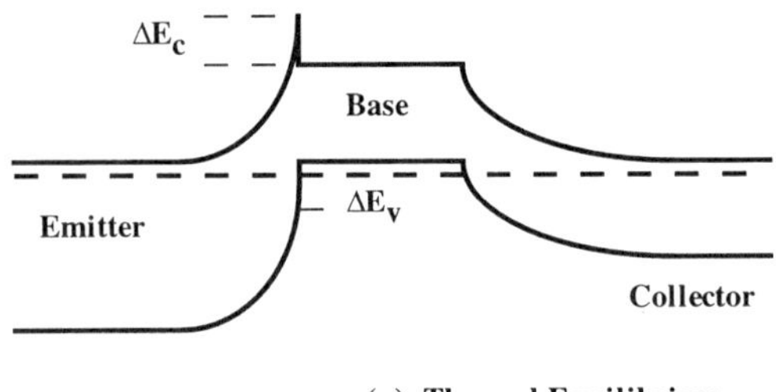

(a) Thermal Equilibrium

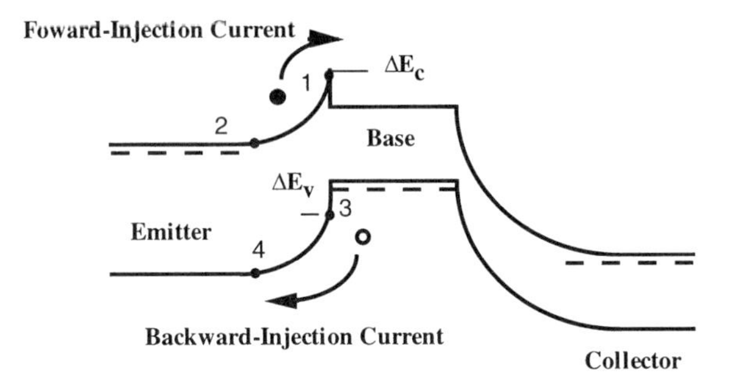

(b) Normal Operation

FIGURE 3-3. Band diagrams of a *Npn* HBT under (*a*) thermal equlibrium and (*b*) in the normal, forward-active mode. This abrupt HBT has a conduction band discontinuity in the base-emitter junction.

The energy-gap difference between the emitter and the base gives HBTs a substantial edge over BJTs. When the base-emitter junction of a BJT is forward-biased (Fig. 3-2), both the electrons forward-injected into the base and the holes back-injected into the emitter experience the same amount of energy barrier. This is a property of the homojunction; there is no design freedom to control the forces acting on the electrons and holes separately and independently of each other. Consequently, the only design parameter available to reduce the undesirable hole back-injection is to increase the doping level in the emitter relative to the base. In a HBT (either abrupt or graded), in contrast, the base-emitter heterojunction causes the holes from the base to experience a much larger energy barrier than the electrons from the emitter. For the abrupt HBT shown in Fig. 3-3, the hole barrier is larger than the electron barrier by a magnitude ΔE_v (since $V_{12} = V_{34}$.) For the graded HBT shown in Fig. 3-4, the hole barrier is larger by $V_{34} - V_{12} = \Delta E_g$. In both types of HBTs, the forces acting on the electrons and holes are different, favoring the electron injection from the emitter into the base to the hole back-injection from the base into the

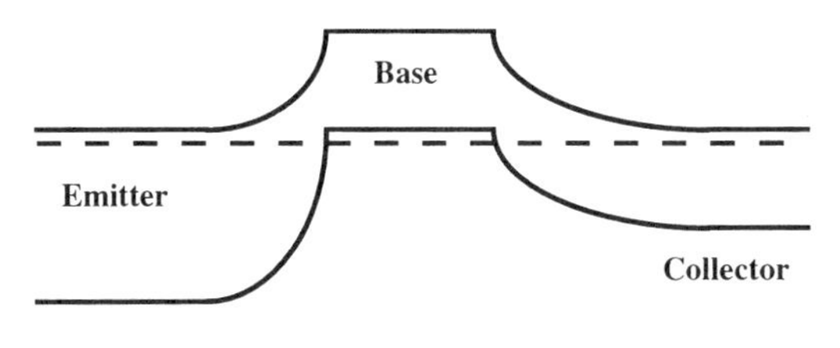

(a) Thermal Equilibrium

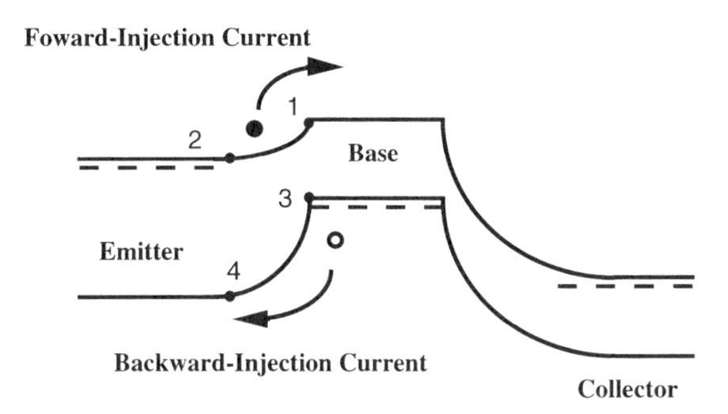

(b) Normal Operation

FIGURE 3-4. Band diagrams of an *Npn* HBT under (*a*) thermal equilibrium and (*b*) in the normal, forward-active mode. This is a graded HBT whose aluminum composition in the base-emitter junction is graded to remove the conduction band discontinuity.

emitter. The base doping can be made larger than the emitter doping without adversely affecting the HBT performance.

We quantify the advantage of HBT compared to BJT by calculating the ratio of the collector current to the base current. We first perform the analysis for a graded HBT whose base and collector currents are limited by the diffusions of carriers in the neutral regions. In this case, both Eqs. (3-1) and (3-2) apply, just as they do to a BJT whose currents are also limited by the diffusions of carriers. Unlike the BJT, however, the intrinsic carrier concentration in the emitter of the HBT is made smaller than that in the base by the use of heteromaterials. If we denote the energy-gap difference of the emitter and the base materials as ΔE_g, then the ratio of I_c/I_{Bp} is

$$\frac{I_C}{I_{Bp}} = \frac{D_{nB}X_E N_E}{D_{pE}X_B N_B}\frac{n_{iB}^2}{n_{iE}^2} = \frac{D_{nB}X_E N_E}{D_{pE}X_B N_B}\exp\left(\frac{\Delta E_g}{kT}\right) \qquad \text{(graded HBT)} \qquad (3\text{-}3)$$

Equation (1-13), which relates the intrinsic carrier concentration to the energy gap, was used in the derivation. This equation illustrates that even if $N_B \gg N_E$, the current ratio can still be made to be $\gg 1$ because of the exponential factor. The degree to which the base doping level can be made higher than the emitter doping depends on the amount of the energy-gap difference and the operating temperature. In a homojunction BJT, $\Delta E_g = 0$ and the exponential factor is unity. N_B has to be smaller than N_E in order for the current ratio to exceed unity.

For an abrupt HBT whose base-emitter junction is an abrupt heterojunction, we learned in Chapter 2 that the forward collector current is not limited by the diffusion through the base. The current transport is affected by the conduction band spike, and the collector current assumes a different relationship from that of Eq. (3-2). If we adopt the accurate expression of I_C in the abrupt HBT the ratio I_C/I_{Bp} will be considerably more complicated than that for the graded HBT. For the sake of establishing a simple expression, the following analysis based on the thermionic-emission concept is used. It assumes that the magnitude of forward current depends on the amount of electrons having enough energy to surmount the potential barrier ϕ_N, which is simply V_{12} in the band diagrams of Figs. 3-3 and 3-4. Therefore,

$$\frac{I_{C.\text{abrupt}}}{I_{C.\text{graded}}} = \frac{\exp(q\phi_{N.\text{abrupt}}/kT)}{\exp(q\phi_{N.\text{graded}}/kT)} = \exp\left(-\frac{\Delta E_c}{kT}\right) \tag{3-4}$$

Although I_C of the abrupt HBT is not exactly that given by Eq. (3-2), the backinjected base current is still limited by diffusion and given by Eq. (3-1). The backinjection current is generally small. It is not likely to be limited by other transport mechanisms, whether it is a graded or an abrupt HBT. Combining Eqs. (3-3) and (3-4), we obtain an I_C/I_{Bp} ratio for the abrupt HBT as

$$\frac{I_C}{I_{Bp}} = \frac{D_{nB}X_E N_E}{D_{pE}X_B N_B}\exp\left(\frac{\Delta E_v}{kT}\right) \quad \text{(abrupt HBT)} \tag{3-5}$$

Example 3-1:

Compare the I_C/I_{Bp} ratios at room temperature of a Si BJT, a graded $Al_{0.35}Ga_{0.65}As$/GaAs HBT, and an abrupt $Al_{0.35}Ga_{0.65}As$/GaAs HBT. All of the transistors have the following parameters: $N_E = 2\times10^{17}$ cm^{-3}, $N_E = 3\times10^{19}$ cm^{-3}, $D_{nB} = 25$ cm^2/V-s, $D_{pE} = 2.5$ cm^2/V-s, $X_E = 2000$ Å, and $X_B = 800$ Å.

For the Si BJT, $\Delta E_g = 0$. From Eq. (3-3),

$$\frac{I_C}{I_{Bp}} = \frac{D_{nB}X_E N_E}{D_{pE}X_B N_B} = \frac{25(2000)(2\times10^{17})}{2.5(800)(3\times10^{19})} = 0.17$$

For the graded $Al_{0.35}Ga_{0.65}As/GaAs$ heterojunction, ΔE_g is calculated from Eq. (1-88) to be $0.35 \times 1.247 = 0.436$ eV. The ratio for the graded HBT is therefore

$$\frac{I_C}{I_{Bp}} = \frac{D_{nB} X_E N_E}{D_{pE} X_B N_B} \exp\left(\frac{\Delta E_g}{kT}\right) = \frac{25(2000)(2 \times 10^{17})}{2.5(800)(3 \times 10^{19})} \exp\left(\frac{0.436}{0.0258}\right) = 3.7 \times 10^6$$

For the abrupt HBT, ΔE_v is the parameter of interest. From Eq. (1-89), ΔE_g for $\xi = 0.35$ is $0.55 \times 0.35 = 0.193$ eV. Therefore,

$$\frac{I_C}{I_{Bp}} = \frac{D_{nB} X_E N_E}{D_{pE} X_B N_B} \exp\left(\frac{\Delta E_v}{kT}\right) = \frac{25(2000)(2 \times 10^{17})}{2.5(800)(3 \times 10^{19})} \exp\left(\frac{0.193}{0.0258}\right) = 301$$

This example illustrates the advantage of a HBT over a BJT. Without the energy-gap difference, the ratio of the useful collector current to the back-injection current is merely 0.17 for the BJT. This renders the device useless. In contrast, both the graded and the abrupt HBTs remain functional, despite the higher base doping in comparison to the emitter.

With the replacement of the emitter material, the HBTs enjoy enhanced high-frequency performance compared to BJTs. HBTs based on the AlGaAs/GaAs and InP/InGaAs material systems both have reported operations up to the 300-GHz range. The idea of the HBT was actually conceived a long time ago. Shortly after the point-contact transistor was invented in 1948, Shockley stated in a U.S Patent #2569347:

> What is claimed is: 1 ... 2. A device as set forth in claim 1 in which one of the separated zones is of a semiconductive material having a wider energy gap than that of the material in the other zones.

The above brief description laid the foundation for the HBT. Even though the concept of the HBT existed for some time and its potential advantages were well recognized, the technologies to bring it to fruition have only recently emerged. At first, attempts were made to grow Ge–Si junctions by the alloying process and Ge–GaAs junctions by the vapor-deposition method. However, the junctions thus made had high dislocation densities, preventing the devices from being practical. The potentials of HBTs have been realized only since the recent advances in the new epitaxial technologies, such as molecular beam epitaxy (MBE) and the metal-organic chemical vapor deposition (MOCVD). These techniques were described in Chapter 1.

§ 3-2 BASE CURRENT COMPONENTS

The d.c. current gain (β) is one of the most important parameters in bipolar transistors. Certain types of circuits, particularly those used for digital or low-noise applications, do not function properly if the current gains are below certain values. Although the transistor current gain is less critical for power amplifier circuits, insight into the various factors affecting the current gain is nonetheless essential to sound device design. The d.c. current gain is easily measurable and sheds light on many properties of the transistor. For example, if it is suspected that somehow during growth the base doping has become low or that somehow during processing the extrinsic base surface passivation ledge has been removed, a measurement of the current gain can usually pinpoint the source of the problem.

The current gain is the ratio of the collector current (I_C) to the base current (I_B). A solid understanding of β rests on knowledge of the different components of the collector and the base currents. When the base-emitter junction is forward-biased during transistor operation, an offset in the Fermi levels of the two regions brings forth current flows from both sides. The holes that back-inject into the emitter give rise to I_{Bp}, the base current component mentioned in the previous section. In order for the transistor to function properly, the number of these holes should be small compared to the number of electrons that forward-inject into the base. In the absence of a base electric field, the electrons travel through the base layer by diffusion. Although some of these carriers are lost inside the base due to recombination, most of them reach the other end and are swept through the collector by the field in the reverse-biased base-collector junction. The carriers that are successfully collected at the collector constitute the collector current. For every electron that is not collected but is instead recombined in the base, a hole is supplied from the base contact to fulfill such a recombination event. These holes flow through the contact, constituting the remaining of the base current. Depending on the location of the recombination, the base current (excluding I_{Bp}) can be classified into four components. As shown in Fig. 3-5, they are (1) surface recombination current in the exposed extrinsic base region ($I_{B,\text{surf}}$), (2) interface recombination current at the base contact ($I_{B,\text{cont}}$), (3) bulk recombination current in the base region ($I_{B,\text{bulk}}$), and (4) space-charge recombination current in the base-emitter space-charge region ($I_{B,\text{scr}}$). The overall base current consists of these four recombination components and the aforementioned back-injection current (I_{Bp}).

The general characteristics of the five base current components are briefly described here. Some of the five base current components' effects on β will be examined further in the following sections. The origin of the first base current component relates to a GaAs property that the free GaAs surface has a high surface recombination velocity approaching 1×10^6 cm/s. Without special design, HBTs fabricated with typical processes have exposed extrinsic base regions, such as that shown in Fig. 3-6*a*. (Figure 3-6*b* is drawn for a device passivated by depleted AlGaAs ledge to contrast the difference between a passivated and an unpassivated device.) Consequently, some minority carriers injected from the emitter recombine with the base majority carriers at the surfaces, resulting in the extrinsic base surface recombination current, $I_{B,\text{surf}}$. This base current component is proportional to the emitter periphery rather

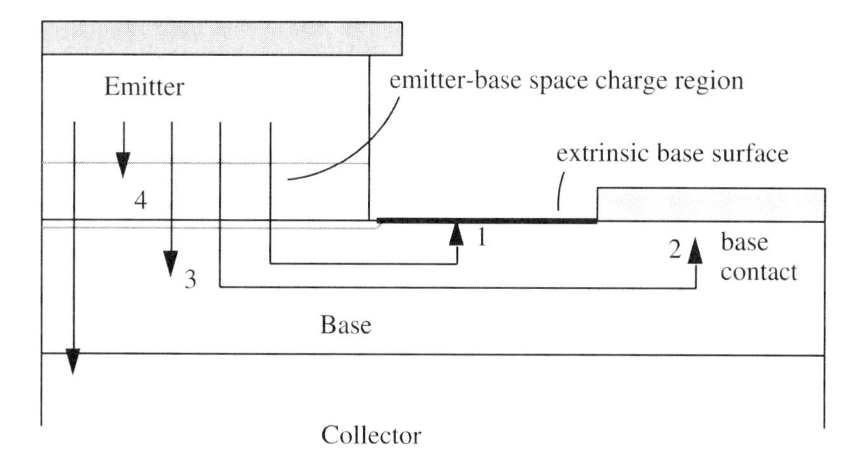

FIGURE 3-5. Schematic diagram showing the locations of the four main base recombination currents in HBTs. Only half of the transistor is shown, since the transistor is symmetrical across the center of the emitter mesa. (From W. Liu, Microwave and d.c. characterizations of *Npn* and *Pnp* HBTs. Ph.D. dissertation, Stanford University, Stanford, CA, 1991.)

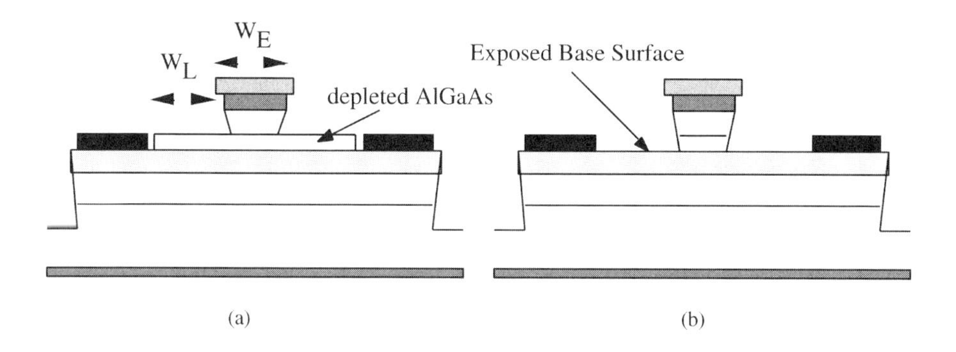

FIGURE 3-6. (*a*) A passivated AlGaAs/GaAs HBT. (*b*) An unpassivated HBT with exposed extrinsic base surfaces. (From W. Liu, Microwave and d.c. characterizations of *Npn* and *Pnp* HBTs. Ph.D. dissertation, Stanford University, Stanford, CA, 1991.)

than the emitter area, unlike the collector current. For small devices whose device perimeter-to-area ratio is large, $I_{B,\mathrm{surf}}$ is a major component of the overall base current, and the current gain is substantially reduced from that of a large device whose perimeter-to-area ratio is small. This is the *emitter size effect*, that the current gain decreases as the emitter area decreases, due to a larger portion of $I_{B,\mathrm{surf}}$ in the overall base current. It is somewhat unfortunate that the term "emitter size effect" is used because it is not very descriptive. Just about any phenomenon that causes a device property to vary as a function of the emitter area qualifies to be called an emitter size effect. In Si BJTs, the term "emitter size effect" bears no relation to $I_{B,\text{surf}}$. It refers to the fact that, as a result of the nonideal diffusion of the base dopant, the base doping concentration is higher near the periphery than in the intrinsic area. This nonuniformity causes a larger portion of the back-injection current to crowd into the perimeter compared to inside the emitter area, hence lowering the current gain.

From a numerical two-dimensional device equation solver, the ideality factor for $I_{B,\text{surf}}$ is found to decrease from about 2 to 1 as I_C increases. Measured ideality factors as small as 1.33 and 1.05 have been determined experimentally for this current in AlGaAs/GaAs and GaInP/GaAs HBTs, respectively. For simplicity, this current's ideality factors is said to be 1, just as it is customary to generalize the space-charge recombination current's ideality factor as 2 even though its actual value is not exactly 2 (it is actually slightly less than 2).

Unlike the GaAs surface, the free InGaAs surface exhibits more ideal characteristics, with a dramatically lower surface recombination velocity of about 1×10^3 cm/s. Therefore, in InP/InGaAs HBTs, $I_{B,\text{surf}}$ is inconsequential, even when the device size is small. Similarly, in silicon BJTs, $I_{B,\text{surf}}$ is hardly ever mentioned, because the extrinsic base surfaces are well passivated by SiO_2.

The base contact recombination current ($I_{B,\text{cont}}$) shares similar characteristics to $I_{B,\text{surf}}$ because both originate from surfaces with high surface recombination velocities. The surface recombination velocity of a metal–semiconductor interface is estimated to be 2×10^7 cm/s. This results from the assumption that there is no reflection of electrons striking the base contact interface. Although this value is higher than that of an unpassivated GaAs surface, $I_{B,\text{cont}}$ is lower than $I_{B,\text{surf}}$ because of the physical locations of the surfaces involved. The excess carrier concentration decreases rapidly from the intrinsic base region toward the extrinsic region. When the base-emitter contact separation is large, the excess carrier concentration at the base contact is nearly zero and $I_{B,\text{cont}}$ is nearly zero. Even in microwave transistors whose base-emitter contact spacing is small (due to base-emitter self-alignment), the number of excess carriers on the contact interface is substantially reduced from its value in the intrinsic region such that $I_{B,\text{cont}} < I_{B,\text{surf}}$. $I_{B,\text{cont}}$, however, can be important if the extrinsic base surface is passivated (so that $I_{B,\text{surf}}$ is made small) while the base-emitter spacing is made small. The magnitude of this current component is proportional to the periphery, just like $I_{B,\text{surf}}$. Therefore, passivated HBTs can still exhibit the emitter size effects if $I_{B,\text{cont}}$ is appreciable. Due to their similarity in their physical origins, we expect the ideality factor of $I_{B,\text{cont}}$ to be the same as that of $I_{B,\text{surf}}$, that is, 1.

Although $I_{B,\text{surf}}$ is relatively uncommon in Si BJTs (except for early devices), $I_{B,\text{cont}}$ can be important in BJTs because a base contact formed on any semiconductor material has a high surface recombination velocity. Despite that $I_{B,\text{cont}}$ causes the current gain to decrease as the perimeter-to-area ratio increases, it is not popular in the Si BJT literature to refer to such as an emitter size effect.

Unlike the previous two components, which appear on surfaces, the base bulk recombination current ($I_{B,\text{bulk}}$) originates in the base bulk layer. There are three major mechanisms through which the electrons and holes recombine to reestablish thermal equilibrium when the equilibrium is disturbed: (1) radiative recombination, (2) Shockley–Read–Hall (SRH) recombination, and (3) Auger recombination. The radiative recombination is a process by which an electron in the conduction band and a hole in the valence band recombine without the assistance of an intermediate state. A photon is released so that the overall energy of the process is conserved. Because momentum is also conserved during the transition, radiative recombination occurs mostly in materials with a direct energy gap. GaAs and III-V materials fall into this category. Si, and indirect-energy-gap material, has a negligible amount of radiative recombination. The SRH recombination describes an event by which an electron makes the jump from the conduction band to the valence band through an intermediate energy level. The energy level (or trap level), residing somewhere in the forbidden energy gap, is created by the impurities in the semiconductor. In this recombination process, an electron is first captured by the traps and is subsequently transferred to the valence band once a hole appears near the trap to recombine with the electron. The energy and momentum are conserved through the creation of phonons. Sometimes the Shockley–Read–Hall recombination is called the Shockley–Hall–Read recombination; we adopt the former terminology.

Another means of disposing of the energy in a single recombination event is to transfer it to another electron in the conduction band. This action, called an Auger process, generates a highly energetic charge carrier that soon shares its energy with other carriers in the band. Auger recombination is most acute in heavily doped regions such as the base layer.

All three recombination mechanisms contribute to the base bulk recombination current. The recombination rates (number of events per unit volume) for radiative recombination (U_{rad}), SRH recombination (U_{SRH}), and auger recombination ($U_{t\text{A}}$) are given by

$$U_{\text{rad}} = B(np - n_i^2) \tag{3-6}$$

$$U_{\text{SRH}} = \frac{pn - n_i^2}{\tau_{\text{SRH}}(n + p + 2n_i)} \tag{3-7}$$

$$U_{\text{A}} = C(n + p)(np - n_i^2) \tag{3-8}$$

where B, τ_{SRH}, and C are the radiative recombination coefficient, SRH recombination lifetime, and Auger recombination coefficient, respectively. These equations are treated as basic equations. We do not intend to derive them, which would require understanding the detailed physical processes associated with the recombination mechanisms. During device operation, the minority electron concentration in the heavily doped p-type base is significant larger than its thermal equilibrium value, which is n_i^2/N_B, where N_B is the base doping. Therefore, n is approximately equal to Δn, which denotes the excess electron concentration compared to its thermal equilibrium concentration. The majority carrier concentration, in contrast, is not significantly perturbed from its thermal equilibrium value. Therefore, $p \approx N_B$ and $n \approx \Delta n \ll N_B$. Substituting these two approximations into the three recombination rate expressions, we obtain

$$U_{\text{rad}} \approx \Delta n(B \cdot N_B) = \frac{\Delta n}{\tau_{\text{rad}}} \tag{3-9}$$

$$U_{\text{SRH}} \approx \frac{pn}{\tau_{\text{SRH}}(p+n)} = \frac{p\Delta n}{\tau_{\text{SRH}}(p+\Delta n)} = \frac{\Delta n}{\tau_{\text{SRH}}} \tag{3-10}$$

$$U_{\text{A}} \approx \Delta n(C \cdot N_B^2) = \frac{\Delta n}{\tau_A} \tag{3-11}$$

The overall recombination rate is $U_{\text{rad}} + U_{\text{SRH}} + U_{\text{A}}$. According to Eq. (1-42), it is convenient to define an effective minority electron recombination lifetime (τ_n) such that the recombination rate is expressed in the following manner:

$$U = U_{\text{rad}} + U_{\text{SRH}} + U_{\text{A}} = \frac{\Delta n}{\tau_n} \tag{3-12}$$

From Eqs. (3-9)–(3-12), τ_n is given by

$$\tau_n = \left(\frac{1}{\tau_{\text{rad}}} + \frac{1}{\tau_{\text{SRH}}} + \frac{1}{\tau_{\text{A}}}\right)^{-1} \tag{3-13}$$

When $I_{B,\text{bulk}}$ is the dominant base current component, the current gain can be expressed as

$$\beta = \frac{I_C}{I_{B,\text{bulk}}} = \frac{\tau_n}{\tau_b} \qquad (I_{B,\text{bulk}} \text{ dominates } I_B) \tag{3-14}$$

where τ_b is the minority carrier transit time across the base. For example, if it takes an electron an average of 10 ns to travel through the base whereas every 1 μs an electron is lost through recombination, then the transistor has a current gain of 1 μs/10 ns = 100. These numbers are typical of Si BJTs. For AlGaAs/GaAs HBTs, the recombination lifetime in the GaAs base material is ~1 ns, which is only 1/1000 that of the Si material. Although the lifetime is significantly shorter, the transit time through the GaAs is also shorter than in Si owing to the higher carrier mobility in GaAs. A well-designed HBT has a τ_b of 0.01 ns; therefore, a $\beta = 100$ is routinely obtainable in A1GaAs/GaAs HBTs as well. The derivation of Eq. (3-14) is presented shortly.

We examine τ_n and τ_b more closely. τ_n is basically a material parameter. Although its value decreases somewhat with increasing base doping, it is not a parameter that can be designed arbitrarily. τ_b, on the other hand, is a parameter that is easily increased or reduced by varying the base thickness (X_B). To determine an expression for τ_b, we first establish the minority carrier profile inside the base layer. The equation governing the electron concentration inside the base layer is the continuity equation given by Eq. (1-67). For now, we consider only the case where the base electric field is zero. The presence of a nonzero base field is examined in §3-5. The time derivative of the electron concentration is zero because we are concerning with steady-state operation. Given these considerations, we rewrite Eq. (1-67) as

$$\frac{d^2n}{dx^2} - \frac{n - n_0}{D_{nB}\tau_n} = 0 \tag{3.15}$$

Two boundary conditions are needed to solve these linear second-order differential equations uniquely. The laws of the junction [Eq. 2-113] state that the number of excess carriers per unit area at the emitter edge of the base layer relates to the junction voltage in an exponential fashion:

$$n(0) = \frac{n_i^2}{N_B} \exp\left(\frac{qV_{BE}}{kT}\right) \tag{3-16}$$

Note that Eq. (2-113) uses a slightly different coordinate system in which the junction edge is taken to be $x = -X_p$. Here, the junction edge is taken to be $x = 0$, as reflected in Eq. (3-16). V_{BE} in Eq. (3-16) is the external base-emitter voltage. N_B and n_i are the base doping and the intrinsic carrier concentration in the base, respectively. This equation assumes that the voltage drops in the bulk emitter and base regions as well as the contacts are negligible. If this were not the case, then V_{BE} should be replaced by the amount of voltage that drops across the junction only. Another boundary condition is established at the base-collector junction. It has a form similar to Eq. (2-113). However, since the base-collector bias is negative so that the base-collector junction is reversed-based, we obtain

$$n(X_B) = \frac{n_i^2}{N_B} \exp\left(\frac{qV_{BC}}{kT}\right) \approx 0 \tag{3-17}$$

For a HBT with $N_B = 3 \times 10^{19}\ \text{cm}^{-3}$, $n_i = 1.79 \times 10^6\ \text{cm}^{-3}$, and $V_{BE} = 1.3\ \text{V}$, $n(0) = 8.2 \times 10^{14}\ \text{cm}^{-3}$. n_0 appearing in Eq. (3-15), the thermal equilibrium concentration, is $n_i^2/N_B = 1 \times 10^{-7}\ \text{cm}^{-3}$. It is conceivable (and in fact true) that throughout most of the base layer, the electron concentration n anywhere is significantly larger than n_0. Therefore, Eq. (3-15) is simplified to

$$\frac{d^2n}{dx^2} - \frac{n}{L_n^2} = 0 \tag{3-18}$$

L_n, the diffusion length of the minority electron in the base, is defined as

$$L_n = \sqrt{D_{nB} \cdot \tau_n} \tag{3-19}$$

A general solution to Eq. (3-18) is $A \exp(x/L_n) + B \exp(-x/L_n)$. Matching the boundary conditions of Eqs. (3-16) and (3-17), we solve for the unknowns A and B. The resulting electron profile is given by

$$n(x) = \frac{n(0)}{\exp(X_B/L_n) - \exp(-X_B/L_n)} \left[\exp\left(\frac{X_B - x}{L_n}\right) - \exp\left(-\frac{X_B - x}{L_n}\right) \right] \tag{3-20}$$

Typically, τ_n in HBT is of the order of 1 ns, while D_{nB} is of the order of 25 cm^2/s. L_n according to Eq. (3-19) is about 1.58 μm. The typical base thickness, in contrast, is only 1000 Å or less. Since $L_n \gg X_B$, we use the fact that $\exp(x) \approx 1 + x$ when $x \ll 1$ to simplify Eq. (3-20) as

$$n(x) = n(0)\left(1 - \frac{x}{X_B}\right) \tag{3-21}$$

In the absence of a base electric field, the excess carrier concentration in the intrinsic base varies linearly with position, as shown in Fig. 3-1*b*.

The collector current is determined by taking the gradient of the minority carrier concentration at the collector end of the base layer. Because the minority carrier concentration varies linearly with distance in the base, the gradient is the same in the entire base and it does not really matter at exactly what location the slope is calculated:

$$I_C = A_E q D_{nB} \left.\frac{dn(x)}{dx}\right|_{x = X_B} = A_E q D_{nB} \frac{n(0)}{X_B} = \frac{A_E q D_{nB}}{X_B} \frac{n_i^2}{N_B} \exp\left(\frac{qV_{BE}}{kT}\right) \tag{3-22}$$

(*Note*: In a strict definition, the collector current should be the negative to that given here. As shown in Eq. (3-21), the electron concentration decreases with increasing x. This means that the electrons diffuse from the left to the right and the electron current points from the right to the left. However, it is customary to work with positive current. Therefore, the collector current is defined as positive if the electron carriers are flowing out of the collector terminal. In other words, the collector current is said to be positive if the current flows into the collector terminal. The definition given here is used consistently in this book unless otherwise noted.)

The base transit time is equal to the total excess charge in the base (Q_B) divided by the collector current that removes these charges:

$$\tau_b = \frac{Q_B}{I_C} = \frac{qA_E \int_0^{X_B} n(x)\, dx}{I_C} = \frac{X_B^2}{2D_{nB}} \tag{3-23}$$

In this equation, the collector current (as well as Q_B) is proportional to the emitter area (A_E). It is not related to the collector area because the excess carrier concentration outside the intrinsic emitter mesa decreases exponentially to nearly zero, as mentioned in the discussion of $I_{B,\text{cont}}$. The carrier gradient established by the carrier concentration $n(0)$ and 0 at the two sides of the base layer exists primarily within the intrinsic emitter mesa. For this same reason, the device area refers to the emitter area, not the base-collector junction area. Likewise, the collector current density is equal to I_C divided by the emitter area.

Example 3-2:

Assume that the base current in an *Npn* HBT is dominated by the base bulk recombination current. Find the current gain if the transistor's base doping level is 1×10^{19} cm^{-3} and the base thickness is 800 Å. Repeat the calculation for a base doping of 1×10^{20} cm^{-3}.

The base layer in an *Npn* HBT is *p*-type, in which the minority carrier is the electron. We need to find out two material parameters from § 1-7. According to Eq. (1-75), the minority electron mobility at a doping level of 1×10^{19} cm^{-3} is

$$\mu_n = 8300\left[1 + \frac{1 \times 10^{19}}{3.98 \times 10^{15} + (1 \times 10^{19})/641}\right]^{-1/3} = 1038 \text{ cm}^2/\text{V-s}$$

The diffusion coefficient, D_{nB}, according to the Einstein relationship, is $1038 \times 0.0258 = 27$ cm^2/s. Together with $X_B = 800$ Å, we find the base transit time from Eq. (3-23) to be

$$\tau_b = \frac{(800 \times 10^{-8})^2}{2 \times 27} = 1.2 \times 10^{-12} \text{ s}$$

The second material constant to be found from § 1-7 is the minority carrier lifetime. It is given by Eq. (1-71):

$$\tau_n = \left[\frac{1 \times 10^{19}}{1 \times 10^{10}} + \frac{(1 \times 10^{19})^2}{1.6 \times 10^{29}}\right]^{-1} = 6.2 \times 10^{-10} \text{ s}$$

The current gain for $N_B = 1 \times 10^{19}$ cm^{-3} is $\tau_n/\tau_b = 520$.

When $N_B = 1 \times 10^{20}$ cm^{-3}, we find that $\mu_n = 970$ cm^2/V-s. D_{nB} is $970 \times 0.0258 = 25$ cm^2/s. The base transit time τ_b is 1.3×10^{-12} s. Comparing this value to the previous one, we find that the transit time is not a sensitive function of the base doping. The recombination lifetime, on the other hand, is a material parameter that depends on the quality of the material. It varies drastically with N_B because the number of recombination traps generally increases with the base doping. From Eq. (1-71), the recombination lifetime is found to be $\tau_n = 1.4 \times 10^{-11}$ s, which is merely 2% of its value at the lower doping level. The current gain is $\tau_n/\tau_b = 10$. This is a small value compared to 520 obtained previously.

Example 3-2 demonstrates that the base doping level is an important parameter determining the current gain of the transistor. As τ_n decreases to low values in heavily doped HBTs, the base recombination current often dominates the overall base current. Growing wafers with extremely high base doping while maintaining a useful current gain presents challenges to material growth.

It was shown that in a quasi-neutral region (such as in the base layer), the recombination rate (number of events per cubic centimeter) is equal to the product of the excess carrier density and the inverse of the carrier lifetime. Hence, the base bulk recombination current is written as

$$I_{B,\text{bulk}} = \frac{A_E \displaystyle\int_0^{X_B} qn(x)\,dx}{\tau_n} = \frac{qn(0)\;X_B}{2\tau_n} A_E \tag{3-24}$$

Since the excess carrier concentration at $x = 0$ (the emitter edge) is proportional to $\exp(qV_{BE}/kT)$, the bulk recombination current has a unity ideality factor. Incidentally, using $I_{B,\text{bulk}}$ of Eq. (3-24) and I_C of Eq. (3-22), we can obtain the current gain given in Eq. (3-14).

Recombination traps are found not just in the base, but all over the semiconductor. The recombination current in the base layer is noteworthy because there are significant excess carriers. In other layers such as the bulk emitter layer, the bulk collector layer, and the reverse-biased base-collector junction, the excess carrier concentration and thus the recombination rate are negligible. However, in the forward-biased base-emitter junction, where both the electrons and holes diffuse from the adjacent emitter and the base layers, the space-charge recombination current occuring within can be important. This current is identical to the recombination current of a *p-n* junction discussed in § 2-6. We shall denote them differently: The space-charge current in the transistor is called $I_{B,\text{scr}}$, while the space-charge recombination current in a diode was called I_{rec}. It was remarked then that I_{rec} in the *p-n* junction is not too important since it is much smaller than the forward-biased diode current I_D. In the context of the transistor, the large diode current is essentially the collector current that flows from the forward-biased emitter-base junction to the collector. The recombination current occurring in the space-charge region is part of the base current, which is a distinct terminal current that is not added on top of the collector current. Therefore, in a transistor the amount of space-charge recombination ($I_{B,\text{scr}}$) is an important part of I_B, whereas in a diode the amount of space-charge recombination current (I_{rec}) is a small portion of I_D.

In § 2-6 we said that the maximum recombination rate occurs where the hole and electron concentrations are equal. We made the approximation that the maximum recombination rate takes place everywhere inside the depletion region. This picture was given for the forward-biased diode, and it should be equally applicable in the forward-biased *p-n* junction of a transistor. We therefore are led to the same conclusion obtained previously, that the recombination current has an ideality factor of ~ 2. Since $I_{B,\mathrm{scr}} \propto \exp(qV_{BE}/2kT)$ whereas other recombination currents are $\propto \exp(qV_{BE}/kT)$, $I_{B,\mathrm{scr}}$ is most significant when the current levels are low.

The carrier lifetime in the GaAs material system is generally shorter than in Si. It is expected that the lifetime in AlGaAs is even shorter because, as a result of the keen reactivity of aluminum, AlGaAs is particularly susceptible to impurity incorporation during material epitaxy. Nevertheless, it is not necessarily true that $I_{B,\mathrm{scr}}$ in AlGaAs/GaAs HBTs exceeds that in Si BJTs in a comparable transistor area. The space-charge recombination current, while being inversely proportional to the carrier lifetime, is also proportional to the intrinsic carrier concentration. AlGaAs, which has a higher energy gap than the other materials, has considerably less intrinsic carrier concentration. Overall, the magnitudes of the space-charge recombination currents in Si BJTs and in AlGaAs/GaAs HBTs are comparable.

The back-injection current is the most distinguishing component between a HBT and a BJT. As discussed in § 3-1, its presence forces homojunction transistors to be designed with a higher emitter doping than the base doping. The energy-gap difference between the base-emitter heterojunction in the HBT effectively eliminates this current component and removes the constraint of the base doping level in relation to the emitter doping. However, at ambient temperatures higher than room temperature, the effectiveness of the energy-gap difference to deter back-injection decreases. The increasing amount of carrier back-injection contributes to β's lowering in HBTs operating at high temperatures. There are other times when this component unexpectedly alters the device characteristics. For instance, if somehow the aluminum concentration is not sufficient during the growth of the emitter, or if somehow the base dopant diffuses into the emitter and thereby makes an AlGaAs homojunction, then the current gain can even be below unity.

The magnitude of I_{Bp} is determined entirely by the diffusion of the carriers in the bulk emitter region. Because the carrier concentration is proportional to $\exp(qV_{BE}/kT)$, I_{Bp} has an ideality factor of unity.

Example 3-3:

Two transistors are measured side by side on a wafer. The only difference is the size of their emitter areas, one large ($100 \times 100\ \mu\mathrm{m}^2$) and one small ($2 \times 30\ \mu\mathrm{m}^2$). The ideality factors of the base currents are measured to be 1.9 for the large device and 1.5 for the small device. What is likely cause for such a discrepancy in the ideality factors?

The ideality factor of 1.9 is close to 2. Hence, the space-charge recombination dominates the overall base current in the large device. We note that $I_{B,\text{scr}}$ is proportional to the emitter area, whereas $I_{B,\text{surf}}$ is proportional to the perimeter. Even if its extrinsic base surface were unpassivated, $I_{B,\text{surf}}$ would be small since a squarish large device has a small perimeter-to-area ratio. This is not true in a small device, whose perimeter-to-area ratio is comparatively larger. Both $I_{B,\text{scr}}$ and $I_{B,\text{surf}}$, which make up the overall base current, are appreciable. Because $I_{B,\text{surf}}$ has an ideality factor of unity, the overall base current ideality factor is lowered from 2, which is the case when $I_{B,\text{scr}}$ dominates. The exact value of the ideality factor depends on the exact contributions from $I_{B,\text{surf}}$ and $I_{B,\text{scr}}$. In this example, apparently $I_{B,\text{surf}}$ and $I_{B,\text{scr}}$ are roughly equal, so the overall ideality factor of 1.5 is an average of 1 and 2.

The following example describes a common mistake when the experiment is not well devised. It points out the importance of designing the correct epitaxial structure.

Example 3-4:

A graduate student invents a novel fabrication technique that he claims can passivate the extrinsic base of HBTs and thereby eliminate $I_{B,\text{surf}}$. To test out his process, he first epitaxially grows a wafer with a base doping of 1×10^{20} cm^{-3}. Then, he fabricates several HBTs on the same wafer. The transistor areas are designed so that the emitter perimeter-to-area ratio varies from one extreme to another. He gets lucky, not dropping his wafer during the entire processing, and is able to yield several devices. During characterization, he finds that all of the transistors have roughly the same value of current gain. He concludes that, since the current gains do not vary even though the perimeter-to-area ratio of the devices changes, his fabrication technique passivates the surface. What is wrong with his conclusion?

As shown in Example 3-2, even if we neglect all other base current components and consider only $I_{B,\text{bulk}}$, the current gain at a base doping of 1×10^{20} cm^{-3} is at most only 10. Suppose that the novel technique does not work and $I_{B,\text{surf}}$ varies from device to device due to the different perimeter-to-area ratios. Even though $I_{B,\text{surf}}$ may vary significantly, the magnitudes are small compared to $I_{B,\text{bulk}}$. The overall base current, which is a sum of $I_{B,\text{bulk}}$ and $I_{B,\text{surf}}$, will therefore not vary much. The measurement simply confirms that the base doping level is high, not that the technique works. In order to assess the effectiveness of the technique, the student should have used a low base doping level.

As a brief summary, we list in Table 3-1 some characteristics associated with the five main base current components in HBTs. These five components are the extrinsic base surface recombination current ($I_{B,\text{surf}}$), the base contact

TABLE 3-1. Characteristics of Base Current Components

Current Component	Ideality Factor	Proportional to	Most Important in Devices with
$I_{B,\text{surf}}$	1	Perimeter	Small geometry
$I_{B,\text{cont}}$	1	Perimeter	Self-aligned base-emitter contacts
$I_{B,\text{bulk}}$	1	Area	Heavy base doping
$I_{B,\text{scr}}$	2	Area	Large number of traps, graded base-emitter junction
I_{Bp}	1	Area	High junction temperature, abrupt base-emitter junction

recombination current ($I_{B,\text{cont}}$), the base bulk recombination current ($I_{B,\text{bulk}}$), the space-charge recombination current ($I_{B,\text{scr}}$), and the back-injection current (I_{Bp}).

§ 3-3 CURRENT GAIN FLATTENING

The HBTs examined in this section are "large" ($50 \times 50\ \mu\text{m}^2$) devices whose base-emitter contact spacing is "wide" ($> 2\ \mu\text{m}$). In a large device, both $I_{B,\text{surf}}$ and $I_{B,\text{cont}}$ are negligible compared to $I_{B,\text{bulk}}$ because the former two components are proportional to the emitter periphery whereas the latter is proportional to the device area. If we focus on room-temperature operation, then I_{Bp} is negligible altogether. Therefore, the two important base current components are the space-charge recombination current and the base bulk recombination current. This situation also occurs in well-designed silicon BJTs in which the back-injection current is made small by doping the emitter much more heavily than the base, and in which the surface recombination currents are made insignificant by having the proper device geometry. Generally, the current gain in silicon BJTs increases initially with I_C and then flattens out as I_C becomes large. For HBTs, however, such gain behavior is observed only when the base-emitter junction is abrupt. Figure 3-7 is a Gummel plot of an abrupt HBT. (A Gummel plot is a plot of the base and collector currents as a function of V_{BE} when $V_{CB} = 0$ V.) Figure 3-7 shows that the collector current increases with V_{BE} with an ideality factor of $\eta = 1.13$, a value which is close to unity, as expected from Eq. (3-2). The base current ideality is $\eta = 1.83(\sim 2)$ at low current levels ($I_C \leq 10^{-8}$ A), indicating that $I_{B,\text{scr}}$ is larger than $I_{B,\text{bulk}}$ at these current levels. However, as I_C increases past 10^{-8} A, the base current ideality factor changes to $\eta = 1.30(\sim 1)$, showing that $I_{B,\text{bulk}}$ dominates the base current within a wide range of operating current levels.

The base current eventually departs from the $\eta = 1.30$ extrapolation curve because a large proportion of V_{BE} drops across the parasitic base and emitter resistances, causing the available junction voltage, V_{BEj}, to decrease. However,

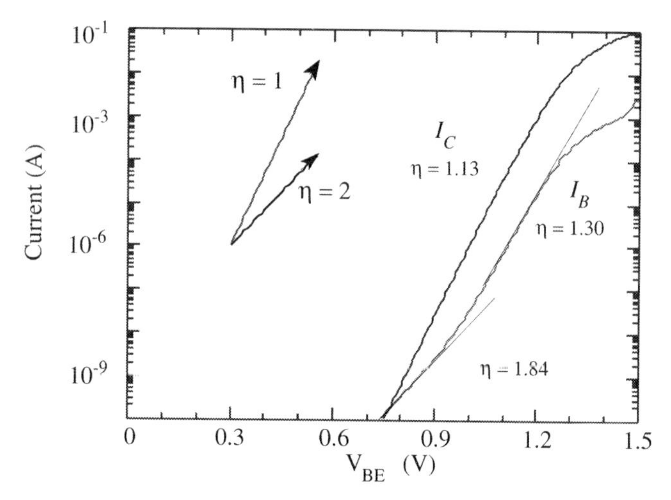

FIGURE 3-7. Measured Gummel plot of an abrupt AlGaAs/GaAs HBT. The Gummel plot is taken with a collector-base bias of 0 V. (From W. Liu, Experimental comparison of base recombination currents in abrupt and graded AlGaAs/GaAs heterojunction bipolar transistors, *Electr. Lett.* **27,** 2115–2116, 1991. © IEE, reprinted with permission)

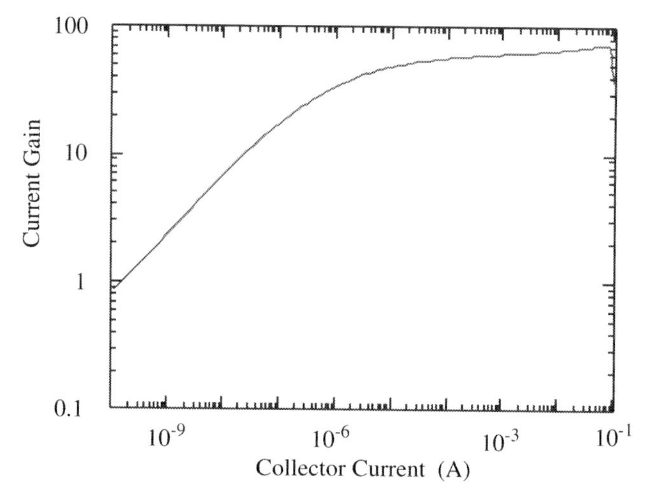

FIGURE 3-8. Measured current gain for the abrupt HBT of Fig. 3-7. $V_{CB} = 0$. (From W. Liu, Experimental comparison of base recombination currents in abrupt and graded AlGaAs/GaAs heterojunction bipolar transistors, *Electr. Lett.* **27,** 2115–2116, 1991. © IEE, reprinted with permission)

even at these high current levels, the I_B ideality factor can still be estimated from the β-versus-I_C plot of Fig. 3-8. The lack of the high-injection effects in HBTs ensures that I_C continues to increase with $\eta = 1.13$ at high current levels. As observed in Fig. 3-8, the current gain nearly saturates at $I_C \geq 10^{-5}$ A, indicating

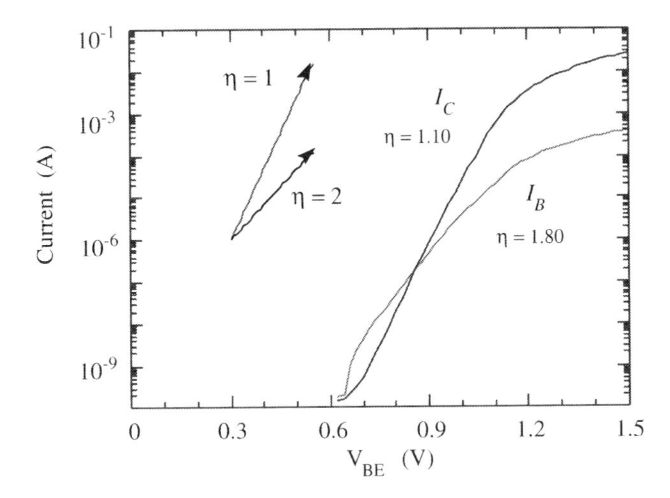

FIGURE 3-9. Measured Gummel plot of a graded AlGaAs/GaAs HBT. (From W. Liu, Experimental comparison of base recombination currents in abrupt and graded AlGaAs/GaAs heterojunction bipolar transistors, *Electr. Lett.* **27,** 2115–2116, 1991. © IEE, reprinted with permission)

that I_C and I_B have similar ideality factors. Since the ideality factors of 1.13 for I_C and 1.30 for I_B are not much different at low current levels, we estimate that the I_B ideality factor at high current levels assumes its low-current values. That is, η for I_B continues to be 1.30 at $I_C \geq 10^{-5}$ A.

Because $I_{B,\text{scr}}$ is proportional to $\exp(qV_{BE}/2kT)$ whereas $I_{B,\text{bulk}}$ is proportional to $\exp(qV_{BE}/kT)$, it makes sense for $I_{B,\text{bulk}}$ to increase at a faster rate than $I_{B,\text{scr}}$ as V_{BE} increases. Although $I_{B,\text{bulk}}$ eventually surpasses $I_{B,\text{scr}}$ in silicon BJTs and abrupt AlGaAs/GaAs HBTs at some large V_{BE} values, $I_{B,\text{bulk}}$ hardly ever exceeds $I_{B,\text{scr}}$ in graded HBTs. Compared to abrupt HBTs, graded HBTs have more significant $I_{B,\text{scr}}$ because at the location where most recombination events take place (the depleted emitter region next to the base), the aluminum concentration is small. The energy gap is smaller than that of the bulk emitter and the intrinsic carrier concentration is relatively higher. In abrupt HBTs, in contrast, the entire depletion region is at AlGaAs with the maximum aluminum mole fraction. The intrinsic carrier concentration is small, leading to a small $I_{B,\text{scr}}$.

Figure 3-9 is a Gummel plot of a HBT that has the same device geometry as that of Fig. 3-7. The epitaxial structures, however, are different. In the HBT shown in Fig. 3-9, the aluminum composition of the last 300 Å of the AlGaAs emitter adjacent to the base is graded parabolically from 30% to 0% at the base. (The significance of parabolic grading was discussed in § 2-4.) In contrast to that of abrupt HBTs, the I_B ideality factor in graded HBTs is nearly $2\,(\eta = 1.80)$ throughout the entire current range. This suggests that the space-charge recombination current in graded HBT dominates over the base bulk recombination current at all operating V_{BE}'s, even though the space-charge recombination

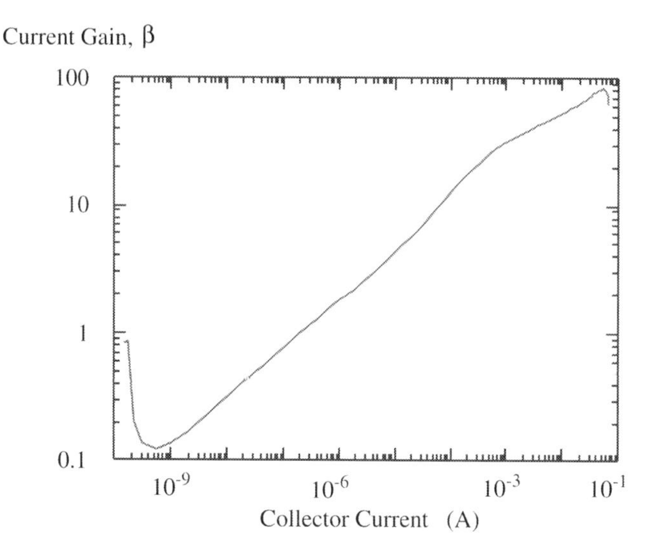

FIGURE 3-10. Measured current gain for the graded HBT of Fig. 3-9. (From W. Liu, Experimental comparison of base recombination currents in abrupt and graded AlGaAs/GaAs heterojunction bipolar transistors, *Electr. Lett.* **27,** 2115–2116, 1991. © IEE, reprinted with permission)

current has an ideality factor value twice as large as that of the base bulk recombination current.

The finding that $I_{B,\text{scr}}$ dominates the base current of graded HBTs is also observed in Fig. 3-10, which shows the measured β versus I_C. β of the graded HBTs continues to increase with I_C and never saturates at a constant value. From the slope of β versus I_C, the I_B ideality factor is estimated to be roughly twice the I_C ideality factor. This nonsaturation of current gain in graded HBTs marks the most striking difference from the current gain in abrupt HBTs, which levels off at moderate collector current levels.

In summary, the space-charge recombination current with $\eta = 2$ usually dominates at low current levels. It continues to dominate at high current levels for graded HBTs. Therefore, the current gain never flattens out as I_C increases. In abrupt HBTs, the base bulk recombination current dominates at high collector current levels. Current gain flattening is then observed, as in Si BJTs.

§ 3-4 SURFACE PASSIVATION

The natural oxide of silicon, the silicon dioxide, passivates the silicon surface effectively. On the other hand, the native oxide grown on GaAs is not stable and is not suitable as a dielectric material. Several methods have been proposed to passivate the GaAs surface. They include a spun-on layer of sulfur-based chemical such as NaS, a spun-on film of Langer film, and an epitaxially grown AlGaAs layer. The first two approaches, though demonstrating some success, require careful processing, such as nitride capping with limited thermal budget to prevent the layer from being dissociated over time. These techniques still require some research before becoming manufacturable. The depleted AlGaAs technique, in contrast, has proven to passivate the free GaAs surface effectively, at least in the extrinsic base area of the HBTs. The AlGaAs/GaAs interface is characterized by a surface recombination velocity of 10^3 cm/s. This value is small enough for the surface to be considered passivated and for $I_{B,\text{surf}}$ to be approximated by zero. The fabrication procedure resulting in the depleted AlGaAs structure shown in Fig. 3-6*b* is described in § 7-1. The passivation ledge needs to be thin enough so that it is fully depleted by a combination of the free-surface Fermi level pinning above and the base-emitter junction below. If the passivation ledge is not fully depleted, an appreciable emitter current flows through the ledge outside of the intrinsic emitter mesa before injecting into the base-emitter junction. Hence, the active device area would be much larger than the designed emitter area. The requirement for the AlGaAs layer to be fully depleted limits the AlGaAs thickness to be about 1000 Å when the emitter doping is 2×10^{17} cm^{-3}. If we assume that the Fermi level of the exposed $Al_\xi Ga_{1-\xi}As$ ($E_g = 1.86$ eV at $\xi = 0.35$) is pinned at midgap, then the surface depletion distance is 820 Å. The remaining is depleted by the *p-n* junction. The passivation ledge should not be too thin, either. It needs to be thick enough to passivate the surface effectively. In the extreme that the ledge has zero thickness, the extrinsic base surface is unpassivated and the surface recombination current surges. The same AlGaAs layer that is depleted in the extrinsic base region acts as the active emitter layer in the intrinsic emitter region. Therefore, when the depleted AlGaAs technique is used to passivate the free surface, the active emitter thickness is subjected to the same limitation described above. Fortunately, an emitter thickness of 500 Å is sufficient for the electrons to thermalize completely before injecting into the base. If the depletion requirement for the emitter limited the thickness to, for example, 20 Å, then significant tunneling of carriers would be possible at the emitter-base junction, rendering the device useless.

Figure 3-11 shows β versus I_C for both the passivated and unpassivated devices of $4 \times 10\,\mu\text{m}^2$. The emitter area is small enough to demonstrate the benefit of the surface passivation. A large device has negligible surface recombination current and the current gain does not depend on whether the surface is passivated or not. Because the two devices are fabricated simultaneously and physically adjacent to each other, the difference between the measured β's is accounted for by the additional $I_{B,\text{surf}}$ of the unpassivated device. Such β improvement in the passivated device is maintained over time, unlike in HBTs passivated by NaS, whose gain decreases gradually after the passivation coating dissociates over time. The current gain difference is most significant at medium to high current levels. At low current levels, both passivated and unpassivated devices exhibit similar current gains. This observation indicates that the current

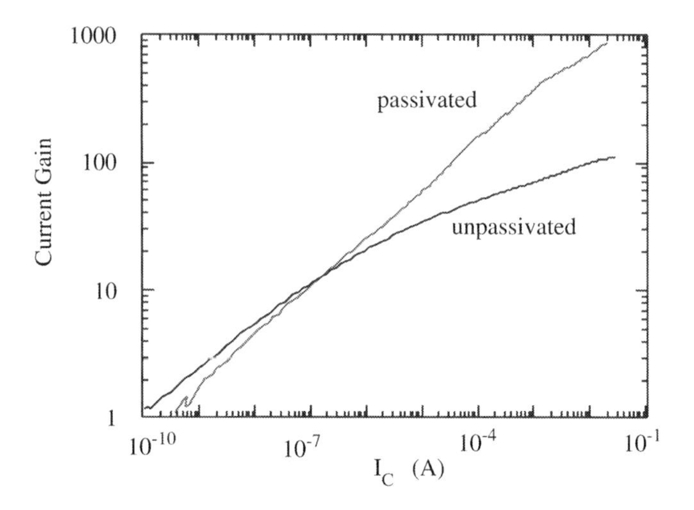

FIGURE 3-11. Measured current gain of passivated and unpassivated HBTs. The transistors have abrupt base-emitter junctions and the area is $4 \times 10\,\mu m^2$. $V_{CB} = 0$. (From W. Liu and J. Harris, Diode ideality factor for surface recombination current in AlGaAs/GaAs heterojunction bipolar transistors, *IEEE Trans. Electr. Dev.* **39,** 2726–2732, 1992. © IEEE, reprinted with permission)

gains at low current levels for both devices are limited by the base-emitter junction space-charge recombination. The unpassivated devices actually have slightly larger current gains than passivated devices. One plausible explanation is the following. In passivated devices, there exists a depleted passivation ledge in addition to the base-emitter space-charge region. It is possible that such an additional depletion region leads to larger space-charge recombination current, resulting in the smaller current gain in passivated devices at low current levels. Because we are interested mostly in how passivation affects the device current gain, in the following discussion we will concentrate on the device characteristics at medium to high current levels.

One method to ascertain the importance of $I_{B,\text{surf}}$ in a given technology is to measure the current gains of several HBTs with various emitter widths (W_E) and/or emitter lengths (L_E). In general,

$$I_B = W_E L_E (J_{B,\text{bulk}} + J_{B,\text{scr}} + J_{Bp}) + 2(W_E + L_E)\, K_{B,\text{surf}} \tag{3-25}$$

where $J_{B,\text{bulk}}$ (A/cm^2), $J_{B,\text{scr}}$(A/cm^2), and J_{Bp}(A/cm^2) are the base bulk recombination current density, the base-emitter space-charge recombination current density, and the base-to-emitter back-injected current density, respectively; $K_{B,\text{surf}}$ (A/cm) is the surface recombination current divided by the emitter periphery; and J_C (A/cm^2) is the collector current density. At room temperature, J_{Bp} is generally negligible due to the energy difference experienced by the carriers

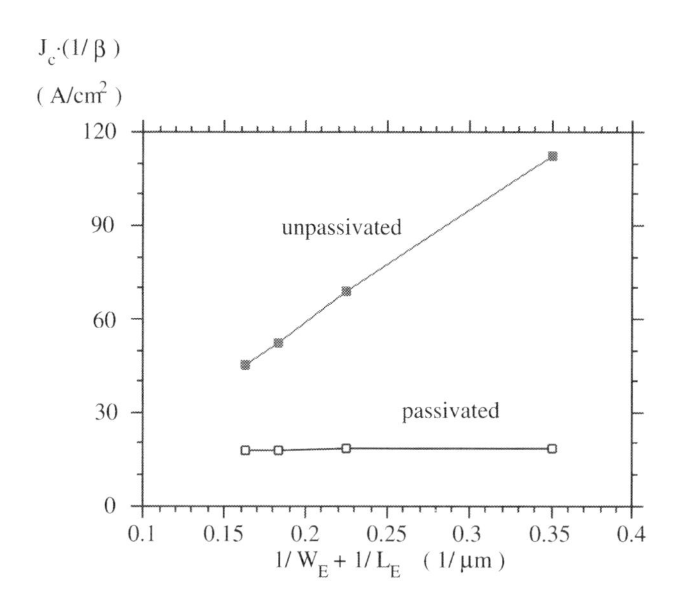

FIGURE 3-12. Measured $J_C(1/\beta)$ versus $(1/W_E + 1/L_E)$ for passivated and unpassivated HBTs. W_E varies from 4, 8, 12, to 16 μm and L_E remains at 10 μm. (From W. Liu and J. Harris, Diode ideality factor for surface recombination current in AlGaAs/GaAs heterojunction bipolar transistors, *IEEE Trans. Electr. Dev.* **39,** 2726–2732, 1992. © IEEE, reprinted with permission)

injected from the emitter and from the base as they traverse the base-emitter junction. Dividing both sides by I_C, we obtain

$$\frac{1}{\beta} = \frac{(J_{B,\text{bulk}} + J_{B,\text{scr}} + J_{Bp})}{J_C} + 2\frac{K_{B,\text{surf}}}{J_C}\left(\frac{1}{W_E} + \frac{1}{L_E}\right) \tag{3-26}$$

Figure 3-12 shows measured $J_C(1/\beta)$ versus $(1/W_E + 1/L_E)$ for both passivated and unpassivated HBTs, with W_E varying from 4, 8, 12, to 16 μm and L_E remaining at 10 μm. All data are taken at $J_C = 1 \times 10^4\ \text{A/cm}^2$. From Eq. (3-26), the intercept of the curves in Fig. 3-12 with the y axis, $J_C(1/\beta)$, represents the sum of $J_{B,\text{bulk}}$, $J_{B,\text{scr}}$ and J_{Bp}, and the slope represents $2K_{B,\text{surf}}$. This figure confirms that whereas $K_{B,\text{surf}}$ is significant for unpassivated devices, it is nearly zero for passivated devices. The intercept of the curve for unpassivated devices is extrapolated to be negative. This is because Eq. (3-26) is derived from an approximation that the emitter area is $W_E \times L_E$ and the emitter periphery is $2W_E + 2L_E$. It is assumed that $K_{B,\text{surf}}$ is uniform along both the emitter width and length, even though the base contacts in these devices do not surround the entire emitter area, but are located on only two sides of the emitter area.

Knowledge of the ideality factors of the base current components enables us to better understand and interpret experimental results. The ideality factors for base bulk recombination and the back-injection currents are ~ 1, since these base current components are directly proportional to the collector current. The ideality factor for the space-charge recombination current in the base-emitter junction is ~ 2. It is not obvious, however, what value the ideality factor should be for the extrinsic base surface recombination current. In a simplified picture it seems to be 2, since there is a depletion region where the surface recombination occurs. However, the dynamics of carrier recombination in the surface depletion region differ from those in the space-charge depletion region. Several experimental and theoretical studies indicate that the ideality factor for the extrinsic base surface recombination current is ~ 1.

It is important to distinguish the extrinsic base-surface recombination current from the junction-surface recombination current. The physical locations at which these two currents take place are illustrated in Fig. 3-13. The ideality factor of the junction-surface recombination current was shown to have a value of ~ 2, whereas we have discussed that the ideality factor is ~ 1 for the extrinsic base-surface recombination current. It should be emphasized that the conclusion that $\eta \approx 2$ for the junction-surface recombination current was obtained from heterojunction laser diodes. In these diodes, the central active layer had lower aluminum content than the barriers at both sides. This feature increased the likelihood of carriers recombining in the junction space-charge region, since the carriers were confined in the sandwiched active layer. An ideality factor of 2 was thus expected because the current conduction in a laser diode structure was not limited by the diffusion current in the quasi-neutral regions of the diode. Rather, it was limited by the current conduction in the sandwiched active layer, where both carriers were proportional to $\exp(qV_a/2kT)$. (V_a is the applied voltage.) This differs from the collector current in HBT structures, which is limited by the diffusion process in the base layer and has an ideality factor value close to unity.

Moreover, the types of devices being measured are different. The measurable current for the diodes was the total current flowing across the junction. Specific recombination current components, such as the extrinsic base-surface recombination current, could not be ascertained from the total diode current. This is because the extrinsic base-surface recombination current may potentially represent only a small fraction of the total diode current. In contrast, a transistor structure is utilized to analyze the extrinsic base-surface recombination current. As the base current consists only of the various recombination current components, the characteristics of the extrinsic base-surface recombination current component can be examined directly with proper device geometry and device processing to retain or remove the passivation ledge.

§ 3-5 BASE QUASI-ELECTRIC FIELD

In a bipolar transistor that has a uniformly doped base layer, the electrons injected from the emitter travel through the base layer by diffusion.

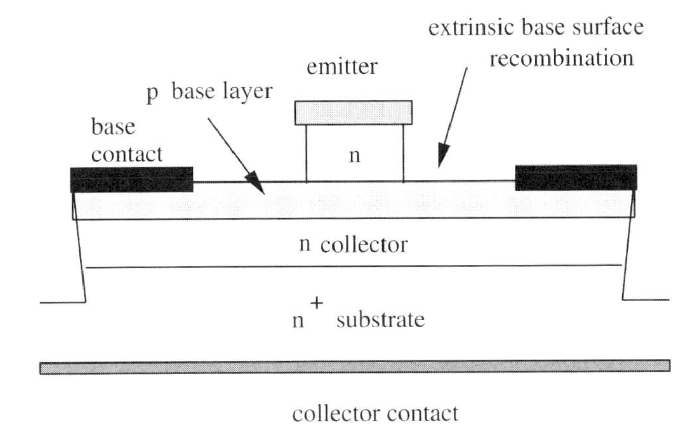

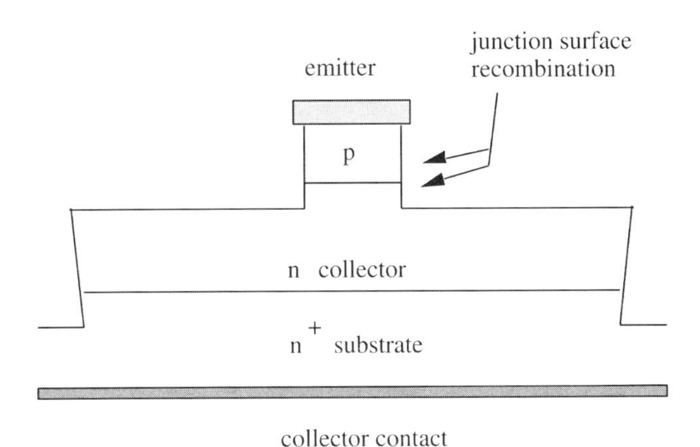

FIGURE 3-13. The physical locations at which the extrinsic surface recombinations occur in HBTs (top) and in diodes (below). (From W. Liu and J. Harris, Diode ideality factor for surface recombination current in AlGaAs/GaAs heterojunction bipolar transistors, *IEEE Trans. Electr. Dev.* **39,** 2726–2732, 1992. © IEEE, reprinted with permission)

When a base electric field is established, the current flow includes a drift component in addition to the diffusion component. Consequently, the base transit time decreases. The carrier lifetime, in contrast, is purely a material parameter in first-order analysis. It is not expected to be modified by the base electric field. With a reduction in τ_b and a relatively constant τ_n, Eq. (3-14) states that the current gain increases due to the base electric field.

In practice, the base field in AlGaAs/GaAs HBTs is established by the grading of aluminum concentration in the base layer. Because of the relative reactiveness of AlGaAs compared to GaAs, the AlGaAs material typically incorporates more impurity during epitaxy growth and has a shorter recombination lifetime. However, compared to the reduction in transit time, such slight modification in the carrier lifetime is a second-order effect. Therefore, as far as this analysis is concerned, the amount of current gain improvement is directly proportional to the transit time reduction.

The transit time in the presence of a constant base electric field is estimated as follows. Since the base electric field is intended to reduce the base transit time, the field direction is such that it aids electron flow from the emitter to the collector. This implies that the electric field points from the collector to the emitter. Adopting the coordinate system that $x = 0$ denotes the emitter side of the base layer and $x = X_B$, the collector side, the electric field is therefore negative. For simplicity, we assume that the electric field is constant. This is a good assumption for both HBTs and Si BJTs. In HBTs, the base electric field is usually established by linearly grading the aluminum (or indium) content in the AlGaAs (or InGaAs) base. In the case of an AlGaAs base, the higher aluminum content is placed near the emitter side to reduce the transit time. In the case of an InGaAs base, the higher indium content is placed near the collector side to reduce the transit time. Because the energy gap is a linear function of x (due to the linear grading), the field is a constant. We also note that since the field in HBTs is not established by conventional electrostatic charges, the field is referred to as *quasi-electric field*. In spite of its name, the quasi-electric field is as good as a conventional electric field in terms of providing drift action to carriers. In BJTs, the base electric field is established by electrostatic charge due to the grading of the base doping (see Example 1-6). The base field is thus a regular field rather than a quasi-electric field. It so happens that the base doping profile as a result of a diffusion process is an exponential function of x (actually, $-x$ to be exact). Therefore, the established field is also nearly constant. This is elaborated shortly.

Due to the electric field, the collector current consists of both the diffusion and the drift components:

$$I_C = -qA_E D_n \frac{dn(x)}{dx} + qA_E \mu_n n(x)\,\varepsilon_B = -qA_E D_n \left[\frac{dn(x)}{dx} - \frac{q\varepsilon_B}{kT} n(x)\right] \qquad (3\text{-}27)$$

I_C is negative because the electrons flow from the emitter to the collector. Einstein's relationship relating D_n and μ_n was applied in the above equation. If we neglect the recombination currents in the base, then I_C is a constant throughout $x = 0$ to X_B. Equation (3-27) represents a linear first-order differential equation, which can always be solved analytically. The approach is to rewrite the differential equation and multiply each term by an unknown function $R(x)$, which is to be determined shortly:

$$R(x)\frac{I_C}{(-qA_E D_n)}\,dx = R(x)\,dn(x) - R(x)\frac{q\varepsilon_B}{kT} n(x)\,dx \qquad (3\text{-}28)$$

$R(x)$ is an integration factor. If we take the derivative of $R(x)n(x)$, we find

$$d(Rn) = R\,dn + n\,dR \tag{3-29}$$

Comparing the above two equations, we find that the right-hand sides of the equations are equivalent if we choose

$$dR(x) = -\frac{q\varepsilon_B}{kT} R(x) \tag{3-30}$$

The solution to this differential equation is obtained by a simple integration:

$$R(x) = \exp\left(-\frac{q\varepsilon_B}{kT} x\right) \tag{3-31}$$

Plugging $R(x)$ back into Eq. (3-28), we obtain

$$d\left[\exp\left(-\frac{q\varepsilon_B}{kT}x\right)n(x)\right] = \exp\left(-\frac{q\varepsilon_B}{kT}x\right)\left(\frac{I_C}{-qA_E D_n}\right)dx \tag{3-32}$$

We note that $n(x = X_B) = 0$. Integrating both sides of the above equation and taking the limit from $x' = X_B$ to $x' = x$, we have

$$\exp\left(-\frac{q\varepsilon_B}{kT}x\right)n(x) = \frac{I_C}{-qA_E D_n}\int_{XB}^{x} \exp\left(-\frac{q\varepsilon_B}{kT}x'\right)dx' \tag{3-33}$$

Carrying out the integration, $n(x)$ is found to be

$$n(x) = \frac{I_C}{qA_E D_n}\frac{kT}{q\varepsilon_B}\left\{1 - \exp\left[-\frac{q\varepsilon_B}{kT}(X_B - x)\right]\right\} \tag{3-34}$$

The carrier profile is shown in Fig. 3-14, with the assumption of $J_C = I_C/A_E = 1 \times 10^4\ \mathrm{A/cm^2}$; $\mu_n = 1000\ \mathrm{V/cm}$; $X_B = 1000\ \text{Å}$. For convenience, the carrier concentration with identical parameters but without the base electric field is also shown for comparison. The charge gradient near the emitter junction is practically zero; hence, the diffusion component is nearly 0. The drift component, however, is large because the number of electrons there is appreciable. As the distance approaches the collector junction, the drift component of the current diminishes and in fact goes to zero right at the collector junction, where the excess carrier density is zero. In order for the collector current to remain constant throughout the base region, the diminished drift component is replenished by the diffusion component. Therefore, though the charge density decreases toward the collector, the gradient of the charge increases.

For convenience, we define a base electric field factor (unitless), κ, as

$$\kappa = \frac{q\varepsilon_B}{kT} X_B = \frac{\Delta E_{gB}}{kT} \tag{3-35}$$

where ΔE_{gB} denotes the bandgap energy difference at the emitter and the collector ends of the base layer whose aluminum (or indium) content is linearly

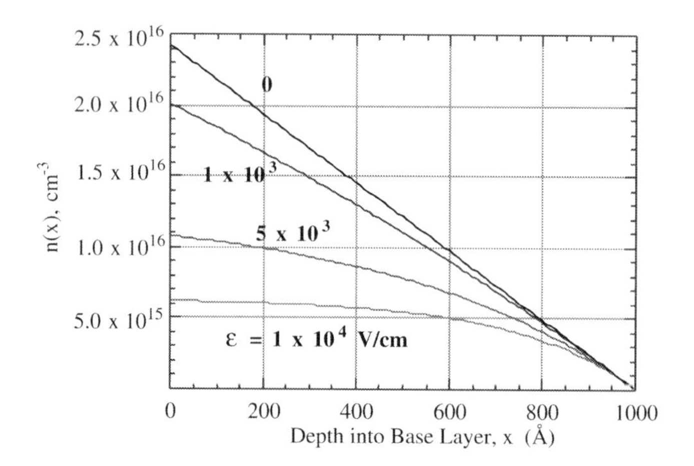

FIGURE 3-14. Minority carrier concentration profile in the base under various magnitudes of the base electric field. As the base electric field increases, the carrier concentration becomes progressively nonlinear.

graded. Therefore, $n(x)$ of Eq. (3-34) simplifies to

$$n(x) = \frac{I_C X_B}{q A_E D_n} \frac{1}{\kappa} \left\{ 1 - \exp\left[-\kappa \left(1 - \frac{x}{X_B} \right) \right] \right\} \tag{3-36}$$

In the absence of base electric field, $\kappa = 0$ and Eq. (3-36) reduces to

$$n(x) = \frac{I_C X_B}{q A_E D_n} \left(1 - \frac{x}{X_B} \right) = n(0) \left(1 - \frac{x}{X_B} \right)$$

which was derived previously for HBTs with a uniform base. It should be noted that $n(0)$ in the above equation is the concentration at the emitter edge when the base field is zero. In general, $n(0)$ at nonzero base electric field is smaller, as shown in Fig. 3-14, which plots $n(x)$ as a function of the field factor.

The carrier transit time is evaluated from the same definition of Eq. (3-23), except that Eq. (3-36) is used for $n(x)$:

$$\tau_b = \frac{Q_B}{I_C} = \frac{A_E \int_0^{X_B} n(x)\, dx}{I_C} = \frac{X_B^2}{2 D_n} \cdot f(\kappa) \tag{3-37a}$$

$$f(\kappa) = \frac{2}{\kappa} \left(1 - \frac{1}{\kappa} + \frac{1}{\kappa} e^{-\kappa} \right) \tag{3-37b}$$

When $\varepsilon_B = 0$, $\kappa = 0$. Expressing the exponential term in $f(\kappa)$ as an infinite summation and taking the limit as $\kappa \to 0$, we find that $f(\kappa) = 1$. Therefore, in the case of a uniform base without a base electric field, Eq. (3-37) simplifies to the familiar form, $\tau_b = X_B^2/2D_n$, agreeing with out analysis in § 3-2.

In BJTs, the base electric field is established by grading the base dopant concentration. Its magnitude is obtained by examining the base current (not the collector current). If again the base recombination currents are neglected, then the current continuity equation for the holes in the base is written as

$$I_B = qD_n \frac{dN_B(x)}{dx} + q\mu_n N_B \varepsilon_B \approx 0 \tag{3-38}$$

The built-in electric field is then

$$\varepsilon_B = -\frac{kT}{q} \frac{1}{N_B(x)} \frac{dN_B(x)}{dx} \tag{3-39}$$

The typical base doping profile in Si BJTs has an exponential dependence on the position, as a result of dopant diffusion. The profile is represented mathematically by

$$N_B(x) = N_B(0) \exp\left(-\frac{\alpha x}{X_B}\right) \qquad \text{where } \alpha = \ln\left[\frac{N_B(0)}{N_B(X_B)}\right] \tag{3-40}$$

Substituting this $N_B(x)$ into Eq. (3-39), we find that the built-in electric field is constant, equal to $kT/q \cdot \alpha/X_B$. In a transistor for which $\alpha = 5$ and $X_B = 1000$ Å, the electric field is 1.29×10^4 V/cm. For this transistor, the parameter κ from Eq. (3-37) is equal to 5, and the current gain is expected to decrease by three-fold compared to the case when the base doping is uniform.

One drawback of using the base doping gradient to set up the base electric field is that the base doping is lighter near the collector junction. Hence, the base sheet resistance is increased compared to that obtainable if the base layer is uniformly doped to the highest possible concentration. It is advantageous to establish the base electric field through energy-gap engineering, a possibility achievable only in heterostructures. In HBTs based on the GaAs material system, a common method to establish a constant base field is to linearly grade the aluminum concentration of the AlGaAs base layer from, for example, 10% at the emitter side to 0% at the collector side. This choice of grading results in a field that points from the collector to the emitter side, thus aiding the electron flow. Figure 3-15 illustrates the band diagram associated with this approach (for graded and abrupt HBTs). The base region, as a result of the quasi-neutrality approximation, does not have space charges. The electric field due to the space charge ($\varepsilon_{\text{Poisson}}$) is zero. Nonetheless, a quasi-electric field exists because of the spatial variation of the electron affinity in the graded AlGaAs base layer. The magnitude is, from Eq. (1-32), equal to the slope of the conduction band edge.

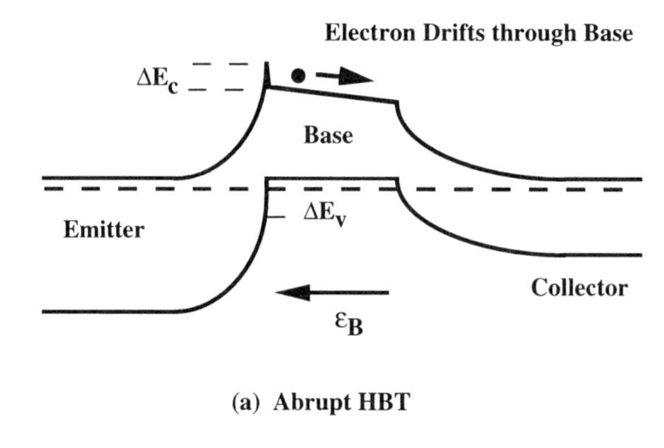

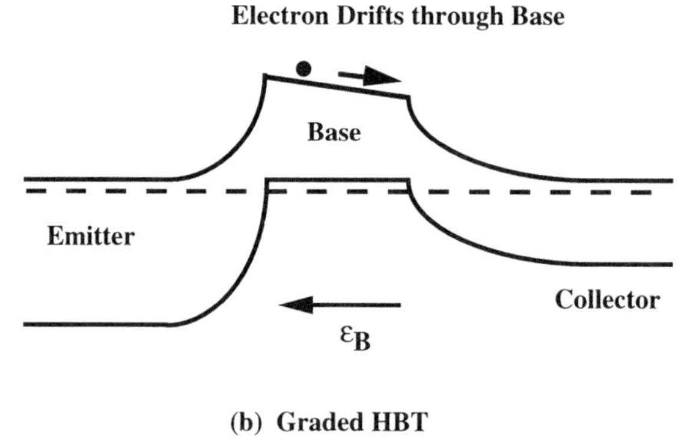

FIGURE 3-15. Band diagrams of (*a*) abrupt and (*b*) graded HBTs having a base quasi-electric field.

Because the grading is linear, ε_B is constant, given by the difference in the energy gaps at the two ends of the base material (ΔE_{gB}) divided by the base thickness:

$$\varepsilon_B = \frac{\Delta E_{gB}}{qX_B} \tag{3-41}$$

Example 3-5:

There are two HBTs, both having heavily doped base layers at the same doping level and the same base thickness of 1000 Å. One of the base layers is purely GaAs. Another is an $Al_\xi Ga_{1-\xi}As$ base layer, with aluminum linearly graded from $\xi = 0$ at the collector side to $\xi = 0.1$ at the emitter side. Assume that the base bulk recombination current is the dominant base current in both transistors. Find the ratio of the current gains of the two transistors.

For the graded AlGaAs base, $\Delta E_{gB} = 0.1 \times 1.247 = 0.125\,\text{eV}$. The base quasi-electric field is therefore $\varepsilon_B = 0.125/(1000 \times 10^{-8}) = 1.25 \times 10^4$ V/cm. The κ value, from Eq. (3-37), is

$$\kappa = \frac{1.25 \times 10^4}{0.0258} 1000 \times 10^{-8} = 4.84$$

The factor $f(\kappa)$ is found from Eq. (3-36) to be

$$f(\kappa) = \frac{2}{4.84}\left(1 - \frac{1}{4.84} + \frac{1}{4.84} e^{-4.84}\right) = 0.33$$

The HBT with a uniform base has zero base electric field and $\kappa = 0$. As mentioned, $f(\kappa)$ is unity for this case. Since the carrier lifetime and diffusion coefficient are the same for these two HBTs, the ratio of their current gains is equal to the inverse ratio of the $f(\kappa)$'s. The current gain of the AlGaAs base HBT is larger than that of uniform base HBT, by a factor of $1/0.33 \approx 3$.

As demonstrated in Example 3-5, a rather high base quasi-electric field of 1.25×10^4 V/cm can be established in practice. This reduces the base bulk recombination current by threefold; hence, it increases the current gain by threefold if $I_{B,\text{bulk}}$ is the dominant base current component. Besides reducing $I_{B,\text{bulk}}$, the base electric field also decreases $I_{B,\text{surf}}$. This is easy to comprehend, since the electric field tends to sweep the minority electrons away from the extrinsic base surfaces toward the collector. As fewer carriers are present on the surface, $I_{B,\text{surf}}$ decreases. As far as $I_{B,\text{scr}}$ is concerned, however, the base electric field is irrelevant, because it does not affect the carrier distribution inside the base-emitter junction. Therefore, designing a base quasi-electric field will result in the most significant gain improvement if the device's base current is dominated by either $I_{B,\text{bulk}}$ or $I_{B,\text{surf}}$. If the device is instead limited by the space-charge recombination current, then the improvement will be trivial.

§ 3-6 EMITTER CROWDING

Typical bipolar transistors are vertical structures with base current flowing horizontally from the base contacts to the center of the emitter region. The finite base resistance causes a finite voltage drop as the base current traverses through. Therefore, the effective base-emitter junction voltage decreases from the edge to the center. Because the amount of carrier injection depends exponentially on the junction voltage at a given position, most of the injection occurs near the edge of the emitter, where the reduction in the available junction voltage due to resistive

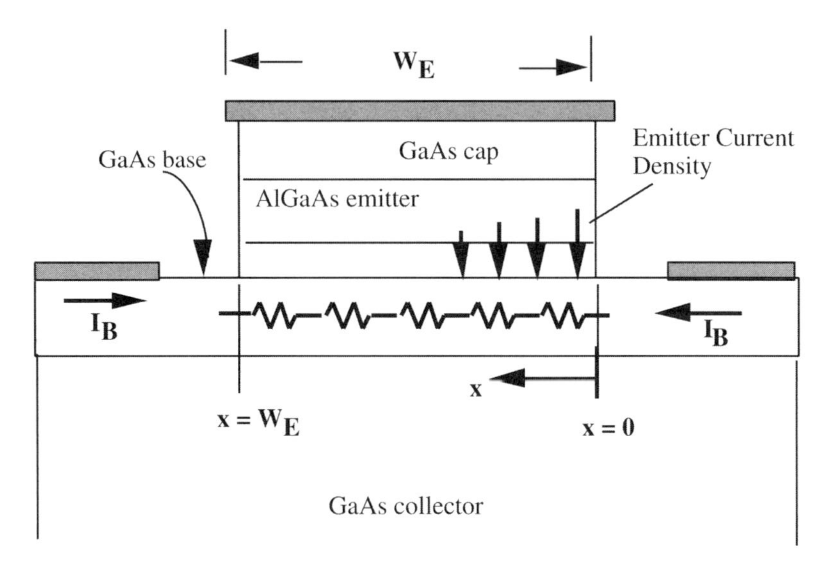

FIGURE 3-16. Essential elements in the analysis of emitter crowding. The region of focus is the intrinsic portion of the base layer.

voltage drop is a minimum. In this phenomenon, called *emitter crowding*, the emitter current flow is nonuniform across the emitter area. Sometimes, emitter crowding is referred to as *base crowding*. Both terms are acceptable, since the current constriction occurs in the base-emitter junction. We adopt the former term.

Figure 3-16 illustrates the essential elements involved in the analysis of the emitter crowding. We note that the boundary conditions are reflective along the middle of the emitter mesa. The following analysis applies to transistor structures in which two base contacts are placed symmetrically besides the emitter mesa.

When I_C increases, the base current supplying the recombined holes in the base also increases. As the base current flows through the base layer, the finite base resistance causes a finite amount of voltage drop. The potential difference between any point in the base layer with respect to the emitter is less than the externally applied base-emitter bias. Although the voltage drop also occurs in the extrinsic base region, we are particularly interested in the intrinsic base region, where the injection of the electrons from the emitter into the collector take place. Because of the exponential dependence of the emitter current density on the available junction voltage, the majority of the emitter current is injected at the periphery of the emitter stripe, where the voltage drop across the base resistance is the smallest. In the extreme case of high base resistance, it is possible that the central portion of the intrinsic emitter ceases contributing to the collector current conduction. The most notable effect is the reduction in the effective emitter width, causing undesirable complications in the analysis of experimental results since the device junction area is not determined by designed emitter width.

The available junction voltage as a function of x (Fig. 3-16) is equal to the externally applied bias minus the voltage drop across the base resistance:

$$V_{BEj}(x) = V_{BEj}(0) - \int_0^x J_{Bi}(x')\,\rho_B\,dx' \tag{3-42}$$

where ρ_B is the base sheet resistivity. The intrinsic base current density (J_{Bi}) is simply equal to the intrinsic base current (I_{Bi}) divided by the product of L_E and X_B. The intrinsic base current density at a given location is related to the emitter current density by

$$J_{Bi}(x) = \frac{1}{X_B}\int_x^{W_E/2} \frac{1}{(\beta+1)} J_E(x')\,dx' \tag{3-43}$$

$J_E(x)$, in turn, is related to the junction voltage by the exponential junction law of Eq. (3-22), which is restated here for convenience:

$$J_E(x) = J_0 \exp\left[\frac{qV_{BEj(x)}}{kT}\right] \tag{3-44}$$

where J_0 denotes the reverse saturation current density of the base-emitter junction. There are three unknown variables in the above three equations: $V_{BE_j}(x)$, $J_E(x)$, and $J_{Bi}(x)$. After substituting Eqs. (3-44) and (3-42) into Eq. (3-43) to eliminate $V_{BEj}(x)$ and $J_E(x)$ and then taking the derivative of the resulting expression twice, we obtain a differential equation:

$$\frac{d^2 J_{Bi}(x)}{dx^2} + \frac{q\rho_B}{kT}\frac{dJ_{Bi}(x)}{dx} J_{Bi}(x) = 0 \tag{3-45}$$

Two boundary conditions are needed to solve uniquely for the intrinsic base current density. At $x = W_E/2$, there is no more intrinsic base current flow due to the reflective boundary condition. Therefore, $J_{Bi}(W_E/2) = 0$. The second boundary condition is established at $x = 0$, for which Eq. (3-43) states that

$$J_{B_i}(0) = \frac{1}{X_B}\frac{1}{(\beta+1)}\int_0^{W_E/2} J_E(x)\,dx \tag{3-46}$$

This integral, when multiplied by L_E, is basically the terminal emitter current flowing through the half-mesa. If we denote the total emitter current as I_E, then

$$J_{Bi}(0) = \frac{1}{X_B L_E} \frac{1}{(\beta + 1)} \frac{I_E}{2} \tag{3-47}$$

The nonlinear second-order differential equation of Eq. (3-45) can be solved with a "p-technique," given in Ref. [1]. We find that

$$J_{Bi}(x) = \frac{kT}{q} \frac{2c}{\rho_B W_E} \tan\left[c\left(1 - \frac{2x}{W_E}\right)\right] \tag{3-48}$$

where the unknown constant c (unitless) is obtained from the remaining boundary condition at $x = 0$. Applying Eq. (3-47) to the above equation, we obtain the following transcendental equation from which c is determined:

$$c \tan c = \frac{q}{kT} \frac{I_E}{X_B} \frac{\rho_B}{4(\beta + 1)} \frac{W_E}{L_E} \tag{3-49}$$

Although in general many solutions of c satisfy Eq. (3-49), the root of interest lies between 0 and $\pi/2$. After c is established from Eq. (3-49). J_{Bi} as a junction of x is found from Eq. (3-48). The analysis is basically complete. However, often we are interested in the emitter current density as a function of distance. We derive $J_E(x)$ by taking derivative of both sides of Eq. (3-43) and then substituting Eq. (3-48):

$$J_E(x) = -X_B(\beta + 1)\frac{dJ_{Bi}(x)}{dx} = \frac{kT}{q} \frac{4X_B(\beta + 1)c^2}{\rho_B W_E \cos^2[c(1 - 2x/W_E)]} \tag{3-50}$$

Perhaps the best figure of merit quantifying the amount of emitter crowding is the effective emitter width (W_{eff}). This is defined as the emitter width that would result in the same current level if current crowding were neglected and the emitter current density were uniform at its edge value. The ratio, W_{eff}/W_E, can be expressed as

$$\frac{W_{\text{eff}}}{W_E} = \frac{2\int_0^{W_E/2} J_E(x)\,dx}{W_E J_E(0)} = \frac{\sin c \cos c}{c} \tag{3-51}$$

where c is determined from Eq. (3-49).

Example 3-6:

A $W_E \times L_E = 4 \times 20\ \mu\text{m}^2$ HBT has a base resistivity of $3 \times 10^{-3}\ \Omega$-cm. The base thickness is 1000 Å. Its current gain is 17.5, find the effective emitter width when the emitter current is 1 mA.

The first step is to evaluate the unknown constant c, which is governed by the transcendental equation of Eq. (3-49):

$$c \tan c = \frac{1 \times 10^{-3}}{10^{-5}} \frac{3 \times 10^{-3}}{4(17.5 + 1)(0.0258)} \frac{20 \times 10^{-4}}{4 \times 10^{-4}} = 0.785$$

The value 0.785 is simply $\pi/4$. It is clear from inspection that $c = \pi/4$ satisfies the above relationship. Substituting $c = \pi/4$ into Eq. (3-51), we find:

$$\frac{W_{\text{eff}}}{W_E} = \frac{\sin(\pi/4)\cos(\pi/4)}{(\pi/4)} = \frac{2}{\pi} = 0.64$$

The effective emitter width is hence $0.64 \times 4\,\mu\text{m} = 2.55\,\mu\text{m}$.

§ 3-7 INTRINSIC BASE RESISTANCE

Besides the amount of emitter utilization, the intrinsic base resistance (R_{Bi}) is also affected by the emitter current crowding. We caution that R_{Bi} is not a simple resistance that obeys Ohm's law in its usual sense, but is instead some sort of an averaged resistance defined from the equivalent power dissipation. Sometimes, the intrinsic base resistance is referred to as the *base spreading resistance*, to distinguish it form the conventional resistance. We will not use the latter term because the term *spreading resistance* is used in this book to describe the resistance resulting from nonuniform conduction from one terminal to the other, as in § 4-4.

Although R_{Bi} is not an ohmic resistance, a closely related resistance is. This resistance arises from a hypothetical situation, in which all carriers constituting the base current are assumed to flow into the base through the cross section at $x = 0$ and exit through the cross section at $x = W_E$ (Fig. 3-16). Because of the transverse flow direction, we call such a hypothetical resistance the *transverse intrinsic base resistance*, $R_{Bi(\text{tvs})}$. It is given by

$$R_{Bi(\text{tvs})} = \rho_B \frac{W_E}{L_E X_B} \tag{3-52}$$

where ρ_B is the resistivity of the base material:

$$\rho_B = \frac{1}{q\mu_p N_B} \tag{3-53}$$

In real device operation, none of the base current carriers exit through the cross section at $x = W_E$ once they enter the cross section at $x = 0$. In fact, if we neglect the space-charge recombination, all of the base current that enters the intrinsic base region eventually disappears in the intrinsic base region through recombination. The carriers gradually decrease in number from the maximum at the edge of the intrinsic region to zero at the center. If we further assume that there is no recombination in the extrinsic base region, then the amount of hole carriers flowing through the cross section at $x = 0$ is the same as that flowing out the base terminal. The current crossing the $x = 0$ cross section is simply the terminal current I_B. The current flowing inside the intrinsic base region, denoted the intrinsic base current (I_{Bi}), is a function of x rather than a constant because of the aforementioned carrier recombination in the intrinsic region. In the absence of emitter crowding, the emitter current density is uniform at any point along the x direction. The amount of carriers undergoing recombination is therefore the same at any position along x. Since these carriers to be recombined are all supplied from the cross section at $x = 0$, I_{Bi} is a "quasi-linear" function of x, decreasing linearly from its maximum value $I_B/2$ at $x = 0$ to zero as $x = W_E/2$, and then increasing linearly to $I_B/2$ at $x = W_E$. (Note, if there were only one base contact, then $I_{Bi}(x)$ would be truly a linear function x that decreases from a value of I_B at $x = 0$ to 0 at $x = W_E$. The following results, which are obtained for the conventional bipolar structure with two base contacts placed symmetrically at the two sides of the emitter mesa, would then not apply.) Since the problem is symmetrical along the $x = W_E/2$ plane, we focus only on the region at $0 \leq x \leq W_E/2$. $I_{Bi}(x)$ is expressed as

$$I_{Bi}(x) = \frac{I_B}{2}\left(1 - \frac{2x}{W_E}\right) \qquad \left(0 \leq x \leq \frac{W_E}{2}\right) \tag{3-54}$$

The intrinsic base current as a function of x is shown schematically in Fig. 3-17 (for all x between 0 and W_E). At any position x, the incremental resistance associated with an incremental distance Δx with a cross-section area of $L_E X_B$ is

$$\Delta R = \rho_B \frac{\Delta x}{L_E X_B} \tag{3-55}$$

The incremental voltage drop across dx is therefore

$$dV_{Bi}(x) = -\frac{I_B}{2}\left(1 - \frac{2x}{W_E}\right)\frac{\rho_B}{L_E X_B}dx \tag{3-56}$$

The minus sign reflects the fact that the voltage decreases with increasing x. Similar to the definition of $I_{Bi}(x)$, $V_{Bi}(x)$ denotes the base voltage with respect to the emitter inside the intrinsic region. At $x = 0$, it is equal to V_B, the base terminal voltage. $V_{Bi}(x)$ is obtained by integrating Eq. (3-56) with the boundary condition at $x = 0$:

$$V_{Bi}(x) = -\frac{I_B}{2}\frac{\rho_B}{L_E X_B}\left(x - \frac{x^2}{W_E}\right) + V_B \tag{3-57}$$

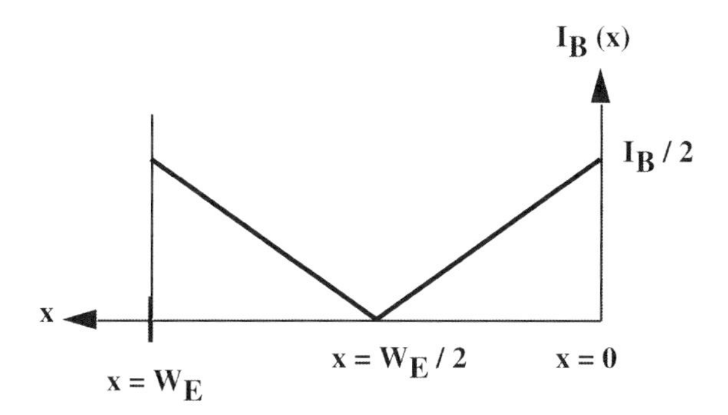

FIGURE 3-17. Intrinsic base current as a function of position away from one side of the intrinsic emitter mesa. The current distribution is symmetrical across the center of the mesa.

The average intrinsic base voltage is equal to the integral of $V_{Bi}(x)$ from $x = 0$ to $W_E/2$, and then divided by $W_E/2$. After carrying out the definition, we find the average value to be

$$\overline{V_{Bi}} = -I_B \frac{\rho_B}{L_E X_B} \frac{W_E}{12} + V_B \tag{3-58}$$

The average intrinsic base resistance (or simply the intrinsic base resistance) inside the half portion of the intrinsic base region from $x = 0$ to $W_E/2$ is

$$R_{Bi(\mathrm{half})} = \frac{V_B - \overline{V_{Bi}}}{I_B/2} = \frac{W_E}{6} \frac{\rho_B}{L_E X_B} = \frac{1}{6} R_{Bi(\mathrm{tvs})} \tag{3-59}$$

The $R_{Bi(\mathrm{half})}$ from the emitter to one base contact is equivalent to one-sixth of $R_{Bi(\mathrm{tvs})}$. Because of symmetry, the other half also has the same average resistance. The situation is illustrated in Fig. 3-18. Because the two averaged resistances are in parallel, the overall intrinsic base resistance defined between the emitter and the base terminals is thus one-half of $R_{Bi(\mathrm{half})}$. That is,

$$R_{Bi} = \frac{R_{Bi(\mathrm{half})}}{2} = \frac{1}{12} R_{Bi(\mathrm{tvs})} \tag{3-60}$$

If the transistor has only one base contact rather than two, the intrinsic base resistance is one-third instead of one-twelfth of the transverse intrinsic base resistance.

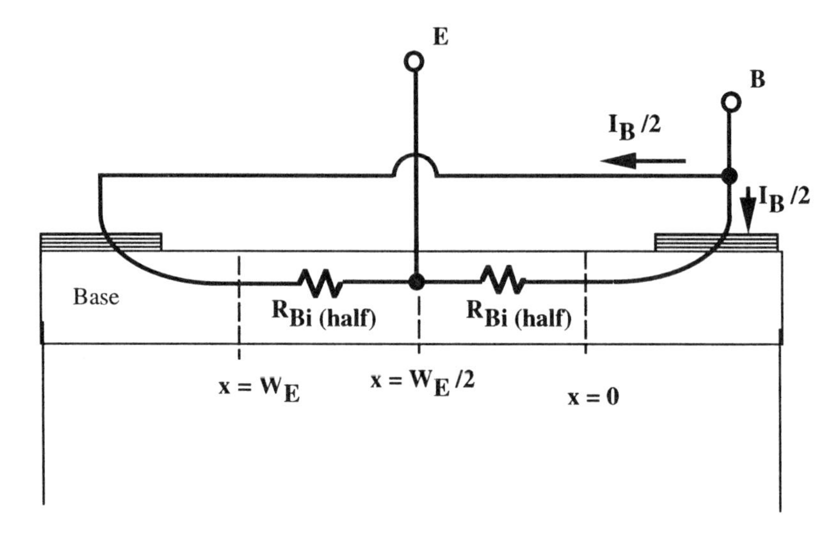

FIGURE 3-18. Schematic circuit diagram showing the parallel paths in the calculation of the intrinsic base resistance.

An alternative but equivalent method of determining the intrinsic base resistance is through consideration of the power. The power dissipation in the intrinsic base region is given by

$$P_{Bi} = \int I_{Bi}^2(x)\,dR = \frac{\rho_B}{L_E X_B}\int_0^{W_E/2} I_{Bi}^2(x)\,dx = \frac{1}{24} I_B^2 R_{Bi(\text{tvs})} \tag{3-61}$$

The intrinsic base resistance (more accurately, the average intrinsic base resistance) in this half of the intrinsic base region is defined by

$$P_{Bi} = I_{Bi}^2(0)\,R_{Bi(\text{half})} = \frac{I_B^2}{4} R_{Bi\,(\text{half})} \tag{3-62}$$

From a comparison of Eqs. (3-61) and (3-62), it is clear that $R_{Bi(\text{half})}$ is equal to one-sixth of $R_{Bi(\text{tvs})}$, just as obtained Eq. (3-58). The intrinsic base resistance in the entire intrinsic base region is expressed in Eq. (3-60).

The analysis has neglected the effects of emitter crowding. Neglecting emitter crowding enables us to assume a predetermined function for $I_B(x)$, as given in Eq. (3-54). When the crowding is significant, the amount of emitter current injection is not constant as assumed previously, but decreases as the distance away from the edges of the emitter mesa increases. The intrinsic base current $I_{Bi}(x)$ is hence not a linear function of x, but decreases in some complex tangential function that was established for the intrinsic base current density.

We again employ power considerations to find R_{Bi} under emitter crowding. Equating the power dissipation definitions given in Eqs. (3-61) and (3-62), we find:

$$P_{Bi} = \frac{\rho_B}{L_E X_B} \int_0^{W_E/2} I_{Bi}^2(x)\, dx = I_{Bi}^2(0)\, R_{Bi(\text{half})} \tag{3-63}$$

Because $I_{Bi}(x) = L_E X_B J_{Bi}(x)$ and $R_{Bi(\text{half})} = R_{Bi}/2$, we find:

$$R_{Bi} = \frac{1}{2}\left[\frac{1}{J_{Bi}(0)}\right]^2 \frac{R_{Bi(\text{tvs})}}{W_E} \int_0^{W_E/2} J_{Bi}^2(x)\, dx \tag{3-64}$$

Applying the $J_{Bi}(x)$ expression from Eq. (3-48), we find the intrinsic base resistance to be

$$R_{Bi} = \frac{R_{Bi(\text{tvs})}}{4}\left[\frac{\tan(c) - c}{c \tan^2(c)}\right] \tag{3-65}$$

When there is no emitter crowding, either because $I_E \approx 0$ or $\rho_B \approx 0$, then $c \approx 0$. From L'Hôpital's rule, we find

$$\lim_{c \to 0} \frac{\tan(c) - c}{c \tan^2(c)} = \frac{1}{3} \tag{3-66}$$

In the absence of emitter crowding, the intrinsic base resistance is thus one-twelfth of transverse intrinsic base resistance, agreeing with Eq. (3-60) obtained under this assumption. Further, the function inside the brackets in Eq. (3-65) decreases below its maximum value of 1/3 as c increases. In other words, the intrinsic base resistance decreases from its maximum value of $R_{Bi(\text{tvs})}/12$ in the presence of emitter current crowding. Usually we treat the decrease in resistance as a good event because resistance implies power dissipation and a loss of gain. It is paradoxical that a reduction in R_{Bi} due to emitter crowding actually degrades the device's power performance. Basically, a reduction of R_{Bi} comes from the fact that a portion of the intrinsic region ceases to function as a transistor. As part of the intrinsic region becomes inactive, the base current flows only in a smaller portion of the intrinsic base, thus lowering the intrinsic base resistance. The inactive portion of the base contributes parasitic capacitances in both the base-emitter and the base-collector junctions. Therefore, the overall device performance is degraded as R_{Bi} decreases.

Example 3-7:

For the HBT of Example 3-6, find the intrinsic base resistances, both with and without consideration of emitter crowding.

From the information of Example 3-6, we know that the $4 \times 20\ \mu m^2$ HBT has a base resistivity of $3 \times 10^{-3}\ \Omega$-cm and a base thickness of 1000 Å. $R_{Bi(tvs)}$, given by Eq. (3-52), is

$$R_{Bi(tvs)} = 3 \times 10^{-3} \frac{4 \times 10^{-4}}{20 \times 10^{-4} \times 10^{-5}} = 60\ \Omega$$

Without considering emitter crowding, the intrinsic base resistance is calculated from Eq. (3-60):

$$R_{Bi} = \frac{1}{12} \times 60\ \Omega = 5\ \Omega$$

This value is calculated with the assumption that two base contacts are placed symmetrically at each side of the emitter mesa. If there is only one base contact, the intrinisic base resistance is four times as large, or 20 Ω.

If emitter crowding is considered, then we take the c value obtained from Example 3-6. Substituting $c = \pi/4$ into Eq. (3-65), we find:

$$R_{Bi} = \frac{60\ \Omega}{4} \left[\frac{\tan(\pi/4) - \pi/4}{(\pi/4)\tan^2(\pi/4)} \right] = 4.1\ \Omega$$

As discussed, the intrinsic base resistance is smaller when emitter crowding is taken into consideration.

§ 3-8 KIRK EFFECT

The threshold current level at which the current gain falls off sets the upper bound to the operating current. For microwave power applications in which high current levels are desired, it is essential to understand the physical mechanisms that set the limit. From the outset, we state that often several mechanisms interact and can simultaneously cause the current gain to fall off. There is not a single physical process that dictates the current gain fall-off in HBTs, a fact that complicates the analysis. However, Kirk effect is generally considered to be the most important factor.

The base-collector junction in a bipolar transistor is reverse-biased to ensure proper transistor operation. Figure 3-19 shows the charge distribution and the field profile inside the junction when the current level is low. The Poisson equation states that

$$\frac{d\varepsilon}{dx} = \frac{q}{\epsilon_s}(p - n + N_d - N_a) \tag{3-67}$$

where N_d and N_a are the donor and acceptor doping levels, respectively. This equation is correct in general, independent of whether the current level is high or low. We first examine the situation when the current is low, such that the mobile electrons constituting the collector current is about 0 (i.e., $n \approx 0$). Since the collector is n-type, the mobile hole concentration is also nearly zero and there is no acceptor impurity ($N_a = p = 0$). The field is established from the ionized donor charges alone. With $N_d = N_C$ (the collector doping), Eq. (3.67) leads to

$$\frac{d\varepsilon}{dx} = \frac{q}{\epsilon_s} N_C \qquad \text{(valid at low current levels)} \tag{3-68}$$

If we denote depletion thickness in the collector as X_{dep}, then the maximum electric field that occurs at $x = 0$ is

$$\varepsilon(0) = \frac{q}{\epsilon_s} N_C X_{\text{dep}} \tag{3-69}$$

We take advantage of the fact that the base doping in HBTs is approximately two to four orders magnitude higher than the collector doping. Because the depletion region resides mostly in the collector side, the area enclosed by the ε-field profile is the sum of the applied reverse base-collector bias (V_{CB}) plus the base-collector junction potential (ϕ_{CB}):

$$V_{CB} + \phi_{CB} = \frac{1}{2}\,\varepsilon(0)\,X_{\text{dep}} \tag{3-70}$$

Because the base in HBTs is heavily doped, we assume that the Fermi level in the base region lies on top of the valence band edge. The base-collector junction potential therefore depends solely on the collector doping:

$$\phi_{CB} = \frac{E_g}{2} + \frac{kT}{q} \ln\left(\frac{N_C}{n_i}\right) \tag{3-71}$$

Combining Eqs. (3-69) and (3-70), we deduce the depletion thickness as

$$X_{\text{dep}} = \sqrt{\frac{2\epsilon_s}{qN_C}(V_{CB} + \phi_{CB})} \tag{3-72}$$

Throughout this book, V_{CB} means the collector voltage with reference to the base voltage. It has a negative value when the base-collector junction of an *Npn* HBT is forward-biased and a positive value when it is reversed-biased.

Example 3-8:

Determine the electric profile for a power HBT, whose collector thickness and collector doping are 1 μm and 1×10^{16} cm^{-3}, respectively. The collector-base reverse bias is $V_{CB} = 4$ V.

According to Eqs. (3-71) and (3-72),

$$\phi_{CB} = \frac{1.424}{2} + 0.0258 \ln\left(\frac{1 \times 10^{16}}{1.79 \times 10^{6}}\right) = 1.29 \text{ V}$$

$$X_{\text{dep}} = \sqrt{\frac{2 \times 13.18 \times 8.85 \times 10^{-14}}{1.6 \times 10^{-19} \times 1 \times 10^{16}}(4 + 1.29)} = 0.88 \ \mu\text{m}$$

The maximum electric field $\varepsilon(0)$ is determined to be 1.3×10^{5} V/cm from Eq. (3-69). The magnitude of the electric field profile is shown in Fig. 3-19. (The actual electric field is negative in this coordinate system.) As dictated by Eq. (3-68), the electric field profile inside the collector is a straight line with a slope proportional to the collector doping.

As shown in Fig. 3-19, the electric field in practically the entire collector exceeds 1×10^{4} V/cm, which is the critical field beyond which the electron carriers travel at a constant saturation velocity (v_{sat}). Such a velocity is much greater than the diffusion velocity, so the carriers travel in the collector mainly by drift (although in the base they travel mainly by diffusion). Further, because the drift velocity is a constant in this high-field region, the mobile electron concentration is constant with respect to position inside the junction. It is given by

$$n(x) = \frac{J_C}{q v_{\text{sat}}} = \text{constant} \qquad (3\text{-}73)$$

where J_C is the collector current density.

When J_C is small, $n(x) \ll N_C$ and Eq. (3-68), which approximates Eq. (3-67), is valid. As the current density increases, the mobile electron concentration calculated from Eq. (3-73) becomes significant. The right-hand side of Eq. (3-67) can no longer be approximated as N_C alone. The net charge concentration should be written as $N_C - n$ (both p and N_a remain roughly zero). This modification has a profound impact on the electric field profile, since the slope of the electric field is directly proportional to the net charge concentration. As $N_C - n$ decreases with increasing J_C, and the slope of the field decreases, as shown in Fig. 3-20*a*. While the current density increases, the junction voltage drop is maintained at $V_{CB} + \phi_{CB} = 4.84$ V and the area under the field profile is unchanged. The simultaneous requirements of the decreasing field slope and the

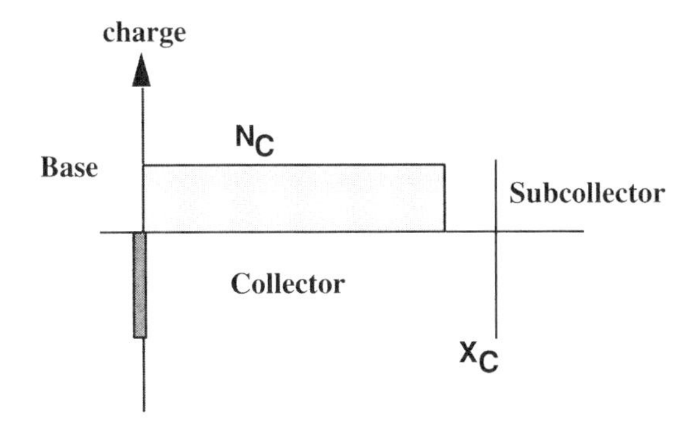

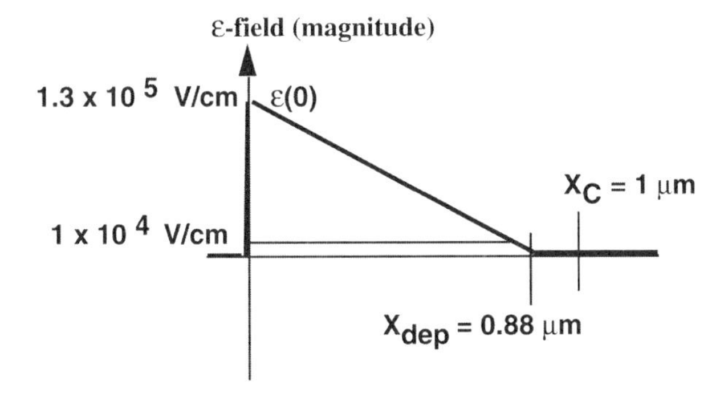

FIGURE 3-19. Charge distribution and field profile inside the collector side of a base-collector junction when the collector current is small. The values are calculated in Example 3-8.

constant area imply that the depletion thickness increases. If we do not consider impact ionization, the depletion thickness would continue to increase until it reaches $x = X_C$, the full collector thickness. The depletion thickness does not extend beyond X_C because the subcollector underneath the collector layer is heavily doped. Further increase of current results in a pentagonal field profile, as shown in Fig. 3-20*b*, replacing the previous triangular profile. As the current density increases a level such that $n = N_C$, the net charge inside the junction becomes zero, and the field profile stays constant with the position inside the junction (slope = 0). This situation, depicted in Fig. 3-20*c*, marks the beginning the *field reversal*. When J_C increases further such that $n > N_C$, the net charge inside the junction becomes negative. The electric field then takes on a negative slope (Fig. 3-20*d*), with a smaller magnitude at the base side of the collector than at the subcollector side. As the trend progresses, the magnitude of the field eventually diminishes to zero (Fig. 3-20*e*). When there is no more field to prevent the holes from "spilling" into the collector, *base pushout* is said to occur. The effective base thickness is the sum of the physical base thickness plus the hole

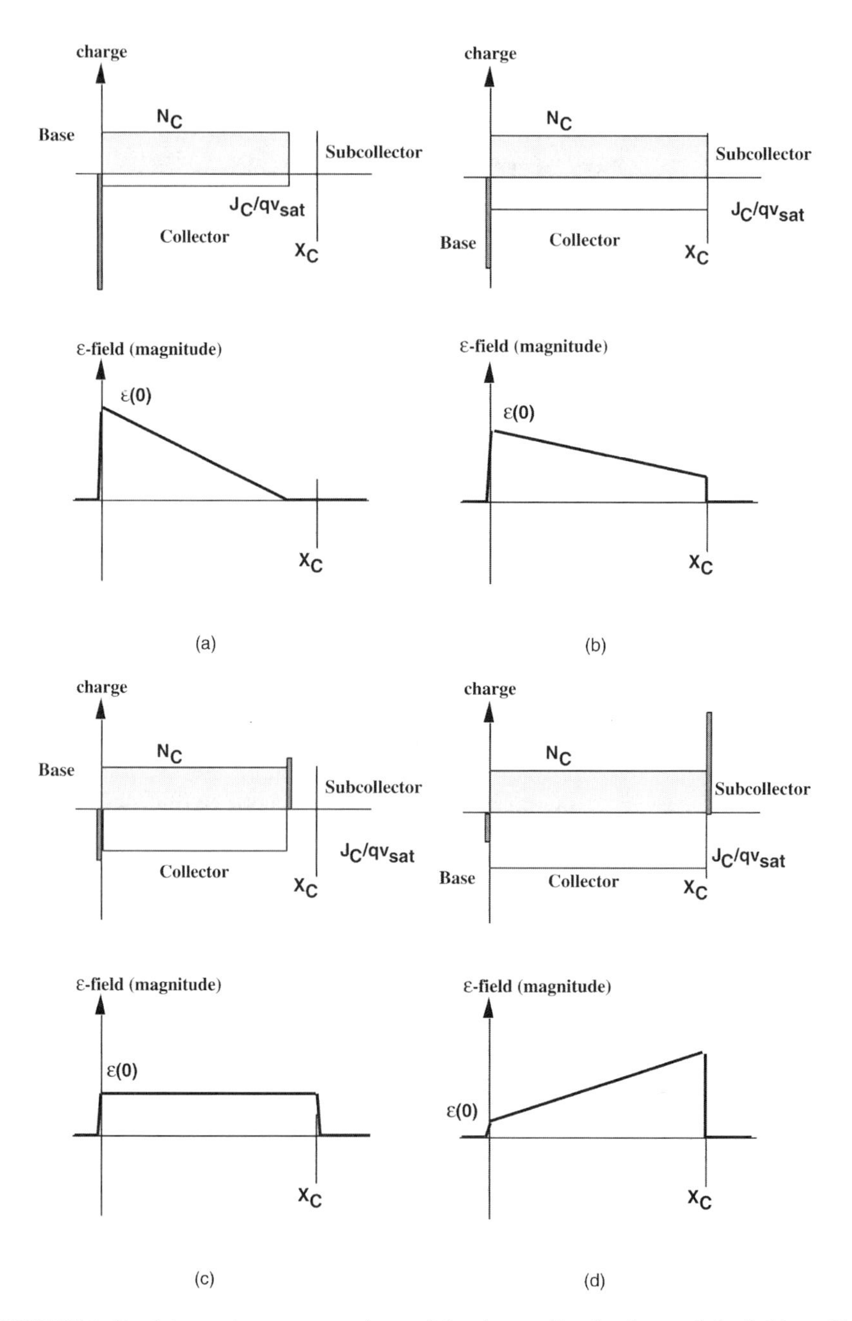

FIGURE 3-20. Schematic representations of the charge distribution and the field profile inside the base-collector junction: (*a*) $I_C \approx 0$; (*b*) I_C is such that the carrier concentration in the collector is $n < N_C$; (*c*) I_C is large enough that $n = N_C$; (*d*) I_C is large enough that $n > N_C$; (*e*) I_C is so large that the electric field at the base-collector junction interface becomes zero. The field reversal occurs in (*d*).

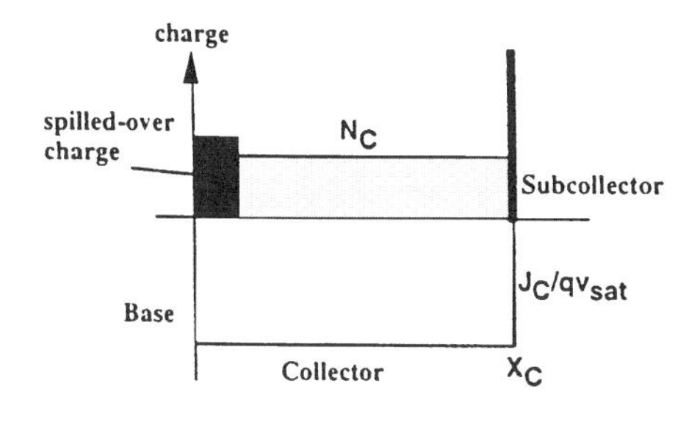

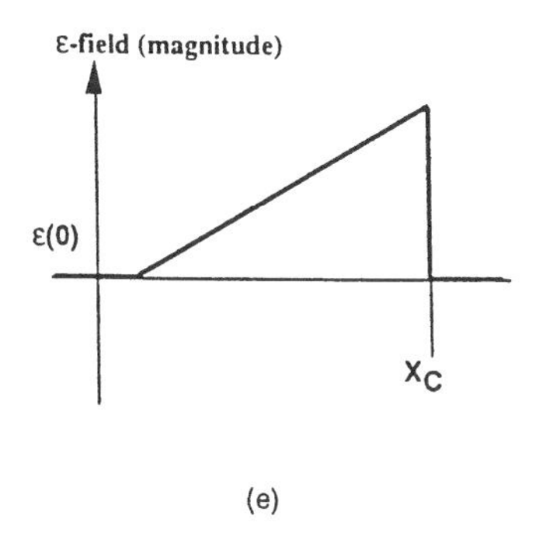

(e)

FIGURE 3-20. *(Continued)*

spilling distance. The current gain decreases as the transit time associated with the thickened base layer increases.

The device characteristics as a result of base pushout are referred to as the *Kirk effects*. The above description suggests that the threshold current due to the Kirk effects increases if the collector doping increases. However, in many applications where the collector doping may not be increased arbitrarily (in order that the operating voltage is greater than a certain value), the Kirk effects become an important mechanism affecting the current gain fall-off in HBTs. For the HBT examined in Example 3-8, which utilizes a collector doping of 1×10^{16} cm^{-3}, the threshold current density is roughly (a more exact quantity will be defined later) $qN_C v_{sat} = 1.3 \times 10^4$ A/cm^2 if v_{sat} is taken to be 8×10^6 cm/s. Clearly, the value of such a threshold current density depends on the magnitude of the saturation velocity. Since the saturation velocity decreases with the ambient temperature, the threshold density due to the Kirk effects is lower at higher temperatures.

We examine the Kirk effect quantitatively, also considering the voltage drop in undepleted collector. As mentioned, the mobile carrier concentration in the base-collector junction is simply $n = J_C/qv_{sat}$. Since $p \approx N_a \approx 0$, Eq. (3-67) simplifies to

$$\frac{d\varepsilon}{dx} = \frac{1}{\epsilon_s}\left(qN_C - \frac{J_C}{v_{sat}}\right) \tag{3-74}$$

Integrating both sides from $x = 0$ to X_{dep} with the boundary condition that $\varepsilon(X_{\text{dep}}) = 0$, we have

$$\varepsilon(x) = \left(qN_C - \frac{J_C}{v_{\text{sat}}}\right)(x - X_{\text{dep}}) \tag{3-75}$$

Since the negative of the area enclosed by the electric field is equal to the potential, we obtain the following by integrating one more time from $x = 0$ to X_{dep}.

$$V_{CBj} + \phi_{CB} = \left(qN_C - \frac{J_C}{v_{\text{sat}}}\right)\left(\frac{X_{\text{dep}}^2}{2}\right) \tag{3-76}$$

The junction voltage is equal to the applied base-collector bias (V_{CB}) minus the resistive voltage drop in the undepleted collector layer:

$$V_{CBj} = V_{CB} - J_C \rho_C (X_C - X_{\text{dep}}). \tag{3-77}$$

where ρ_C is the resistivity of the undepleted collector material and X_C is the designed collector thickness. Equations (3-76) and (3-77) are combined to solve for the depletion thickness as a fucntion of external bias and current density:

$$X_{\text{dep}} = \sqrt{\frac{2\epsilon_s(V_{CB} + \phi_{CB})}{qN_C}} \left(1 - \frac{J_C}{J_2}\right)^{1/2} \left(1 - \frac{J_C}{J_1}\right)^{-1/2} \tag{3-78}$$

where

$$J_1 = qv_{\text{sat}} N_C \tag{3-79}$$

$$J_2 = \frac{(V_{CB} + \phi_{CB})}{\rho_C (X_C - X_{\text{dep}})} \tag{3-80}$$

ϕ_{CB} is the built-in potential across the junction in equilibrium. The relative magnitudes of J_1 and J_2 dictate whether the Kirk effects or quasi-saturation takes place under a given bias condition. If $J_1 > J_2$, X_{dep} calculated from Eq. (3-78) decreases as J_C increases. Physically, this means that the resistive voltage drop is significant, hence depriving the available junction voltage to sustain a given depletion thickness. This sets the stage for *quasi-saturation*, in which the actual base-collector junction is slightly forward-biased even though the externally applied volatage is reverse-biased. If instead $J_2 > J_1$, then Eq. (3-78) states that X_{dep} increases with J_C. Although a lot more of the current increase is needed to eventually trigger the base pushout (or the Kirk effects), the fact that $X_{\rm dep}$ is expanding (rather than shrinking) toward X_C indicates that the Kirk effects will eventually occur and that quasi-satuartion is out of the picture. For typical modern microwave transistors, $J_2 \gg J_1$. Therefore, when HBTs are operated at high current levels, generally the Kirk effects take place instead of the quasi-saturation.

Example 3-9:

Find J_2 for both a low doping case ($N_C = 1 \times 10^{16}$ cm^{-3}) and a high doping case ($N_C = 2 \times 10^{17}$ cm^{-3}), assuming that μ_n is about 5000 cm^2/V-s for the light doping and 4000 cm^2/V-s for the heavy doping. The collector thickness is 1 μm for both cases.

Let us first consider the low doping case, in which $N_C = 1 \times 10^{16}$ cm^{-3}. For this doping level, ϕ_{CB} according to Eq. (3-71) is 1.29 V, as calculated in Example 3-8. ρ_C is calculated in a similar equation used for the base layer, Eq. (3-53):

$$\rho_C = \frac{1}{q\mu_n N_C} = \frac{1}{(1.6 \times 10^{-19})(5000)(1 \times 10^{16})} = 0.125\ \Omega\text{-cm}$$

Even in the absence of any externally applied V_{CB} and assuming that $X_{\rm dep} = 0$, J_2 still has a high value:

$$J_2 = \frac{0 + 1.29}{0.125(1 \times 10^{-4} - 0)} = 1.03 \times 10^5\ \text{A/cm}^2$$

This value is larger than $qN_C v_{\rm sat}$ at $N_C = 1 \times 10^{16}$ cm^{-3}.

At the other extreme of $N_C = 2 \times 10^{17}$ cm^{-3}, ϕ_{CB} is calculated to be about 1.37 V. ρ_C determined from the above procedure is 0.0156 Ω-cm. Even when $V_{CB} = 0$ and $X_{\rm dep} = 0$, J_2 retains a large value of $1.37/0.0156/1 \times 10^{-4} = 8.8 \times 10^5$ A/cm^2, which is much greater than $qN_C v_{\rm sat}$ when $N_C = 2 \times 10^{17}$ cm^{-3}.

Perhaps the only time that J_1 can exceed J_2 in HBTs happens when there are processing problems such that the collector contact resistance is much higher than expected, or the grown collector epitaxy somehow has excessive impurity such that the mobility is low. In gerneral, quasi-saturation is hardly observed in HBTs.

It is desirable to find out the depletion thickness given a certain current flow and bias condition. The depletion thickness is not obtained straightforwardly from Eq. (3-78) because one of its parameters (J_2) is itself dependent on the depletion thickness. A trial-and-error approach is to calculate the depletion thickness at negligible current level using Eq. (3-72), from which J_2 is first estimated from Eq. (3-80). A first-order guess of $X_{\rm dep}$ is then obtained from Eq. (3-78). This thickness is then fed into Eq. (3-80) again to establish a more accurate J_2, from which a second-order estimate of X_{dep} is calculated from Eq. (3-78). This process is repeated until sufficient convergence is obtained. Another method, however, takes the advantage of the fact that X_{dep} can be expressed in a quadratic equation. Plugging Eq. (3-80) directly into Eq. (3-78), we obtain

$$X_{dep}^2 = \frac{2\epsilon_s(V_{CB} + \phi_{CB})}{qN_C}\left(1 - \frac{J_C}{J_1}\right)^{-1} - \frac{2\epsilon_s\rho_C J_C}{qN_C}\left(1 - \frac{J_C}{J_1}\right)^{-1}(X_C - X_{dep}) \tag{3-81}$$

Solving the quadratic equation and taking the positive root, we obtain an expression for the depletion thickness in the collector as a function of the external bias and the collector current density:

$$X_{dep} = \frac{\epsilon_s\rho_C J_C}{qN_C}\left(1 - \frac{J_C}{J_1}\right)^{-1} + \left[\left(\frac{\epsilon_s\rho_C J_C}{qN_C}\right)^2\left(1 - \frac{J_C}{J_1}\right)^{-2} + \frac{2\epsilon_s(V_{CB} + \phi_{CB})}{qN_C}\left(1 - \frac{J_C}{J_1}\right)^{-1} + \frac{\epsilon_s\rho_C J_C X_C}{2qN_C}\left(1 - \frac{J_C}{J_1}\right)^{-1}\right]^{1/2} \tag{3-82}$$

This equation is applicable as long as $J_C < J_1$.

As current continues to increase, X_{dep} continues to expand, eventually reaching the collector/subcollector interface. The current density at which the collector is fully depleted is denoted as J_{fd}. It is obtained from the condition that $X_{dep} = X_C$, with the help of Eq. (3-78):

$$J_{fd} = (V_2 - V_{CB})\left(\frac{V_2 + \phi_{CB}}{J_1}\right)^{-1} \tag{3-83}$$

V_2 is defined as the applied base-collector bias that totally depletes X_C when $J_C \equiv 0$.

$$V_2 = \frac{qN_C X_C^2}{2\epsilon_s} - \phi_{CB} \tag{3-84}$$

After this current density is reached, no more layer can be depleted to sustain the voltage drop because the subcollector is heavily doped, yet the slope of the electric field continues to decrease as J_C increases (Eq. 3-74). Therefore, once J_C exceeds J_{fd}, $\varepsilon(X_{dep})$ is no longer equal to 0 as assumed in Eq. (3-75). The electric field ceases to be triangular in shape, but instead takes on a pentagonal shape as shown in Fig. 3-20. At the current density $J_C = qN_C v_{sat}$, the field slope decreases to zero and the field profile is constant inside the junction with $\varepsilon(0) = \varepsilon(X_{\mathrm{dep}})$. As the collector current increases even further, the field reversal takes place and $\varepsilon(0)$ ultimately decreases to zero at the Kirk current density (J_{Kirk}). J_{Kirk} is calculated from Fig. 3-20*d*; with the condition that $\varepsilon(0) = 0$, the total potential enclosed by the electric field is

$$V_{CB} + \phi_{CB} = \frac{1}{\epsilon_s}\left(\frac{J_C}{v_{\mathrm{sat}}} - qN_C\right)\frac{X_C^2}{2} \tag{3-85}$$

Therefore,

$$J_{\mathrm{Kirk}} = \left(1 + \frac{V_{CB} + \phi_{CB}}{V_2 + \phi_{CB}}\right) qN_C v_{\mathrm{sat}} \tag{3-86}$$

The maximum current is proportional to the collector doping and the electron saturation velocity. There are some dependences on the collector thickness through the dependence of V_2, but this is usually of relatively minor importance.

Example 3-10:

Find J_{Kirk} for the HBT of Example 3-8. Its collector thickness is 1 μm and its collector doping is 1×10^{16} cm^{-3}. The bias V_{CB} is 4 V. Assume that the saturation velocity is 8×10^6 cm/s.

The built-in voltage ϕ_{CB} was already determined in Example 3-8 for $N_C = 1 \times 10^{16}$ cm^{-3}. It was found to be 1.29 V. We determine V_2 from Eq. (3-84):

$$V_2 = \frac{(1.6 \times 10^{-19})(1 \times 10^{16})(1 \times 10^{-4})^2}{2(13.18)(8.85 \times 10^{-14})} - 1.29 = 5.6 \text{ V}$$

We find J_{Kirk} in accordance with Eq. (3-86):

$$J_{\mathrm{Kirk}} = \left(1 + \frac{4 + 1.29}{5.6 + 1.29}\right)(1.6 \times 10^{-19})(1 \times 10^{16})(8 \times 10^6) = 2.3 \times 10^4 \text{ A/cm}^2$$

Example 3-11:

The depletion thickness of the HBT in Example 3-8 was found to be 0.88 μm when $J_C = 0$. Find the depletion thickness when $J_C = J_{\mathrm{fd}}/2$, assuming that $J_2 \gg J_1$. The collector thickness in 1 μm and the collector doping is 1×10^{16} cm^{-3}. The bias V_{CB} is 4 V, and the saturation velocity is 8×10^6 cm/s. V_2 was found in Example 3-10 to be 5.6 V, and $\phi_{CB} = 1.29$ V.

If $J_2 \gg J_1$, then Eq. (3-78) can be simplified to

$$X_{\text{dep}} = \sqrt{\frac{2\epsilon_s(V_{CB} + \phi_{CB})}{qN_C}} \left(1 - \frac{J_C}{J_1}\right)^{-1/2}$$

$$= X_{\text{dep}}(J_C = 0)\left(1 - \frac{J_C}{J_1}\right)^{-1/2}$$

We calculate J_1 and J_{fd} first, from Eq. (3-79) and Eq. (3-83), respectively:

$$J_1 = (1.6 \times 10^{-19})(1 \times 10^{16})(8 \times 10^6) = 1.3 \times 10^4 \text{ A/cm}^2$$

$$J_{\text{fd}} = (5.6 - 4)\left(\frac{5.6 + 1.29}{1.3 \times 10^4}\right)^{-1} = 3 \times 10^3 \text{ A/cm}^2$$

At $J_C = J_{\text{fd}}/2 = 1.5 \times 10^3$ A/cm^2, X_{dep} is

$$X_{\text{dep}} = 0.88\left(1 - \frac{1.5 \times 10^3}{1.3 \times 10^4}\right)^{-1/2} = 0.94 \text{ μm}$$

The moment J_C reaches J_{Kirk} is not necessarily the same moment that the current gain falls off. The conventional description about what happens after $J_C = J_{\text{Kirk}}$ states that, because the base thickness increases after the base pushout, the base transit time increases and the current gain decreases. In this line of thinking, if the transistor's gain were dominated by surface recombination and the bulk recombination were only a minor part of the overall base current, then the current gain would not fall off immediately despite the onset of the base pushout. For simplicity, however, we take J_{Kirk} to be the threshold current density above which β falls off.

§ 3-9 AVALANCHE BREAKDOWN

The previous section relates to high-current effects. The described transistors are operated with relatively low values of collector-emitter bias. If the V_{CE} of these transistors increases to extreme values, either one or both of two phenomena take place. In the case when the transistor has more than one emitter finger and operates at relatively high current levels, the collapse of current gain marked by a sudden drop of I_C occurs as V_{CE} increases. The collapse of current gain is the subject of discussion in § 3-10. The other phenomenon occurs more readily in one-finger HBTs or multifinger HBTs operating at low current levels. It is the avalanche breakdown of the base-collector junction during which the collector current increases drastically. In this case, the current gain surges, in sharp

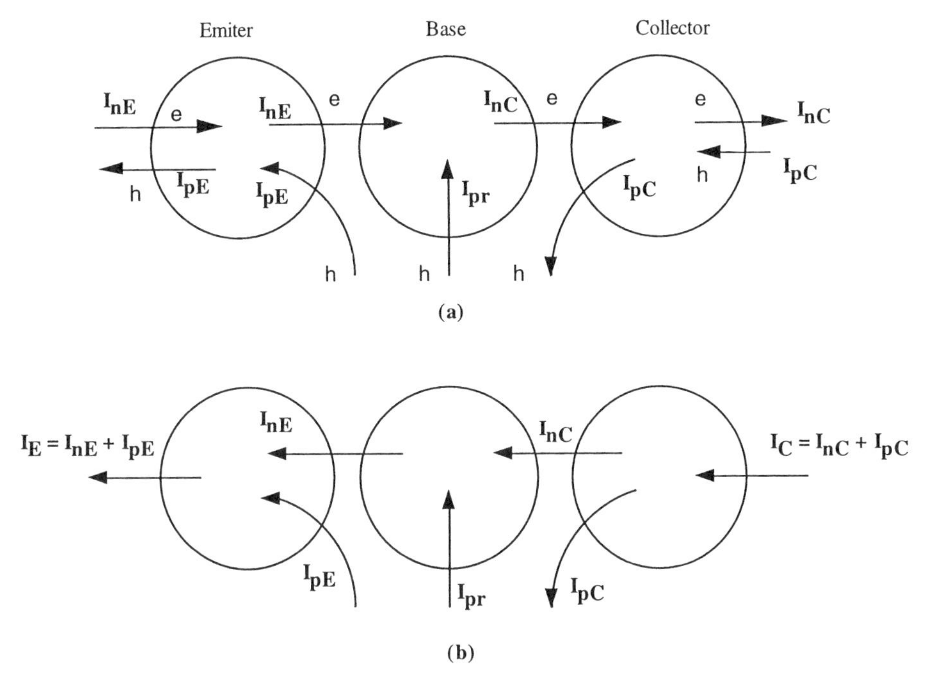

FIGURE 3-21. Various current components in a HBT: (*a*) the arrow points to the direction of the carrier flow; (*b*) the arrow points to the direction of the current flow.

contrast to the collapse of current gain The avalanche breakdown places a limit on the maximum operating voltage of bipolar transistors.

We have analyzed the base current components without presenting a formal treatise on the so-called emitter injection efficiency and base transport factor. The introduction of avalanche breakdown here serves as a convenient place to introduce these concepts. The emitter injection efficiency (which characterizes the amount of back-injection), the base transport factor (which characterizes the amount of bulk recombination), together with the avalanche multiplication factor (which characterizes the avalanche breakdown), completely determines the device current gain when leakage currents are neglected. Rather than focusing on the current gain directly, we first analyze the current transfer ratio, α, which is the ratio of the collector current to the emitter current.

The various current components that determine α are shown in Fig. 3-21. Figure 3-21*a* emphasizes the direction of the carrier flow, as marked by the arrows. The arrow direction is the same as the current flow for the hole current, but points to the opposite direction for the electron current, as shown in Fig. 3-21*b*. When the base-emitter junction is forward-biased, both the desired electron current injected from the emitter (I_{nE}) and the back-injection of holes from the base (I_{pE}) flow through the junction. These currents flow through the emitter terminal; therefore,

$$I_E = I_{pE} + I_{nE} \tag{3-87}$$

The emitter injection efficiency, Γ, is the ratio of the desired forward current to the overall emitter current:

$$\Gamma = \frac{I_{nE}}{I_{pE} + I_{nE}} \tag{3-88}$$

The back-injection current I_{pE}, though it exits through the emitter terminal, originates from the base terminal and is also part of the base current. In fact, it is identical to I_{Bp} described in § 3-2. I_{nE}, in contrast, originates from the emitter and flows into the base. A typical value of Γ for Si BJTs is 0.99. Γ is practically equal to 1 for HBTs because of the additional energy barrier experienced by the back-injecting holes (I_{pE}) compared to the forward-injecting electrons (I_{nE}).

Although most of the electrons diffuse through the base and enter the collector region, a small fraction of the electrons are lost in the base due to recombination. The recombination causes I_{nC} to be slightly smaller than I_{nE}. For each electron that is lost, a hole coming from the base contact must be supplied into the base region to enable such recombination event. Therefore,

$$I_{pr} = I_{nE} - I_{nC} \tag{3-89}$$

where I_{pr} is the sum of the recombination currents that take place in the base layer, equal to the sum of $I_{B,\text{bulk}}$, $I_{B,\text{surf}}$, and $I_{B,\text{cont}}$ described in § 3-2. The base transport factor (α_T) characterizes the amount of carriers being lost. It is defined as

$$\alpha_T = \frac{I_{nC}}{I_{nE}} \tag{3-90}$$

The magnitude of α_T is typically about 0.99 for both BJTs and HBTs.

After traveling through the base, the electrons constituting I_{nC} enter the collector region, where there is a high electric field. If the magnitude of the field exceeds a certain threshold, the electrons acquire enough energy to generate electron-hole pairs upon impacting the lattice atoms. Because the field is in the direction from the subcollector to the base, the generated holes are swept toward the base while the generated electrons move with the initiating electrons toward the subcollector. Before reaching the subcollector layer, the generated electrons pick up energy from the field and generate further electron-hole pairs upon colliding with the lattice atoms. This results in a chain reaction, similar to the avalanche process and from which the name avalanche breakdown originated. Unlike the recombination, which slightly decreases the electron population in the base, the avalanche multiplication dramatically increases the number of electrons collected at the subcollector. The amount of carrier multiplication is characterized by the avalanche multiplication coefficient, M:

$$M = \frac{I_C}{I_{nC}} \tag{3-91}$$

M has a value greater than unity. If we multiply Γ, α_T, and M together, we find that

$$\Gamma \alpha_T M = \frac{I_{nE}}{I_{nE} + I_{pE}} \cdot \frac{I_{nC}}{I_{nE}} \cdot \frac{I_C}{I_{nC}} = \frac{I_C}{I_E} \equiv \alpha \tag{3-92}$$

The *current gain transfer ratio* is simply the product of the emitter injection efficiency, the base transport factor, and the avalanche multiplication coefficient. The base current, equal to $I_E - I_C$, is expressed as

$$I_B = \left(\frac{I_E - I_C}{I_C}\right) I_C = \left(\frac{1-\alpha}{\alpha}\right) I_C = \left(\frac{1 - \Gamma \alpha_T M}{\Gamma \alpha_T M}\right) I_C \tag{3-93}$$

It is instructive to find I_B by directly summing the individual components of the base current rather than by equating it to $I_E - I_C$. From the charge conservation in the base region (the central circled region of Fig. 3-21), we write

$$I_B = I_{pE} + I_{pr} - I_{pC} \tag{3-94}$$

I_{pC}, consisting of the holes generated in the avalanche process, flows out the base terminal. It is therefore a *negative* component of the base current. If we focus instead on the collector region (the rightmost circled region), the consideration of charge conservation dictates that I_{pC} is a positive component of I_C. That is, the collector terminal current is

$$I_C = I_{nC} + I_{pC} \tag{3-95}$$

The three base current components can then be determined from the following:

$$I_{pC} = I_C - I_{nC} = \frac{M-1}{M} I_C \tag{3-96}$$

$$I_{pr} = I_{nE} - I_{nC} = \frac{I_{nC}}{\alpha_T} - I_{nC} = \frac{I_C}{\alpha_T M} - \frac{I_C}{M} = \frac{1 - \alpha_T}{\alpha_T M} I_C \tag{3-97}$$

$$I_{pE} = I_E - I_{nE} = \frac{I_C}{\Gamma \alpha_T M} - \frac{I_C}{\alpha_T M} = \frac{1 - \Gamma}{\Gamma \alpha_T M} I_C \tag{3-98}$$

It is easy to verify that substituting these expressions into Eq. (3-94) leads to Eq. (3-93).

Sometimes, the base current is said to be equal to

$$I_B \approx \frac{I_C}{\beta_0} - (M-1)I_C \tag{3-99}$$

where β_0 is the current gain of the transistor when it is biased at low voltages such that $M = 1$. That is, $\beta_0 = (\Gamma\alpha_T)/(1 - \Gamma\alpha_T)$. Equation (3-99) is appealing because it seems to state that the base current is equal to $(I_{pr} + I_{pE})$ minus I_{pC}. However, from Eqs. (3-97) and (3-98), $I_{pr} + I_{pE}$ is really equal to $(1 - \Gamma\alpha_T)/(\Gamma\alpha_T M)$, not exactly equal to the first term of Eq. (3-99). Further, I_{pC} is really equal to $(M-1)/MI_C$, not exactly equal to the second term of Eq. (3-99). Equation (3-99) is useful as an approximation to the actual base current expression (at least conceptually). It clarifies the origin of the *current-reversal* phenomenon observed in bipolar transistors, which is observable when I_B is plotted against V_{CE} at a constant V_{BE}. Initially, when V_{CE} is small, M is about unity (no impact ionization). The base current thus consists solely of the first component and I_B is positive. As V_{CE} increases, M increases to a value above unity due to avalanche breakdown, thus increasing the magnitude of the second base current component. As V_{CE} increases past a certain value, the second component eventually exceeds the first and I_B changes sign.

We are interested in expressing the three basic parameters in terms of more fundamental device parameters. Both Γ and α_T can be written in terms of expressions developed previously. The expression for M will be developed shortly. According to Eqs. (3-1) and (3-22), we write that, for graded HBTs,

$$I_{pE} = I_{Bp} = \frac{qA_E D_{pE}}{X_E N_E} n_i^2 \exp\left(-\frac{\Delta E_g}{kT}\right) \exp\left(\frac{qV_{BE}}{kT}\right) \tag{3-100}$$

$$I_{nE} = I_C = \frac{qA_E D_{nB}}{X_B N_B} n_i^2 \exp\left(\frac{qV_{BE}}{kT}\right) \tag{3-101}$$

In both of these equations, n_i refers to the intrinsic carrier concentration of the GaAs base. The expressions for abrupt HBTs can be obtained similarly. We will continue to use I_{pE} instead of I_{Bp} to conform with the discussion in this section. We also use I_{nE} instead of I_C. It is obvious now that the I_C expression given in Eq. (3-22) is truly the collector current only if both I_{pr} and I_{pC} are zero (i.e., $\alpha_T = M = 1$). Γ, using the definition of Eq. (3-88), is

$$\Gamma = \frac{1}{1 + [(X_B N_B D_{pE})/(X_E N_E D_{nB})]\ \exp(-\Delta E_g/kT)} \tag{3-102}$$

This is a common expression. We caution that such an expression accounts only for the back-injection current, but not the space-charge recombination current that also originates from the base and enters the emitter region. I_{pE} is therefore equal to the sum of that given in Eq. (3-100) and the space-charge recombination, which, according to the discussion in §3-2, can be expressed qualitatively as

$$I_{B,\,\mathrm{scr}} = I_0 \exp\left(\frac{qV_{BE}}{2kT}\right) \tag{3-103}$$

For each hole recombined in the process, an electron is supplied from the emitter contact. Thus, I_{nE} also increases by the same amount of current as given in Eq. (3-101). Γ, with definition of Eq. (3-88), is

$$\Gamma = \{1 + [(X_B N_B D_{pE})/(X_E N_E D_{nB})] \exp(-\Delta E_g/kT) + [(X_B N_B)/(qA_E D_{nB} n_i^2)] I_0 \exp(-qV_{BE}/2kT)\}^{-1} \tag{3-104}$$

The resulting complicated expression is a reason why $I_{B,\,\mathrm{scr}}$ is often not considered in writing down the correct Γ expression.

In the discussion of Eqs. (3-14) and (3-23), the bulk recombination current is treated as the only source of the base current. We can write:

$$\frac{I_{nE}}{I_{pr}} = \frac{2D_n\tau_n}{X_B^2} \tag{3-105}$$

Note that β in Eq. (3-14) is replaced by the ratio of I_{nE} to I_{pr}. Equation (3-14) is correct if we are concerned only with the base bulk recombination current. When other base current components are considered, the ratio should more properly be taken as I_{nE}/I_{pr}. β and I_{nE} are equivalent only if Γ and M are exactly 1.

α_T from the definition of Eq. (3-90) is

$$\alpha_T = \frac{I_{nE} - I_{pr}}{I_{nC}} = 1 - \frac{X_B^2}{2D_n\tau_n} \tag{3-106}$$

This equation for α_T implicitly assumes that the majority of the base recombination current (I_{pr}) is $I_{B,\,\mathrm{bulk}}$. In HBTs, where $I_{B,\,\mathrm{surf}}$ and $I_{B,\,\mathrm{cont}}$ are sometimes significant, the α_T expression given above would be incorrect. An accurate expression accounting for the two surface recombination currents is difficult to obtain because of the two-dimensional nature involved in the problem.

We are now ready to present an expression for M, which is the ratio of I_C (composed of both the electron and hole currents) to I_{nC} (composed only of the electron current injected from the base). We note that some popular equations of M are obtained without considering the following fact. When an electron moves in the high-field region, it has to drift a certain distance before it acquires enough energy to initiate an impact ionization. The distance and energy are referred to as the *dead space* and the *threshold energy*, respectively. Here, we adopt a formula for M obtained from an extensive experimental study, which accounts for the so-called dead space effects.

$$1 - \frac{1}{M} = \int_{X_{\text{th}}}^{X_{\text{dep}}} \alpha_e[\varepsilon(x)] \exp\left(- \int_{X_{\text{th}}}^{x} \{\alpha_e[\varepsilon(x')] - \alpha_h[\varepsilon(x')]\}\, dx' \right) dx \times \exp\left[\int_0^{X_{\text{th}}} \alpha_h[\varepsilon(x)]\, dx \right] \tag{3-107}$$

where x is the distance measured from the base side of the base-collector junction; $\varepsilon(x)$ is the electric field profile in the junction; X_{dep} is the base-collector junction depletion thickness; X_{th} is the dead space thickness within which carriers have not acquired enough energy to ionize the lattice atoms; and α_e and α_h denote the electron and hole ionization rates, respectively. The dead space thickness is defined as the distance within which an electron gains enough energy (the threshold energy, E_{th}) before it can initiate impact ionization:

$$E_{\text{th}} = q \int_0^{X_{\text{th}}} \varepsilon(x)\, dx \tag{3-108}$$

The threshold energy was determined experimentally to be 1.7 eV for GaAs. The ionization coefficients are expressed as

$$\alpha_e[\varepsilon(x)] = 1.899 \times 10^5 \exp\left[- \left(\frac{5.75 \times 10^5}{\varepsilon(x)} \right)^{1.82} \right] \text{(1/cm)} \tag{3-109}$$

$$\alpha_h[\varepsilon(x)] = 2.215 \times 10^5 \exp\left[- \left(\frac{6.57 \times 10^5}{\varepsilon(x)} \right)^{1.75} \right] \text{(1/cm)} \tag{3-110}$$

where $\varepsilon(x)$ is in volts per centimeter. With Eqs. (3-107)–(3-110), M can be calculated once the electric profile is known. In the most general case, the doping of the collector layer is arbitrary, rendering the determination of $\varepsilon(x)$ difficult. However, if we restrict ourselves to the most common collector design, in which the collector doping is constant, we can easily calculate the electric profile.

We assume that the following parameters are known: the collector thickness (X_C), the collector doping (N_C), the external base-collector bias (V_{CB}). From the Poisson equation (Eq. 3-68),

$$\varepsilon(x) = \varepsilon(0) - \frac{qN_C}{\epsilon_s} x \tag{3-111}$$

where $\varepsilon(0)$ is the peak electric field, which is located at the base edge of the base-collector junction ($x = 0$). Its value depends on whether the applied V_{CB} is large enough to fully deplete the collector: For

$$\frac{1}{2\epsilon_s} qN_C X_C^2 \geq V_{CB} + \phi_{CB} \qquad \text{(collector not fully depleted)}$$

$$\varepsilon(0) = \left[\frac{2}{\epsilon_s} qN_C(V_{CB} + \phi_{CB})\right]^{1/2}$$

For

$$\frac{1}{2\epsilon_s} qN_C X_C^2 < V_{CB} + \phi_{CB} \qquad \text{(collector fully depleted)}$$

$$\varepsilon(0) = \frac{1}{X_C}\left(V_{CB} + \phi_{CB} - \frac{1}{2\epsilon_s} qN_C X_C^2\right) + \frac{1}{\epsilon_s} qN_C X_C \tag{3-112}$$

These expressions are derived with the one-sided depletion approximation, valid for typical HBTs whose base is much more heavily doped than the collector. It also neglects the presence of the mobile carriers due to the collector current.

M is calculated from Eq. (3-107), whose parameters are obtained from Eqs. (3-109) and (3-112). This concludes the model to determine M. Together with Γ and α_T from Eqs. (3-88) and (3-90), the current gain of a common-emitter transistor can be found:

$$\text{Common-emitter current gain:} \qquad \beta = \frac{I_C}{I_B} = \frac{\alpha_T \Gamma M}{1 - \alpha_T \Gamma M} \tag{3-113}$$

The current gain in a common-base transistor is defined as I_C/I_E. It is equal to the current gain transfer ratio described in Eq. (3-92):

$$\text{Common-base current gain:} \qquad \alpha = \frac{I_C}{I_E} = \alpha_T \Gamma M \tag{3-114}$$

Example 3-12:

An HBT has a heavily doped base layer of 8×10^{19} cm^{-3}, with a thickness of 800 Å. Its current gain is completely limited by the base bulk recombination. Find the current gain when the transistor is biased in the common-emitter configuration. What is the current gain when the transistor is configured in common-base? Assume that $M = 1.05$.

Since the current gain is completely limited by the base bulk recombination, Γ, which measures the amount of I_{Bp}, is equal to 1. From Eqs. (1-71) and (1-75), we find that, for $N_B = 8 \times 10^{19}\ \text{cm}^{-3}$, the minority lifetime and mobility are

$$\tau_n = \left(\frac{8 \times 10^{19}}{1 \times 10^{10}} + \frac{6.4^{39}}{1.6 \times 10^{29}}\right)^{-1} = 2.1 \times 10^{-11}\ \text{s}$$

$$\mu_n = 8300\left(1 + \frac{8 \times 10^{19}}{3.98 \times 10^{15} + 8 \times 10^{19}/641}\right)^{-1/3} = 970\ \text{cm}^2\ \text{V-s}$$

D_n, from the Einstein relationship, is $25\ \text{cm}^2/\text{s}$. The base transport factor according to Eq. (3-106) is

$$\alpha_T = 1 - \frac{(800 \times 10^{-8})^2}{2(25)(2.1 \times 10^{-11})} = 0.94$$

With $\Gamma = 1$, $\alpha_T = 0.94$ and $M = 1.05$, the current gain is found from Eq. (3-113) as

$$\beta = \frac{0.94 \times 1 \times 1.05}{1 - 0.94 \times 1 \times 1.05} = 76$$

The above model of current gains allows us to calculate the avalanche breakdown voltages. The breakdown voltage for a common-emitter transistor, termed BV_{CE}, is equal to the applied external V_{CE} such that $\beta \to \infty$. Likewise, the breakdown voltage for a common-base transistor, termed BV_{CB}, is equal to the applied external V_{CB} such that $\alpha \to \infty$. In other words, BV_{CE} and BV_{CB} are the voltages at which M increases from unity to $1/(\Gamma\alpha_T)$ and ∞, respectively. The breakdown voltages vary with the current levels (mainly because Γ and α_T vary with I_C, though M also vary with I_C). The desire to yield breakdown voltages independent of the bias current leads to the introduction of two parameters: BV_{CEO} and BV_{CBO}. BV_{CEO} applies to the common-emitter configuration, denoting the breakdown voltage when the input base current is zero. For the common-base configuration, BV_{CBO} denotes the breakdown voltage when the input emitter current is zero.

Because typically $\Gamma \approx \alpha_T \approx 0.95$, M needs to be only ≈ 1.11 before the avalanche breakdown in a common-emitter transistor is initiated. Compared to the value of ∞ to initiate the breakdown in a common-base transistor, the relatively small value of 1.11 immediately suggests that $BV_{CEO} \ll BV_{CBO}$. Experimentally, these two figures of merits are identified as the voltages at which the measured collector currents increases "rapidly." We cannot ascertain their exact values because M is not a directly measurable quantity. Surely the definition based on eyeing I_C is somewhat arbitrary, because what appears to be a rapid increase of I_C to one person is a slow increase to others. Two people may read out different values of breakdown voltages, especially when the measurement

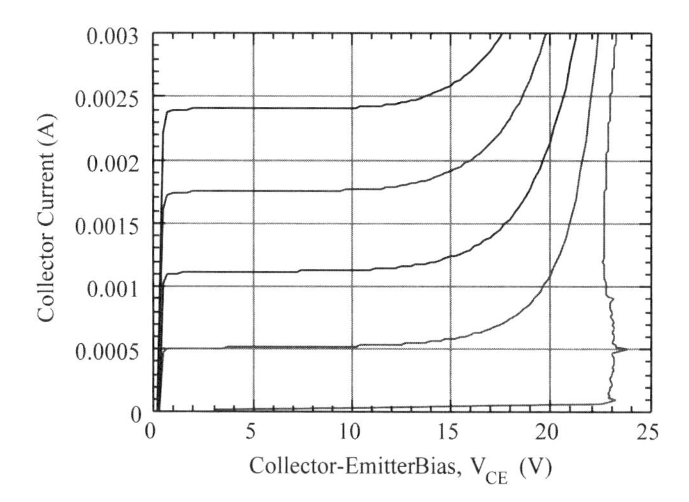

FIGURE 3-22. Measured *I–V* characteristics exhibiting avalanche breakdown behavior. The transistor is biased in the common-emitter configuration.

display (in a curve tracer or a parametric analyzer) plots the currents on different scales. However, the reported breakdown voltages are generally within 0.5 V of each other.

Figure 3-22 shows the measured current–voltage characteristics of the avalanche breakdown when the transistor is biased in the common-emitter configuration. Common-emitter *I–V* characteristics plot I_C as a function of V_{CE}, with the input base current fixed at a certain level. Common-base *I–V* characteristics plot I_C as a function of V_{CB}, with the input emitter current fixed at a certain level. From Fig. 3-22, the collector current is initially low due to the relatively low base current level. As V_{CE} increases to greater that 20 V, avalanche breakdown occurs and the current increases substantially. This marks the BV_{CEO} value. Sometimes, with further increase in the collector current, the measured V_{CE} decreases slightly from the BV_{CEO} value (such as the lowest curve of Fig. 3-22). This is explained by noting that in general the transistor current gain increases with current (§3-3). More specifically, the base current at low I_C levels is dominated by the space-charge recombination current, which has an ideality factor of 2. As I_C increases, the relative magnitude of $I_{B,\text{scr}}$ compared to I_C decreases, resulting in the increase in Γ (see Eq. 3-104). Since the requirement for the breakdown is $\alpha_T \Gamma M = 1$ for a transistor biased in the common-emitter configuration, a smaller value of M is needed to satisfy the requirement when Γ is larger. Consequently, the required V_{CE} bias to continue the trend of surging I_C is smaller than BV_{CEO}.

Figure 3-23 exhibits the measured characteristics when the transistor is biased in the common-base configuration. The breakdown voltage in the common-base configuration is larger than that measured in the common-emitter configuration.

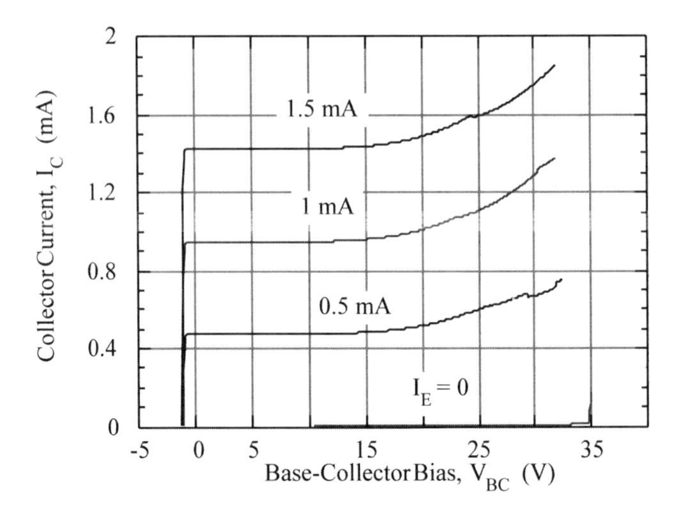

FIGURE 3-23. Avalanche breakdown behavior when the transistor is biased in the common-base configuration.

In particular, BV_{CBO} identified from Fig. 3-23 is 35 V, which is nearly twice that of BV_{CEO} identified from Fig. 3-22. It is also clear that the current gain in the common-base configuration (I_C/I_E) is slightly less than unity.

§ 3-10 COLLAPSE OF CURRENT GAIN

The *collapse of current gain* is a sudden decrease of current gain observed when the transistor is operated at high power levels. Figure 3-24 illustrates the common-emitter I-V characteristics measured from a two-finger HBT. Two distinct operating regions are evident, depending on the magnitude of the collector-emitter bias. At smaller V_{CE} values, the collector current decreases gradually with V_{CE}, exhibiting a negative differential resistance (NDR). The region is called either the NDR or the normal operating region. The cause of NDR, which is relevant to the understanding of collapse, is briefly described as follows. In *Npn* HBTs, the base holes back-injected into the emitter experience a larger energy barrier than the emitter electrons forward-injected into the base. The ratio of the desirable forward injection to the undesirable backward injection is proportional to a factor $\exp(\Delta E_a/kT)$, where ΔE_a is the effective energy barrier difference for the electrons and the holes. It is equal to ΔE_g for graded HBTs and ΔE_v for abrupt HBTs (§ 3-1). At room temperature, this ratio is large and the emitter injection efficiency is roughly equal to unity. However, as V_{CE} increases, the power dissipation in the HBT increases, gradually elevating the junction temperature above the ambient temperature. The emitter injection

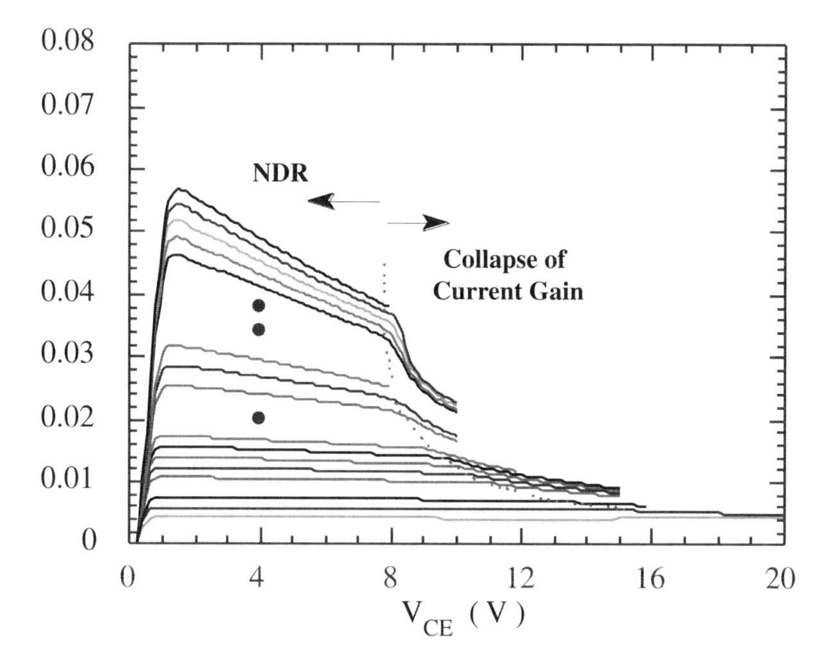

FIGURE 3-24. Measured common-emitter I-V characteristics of a two-finger HBT with a total emitter area of $2 \times 60\ \mu m^2$. The base current is constant during each scan of the curve. The dotted line is the collapse loci indicating the operating conditions when the collapse occurs. (From W. Liu, S. Nelson, D. Hill and A. Khatibzadeh [1993]. Current gain collapse in microwave multi-finger heterojunction bipolar transistors operated at very high power density. *IEEE Trans. Electr. Dev.* **40,** 1917–1927. © IEEE, reprinted with permission)

efficiency, and hence the current gain, gradually decreases with increasing V_{CE}. Since Fig. 3-24 plots measured I_C for constant applied I_B, the decreasing values of β directly results in the observed gradual decreasing of I_C.

As V_{CE} increases even more ($V_{CE} \geq 8$ V at $I_C > 0.02$ A in Fig. 3-24), NDR is suddenly replaced by the collapse phenomenon, as marked by an abrupt, dramatic lowering of I_C. The *collapse loci*, the I_C as a function of V_{CE} at which collapse occurs, are shown as a dotted curve in Fig. 3-24. We first present a qualitative description of the cause of the collapse. Quantitative analysis of the I-V characteristics using a model based on this qualitative description will then be discussed.

When several identical transistors are connected together to common emitter, base, and collector electrodes, each transistor is expected to conduct the same amount of collector current for any biases. However, as theoretically shown and experimentally demonstrated in the later part of this section, equal conduction occurs only when the power dissipation is low to moderate, so that the junction temperature rise above the ambient temperature is small. At high V_{CE} and/or high I_C operation, where the transistor is operated at elevated temperatures, one transistors starts to conduct more current than the others. Eventually, one transistor conducts all the current while the others become electrically inactive. This current domination results because of an intrinsic transistor property that, as the junction temperature increases, the bias required to turn on some arbitrary current level decreases. This property is illustrated by an empirical expression relating I_C, V_{BE}, and T:

$$I_C = I_{CO} \cdot \exp\left\{\frac{q}{\eta k T_A} \cdot [V_{BEj} - \phi \cdot (T - T_A)]\right\} \tag{3-115}$$

where I_{CO} is collector saturation current, V_{BEj} is the bias across the base-emitter junction, η is the collector current ideality factor, T_A is the ambient temperature, and T is the actual junction temperature. The degree of the change of the turn-on voltage in response to junction temperature change is characterized by ϕ, which is named as the *thermal-electrical feedback coefficient*. Likewise, Eq. (3-115) is referred to as the *thermal-electrical feedback equation*. This equation is an approximation of a more accurate expression. However, we use this equation because it highlights the feedback action of the increased self-heating on the collector current. When the junction temperature exceeds the ambient temperature, the transistor needs an amount of $\phi \cdot \Delta T$ less voltage to turn on the same amount of I_C.

A multifinger HBT can be viewed as consisting of several identical sub-HBTs, with their respective emitter, base, and collector leads connected together. From Eq. (3-115), if one finger (i.e., one sub-HBT) becomes slightly warmer than the others, then its base–emitter junction turn-on voltage becomes slightly lower. Consequently, this particular finger conducts more current for a given fixed base–emitter voltage. This increased collector current, in turn, increases the power dissipation in the junction, raising the junction temperature even further. Collapse occurs when the junction temperature in one finger of the entire device becomes much hotter than the rest of the fingers, so that the feedback action of increased collector current with junction temperature quickly leads to the fact that just one particular finger conducts the entire device current. Since the transition from uniform current conduction to one finger domination occurs suddenly, an abrupt, dramatic current gain lowering due to the lowered emitter injection efficiency at high temperatures is observed. Therefore, the fundamental cause of both NDR and collapse is the current gain drop with increasing temperature. Their difference, however, lies in the degree of temperature increase as V_{CE} increases. In the NDR region, all fingers share relatively the same amount of current and the junction temperatures increase gradually with V_{CE}. In contrast, in the collapse region, as the device power is entirely dissipated in one finger and the junction temperature surges rapidly, the current gain suddenly plummets.

The gain collapse does not cause the HBTs to fail immediately. The I-V characteristics shown in Fig. 3-24 are repeatedly measured as the device is biased into and out of the collapse. Because the collapse is reversible, the same device can be measured for several experiments. This is to be contrasted with thermal runaway typically reported in silicon bipolar junction transistors (Si BJTs) which causes the devices to fail irreversibly. The gain collapse does not exhibit hysteresis, either. If an I-V curve is measured by increasing V_{CE} from, say, 0 to 20 V, the same curve is measured when V_{CE} is instead decreased from 20 to 0 V.

At low to moderate junction temperatures, the collector current is shared equally by the fingers. However, when a critical condition is reached, the collector current suddenly flows through only one finger. We now present a simple theoretical model of the collapse phenomenon, seeking an equation governing such a critical condition. Both the electrical and thermal models of the multifinger HBT are schematically shown in Fig. 3-25, where R_E and R_{th} denote the per-finger values of the emitter resistance and the thermal resistance, respectively. (Suppose the overall electrical and thermal resistances of the two-finger HBT are 2 Ω and 750°C/W, respectively, then the per-finger values are 4 Ω and 1500°C/W, respectively.) R_{Ex} and R_{thx}, by contrast, respectively denote the resistance and the thermal resistance common to all the fingers. For simplicity, the following derivation concerns with an HBT with two fingers. If under some bias condition the thermal instability has occurred and the currents flowing through the fingers are I_{C1} and I_{C2}, then from Eq. (3-115),

$$V_{BEj1} = \frac{\eta k T_A}{q} \ln\left(\frac{I_{C1}}{I_{CO}}\right) - \phi \cdot \Delta T_1 \tag{3-116a}$$

$$V_{BEj2} = \frac{\eta k T_A}{q} \ln\left(\frac{I_{C2}}{I_{CO}}\right) - \phi \cdot \Delta T_2 \tag{3-116b}$$

ΔT, the increase in junction temperature with respect to the ambient temperature, is written from the inspection of the thermal model of Fig. 3-25 as:

$$\Delta T_1 = (R_{\text{th}} + R_{\text{thx}}) \cdot I_{C1} V_{CE} + R_{\text{thx}} \cdot I_{C2} V_{CE} \tag{3-117a}$$

$$\Delta T_2 = R_{\text{thx}} \cdot I_{C1} V_{CE} + (R_{\text{th}} + R_{\text{thx}}) \cdot I_{C2} V_{CE} \tag{3-117b}$$

From the Kirchoff's voltage loop, we obtain a relationship between the junction voltages of the two subtransistors:

$$V_{BEj1} - V_{BEj2} = I_{C2} \cdot R_E - I_{C1} \cdot R_E \tag{3-118}$$

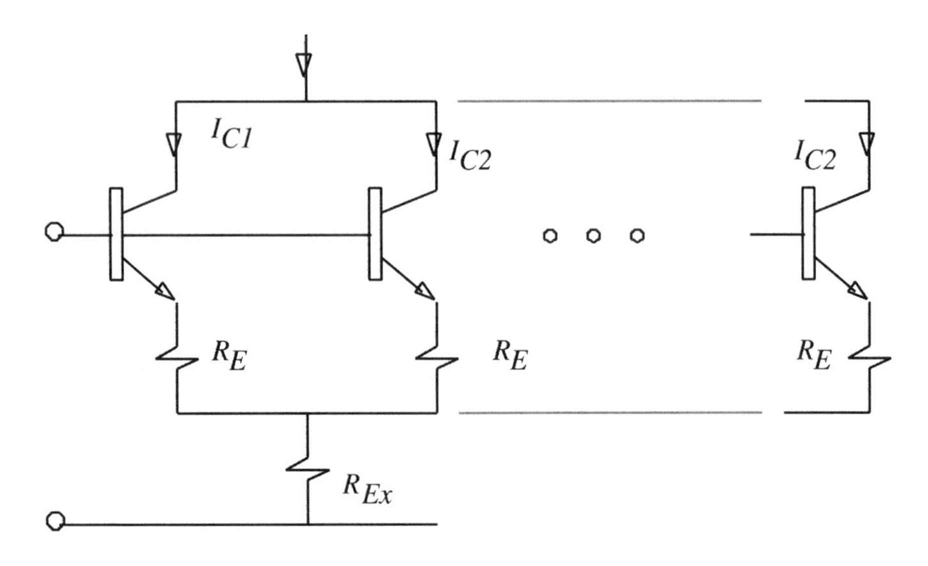

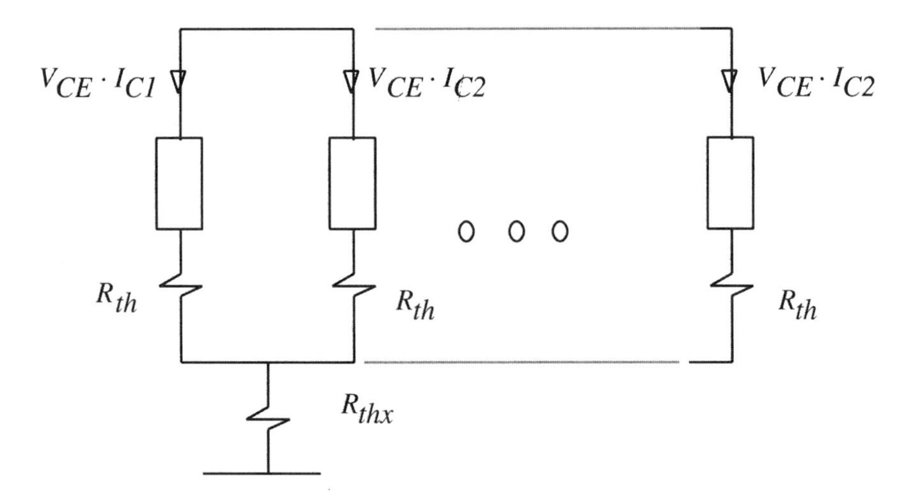

FIGURE 3-25. Electrical and thermal models of a multifinger HBT that can be viewed as consisting of several identical sub-HBTs. (From W. Liu, S. Nelson, D. Hill and A. Khatibzadeh [1993]. Current gain collapse in microwave multi-finger heterojunction bipolar transistors operated at very high power density. *IEEE Trans. Electr. Dev.* **40,** 1917–1927. © IEEE, with permission.)

Substituting the ΔT's of Eq. (3-117) into Eq. (3-116), which in turn is substituted into Eq. (3-118), we obtain an equation relating I_{C1} and I_{C2}:

$$I_{C2} = \frac{\eta k T_A}{q} \ln\left(\frac{I_{C1}}{I_{C2}}\right) \frac{1}{\left(1 - \frac{I_{C1}}{I_{C2}}\right)(R_E - \phi \cdot R_{th} \cdot V_{CE})} \tag{3-119}$$

This equation is derived under the assumption that the transistor has entered the collapse region of operation. We know that right before the collapse, $I_{C1} = I_{C2}$. Therefore, if we take the limit of $I_{C1} \to I_{C2}$ in Eq. (3-119), we can determine the current level at which the transistor just enters the collapse region of operation:

$$I_{\text{Collapse}} = \lim_{x \to 1} \frac{\eta k T_A}{q} \frac{1}{(R_E - \phi \cdot R_{th} \cdot V_{CE})} \frac{\ln(x)}{(1 - x)}, \quad \text{where } x = \frac{I_{C1}}{I_{C2}} \tag{3-120}$$

I_{Collapse} is the critical current above which the collapse of current gain occurs. We stress that I_{Collapse} is a per-finger value. For example, if I_{Collapse} is found to be 2 mA for a three-finger HBT, then the total collector current at which the collapse of current gain occurs is 6 mA. Likewise, the values of R_E and R_{th} also denote the per-finger values.

From L'Hôpital's rule, Eq. (3-120) is simplified:

$$I_{\text{Collapse}} = \frac{\eta k T_A}{q} \frac{1}{R_{th} \cdot \phi \cdot V_{CE} - R_E} \tag{3-121}$$

We generalize the above result derived from a two-finger HBT to a multifinger HBT. When the individual device current is below the critical current level (or when I_{Collapse} is negative), all fingers share the same amount of current. Above this critical current level, one finger conducts most of the current, whereas the rest of the fingers share the remaining current equally. Let I_{C1} be the collector current for the finger that becomes hotter than other fingers during the collapse, I_{C2} be the collector current flowing through each of the remaining fingers, and I_{CT} be the total collector current. Then, the preceding description is mathematically expressed as the following, with N being the number of fingers of the transistor:

For $I_{CT} \leq N \cdot I_{\text{Collapse}}$, or $I_{\text{Collapse}} \leq 0$,

$$I_{C1} = I_{C2} = \frac{1}{N} I_{CT} \tag{3-122a}$$

and for $I_{CT} > N \cdot I_{\text{Collapse}}$

$$I_{C1} = \frac{(\eta k T_A/q) \ln (I_{C2}/I_{C1})}{[(I_{C2}/I_{C1}) - 1](R_{th} \cdot \phi \cdot V_{CE} - R_E)} \tag{3-122b}$$

$$I_{C2} = \frac{1}{(N-1)} \cdot (I_{CT} - I_{C1}) \tag{3-122c}$$

Example 3-13:

What is the device current at the onset of the gain collapse for a two-finger HBT? The transistor has a thermal resistance of 630°C/W per finger, and each finger has an area of $2 \times 50\,\mu\text{m}^2$. The measured collector turn-on voltage is found to decrease 24 mV for every 20°C increase in junction temperature. Its overall emitter resistance is 1.5 Ω. The collector current ideality factor is 1.0 and $T_A = 300$ K. The bias voltage is 8 V.

$R_{\text{th}} = 630$°C/W for a 2×50-μm² finger. (The overall device thermal resistance is about 315°C/W.) The thermal-electrical feedback coefficient is 24 mV/20°C = 1.2 mV/°C. The emitter resistance per finger is $1.5\,\Omega \times 2 = 3\,\Omega$. The onset of collapse occurs at

$$I_{\text{Collapse}} = 1.0 \cdot 0.0258 \frac{1}{630 \cdot 1.2 \times 10^{-3} \cdot 8 - 3} = 8.46 \times 10^{-3} A$$

I_{Collapse} is a per-finger value. The total collector current at the onset of the gain collapse is 2×8.46 mA = 17 mA.

An inspection of Eq. (3-121) shows that I_{Collapse} can be maximized by increasing the emitter resistance in the individual finger. When the transistor is fabricated with the most simplistic process, the emitter resistance of the transistor is solely due to the contact resistance and the epitaxial resistance of the emitter layer. These two components of the emitter resistance are fixed after the transistor design and process are chosen. The intentionally added resistance to increase I_{Collapse} is referred to as the *ballasting resistance*. Likewise, a ballasted HBT thus refers to a HBT in which intentional resistance is added externally to the device. An unballasted HBT refers to one in which the overall emitter resistance is due purely to the contact and epitaxial resistances, without an externally added resistance. The term *ballasting* generally has a positive connotation since it is used to prevent undesirable thermal instability in bipolar transistors. However, if there is an excessive amount of ballasting resistance, or intentionally added resistance, then the transistor's high speed performance may degrade. The value of the ballasting resistance is therefore a trade-off between high-frequency performance and thermal stability. In a typical HBT finger with 2×50-μm² area, the contact resistivity is about 0.5 Ω when the contact resistivity is $5 \times 10^{-7}\,\Omega \cdot \text{cm}^2$. Because of the high mobility in n-GaAs emitter layers, the typical internal epitaxial resistance is practically 0 Ω. In typical applications for which 3 to 10 Ω/finger of R_E is required to ensure the operating current level to be below I_{Collapse}, then the ballasting resistance should be between 2.5 and 9.5 Ω/finger.

Adding ballasting resistance requires additional processing steps, such as the deposition of a resistive material on the wafer and the formation of an electrical contact between the resistive material and the transistor finger. The typical material for the ballasting resistance is tantalum nitride (TaN), whose sheet resistance is around 50 Ω/sq. when it is about 1000 Å thick. Nickel chrome (NiCr) is sometimes used; however, its resistance has a higher temperature coefficient. That is, its resistance value tends to vary more with temperature than TaN. A ballasted HBT with an external TaN resistor is schematically shown in Fig. 3-26, which is an example of emitter-ballasted HBT. The ballasting resistance also can be placed in the base terminal and may be more effective in preventing thermal instability.

Equation (3-122) exhibits no dependence on R_{Ex} and R_{thx}. The condition for the onset of collapse purely depends on the individual electrical resistance and the thermal resistance rather than the common resistances. This is an important point in the design layout of the ballasting resistor.

Some reasons why the collapse phenomenon is not reported in silicon bipolar transistors are as follows. Unlike in GaAs HBTs, the current gain typically increases with temperature in silicon bipolar transistors. When the critical current condition is reached, one finger dominates the transistor action and becomes hot. Since the current gain in Si bipolar transistor increases rather than decreases with temperature, the increased current in that finger will draw more and more current. This is to be contrasted to the collapse shown in Fig. 3-24, where I_C decreases after the critical condition is achieved. Therefore, for Si bipolar transistors, reaching the critical condition means the start of a *thermal runaway* in which more and more current flows through the hot spot until the device burns out. In addition, the substrate Si bipolar transistors are doped and have larger thermal conductivity than that of the semi-insulating GaAs substrate used in this investigation. Therefore, it is conceivable that the self-heating in Si transistors is not serious and that each finger can dissipate heat efficiently. In a brief summary, the critical condition of Eq. (3-121) applies equally well to both Si BJTs and HBTs. It is a condition of thermal instability, which is inevitable in any bipolar transistor whose base–emitter junction exhibits the thermal-electric feedback property. However, their responses toward such thermal instability are completely different. In Si BJTs whose current gain increases with temperature, the transistors exhibit thermal runaway. In AlGaAs/GaAs HBTs whose current gain lowers with temperature, the transistors exhibit the collapse of current gain.

We describe the major reason why the current gain increases with temperature in Si BJTs, in contrast to the decreasing trend observed in HBTs. As mentioned at the beginning of this section, the ratio of the undesirable hole back injection to the forward electron injection is proportional to $\exp(-\Delta E_a/kT)$, where ΔE_a is the effective energy barrier difference experienced by the holes with respect to the electrons. For HBTs whose energy gap in the emitter is larger than in the base, ΔE_a is positive. In Si BJTs, despite the fact that both the emitter and the base are made of the same material, the heavy doping in the emitter makes the energy gap lower than in the base. Therefore, ΔE_a is negative in Si BJTs and positive in HBTs. The dependences of the current gains on temperature therefore reveal opposite trends for the two types of transistors.

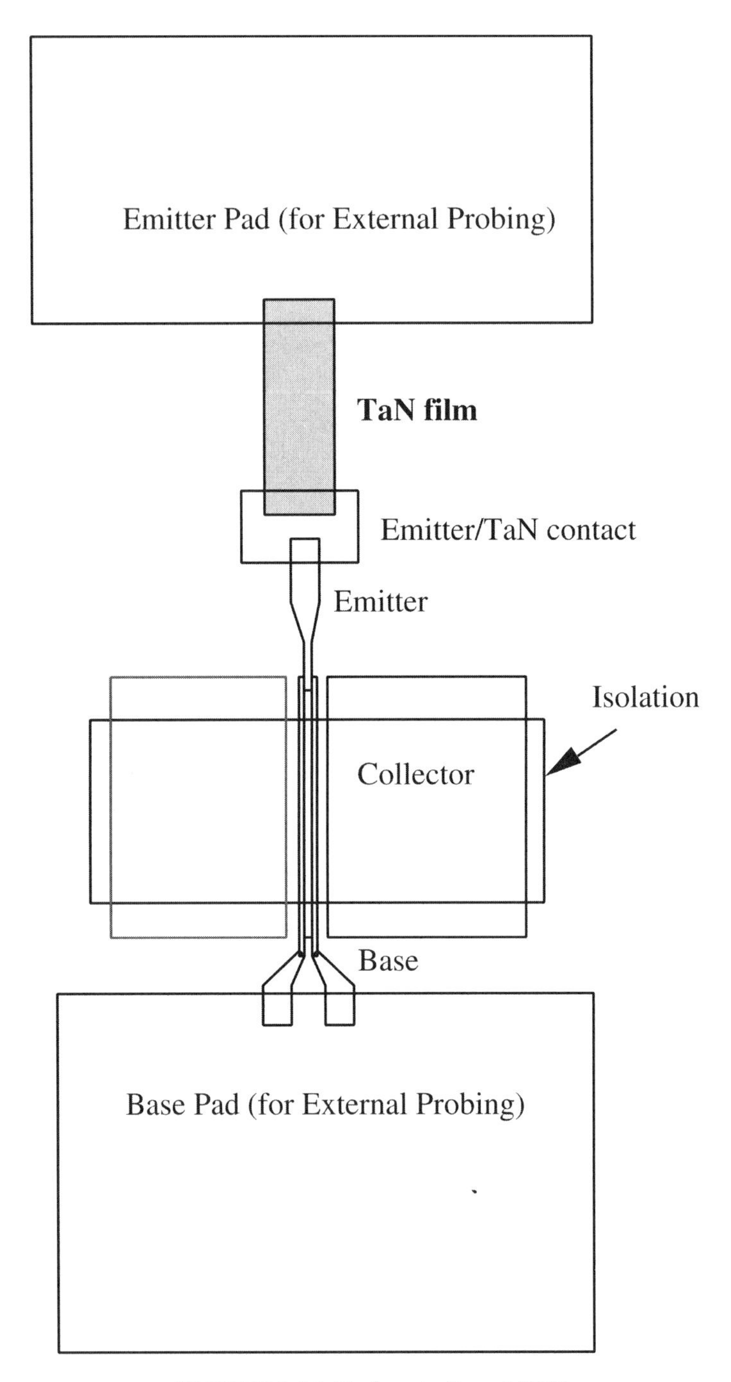

FIGURE 3-26. Emitter-ballasted HBT.

REFERENCE

1. Liu, W. (1988). *Handbook of III-V Heterojunction Bipolar Transistors.* New York: Wiley & Sons. pp. 223–224.

PROBLEMS

1. Consider an abrupt $Al_{0.35}Ga_{0.65}As$/GaAs HBT whose emitter area is so large that the surface recombination currents are negligible. Assume that the space-charge recombination current is negligible compared to I_{Bp} and $I_{B,\text{bulk}}$. The base-emitter heterojunction is abrupt. The emitter doping level is 2×10^{17} cm^{-3}, the emitter thickness is 2000 Å, and the base thickness is 800 Å.
   - **a.** Plot the current gain as a function of temperature when the base doping level is 5×10^{18} cm^{-3}. Neglect the temperature dependence of $I_{B,\text{bulk}}$ and assume it to be constant at its room-temperature value. Use the material parameters of § 1-7.
   - **b.** Plot the current gain as a function of temperature when the base doping level is 1×10^{20} cm^{-3}. Again neglect the temperature dependence of $I_{B,\text{bulk}}$ and assume it to be constant at its room-temperature value.
2. Rework Problem 1, except that the base–emitter heterojunction is graded.
3. A 10×30 μm^2 HBT has a current gain of 30 and a 2×30 μm^2 HBT has a current gain of 10. Estimate the current gain for a HBT whose emitter area is 5×30 μm^2. The current gains are measured at room temperature so that I_{Bp} is negligible. They are measured at high current levels such that $I_{B,\text{scr}}$ is negligible.
4. Consider a HBT in which only $I_{B,\text{scr}}$ and $I_{B,\text{bulk}}$ are of importance. Suppose $I_{B,\text{scr}} = I_{B,\text{bulk}}$ at one particular applied bias V_a. If somehow a base electric field of 2×10^4 V/cm is now established across the 1000 Å base, find the percentage of the change in current gain.
5. A graduate student claims that he has discovered a GaAs surface passivation technique. He measures two HBTs, one with an area of 100×100 μm^2, and another 200×200 μm^2. The base doping is 5×10^{18} cm^{-3} and the base thickness is 1000 Å. He finds the difference in the current gains of the two transistors is only 1%. Is his claim about his passivation technique valid?
6. There are many expressions for the transistor current gain when I_B is dominated by $I_{B,\text{bulk}}$. Start with the differential equation given by Eq. (3-18). Write the general solution in the form of $A\sinh(x/L_n) + B\cosh(x/L_n)$.
   - **a.** Apply the boundary conditions given in Eqs. (3-16) and (3-17). Solve for A and B and show that

$$n(x) = \frac{n(0)}{\sinh(X_B/L_n)} \sinh\left(\frac{X_B - x}{L_n}\right)$$

   - **b.** Show that the above solution is identical to that given in Eq. (3-20).

c. Show that Q_B is given by the following:

$$Q_B = qA_E \int_0^{X_B} n(x)\,dx = qA_E n(0)\, L_n \left[\cosh\left(\frac{X_B}{L_n}\right) - 1\right] \frac{1}{\sinh(X_B/L_n)}$$

d. Show that I_C is

$$I_C = \frac{qA_E D_n n(0)}{L_n} \frac{1}{\sinh(X_B/L_n)}$$

Using the fact that $\sinh(x) \approx x$ when x is small, show that this expression is simplified to Eq. (3-22) if $X_B/L_n \ll 1$.

e. Show that the base transit time is given by

$$\tau_b = \frac{2L_n^2}{D_n} \sinh^2\left(\frac{X_B}{2L_n}\right)$$

Using the fact that $\sinh(x) \approx x$ when x is small, show that this expression leads to Eq. (3-23) when X_B/L_n is small, as expected.

f. Show that a general expression for the d.c. current gain (when $I_{B,\text{bulk}}$ dominates) is

$$\beta = \frac{1}{2\sinh^2(X_B/2L_n)}$$

7. Another expression of the current gain is derived with the following.

a. Apply the part **(a)** result of Problem 6. Show that the emitter current is given by

$$I_E = \frac{qA_E D_n n(0)}{L_n} \coth\left(\frac{X_B}{L_n}\right)$$

b. Use the collector current expression from Problem 6. Show that $I_B = I_E - I_C$ is given by

$$I_B = \frac{qA_E D_n n(0)}{L_n} \sinh^2\left(\frac{X_B}{2L_n}\right) \frac{2}{\sinh(X_B/L_n)}$$

c. Divide I_C given in Problem 6 by I_B obtained just now. Show that the current gain is the same as that given in Problem 6.

d. Assume that the bulk recombination current dominates the base current of a HBT whose base doping is $3 \times 10^{19}\ \text{cm}^{-3}$ and 800 Å thick. What is the current gain?

8. Perform a PISCES simulation (or any other two-dimensional semiconductor device equation solver) for an $Al_{0.35}Ga_{0.65}As/GaAs$ abrupt HBT with $N_E = 5 \times 10^{17}\,cm^{-3}$, $N_B = 5 \times 10^{17}\,cm^{-3}$, and $N_C = 1 \times 10^{16}\,cm^{-3}$. If the simulation program cannot work on heterojunction devices, then work on a GaAs homojunction BJT with $N_E = 5 \times 10^{18}\,cm^{-3}$, $N_B = 5 \times 10^{17}\,cm^{-3}$. X_E, X_B, and X_C are 2000 Å, 1000 Å, and 5000 Å, respectively. The transistor has an emitter width of 2 μm. The spacing between the base contact and emitter contact is 2 μm, and the base contact width is 2 μm. The extrinsic base surface is unpassivated, with a surface recombination velocity of 1×10^6 cm/s.

a. Set up a grid and determine the current gain.

b. Divide the grid spacing by 2. Simulate the current gain. Is the current gain a function of the grid spacing?

9. Modify the following PISCES input deck to obtain a Gummel plot. What is the transistor structure specified in the file, such as the layer thickness and doping level? Is it a HBT?

```
Mesh     rectangular nx=52 ny=38
x.m      n=1 l=0 r=1
x.m      n=3 l=0.2 r=1
x.m      n=15 l=1.0 r=0.8
x.m      n=18 l=1.2 r=1.25
x.m      n=28 l=1.8 r=1
x.m      n=35 l=2.0 r=0.8
x.m      n=42 l=2.2 r=1.25
x.m      n=52 l=2.5 r=1
y.m      n=1 l=-0.2 r=1
y.m      n=9 l=0.0 r=1
y.m      n=13 l=0.05 r=1.25
y.m      n=17 l=0.1 r=0.8
y.m      n=24 l=0.3 r=1.25
y.m      n=31 l=0.5 r=0.8
y.m      n=38 l=0.9 r=1.25
region   num=1 ix.lo=35 ix.hi=52 iy.lo=1 iy.hi=9 gaas
region   num=2 ix.lo=15 ix.hi=52 iy.lo=9 iy.hi=17 gaas
region   num=3 ix.lo=15 ix.hi=52 iy.lo=17 iy.hi=31 gaas
region   num=4 ix.lo=1 ix.hi=52 iy.lo=31 iy.hi=38 gaas
region   num=5 ix.lo=1 ix.hi=35 iy.lo=1 iy.hi=9 nitride
region   num=6 ix.lo=1 ix.hi=15 iy.lo=9 iy.hi=31 nitride
elec     num=1 ix.lo=35 ix.hi=52 iy.lo=1 iy.hi=1
elec     num=2 ix.lo=15 ix.hi=34 iy.lo=9 iy.hi=9
elec     num=3 ix.lo=1 ix.hi=14 iy.lo=31 iy.hi=31
dop      reg=1 unif conc=5e19 n.type
dop      reg=2 unif conc=5e18 p.type
```

```
dop         reg=3 unif conc=5e16 n.type
dop         reg=4 unif conc=2e18 n.type
symb        newton carriers=0
solve       init
symb        newton carriers=2
solve       v3=0.0
solve       v3=0.2
solve       v3=0.5
solve       v3=1.0
solve       v3=2.0
solve       v3=3.0
solve       v3=4.0
solve       electrode=2 v2=0.0 vstep=0.1 nstep=15
solve       v2=1.525
solve       v2=1.55
solve       v2=1.575
solve       v2=1.6 save.bia
plot.2d     boundary
contour     flowline
end
```

10. Determine c if $c\tan(c)$ given in Eq. (3-49) is equal to 0.3. The unknown c is between 0 and $\pi/2$.

a. First, calculate $c\tan(c)$ assuming that $c = 0$ and $\pi/2$. It is fairly easy to find that $c\tan(c) = 0$ when $c = 0$ and $c\tan(c) = \infty$ when $c = \pi/2$.

b. Find $c\tan(c)$ when $c = \pi/4$. Verify that it is equal to 0.785.

c. Since 0.3 is below 0.785, we now know that the answer for c resides between 0 and $\pi/4$. Guess the next value to be $\pi/8$. What is the result of $c\tan(c)$ with this guessed value?

d. Let us suppose that the result calculated with the guess value is x, as calculated in **(c)**. Check to see if $x < 0.3$ or $x > 0.3$. If $x > 0.3$, use $c = \pi/16$ to be the next guessed value. If $x < 0.3$, use $c = 3\pi/16$ to be the next guessed value.

e. Repeat this process three more times. Does the guessed value of c now fairly satisfy the relationship $c\tan(c) = 0.3$?

11. A $2 \times 50\ \mu m^2$ HBT has a base thickness of 800 Å and a base doping of $3 \times 10^{19}\ cm^{-3}$. Assume that its current gain is limited by the base bulk recombination. The operating base current is 0.91 mA.

a. What is the base current distribution inside the emitter mesa?

b. What is the base current at 0.25 μm from one edge of the emitter mesa?

c. What is the effective emitter width?

12. Find the d.c. intrinsic base resistance of the HBT of Problem 11, both with and without consideration of emitter crowding.

13. Calculate the depletion thickness in the collector for a HBT whose collector doping is 1×10^{16} cm^{-3} and whose collector thickness is 1 μm. Assume that ϕ_{CB} is 1 V and $V_{CB} = 5$ V. The operating current density is 50% of J_{fd}, the current density at which the collector is fully depleted. Use material parameters of § 1-7. Calculate J_2.

14. Determine the electric field at the depletion edge in the collector (toward the base side rather than toward the subcollector side), at collector current densities of $J_C = 0$ and $J_C = J_{\text{Kirk}}$. The collector doping is 4×10^{16} cm^{-3} and the collector thickness is 4000 Å. Assume that ϕ_{CB} is 1 V and $V_{CB} = 2$ V.

15. A graded $Al_{0.35}Ga_{0.65}As$/GaAs HBT has the following parameters: $N_E = 5 \times 10^{17}$ cm^{-3}, $X_E = 1000$ Å, $N_B = 3 \times 10^{19}$ cm^{-3}, and $X_B = 800$ Å. Consider only the base bulk recombination and the back-injection current. The collector current is 20 mA.

a. Find α_T and Γ. Assuming that $M = 1$, find the base currents when the transistor is biased in the common-emitter configuration and in the common-base configuration.

b. Suppose now that $M = 1.01$. What are the current gains if the transistor is biased in the common-emitter configuration and in the common-base configuration?

16. The fingers of a two-finger HBT are so far apart that the mutual thermal coupling between them is negligible. Each finger has an area of 2×20 μm^2, and that the thermal resistance for a 2×50 μm^2 finger is 630°C/W.

a. Using the area ratio, estimate the thermal resistance of each of the fingers. What is the overall thermal resistance of the two-finger device?

b. At $V_{CE} = 10$ V, the transistor enters the collapse region of operation. The two finger currents are measured to be 10 mA and 1 mA. Determine the emitter resistance per finger. Assume that $\eta = 1$, $T_A = 300$ K, and $\phi = 1.2$ mV/°C.

CHAPTER 4

HBT HIGH-FREQUENCY PROPERTIES

§ 4-1 INTRINSIC COMMON-BASE y-PARAMETERS

The high-frequency of a HBT, even in the common-emitter configuration, can be expressed in its common-base y-parameters. The derivation of f_T and f_{max} depends largely on the availability of these y-parameters. Figure 4-1 illustrates a HBT in a small-signal operation, biased in the common-base configuration. The small-signal voltages v_{be} and v_{cb} are superimposed on the d.c. biases V_{BE} and V_{CB}, respectively. The overall applied voltages are $v_{BE} = V_{BE} + v_{be}$ and $v_{CB} = V_{CB} + v_{cb}$. The resulting emitter and collector currents are $I_E + i_e$, and $I_C + i_c$, respectively, where I_E and I_C are the values corresponding to the d.c. operating point when the applied voltages are V_{BE} and V_{CB}. In general, the overall emitter and collector currents are some functions of the base–emitter and base–collector voltages:

$$i_E = I_E + i_e = f_1(v_{BE}, v_{CB}) \tag{4-1}$$

$$i_C = I_C + i_c = f_2(v_{BE}, v_{CB}) \tag{4-2}$$

From Taylor's expansions of the functions about their d.c. biasing points, we obtain:

$$i_E \approx f_1(V_{BE}, V_{CB}) + \left.\frac{\partial f_1}{\partial v_{BE}}\right|_{V_{CB}} v_{be} + \left.\frac{\partial f_1}{\partial v_{CB}}\right|_{V_{BE}} v_{cb} \tag{4-3}$$

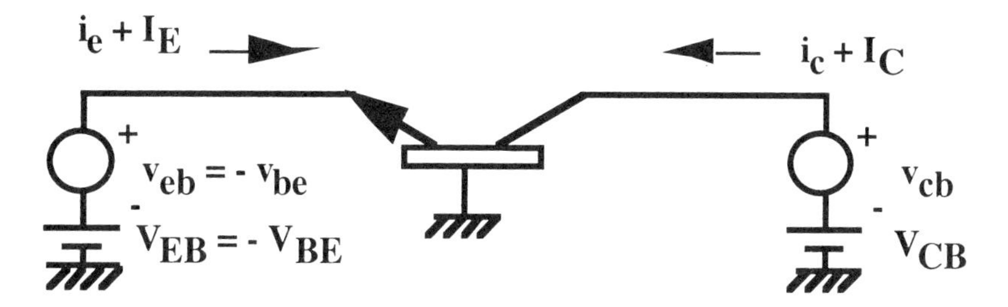

FIGURE 4-1. Voltage and current symbols used to describe a HBT in the small-signal operation.

$$i_C \approx f_2(V_{BE}, V_{CB}) + \left.\frac{\partial f_2}{\partial v_{BE}}\right|_{V_{CB}} v_{be} + \left.\frac{\partial f_2}{\partial V_{CB}}\right|_{V_{BE}} v_{cb} \tag{4-4}$$

Since $I_E = f_1(V_{BE}, V_{CB})$ and $I_C = f_2(V_{BE}, V_{CB})$, we have:

$$\begin{bmatrix} i_e \\ i_c \end{bmatrix} = \begin{bmatrix} y_{ee} & y_{ec} \\ y_{ce} & y_{cc} \end{bmatrix} \begin{bmatrix} v_{be} \\ v_{cb} \end{bmatrix} = \begin{bmatrix} \left.\frac{\partial i_E}{\partial v_{BE}}\right|_{V_{CB}} & \left.\frac{\partial i_E}{\partial v_{CB}}\right|_{V_{BE}} \\ \left.\frac{\partial i_C}{\partial v_{BE}}\right|_{V_{CB}} & \left.\frac{\partial i_C}{\partial v_{CB}}\right|_{V_{BE}} \end{bmatrix} \begin{bmatrix} v_{be} \\ v_{cb} \end{bmatrix} \tag{4-5}$$

Alternatively, we know that, if the input small-signal voltages vary sinusoidally at a given frequency, then the small-signal variation in the device parameters are also sinusoidal at the same frequency. Sometimes, it is convenient to specifically leave out the time variation and express the small-signal parameter as a product of its "phasor" and the time dependence. For example, for the small-signal biasing voltages, we can express

$$v_{be} = \widetilde{v_{be}} e^{j\omega t} \tag{4-6}$$

$$v_{cb} = \widetilde{v_{cb}} e^{j\omega t} \tag{4-7}$$

where ω is the radian frequency of the input sinusoidal excitation. The phasors $\widetilde{v_{be}}$ and $\widetilde{v_{cb}}$ contain the phase and magnitude information of v_{be} and v_{cb}, respectively, but neglects the sinusoidal time dependence. In a similar fashion, the resulting small-signal variations in the currents are expressed as:

$$i_e = \tilde{i}_e e^{j\omega t} \tag{4-8}$$

$$i_c = \tilde{i}_c e^{j\omega t} \tag{4-9}$$

Therefore, the y-parameters of Eq. (4-5) can be expressed as:

$$\begin{bmatrix} \tilde{i}_e \\ \tilde{i}_c \end{bmatrix} = \begin{bmatrix} y_{ee} & y_{ec} \\ y_{ce} & y_{cc} \end{bmatrix} \begin{bmatrix} \widetilde{v_{be}} \\ \widetilde{v_{cb}} \end{bmatrix} = \begin{bmatrix} \left.\dfrac{\partial i_E}{\partial v_{BE}}\right|_{V_{CB}} & \left.\dfrac{\partial i_E}{\partial v_{CB}}\right|_{V_{BE}} \\ \left.\dfrac{\partial i_C}{\partial v_{BE}}\right|_{V_{CB}} & \left.\dfrac{\partial i_C}{\partial v_{CB}}\right|_{V_{BE}} \end{bmatrix} \begin{bmatrix} \widetilde{v_{be}} \\ \widetilde{v_{cb}} \end{bmatrix} \tag{4-10}$$

Our goal is to express the y-parameters in terms of physical parameters, such as diffusion coefficients, doping levels, and layer thicknesses. Figure 4-2, which will be used for the development of the y-parameters, shows the schematic diagram of the small-signal current flows in a normal operation.

y_{ee} y_{ee} is equal to the partial derivative of i_e (or $\tilde{i}_e$) with respect to v_{be} (or $\widetilde{v_{be}}$) when V_{CB} is kept constant. To determine its functional form, we first need to find an expression for the emitter current as a function of the applied biases. When v_{be} is applied on top of V_{BE}, two components of small-signal currents giving rise to the overall i_e result. The first component is the capacitive current flowing through the base–emitter junction capacitance, and the second component is the incremental current due to a change in the minority carrier concentration in the base. The overall i_e is expressed as:

$$i_e = i_{e1} + i_{e2} \tag{4-11}$$

We analyze i_{e1} first. When v_{be} is applied on top of V_{BE}, the depletion thickness of the base–emitter junction decreases or increases depending on whether the magnitude of v_{be} is positive or negative, respectively. The situation when v_{be} is positive is schematically shown in Fig. 4-3, which applies to a p–n junction or a heterojunction with arbitrary doping profiles. In the case of heterojunction, we neglect the difference between the dielectric constants between the emitter and the base materials. Suppose that the magnitude of v_{be} is Δv_{be}. The increase in the

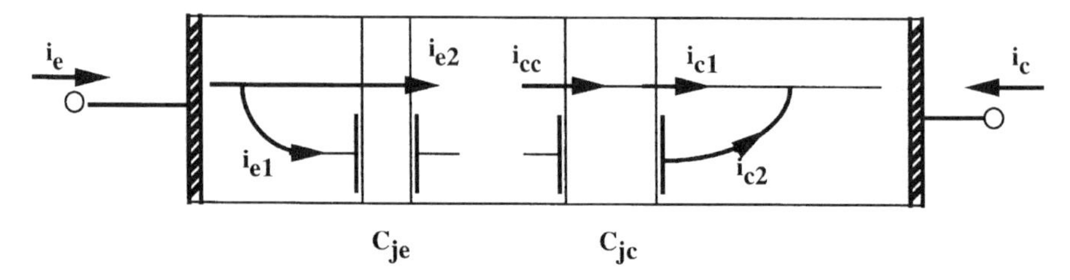

FIGURE 4-2. Various components of the small-signal emitter and collector currents in a common-base HBT.

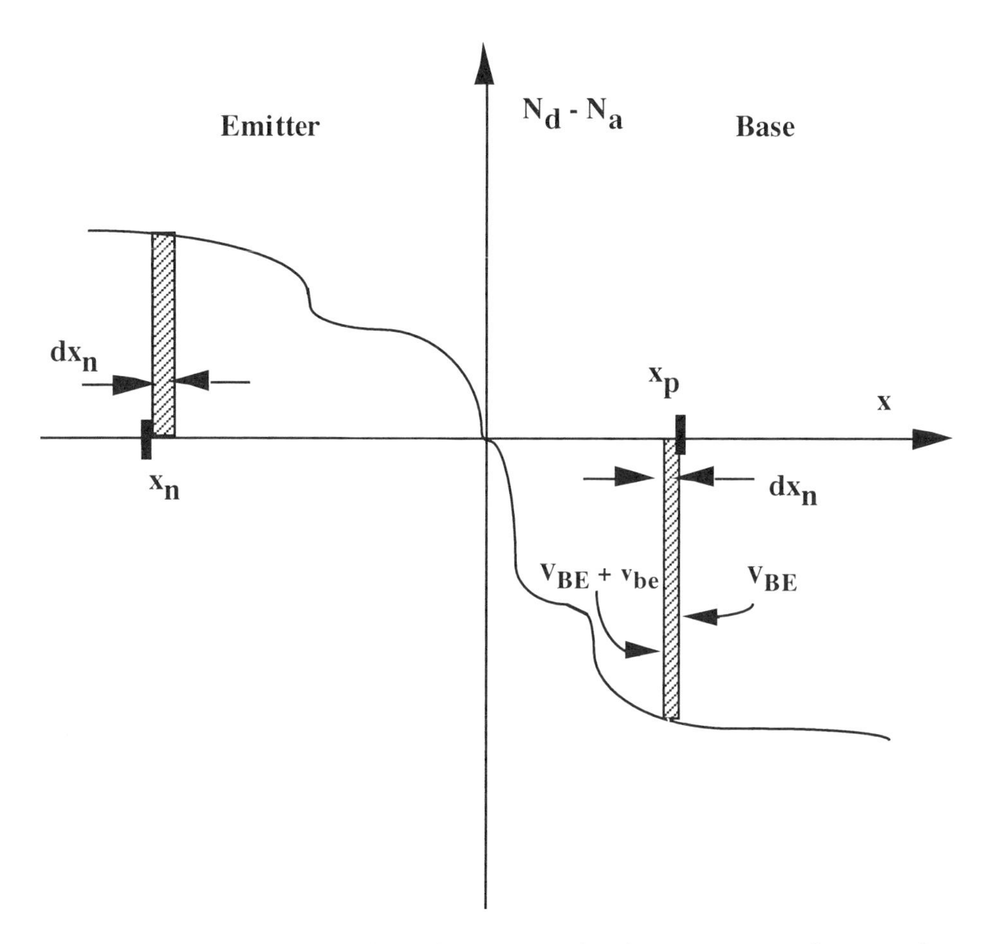

FIGURE 4-3. Change of mobile carrier concentrations in response to an incremental voltage applied to a p-n junction.

bias voltage results in an increase in the electric field (see Fig. 4-4), which, from the Poisson equation, is equal to:

$$\Delta \varepsilon = \frac{q}{\epsilon_s} N_A(X_p) \Delta x \tag{4-12}$$

From Fig. 4-4, which plots the electric field variation, the change in electric field is related to the change in the bias voltage by:

$$\Delta v_{be} = X_{\text{dep}E} \Delta \varepsilon \tag{4-13}$$

where $X_{\text{dep}E}$, the full depletion thickness of the base–emitter junction, is equal to the depletion thickness at the base side (X_p) plus the depletion thickness at the

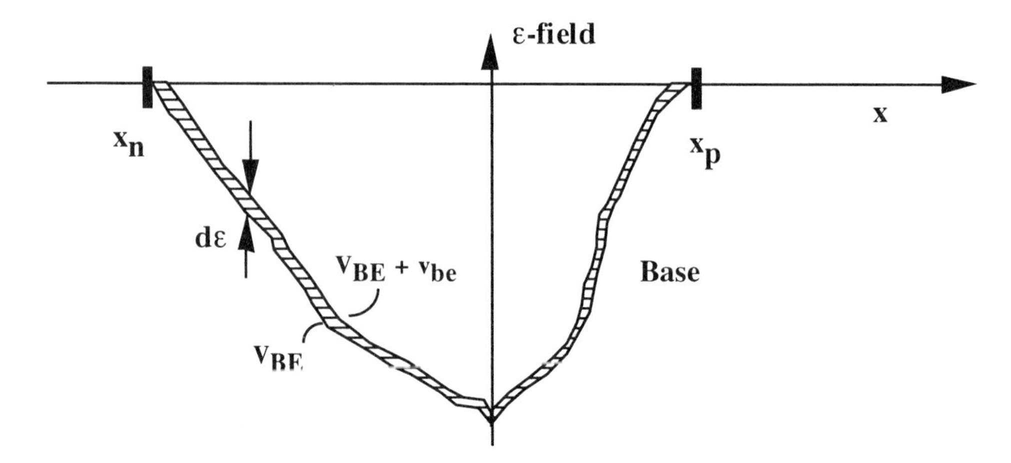

FIGURE 4-4. Change of electric field due to the change of carrier concentration depicted in Fig. 4-3.

emitter side (X_N). (X_{dep}, without the subscript E, is used exclusively to mean the base–collector junction depletion thickness.) In the base–emitter heterojunction of a HBT whose base is much more heavily doped than the emitter, $X_p \to 0$ and $X_{\text{dep}E} \sim X_N$. Similar to the derivation of the depletion thickness in the collector in §3-8, $X_{\text{dep}E}$ is given as:

$$X_{\text{dep}E} = \sqrt{\frac{2\epsilon_s}{qN_E}(\phi_{BE} - V_{BE})} \tag{4-14}$$

where ϕ_{BE} is the built-in potential of the base–emitter junction. For a homojunction, ϕ_{BE} is similarly obtained from the formula given for ϕ_{CB} in Eq. (3.71). For a N–p AlGaAs/GaAs heterojunction, ϕ_{BE} is approximately equal to the energy gap in the AlGaAs material subtracted from the valence band gap discontinuity between the AlGaAs/GaAs heterojunction.

$$\phi_{BE} = E_g|_{AlGaAs} - \Delta E_v|_{AlGaAs/GaAs} \qquad (N\text{-}p \text{ junction}) \tag{4-15}$$

This formulation assumes that the Fermi level is on the conduction band edge in the N-type AlGaAs and on the valence band edge in the p-type GaAs. Such an assumption is valid in typical HBTs whose emitter and base doping levels either exceed or roughly equal the conduction band and the valence band densities of states, respectively. ($N_E \sim 5 \times 10^{17}$ cm^{-3}; $N_B \sim 3 \times 10^{19}$ cm^{-3}, and $N_c \sim 5 \times 10^{17}$ cm^{-3}; $N_v \sim 5 \times 10^{18}$ cm^{-3}.)

Because of the positive magnitude of Δv_{be}, the depletion thickness decreases, meaning that that there is a net hole current flow to compensate the exposed negative impurity charges. Suppose that the amount of hole charges is ΔQ. Then, with Eqs. (4-12) to (4-14), we have:

$$\Delta Q = qA_E N_A(X_p)\Delta x = A_E \frac{\epsilon_s}{X_{\text{dep}E}} \Delta v_{be} \tag{4-16}$$

We express in phasor form the small-signal sinusoidal charge variation due to a sinusoidal small-signal $\widetilde{v_{be}}$:

$$\tilde{Q}e^{j\omega t} = A_E \frac{\epsilon_s}{X_{\text{dep}E}} \widetilde{v_{be}} e^{j\omega t} \tag{4-17}$$

If we define C_{je} (the base–emitter junction capacitance) as:

$$C_{je} = A_E \frac{\epsilon_s}{X_{\text{dep}E}} \tag{4-18}$$

then the small-signal current i_{e1}, which is equal to the time derivative of the small-signal charge variation of Q, is:

$$i_{e1} = \frac{d\tilde{Q}e^{j\omega t}}{dt} = j\omega C_{je} \widetilde{v_{be}} e^{j\omega t} \tag{4-19}$$

Equation (4-19) concludes our determination of i_{e1}. It states that one component of emitter current is the product of $j\omega C_{je}$ and v_{be}. This derivation, particularly the depletion thickness equation (Eq. 4-14) and those afterward, used a so-called *depletion approximation.* The approximation assumes that the mobile carrier concentrations are significantly smaller than the doping levels at either side of the junction. This assumption is always fulfilled in a reverse-biased junction since the junction current is small. However, it becomes increasingly erroneous as the junction is heavily forward-biased, with a large magnitude of current passing through. Because the base–emitter junction is a forward-biased junction, C_{je} expressed in Eq. (4-18) serves only as an approximation. More exact expressions accounting for the presence of mobile carriers are more complicated.

The second small-signal emitter current component results from the perturbation of base minority carrier concentration due to v_{be}. If we limit ourselves to one dimension and neglect the base quasi-electric field, then the time-varying current continuity equation is given by Eq. (1-46):

$$\frac{\partial^2 n(x,t)}{\partial x^2} - \frac{n}{D_n \tau_n} = \frac{1}{D_n}\frac{\partial n}{\partial t} \tag{4-20}$$

The boundary conditions at the two ends of the base layer are:

$$n(x, t)|_{x=0} = \frac{n_i^2}{N_B} \exp\left[\frac{q}{kT}(V_{BE} + v_{be})\right] \tag{4-21}$$

$$n(x, t)|_{x=X_B} = 0 \tag{4-22}$$

Equation (4-21) is an expression of the junction law discussed in Eq. (2-113). Because the small-signal v_{be} is much smaller than the d.c. bias V_{BE}, $n(x)$ at the emitter side is approximated as:

$$n(x, t)|_{x=0} \approx \frac{n_i^2}{N_B} \exp\left(\frac{q}{kT} V_{BE}\right) + \frac{n_i^2}{N_B} \exp\left(\frac{q}{kT} V_{BE}\right) \cdot \frac{q\widetilde{v_{be}} e^{j\omega t}}{kT} \tag{4-23}$$

The second boundary condition, Eq. (4-22), follows from the fact that V_{CB} is applied to reverse-bias the base–collector junction. Since the magnitude of v_{cb} is small compared to V_{CB}, the carrier concentration remains at zero without showing any time dependence.

The solution of $n(x, t)$ is simplified by expressing $n(x, t)$ as a sum of a d.c. component (n_{dc}) and an a.c. component (n_{ac}):

$$n(x, t) = n_{dc}(x) + n_{ac}(x, t) = n_{dc}(x) + \tilde{n}(x)\, e^{j\omega t} \tag{4-24}$$

Plugging Eq. (4-24) into the differential equation Eq. (4-20), as well as the two boundary conditions (Eqs. 4-21 and 4-22), we find two differential equations, each with its own boundary conditions:

$$\frac{d^2 n_{dc}(x)}{dx^2} - \frac{n_{dc}}{D_n \tau_n} = 0 \qquad n_{dc}(0) = \frac{n_i^2}{N_B} \exp\left(\frac{q}{kT} V_{BE}\right) \qquad n_{dc}(X_B) = 0 \tag{4-25}$$

$$\frac{\partial^2 \tilde{n}}{\partial x^2} - \left(\frac{1}{D_n \tau_n} + \frac{j\omega}{D_n}\right)\tilde{n} = 0 \qquad \tilde{n}(0) = \frac{n_i^2}{N_B} \exp\left(\frac{q}{kT} V_{BE}\right) \cdot \frac{q\widetilde{v_{be}}}{kT} \qquad \tilde{n}(X_B) = 0 \tag{4-26}$$

We are interested in solving for the small-signal emitter current due to v_{be}. From this standpoint, we might be tempted to solve Eq. (4-26) only; however, it will prove useful to solve for the d.c. carrier concentration as well. The solution of Eq. (4-25) is simplified if we define:

$$\zeta_{dc} = \sqrt{\frac{1}{D_n \tau_n}} = \frac{1}{L_n} \tag{4-27}$$

L_n is the diffusion length. It is a characteristic length of amount of minority carrier diffusion prior to recombination. In a bipolar transistor design, the base thickness should be much thinner than L_n, so that the bulk recombination does not significantly reduce the transistor current gain. As an example, the typical values of D_n and τ_n for an HBT are 25 cm^2/V-s and 1 ns, respectively. L_n for this transistor is thus 1.58 μm, which is about 10 times the typical base thickness of 1000 Å. With the definition of ζ_{dc}, the linear differential equation in Eq. (4-25) is solved with a straightforward integration. The solution satisfying both boundary conditions is:

$$n_{\rm dc}(x) = \frac{n_i^2}{N_B} \exp\left(\frac{q}{kT} V_{BE}\right) \cdot \frac{\sinh \zeta_{\rm dc}(X_B - x)}{\sinh \zeta_{\rm dc} X_B} \tag{4-28}$$

We just mentioned that $L_n \gg X_B$. Alternatively, $\zeta_{\rm dc} X_B$ approaches zero. Since $\sinh(x) \approx x$ and $\cosh(x) \approx 1$ when x is small, we simplify Eq. (4-28) as:

$$n_{\rm dc}(x) \approx \frac{n_i^2}{N_B} \exp\left(\frac{q}{kT} V_{BE}\right) \cdot (1 - x) \tag{4-29}$$

That is, the carrier concentration decreases linearly from its maximum value at the emitter edge to zero at the collector edge of the base. This fact has been used in the discussion of § 3-2. The magnitude of the emitter current entering at $x = 0$ is:

$$I_E = -qD_n A_E \left.\frac{dn_{\rm dc}}{dx}\right|_{x=0} = qD_n A_E \frac{n_i^2}{N_B} \exp\left(\frac{q}{kT} V_{BE}\right) \tag{4-30}$$

We now seek the small-signal carrier concentration by solving Eq. (4-26). In this case, it is convenient to define a variable $\zeta_{\rm ac}$ similar to $\zeta_{\rm dc}$:

$$\zeta_{\rm ac} = \sqrt{\frac{1 + j\omega\tau_n}{D_n \tau_n}} \tag{4-31}$$

Applying the boundary conditions, we obtain $\tilde{n}$ as:

$$\tilde{n}(x) = \frac{q\widetilde{v_{be}}}{kT} \cdot \frac{n_i^2}{N_B} \exp\left(\frac{q}{kT} V_{BE}\right) \cdot \frac{\sinh \zeta_{\rm ac}(X_B - x)}{\sinh \zeta_{\rm ac} X_B} \tag{4-32}$$

Although $\zeta_{\rm dc} \cdot X_B$ is small and the hyperbolic sine function appearing in the $n_{\rm dc}(x)$ expression is simplified into a linear function, $\zeta_{\rm ac} \cdot X_B$ may or may not be small. As shown in Eq. (4-31), $\zeta_{\rm ac}$ has a frequency dependence in the numerator. Because we are at times concerned with the transistor's high-frequency response (such as at frequencies near f_T), we do not assume $\zeta_{\rm ac} \cdot X_B$ to be small, and $\tilde{n}$ of

Eq. (4-32) cannot be simplified in a manner similar to that done in the d.c. case. The magnitude of the small-signal emitter current is equal to:

$$\widetilde{i_{e2}} = -qD_nA_E\frac{d\tilde{n}}{dx}\bigg|_{x=0} = qD_nA_E\frac{q\widetilde{v_{be}}}{kT}\cdot\frac{n_i^2}{N_B}\exp\left(\frac{q}{kT}V_{BE}\right)\zeta_{ac}\coth(\zeta_{ac}X_B) \tag{4-33}$$

Applying the definition of Eq. (4-30), we then have:

$$\tilde{i}_{e2} = I_E\frac{q\widetilde{v_{be}}}{kT}\zeta_{ac}X_B\coth(\zeta_{ac}X_B) \tag{4-34}$$

With Eqs. (4-19) and (4-34), the small-signal y_{ee} parameter is:

$$y_{ee} = \frac{\tilde{i}_e}{\widetilde{v_{be}}}\bigg|_{\widetilde{v_{cb}}=0} = g_e\zeta_{ac}X_B\coth(\zeta_{ac}X_B) + j\omega C_{je} \tag{4-35}$$

where g_e is the emitter conductance given by:

$$g_e = \frac{qI_E}{\eta kT} \tag{4-36}$$

We have been assuming that the emitter current ideality factor η is unity.

In an attempt to simplify Eq. (4-35) we define a parameter, ω_0, equal to the inverse of the base transit time:

$$\omega_0 = \frac{2D_n}{X_B^2} \tag{4-37}$$

Because X_B/L_n is small, we approximate $\xi_{ac}\cdot X_B$ as:

$$\xi_{ac}X_B = \left[\left(\frac{X_B}{L_n}\right)^2 + j\frac{2\omega}{\omega_0}\right]^{1/2} \approx \left(j\frac{2\omega}{\omega_0}\right)^{1/2} \tag{4-38}$$

Consequently, y_{ee} of Eq. (4-35) is approximated as:

$$y_{ee} = g_e\left(j\frac{2\omega}{\omega_0}\right)^{1/2}\coth\left(j\frac{2\omega}{\omega_0}\right)^{1/2} + j\omega C_{je} \tag{4-39}$$

The first term at the right-hand side of Eq. (4-39), after being normalized with g_e, is plotted in Fig. 4-5. Both the real and imaginary parts are shown. It is seen that

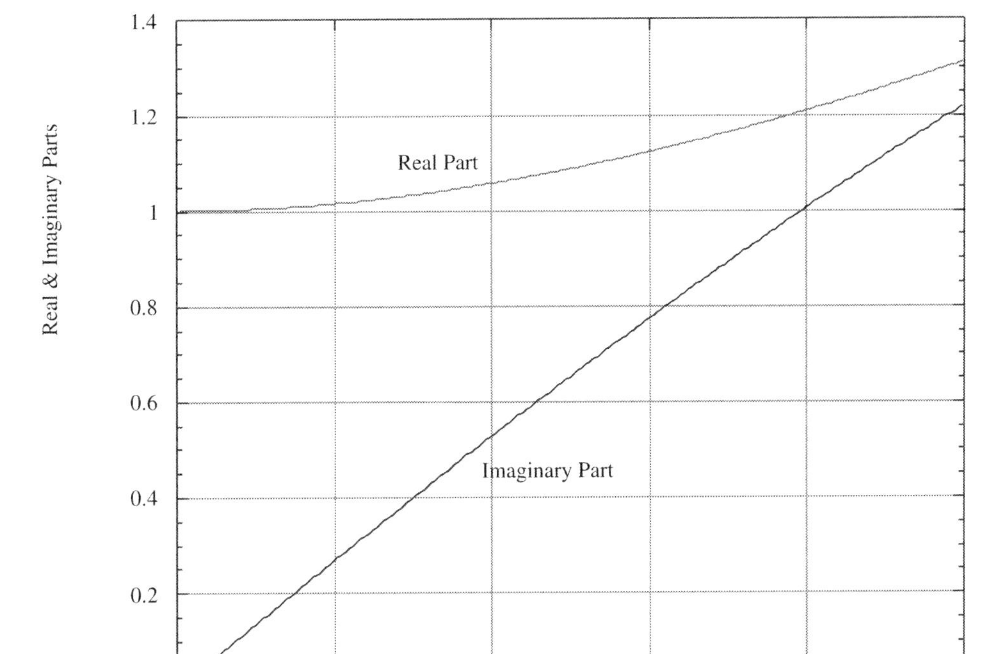

FIGURE 4-5. Real and imaginary parts of y_{ee}/g_e given in Eq. (4-39).

the real part is nearly 1, whereas the imaginary part is approximated by 2/3. (ω/ω_0) at low frequencies. Therefore, y_{ee} can be written as:

$$y_{ee} = g_e\left(1 + j\frac{2\omega}{3\omega_0}\right) + j\omega C_{je} \tag{4-40}$$

A different, but equivalent, expression of y_{ee} is:

$$y_{ee} = g_e + j\omega(C_D + C_{je}) \tag{4-41}$$

where C_D, the diffusion capacitance, is given by:

$$C_D = g_e\frac{X_B^2}{3D_n} \tag{4-42}$$

Because C_D of Eq. (4-42) was derived form Eq. (4-40), which is a low-frequency approximation, the expression for C_D applies to low frequencies. At frequencies approaching f_T, the exact equation of y_{ee} (Eq. 4-39) cannot be simplified into a parallel RC-lumped representation as in Eq. (4-40). Therefore, the term diffusion capacitance is no longer meaningful. Sometimes, without solving for the minority carrier concentration explicitly, the diffusion capacitance at low frequencies is obtained in the following manner. The minority carrier profile is first approximated to be a straight line, as shown in Fig 4-6. Point A is the carrier concentration due to the d.c. bias, V_{BE}, and point B is the carrier concentration due to both the d.c. bias and the small-signal whose magnitude is Δv_{be}. The shaded region, therefore, is the amount of charges due to the presence of Δv_{be}. The incremental charge due to Δv_{be} is the area of the shaded triangle:

$$\Delta Q = \frac{qA_E X_B}{2} \Delta n \tag{4-43}$$

Δn is inferred from the second term of Eq. (4-23). Therefore, Eq. (4-43) reduces to:

$$\Delta Q = \frac{qA_E X_B}{2} \frac{n_i^2}{N_B} \exp\left(\frac{q}{kT} V_{BE}\right) \cdot \frac{q \Delta v_{be}}{kT} \tag{4-44}$$

The diffusion capacitance is the incremental charge over the incremental voltage:

$$C_D = \frac{\Delta Q}{\Delta v_{be}} = \frac{qI_E}{kT} \cdot \frac{X_B^2}{2D_n} = g_e \cdot \frac{X_B^2}{2D_n} \tag{4-45}$$

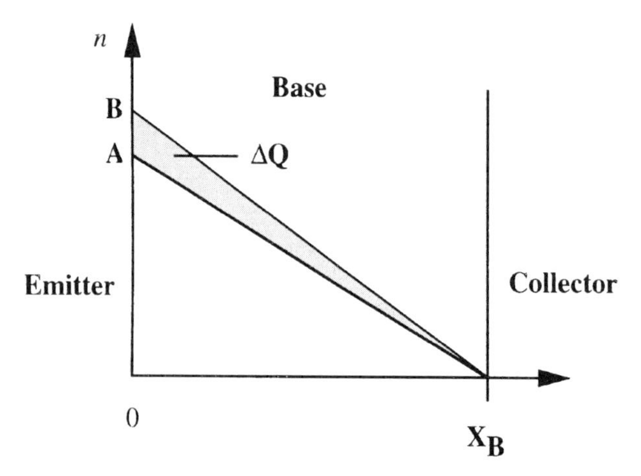

FIGURE 4-6. Incremental of stored base minority charge due to an incremental bias applied across the base–emitter junction.

This result differs from that of Eq. (4-42) by a factor of 3/2. The fundamental cause of the difference is subtle. Equation (4-42) was derived for a transistor biased in the common-base configuration whereas the definition of Eq. (4-45) implicitly assumes that the transistor is in the common-emitter configuration. This statement requires some explanation. When a transistor is in the common-emitter configuration, the input-port terminal current of interest is the base current. Consider first the static situation. The charges contained in the base are those within the triangle formed by point A and the base thickness. It is the base current, not the emitter current, which is proportional to the integral charge in the base. In the small-signal operation, the differential change of the integral charge in the base with respect to time likewise is the small-signal base current. The small-signal emitter current relates to the differential change of carrier density at the emitter edge of the base; it does not relate to the differential change of the integral charge in the base. After we finish deriving the expressions for all other common-base parameters, we show that in the more formal derivation (§4-2), the diffusion capacitance in the common-emitter configuration has the form of Eq. (4-45).

We mentioned that i_e in response to v_{be} consists of two components and y_{ee} was derived accordingly. Strictly speaking, this is an assumption that is correct only for HBTs whose emitter–injection efficiency approaches unity. For silicon BJTs, i_e includes a third component due to the carrier back-injection. We will not derive this component, but merely state the functional form of such emitter current component, which is significant in Si BJTs:

$$i_{e3} = (1 - \Gamma) I_E \cdot (1 + j\omega\tau_E)^{1/2} \frac{qv_{be}}{kT} \tag{4-46}$$

where τ_E is the recombination lifetime in the emitter. Γ is the d.c. emitter injection efficiency (Eq. 3-88). Therefore, for silicon bipolar junction transistors (Si BJTs), y_{ee} should be expressed as:

$$y_{ee} = g_e\left(1 + j\frac{2\omega}{3\omega_0}\right) + j\omega C_{je} + g_e(1 - \Gamma)\cdot(1 + j\omega\tau_E)^{1/2} \tag{4-47}$$

Example 4-1:

An abrupt $Al_{0.35}Ga_{0.65}As/GaAs$ HBT is biased in the common-base configuration. It has $N_B = 3 \times 10^{19}$ cm^{-3}, $X_B = 800$ Å, $N_E = 5 \times 10^{17}$ cm^{-3}. Assume that $D_n = 25$ cm^2/V-s in the base layer. It is operated at a current density of 10^3 A/cm^2, which corresponds to a bias voltage of 1.43 V. Its emitter area is 2×8 μm^2. Compare the magnitude of C_{je} and C_D. Assume that the ideality factor of the HBT is 1.23.

The built-in potential is found from Eq. (4-15):

$$\phi_{BE} = 1.424 + 1.247 \cdot 0.35 - 0.55 \cdot 0.35 = 1.668\ V$$

The current density is low enough such that the depletion approximation embodied in Eq. (4-14) is fair. So,

$$X_{\mathrm{dep}E} = \sqrt{\frac{2 \cdot 1.159 \times 10^{-12}}{1.6 \times 10^{-19} \cdot 5 \times 10^{17}}(1.688 - 1.43)} = 2.62 \times 10^{-6}\ \mathrm{cm}$$

The depletion capacitance of the base–emitter junction is calculated from Eq. (4-18):

$$C_{je} = 2 \cdot 8 \times 10^{-8} \frac{1.159 \times 10^{-12}}{2.62 \times 10^{-6}} = 7.1 \times 10^{-14}\ \mathrm{F}$$

Since the transistor is in the common-base configuration, the diffusion capacitance is that given by Eq. (4-42). The transconductance g_e is given by Eq. (4-36). We approximate the emitter current by the collector current, which is $J_C \cdot A_E = 1.6 \times 10^{-4}$ A. Hence,

$$g_e = \frac{1.6 \times 10^{-4}}{1.23 \cdot 0.0258} = 5.0 \times 10^{-3} \frac{1}{\Omega}$$

From Eq. (4-42)

$$C_D = 5.0 \times 10^{-3} \frac{(800 \times 10^{-8})^2}{3.25} = 4.3 \times 10^{-15}\ \mathrm{F}$$

Since g_e is proportional to I_C, it is possible that C_D approaches C_{je} at large biases.

y_{ce} y_{ce} is defined as the partial derivative of i_c with respect to v_{be} when V_{CB} is kept constant. From Fig. 4-2, which shows the major small-signal current flows, y_{ce} is:

$$y_{ce} = \left.\frac{\tilde{i}_c}{\tilde{v}_{be}}\right|_{v_{cb}=0} = \left.\frac{\tilde{i}_c}{\tilde{i}_{cc}} \cdot \frac{\tilde{i}_{cc}}{\tilde{v}_{be}}\right|_{\tilde{v}_{cb}=0} \tag{4-48}$$

From a cursory examination of Fig. 4-2, i_c and i_{cc} appear to be identical. It seems that, assuming negligible avalanche multiplication, whatever current flowing out of the base should equal the current flowing out of the collector. This simplistic thinking is correct at the static situation, but fails at high frequencies. It takes a finite amount of time for a carrier to travel through the collector, resulting in a phase delay between i_c and i_{cc}. We first determine the relationship between i_{cc} and the applied v_{be}. We then determine the ratio i_c to i_{cc} to complete the finding of y_{ce}.

Unlike i_e, which originates from two distinctive locations, i_{cc} comes purely from the modified carrier concentration inside the base. There is no associated capacitive current from the base–collector junction capacitance because the base–collector bias is maintained constant in the y_{ce} calculation. Therefore, the magnitude of the current is,

$$\widetilde{i_{cc}} = -qD_nA_E \left.\frac{d\tilde{n}}{dx}\right|_{x=X_B} \tag{4-49}$$

i_{cc} is obtained by evaluating the derivative at $x = X_B$, rather than $x = 0$, as for i_{e2}. From Eq. (4-32), we obtain:

$$\left.\frac{\widetilde{i_{cc}}}{\widetilde{v_{be}}}\right|_{\widetilde{v_{cb}}=0} = -\alpha_{T_0} g_e \zeta_{ac} X_B \operatorname{csch}(\zeta_{ac} X_B) \tag{4-50}$$

where ζ_{ac} was defined in Eq. (4-31). The functional form is simple, yet difficult to manipulate. Just as we approximate y_{ee} in the form of Eq. (4-40), we approximate Eq. (4-50) at low frequencies as:

$$\left.\frac{\widetilde{i_{cc}}}{\widetilde{v_{be}}}\right|_{v_{cb}=0} \approx -\alpha_{T_0} g_e \left(j\frac{2\omega}{\omega_0}\right)^{1/2} \operatorname{csch}\left(j\frac{2\omega}{\omega_0}\right)^{1/2} \approx -\alpha_{T_0} g_e \left(1 - j\frac{\omega}{3\omega_0}\right) \tag{4-51}$$

α_{T_0} is the d.c. base transport factor (neglecting the surface recombination because we are considering a one-dimensional problem):

$$\alpha_{T_o} = 1 - \frac{X_B^2}{2L_n^2} \tag{4-52}$$

There is another approach in the derivation of i_{cc}/v_{be}, by seeking the frequency response of the a.c. base transport factor. The a.c. base transport factor has a similar definition as the d.c. transport factor; it relates the current coming out of the collector side of the base to the current incoming from the emitter side of the base:

$$\alpha_{Tac} = -\left.\frac{i_{cc}}{i_{e2}}\right|_{v_{cb}=0} = -\left.\frac{d\tilde{n}(x)}{dx}\right|_{x=X_B} \div \left.\frac{d\tilde{n}(x)}{dx}\right|_{x=0} \tag{4-53}$$

A negative sign is inserted so that α_{Tac} is positive, in accordance with usual physical meaning. The a.c. carrier concentration in the base was found in Eq. (4-32). Carrying out the definition, we find:

$$\alpha_{Tac} = \frac{\operatorname{csch} \zeta_{ac} X_B}{\coth \zeta_{ac} X_B} \tag{4-54}$$

The frequency response of this function is quite complicated. Many forms of approximation exist, each addressing the accuracy in a different frequency domain. For the analysis here, we adopt one that is accurate at lower frequencies, a region that is consistent with the derivation of Eq. (4-51). α_{Tac} of Eq. (4-54) is given by:

$$\alpha_{Tac} = \frac{\alpha_{To}}{1 + j(\omega/\omega_\alpha)} \tag{4-55}$$

where ω_α is the radian frequency at which $|\alpha_{Tac}|$ decreases to 0.707 times its low-frequency value. An accurate calculation of α_{Tac} yields the following:

$$\omega_\alpha = \frac{2.43 D_n}{X_B^2} \approx 1.2\omega_o \tag{4-56}$$

However, for simplicity, we approximate ω_α as ω_o in this section. Therefore, i_{cc}/v_{be} is obtained from the following:

$$\left.\frac{\widetilde{i_{cc}}}{\widetilde{v_{be}}}\right|_{\widetilde{v_{be}}=0} \approx \frac{\widetilde{i_{cc}}}{\widetilde{i_{e1}}} \cdot \frac{\widetilde{i_{el}}}{\widetilde{v_{be}}} = \frac{-\alpha_{To}}{1 + j(\omega/\omega_o)} g_e \left(1 + j\frac{2\omega}{3\omega_o}\right) = -\alpha_{To} g_e \left(1 - j\frac{\omega}{3\omega_o}\right) \tag{4-57}$$

The direct derivation resulting in Eq. (4-51) was sufficient to find i_{cc}/v_{be}. Although the derivation through α_{Tac} is rather circuitous, the use of α_{Tac} helps us understand some ideas to be presented in § 4-3.

We now seek the relationship between i_c and i_{cc}. It turns out that the derivation is quite laborious. According to Ref. [1],

$$\frac{i_c(t)}{i_{cc}(t)} = \frac{1 - \exp(j\omega\tau_m)}{j\omega\tau_m} \tag{4-58}$$

where

$$\tau_m = \frac{X_{\text{dep}}}{v_{\text{sat}}} \tag{4-59}$$

An alternative form of Eq. (4-58) is obtained by some algebraic manipulation:

$$\begin{aligned} \frac{i_c(t)}{i_{cc}(t)} &= \left[\frac{\exp(j\omega\tau_m/2) - \exp(-j\omega\tau_m/2)}{2}\right] \frac{\exp(-j\omega\tau_m/2)}{j\omega\tau_m/2} \\ &= \frac{\sin(\omega\tau_m/2)}{\omega\tau_m/2} \exp(-j\omega\tau_m/2) \end{aligned} \tag{4-60}$$

It turns out that the magnitude of i_c/i_{cc}, equal to $\sin(\omega\tau_m/2)/(\omega\tau_m/2)$, is nearly unity even at high frequencies. Therefore, the ratio is well approximated by:

$$\frac{i_c(t)}{i_{cc}(t)} \approx \exp\left(-\frac{j\omega\tau_m}{2}\right) \approx 1 - j\omega\tau_m/2 \tag{4-61}$$

Depending on whichever form is easier for the algebraic manipulation, y_{ce} as given in Eq. (4-48) is written as either of the following equations:

$$y_{ce} = -\alpha_{To}g_e\left[1 - j(1-m)\frac{\omega}{\omega_o}\right]\exp\left(-j\frac{\omega\tau_m}{2}\right) \tag{4-62a}$$

$$y_{ce} = -\alpha_{To}g_e\left[1 - j(1-m)\frac{\omega}{\omega_o} - j\frac{\omega\tau_m}{2}\right] \tag{4-62b}$$

In writing down Eq. (4-62b), we used the single-pole approximation and neglected terms involving ω^2.

Example 4-2:

Consider the HBT of Example 3-11, which was biased at $J_C = J_{fd}/2 = 1.5 \times 10^3$ A/cm^2. The collector thickness is 1 μm, and the collector doping is 1×10^{16} cm^{-3}. $V_{CB} = 4$ V. What would be the ratio of i_c/i_{cc} in this transistor at 10 GHz?

According to the example, X_{dep} at this operating current level is 0.94 μm, although X_{dep} is only 0.88 μm at $J_C = 0$ (Example 3-8). According to Eq. (4-59):

$$\tau_m = \frac{0.94 \times 10^{-4}}{8 \times 10^6} = 1.175 \times 10^{-11} \text{ s}$$

The product $\omega\tau_m$ is thus $2\pi \cdot 10^{10} \cdot 1.175 \times 10^{-11} = 0.74$ (unitless). The current ratio is given by Eq. (4-60):

$$\frac{i_c(t)}{i_{cc}(t)} = \frac{\sin(0.74/2)}{0.74/2}\exp\left(-j\frac{0.74}{2}\right) = 0.98\exp\left(-j\frac{0.74}{2}\right)$$

$$= 0.98(0.93 - j0.36) = 0.91 - j0.35$$

This calculation demonstrates that the approximation of Eq. (4-60) by Eq. (4-61) is accurate.

y_{ec} y_{ec} is equal to the partial derivative of i_e with respect to v_{cb} when V_{BE} is kept constant. Because the base–collector junction is reverse-biased, a slight change in

v_{cb} on top of V_{CB} does not alter significantly the base minority carrier concentration at the collector side. That is, the concentration there remains nearly zero. Furthermore, since V_{BE} is kept constant, the base minority carrier concentration in the emitter side also is fixed constant. With the boundary conditions unaltered, the carrier profile does not change in spite of the perturbation of v_{be}. Some refinement can be made to the description, so that a nonzero (though still small) value is yielded. This refinement relates to the so-called *Early effect*. Basically, when a positive v_{cb} small-signal is applied, the neutral base thickness decreases slightly because of the applied v_{cb} causes the depletion region in the base–collector junction to expand. Because the base doping is more heavily doped than in the collector, most of the depletion region expansion occurs in the collector. Nonetheless, a finite portion of the depletion expansion also occurs in the base side. As the neutral base thickness shrinks in thickness, the emitter current increases and y_{ec} is nonzero.

An easily observable manifestation of the Early effect is found in the I-V characteristics. As shown schematically in Fig. 4-7, the collector current increases gradually with V_{CE}, instead of exhibiting ideally flat characteristics as V_{CE} increases. The increasing base–collector bias depletes further and further into the neutral base region, causing I_C to continue to rise. The I-V characteristics of Fig. 4-7 are exemplary of the silicon bipolar transistors. Similar characteristics are hardly observed in HBTs because of several reasons. First, the base

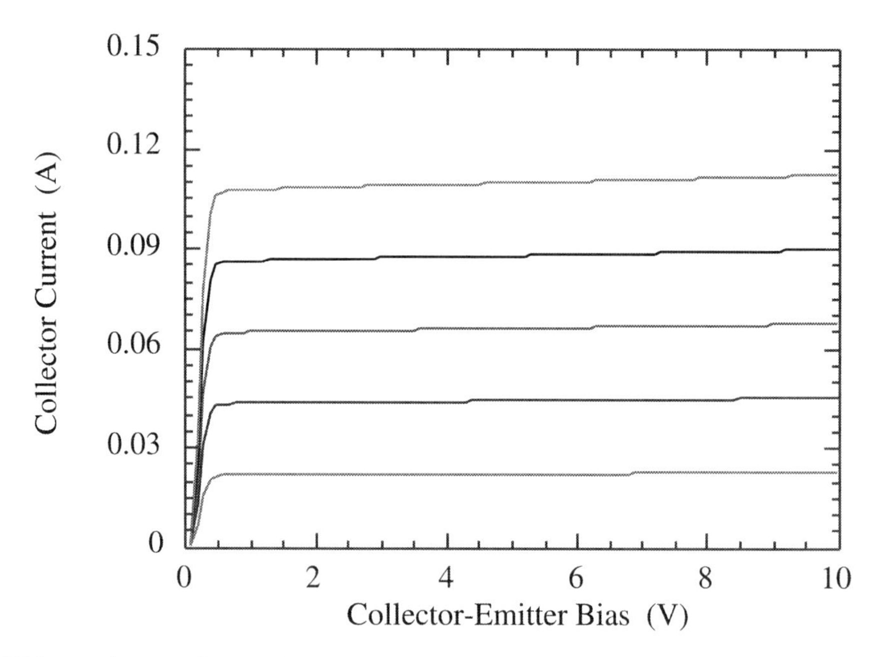

FIGURE 4-7. I-V characteristics exhibiting a certain degree of Early effects. The collector current increases slightly with increasing collector–emitter bias.

doping in HBTs is more heavily doped, unlike in Si BJTs, whose base doping has to be lighter than its emitter doping. Therefore, the reduction in the neutral base thickness is simply too small to cause a noticeable amount of increase in I_C. Second, unlike Si BJTs, which have lower device thermal resistance (because silicon has a higher thermal, conductivity than GaAs), HBTs typically operate at higher junction temperature as V_{CE} increases. Since the current gain decreases in AlGaAs/GaAs HBTs, the collector current actually decreases with V_{CE}, exhibiting a negative differential resistance, as discussed in §3-10.

We therefore approximate y_{ec} as zero for HBTs:

$$y_{ec} \approx 0 \qquad \text{(for HBTs)} \tag{4-63}$$

For Si BJTs, however, the Early effect is significant enough that y_{ec} should not be approximated as zero. The derivation is straightforward, but lengthy. We quote the result without derivation since it is not critical to HBTs:

$$\begin{aligned} y_{ec} &= -\alpha_{To} g_e \Xi \zeta_{ac} \operatorname{csch} \zeta_{ac} \qquad \text{(for Si BJTs)} \\ &= -\alpha_{To} g_e \Xi \left(j\frac{2\omega}{\omega_o} \right)^{1/2} \operatorname{csch} \left(j\frac{2\omega}{\omega_o} \right)^{1/2} \approx -\alpha_{To} g_e \Xi \left(1 - j\frac{\omega}{3\omega_o} \right) \end{aligned} \tag{4-64}$$

where Ξ, the Early effect coefficient, is defined as:

$$\Xi = -\frac{kT}{q}\frac{1}{X_B}\left(\frac{dX_B}{dV_{CB}}\right) \tag{4-65}$$

Ξ is unitless. It is a positive coefficient; a negative sign is inserted in the definition because dX_B/dV_{CB} is negative (the base thickness decreases as V_{CB} increases). As mentioned, for HBTs, whose base doping is high, $\Xi \approx 0$. In the relatively rare events when the base doping is low, Eqs. (4-64) and (4-65) apply equally to HBTs instead of Eq. (4-63).

y_{cc} y_{cc} is equal to the partial derivative of i_c with respect to v_{cb} when V_{BE} is kept constant. As in the discussion of y_{ec}, there is a small change in i_c due to the Early effect when v_{cb} is varied. Unlike i_e, however, i_c includes another component in addition to the Early effect. It is the current component associated with charging up the base–collector junction capacitance (C_{jc}), in an analogous manner as the fact that i_e contains a component from C_{je} when v_{be} is varied. We again do not derive the component due to the Early effect, but just state the result of the overall y_{cc} as:

$$\begin{aligned} y_{cc} &= g_e \Xi \zeta_{ac} \coth \zeta_{ac} + j\omega C_{jc} \\ &= g_e \Xi \left(j\frac{2\omega}{\omega_o} \right)^{1/2} \coth \left(j\frac{2\omega}{\omega_o} \right)^{1/2} + j\omega C_{jc} \approx g_e \Xi \left(1 + j\frac{2\omega}{3\omega_o} \right) + j\omega C_{jc} \end{aligned} \tag{4-66}$$

The Early effect component usually is negligible in HBTs, but the second component has important consequences for the HBT high-frequency performance. As with the derivation of C_{je} in Eqs. (4-12) to (4-18), we express C_{jc} as:

$$C_{jc} = A_C \frac{\epsilon_s}{X_{\text{dep}}} \tag{4-67}$$

A_C is the base–collector junction area, which is typically much larger than A_E, the active device area. X_{dep} is the base-collector depletion width. At low current levels, it is given by the depletion approximation:

$$X_{\text{dep}} = \sqrt{\frac{2\epsilon_s}{qN_C}(\phi_{CB} + V_{CB})} \tag{4-68}$$

At current density levels exceeding $qN_C v_{\text{sat}}$, the depletion thickness is greatly modified by the current and Eq. (4-68) is not accurate. Although the general capacitance expression of Eq. (4-67) remains correct, X_{dep} should be replaced by Eq. (3-78). At even higher current density levels [exceeding J_{Kirk} of Eq. (3-86)], so that base pushout occurs, the exact formulation of the junction capacitance becomes more complicated.

We neglect the contribution from the Early effect and y_{cc} in Eq. (4-66) is approximated as:

$$y_{cc} \approx j\omega C_{jc} \qquad \text{(for HBTs)} \tag{4-69}$$

In summary,

$$[y]_b = \begin{bmatrix} y_{ee} & y_{ec} \\ y_{ce} & y_{cc} \end{bmatrix} = \begin{bmatrix} g_e\left[1 + jm\dfrac{\omega}{\omega_o}\right] + j\omega C_{je} & 0 \\ -\alpha_{To} g_e\left[1 - j(1-m)\dfrac{\omega}{\omega_o}\right]\exp\left(-j\dfrac{\omega\tau_m}{2}\right) & j\omega C_{jc} \end{bmatrix} \tag{4-70a}$$

or, equivalently,

$$[y]_b = \begin{bmatrix} y_{ee} & y_{ec} \\ y_{ce} & y_{cc} \end{bmatrix} = \begin{bmatrix} g_e\left[1 + jm\dfrac{\omega}{\omega_o}\right] + j\omega C_{je} & 0 \\ -\alpha_{To} g_e\left[1 - j(1-m)\dfrac{\omega}{\omega_o} - j\dfrac{\omega\tau_m}{2}\right] & j\omega C_{jc} \end{bmatrix} \tag{4-70b}$$

The value of m is generally 2/3, as derived. To obtain a good fit, sometimes m is varied between 0 and 1. For example, sometimes m is taken to be 5/6 or 1.

§ 4-2 INTRINSIC COMMON-EMITTER y-PARAMETERS

The common-base y-parameters, although easily obtained from fundamental device physics, are not too useful as they stand. After all, most transistors are biased in the common-emitter configuration. It is important to derive the common-emitter y-parameters from the known common-base parameters.

In a common-emitter (or common-base) configuration, the emitter (or the base) node of the transistor is used as the reference node. If we do not choose a particular node of the transistor as the common terminal, we can refer the base, emitter, and collector voltages to a node external to the transistor. In the configuration shown in Fig. 4-8, we write:

$$i_e = y_{ee}v_e + y_{eb}v_b + y_{ec}v_c \tag{4-71a}$$

$$i_b = y_{be}v_e + y_{bb}v_b + y_{bc}v_c \tag{4-71b}$$

$$i_c = y_{ce}v_e + y_{cb}v_b + y_{cc}v_c \tag{4-71c}$$

Although there are a total of nine y-parameters expressed, they are linearly dependent. Once four of the linearly independent y-parameters are known, the other five parameters are readily obtainable. For example, if we assume that $v_e = v_b = v_c = v$, as shown in Fig. 4-9, then from Eq. (4-71a), we have:

$$i_e = (y_{ee} + y_{eb} + y_{ec}) \cdot v \tag{4-72}$$

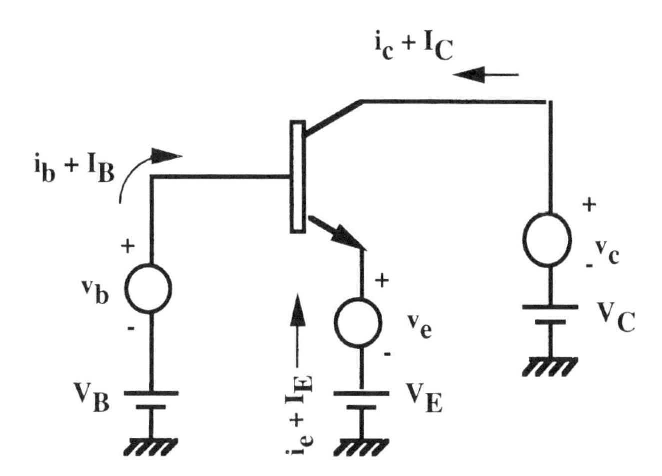

FIGURE 4-8. Small-signal bias condition for an HBT. The reference node resides outside of the emitter, base, and collector nodes.

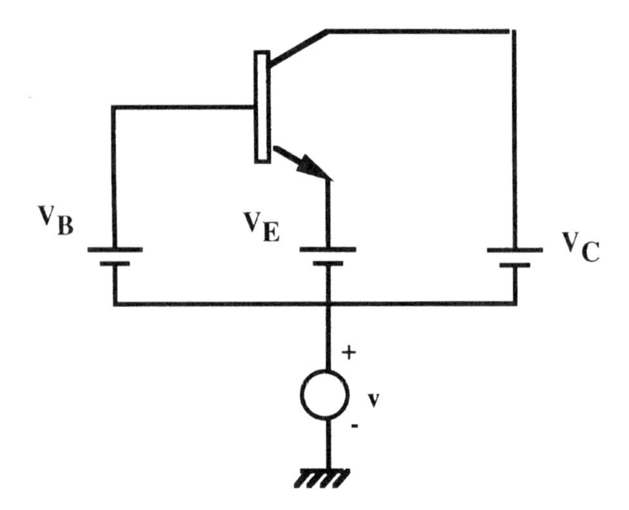

FIGURE 4-9. Illustration when all three terminals of the HBT of Fig. 4-8 are biased to the same small-signal voltage.

However, because there is no small-signal voltage difference across any two of the terminals, all terminal currents must be zero. A similar argument is extended to i_b and i_c. Therefore,

$$y_{ee} + y_{eb} + y_{ec} = 0 \tag{4-73a}$$

$$y_{be} + y_{bb} + y_{bc} = 0 \tag{4-73b}$$

$$y_{ce} + y_{cb} + y_{cc} = 0 \tag{4-73c}$$

Furthermore, from the Kirchoff current law, the small-signal current must add up to zero:

$$i_e + i_b + i_c = 0 \tag{4-74}$$

If we have a situation wherein $v_e \neq 0$, but $v_b = v_c = 0$, then Eq. (4-74), combined with Eq. (4-71a) lead to Eq. (4-75a). Equations (4-75b) and (4-75c) are similarly obtained with other bias conditions:

$$y_{ee} + y_{be} + y_{ce} = 0 \tag{4-75a}$$

$$y_{eb} + y_{bb} + y_{cb} = 0 \tag{4-75b}$$

$$y_{ec} + y_{bc} + y_{cc} = 0 \tag{4-75c}$$

In the common-emitter configuration, we tie the emitter node to the reference node, resulting in a configuration as shown in Fig. 4-10*a*. The emitter voltage, v_e,

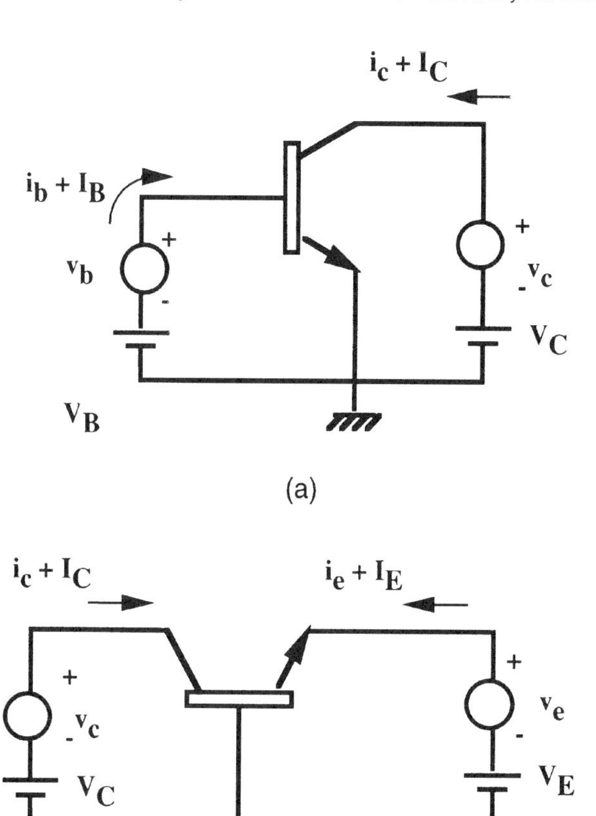

FIGURE 4-10. (*a*) HBT biased in the common-emitter configuration. (*b*) Common-base configuration.

ceases to be a variable and has the value 0. The common-emitter y-parameters are therefore written as:

$$[y]_e = \begin{bmatrix} y_{bb} & y_{bc} \\ y_{cb} & y_{cc} \end{bmatrix} \tag{4-76}$$

Similarly, the common-base configuration is obtained by tying the base node to the reference node, as shown in Fig. 4-10*b*. The common-base y-parameters are:

$$[y]_b = \begin{bmatrix} y_{ee} & y_{ec} \\ y_{ce} & y_{cc} \end{bmatrix} \tag{4-77}$$

In the previous section, we have found the common-base parameters, y_{ee}, y_{ec}, y_{ce}, and y_{cc}. We now try to determine the common-emitter y-parameters based on these parameters. From Eq. (4-73b).

$$y_{bb} = -y_{eb} - y_{cb} \tag{4-78}$$

Moreover, from Eqs. (4-73a) and (4-73c),

$$y_{eb} = -y_{ee} - y_{ec} \quad \text{and} \quad y_{cb} = -y_{ce} - y_{cc} \tag{4-79}$$

Combining Eqs. (4-78) and (4-79) leads to:

$$y_{bb} = y_{ee} + y_{ec} + y_{ce} + y_{cc} \tag{4-80}$$

Similarly, from Eqs. (4-75c) and (4-73c):

$$y_{bc} = -y_{cc} - y_{ec} \tag{4-81}$$

$$y_{cb} = -y_{ce} - y_{cc} \tag{4-82}$$

Therefore, the common-emitter y-parameters in terms of the common-base parameters are:

$$[y]_e = \begin{bmatrix} \sum y & -y_{cc} - y_{ec} \\ -y_{cc} - y_{ce} & y_{cc} \end{bmatrix} \tag{4-83}$$

where Σy, the sum of the common-base y-parameters, is:

$$\sum y = y_{ee} + y_{ec} + y_{ce} + y_{cc} \tag{4-84}$$

Example 4-3:

A common-base HBT has the following parameters: $y_{ee} = (20.5 - j1)\,\text{mS}$; $y_{ec} = (-0.015 - 0.04)\,\text{mS}$; $y_{ce} = (-20 + j3)\,\text{mS}$; and $y_{cc} = (0.02 + j0.06)\,\text{mS}$. Find its common-emitter parameters.

From Eq. (4-83), we find that:

$$y_{11e} = y_{ee} + y_{ec} + y_{ce} + y_{cc} = 0.505 + j2.02\,\text{mS}$$

$$y_{12e} = -y_{ec} - y_{cc} = -0.005 - j0.056\,\text{mS}$$

$$y_{21e} = -y_{ce} - y_{cc} = 20.02 - j3.06\,\text{mS}$$

$$y_{22e} = y_{cc} = 0.02 + j0.06\,\text{mS}$$

In terms of the physical parameters, $[y]_e$ is obtained from either Eq. (4-70a) or (4-70b):

$$[y]_e = \begin{bmatrix} g_e(1-\alpha_{To}) + j\omega(C_{je}+C_{jc}) + jg_e\dfrac{\omega\tau_m}{2} + jg_e\dfrac{\omega}{\omega_o} & -j\omega C_{jc} \\ \alpha_{To}g_e\left[1 - j(1-m)\dfrac{\omega}{\omega_o}\right]\cdot\exp\left(\dfrac{\omega\tau_m}{2}\right) - j\omega C_{jc} & j\omega C_{jc} \end{bmatrix} \quad (4\text{-}85a)$$

or

$$[y]_e = \begin{bmatrix} g_e(1-\alpha_{To}) + j\omega(C_{je}+C_{jc}) + jg_e\dfrac{\omega\tau_m}{2} + jg_e\dfrac{\omega}{\omega_o} & -j\omega C_{jc} \\ \alpha_{To}g_e\left(1 - j(1-m)\dfrac{\omega}{\omega_o} - j\dfrac{\omega\tau_m}{2}\right) - j\omega C_{jc} & j\omega C_{jc} \end{bmatrix} \quad (4\text{-}85b)$$

As mentioned in the previous section, m has a value of 2/3 from a theoretical calculation, but can be optimized to give the best fit.

The term $j\cdot g_e\cdot\omega/\omega_o$ is equal to $j\omega C_D$ where C_D is given by Eq. (4-45). This confirms that C_D of Eq. (4-45) applies to the common-emitter configuration, whereas C_D of Eq. (4-42) applies to the common-base configuration.

Example 4-4:

Determine y_{21} for the HBT discussed in Example 4-2 at 10 GHz. The transistor is biased in the common-emitter configuration. Assume $\alpha_{To} = 0.99$; $X_B = 800$ Å, $D_n = 25$ cm^2/V-s; $\eta = 1.23$; $A_E = 60\ \mu\text{m}^2$; $A_C = 180\ \mu\text{m}^2$.

From the previous example, $\tau_m = 1.175\times10^{-11}$ s and $X_{\text{dep}} = 0.94\ \mu$m. The base-collector junction capacitance is given by Eq. (4-67):

$$C_{jc} = 180\times10^{-8}\,\frac{1.159\times10^{-12}}{0.94\times10^{-4}} = 2.2\times10^{-14}\ \text{F}$$

Further, the operating current density is $J_C = J_{fd}/2 = 1.5\times10^3$ A/cm^2, which corresponds to a collector current of 9×10^{-4} A. The transconductance, from Eq. (4-36), is:

$$g_e = \frac{9\times10^{-4}}{1.23\cdot0.0258} = 2.8\times10^{-2}\,\frac{1}{\Omega}$$

The frequency ω_o is given by Eq. (4-37):

$$\omega_o = \frac{2.25}{(800\times10^{-8})^2} = 7.81\times10^{11}\ \text{rad}$$

From Eq. (4-85b), $y_{21} = -y_{cc} - y_{ce}$ is:

$$y_{21} = \alpha_{To} g_e \left[1 - j(1-m)\frac{\omega}{\omega_o} - j\frac{\omega \tau_m}{2} \right] - j\omega C_{jc}$$

$$= 0.99 \cdot 2.8 \times 10^{-2} \left[1 - j\left(1 - \frac{2}{3}\right) \frac{2\pi \cdot 10^{10}}{7.81 \times 10^{11}} - j\frac{2\pi \cdot 10^{10}\, 1.175 \times 10^{-11}}{2} \right]$$

$$- j\, 2\pi \cdot 10^{10} \cdot 2.2 \times 10^{-14}$$

$$= 0.0277 - j\, 0.111\, \Omega^{-1}$$

§ 4-3 HYBRID-π MODEL

We establish the common-base and the common-emitter models based on the y-parameters. For the intrinsic transistor representation to be discussed presently, the external resistances are not included. A model including such parasitics will be presented in the later part of this section. In general, the y-parameters for a linearized two-port are written in the following matrix form:

$$[y] = \begin{bmatrix} y_{11} & y_{12} \\ y_{21} & y_{22} \end{bmatrix} \tag{4-86}$$

The y-parameters obey the following relationships:

$$i_1 = y_{11} v_1 + y_{12} v_2 \tag{4-87a}$$

$$i_2 = y_{21} v_1 + y_{22} v_2 \tag{4-87b}$$

where i_1, i_2 are the port currents and v_1, v_2 are the port voltages. An equivalent circuit of the two-port that obeys Eq. (4-87) is shown in Fig. 4-11. An alternative

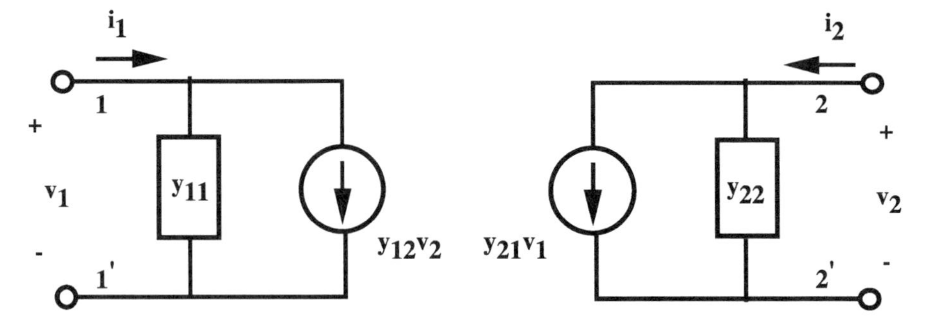

FIGURE 4-11. Two-port circuit representation. The small-signal current and voltages obey Eq. (4-87).

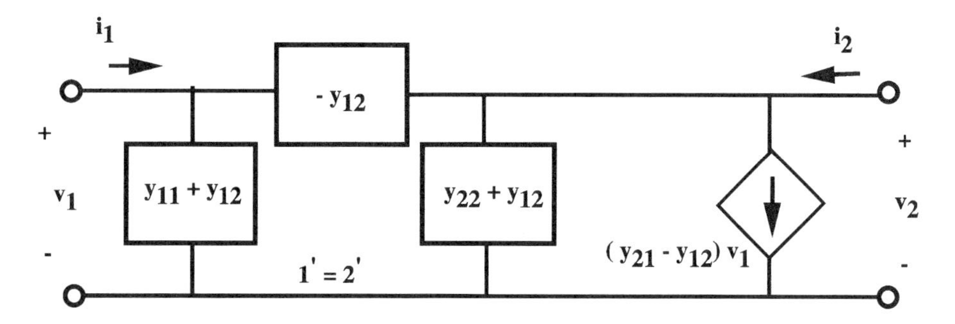

FIGURE 4-12. Rearrangement of the two-port representation of Fig. 4-11. The two models are mathematically identical, despite the obvious differences in the physical appearance.

equivalent circuit using only one dependent current source is shown in Fig. 4-12. The transformation assumes implicitly that the terminals 1′ and 2′ are at the same potential. This requirement is satisfied in the cases at hand because either the base or the emitter are common to both ports. Because the circuit branches of Fig. 4-12 connected by the elements $(y_{11} + y_{12})$, $-y_{12}$, and $(y_{22} + y_{12})$ form a shape that resembles the Greek letter π, it is referred to as the π-model. Using the $[y]_e$ given in Eq. (4-85a) or Eq. (4-85b), we calculate the branch elements found in Fig. 4-12 as:

$$y_{11e} + y_{12e} = g_e(1 - \alpha_{To}) + j\omega C_{je} + j g_e \frac{\omega\tau_m}{2} + j g_e \frac{\omega}{\omega_o}$$

$$= \frac{g_e}{\beta_{dc}} + j\omega\left(C_{je} + \frac{g_e}{\omega_o}\right) + j g_e \frac{\omega\tau_m}{2} \tag{4-88a}$$

$$-y_{12e} = j\omega C_{jc} \tag{4-88b}$$

$$y_{22e} + y_{12e} = 0 \tag{4-88c}$$

$$y_{21e} - y_{12e} = \alpha_{To} g_e \left[1 - j(1 - m)\frac{\omega}{\omega_0}\right] \exp\left(\frac{\omega\tau_m}{2}\right) \tag{4-88d}$$

Based on these equations and the model shown on Fig. 4-12, we represent the common-emitter HBT by a small-signal model shown in Fig. 4-13. We also added the intrinsic base resistance to complete the transistor model. If we neglect $g_e \cdot \omega\tau_m/2$ in the input capacitance and $(1 - m) \cdot \omega/\omega_o$ in the current generator, then the π-model becomes identical to the familiar small-signal model shown in Fig. 4-14.

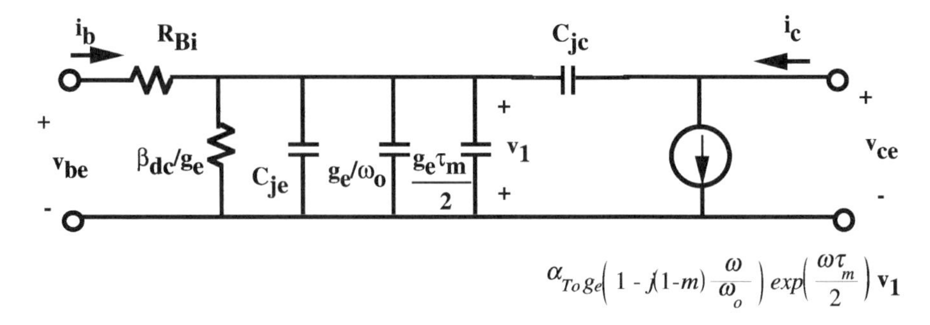

FIGURE 4-13. A small-signal HBT model based on the circuit representation of Fig. 4-12.

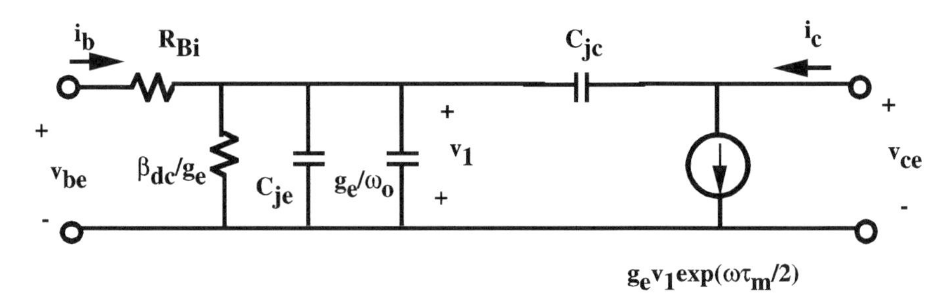

FIGURE 4-14. A simplified equivalent circuit.

Sometimes a T-model is used instead of the π-model. It is called a T-model, again because of its shape, with the branches of the circuit elements arranged in a way resembling the letter T, as shown in Fig. 4-15. It features the use of $\alpha \cdot i_{e2}$ as the collector current source, where α is the modified current transfer ratio. It is the product of the a.c. current gain times the phase delay through the depleted collector region:

$$\alpha = \left.\frac{i_c}{i_{e2}}\right|_{v_{cb}=0} = \frac{i_{cc}}{i_{e2}} \cdot \frac{i_c}{i_{cc}} = \alpha_{Tac} \cdot \exp(-j\omega\tau_m/2) = \frac{\alpha_{To}\exp(-j\omega\tau_m/2)}{1 + j\omega/\omega_0} \tag{4-89}$$

The definition of α makes sense because the collector current (not counting the capacitive current through C_{jc}) is $\alpha \cdot i_{e2}$. We emphasize that i_{e2} is only a part of the total emitter current, excluding the capacitive current through C_{je}. If $\alpha \cdot i_e$ were instead used as the current source, then the carriers flowing through the base–emitter capacitance would be mistakenly assumed to diffuse through the base and be collected in the collector.

There is an important change in the value of a π-model element when it is replaced by the T-model. The input resistor value of $[g_e/(1 - \alpha_{To})]^{-1}$ in the

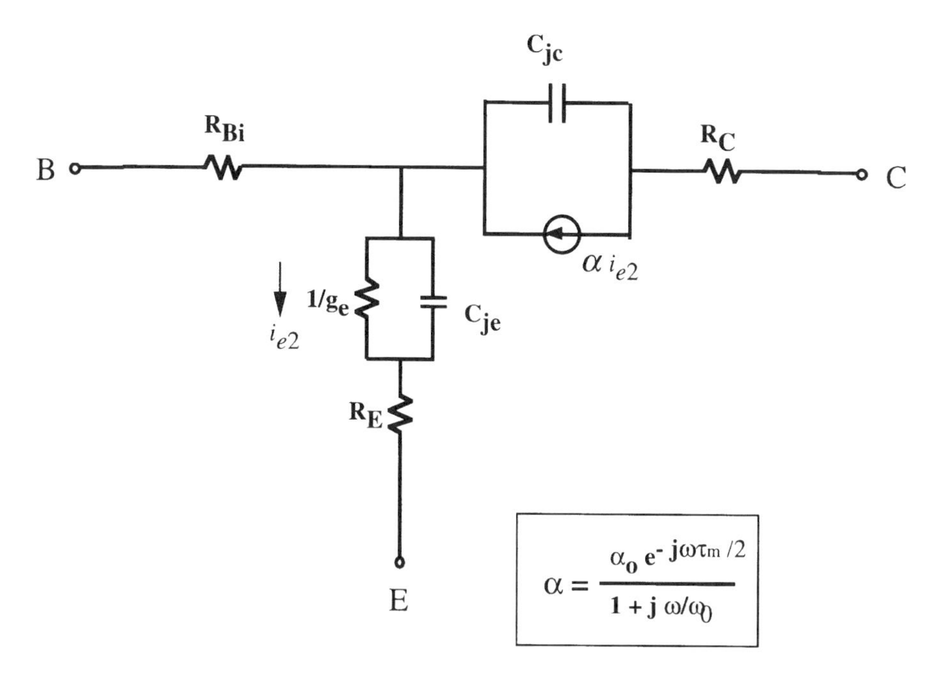

FIGURE 4-15. An alternative model to the hybrid π-model of Fig. 4-14: the T-model.

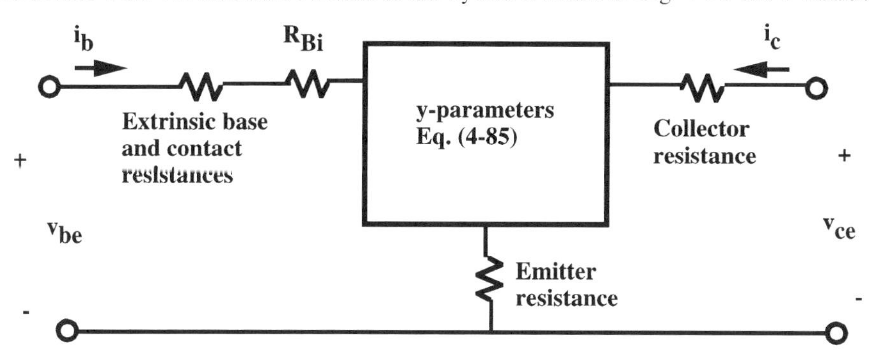

FIGURE 4-16. Addition of terminal resistances to an intrinsic transistor model.

π-model becomes $(g_e)^{-1}$ in the T-model. In the π-model, the amount of current flowing through the resistor is the base current. In contrast, the way the T-model elements are connected is such that the current flowing through the input resistor is the sum of i_b and $\alpha \cdot i_{e2}$. For v_{be} to maintain the same value in either representation, the resistance values must change in response to the change in the current flowing through the resistors.

The models presented so far are for intrinsic transistors, including the capacitive elements such as C_{je} and C_{jc}, but excluding the parasitic resistances. To include the latter, we simply attach resistors to appropriate terminals as shown in Fig. 4-16. Note that we move the intrinsic base resistance outside of the inner two-port, which is defined by the y-parameters discussed previously. The intrinsic base resistance can be lumped with the extrinsic base resistance as well as the base contact resistance.

Since the device properties are often expressed in terms of z- (impedance), and h- (hybrid) parameters, we describe them briefly. For the y-parameter discussed extensively in § 8-2, i_1 and i_2 are the independent variables describing a network with ports 1 and 2. For z-parameters, the independent variables are v_1 and v_2. For h-parameters, they are i_1 and v_2. For convenience, we write out the relevant relationships:

$$\begin{bmatrix} i_1 \\ i_2 \end{bmatrix} = \begin{bmatrix} y_{11} & y_{12} \\ y_{21} & y_{22} \end{bmatrix} \begin{bmatrix} v_1 \\ v_2 \end{bmatrix} \tag{4-90}$$

$$\begin{bmatrix} v_1 \\ v_2 \end{bmatrix} = \begin{bmatrix} z_{11} & z_{12} \\ z_{21} & z_{22} \end{bmatrix} \begin{bmatrix} i_1 \\ i_2 \end{bmatrix} \tag{4-91}$$

$$\begin{bmatrix} v_1 \\ i_2 \end{bmatrix} = \begin{bmatrix} h_{11} & h_{12} \\ h_{21} & h_{22} \end{bmatrix} \begin{bmatrix} i_1 \\ v_2 \end{bmatrix} \tag{4-92}$$

We have determined y's for both the common-base and the common-emitter configurations. Once these parameters are known, the z- and the h-parameters are readily determined. For example, let us find z_{11} in terms of the y-parameters. From Eq. (4-91), z_{11} is the ratio of v_1 over i_1 when $i_2 = 0$, Eq. (4-90) implies that:

$$v_2 = -\frac{y_{21}}{y_{22}} v_1 \tag{4-93}$$

Substituting this result into Eq. (4-90), it is found that:

$$i_1 = y_{11} v_1 - y_{12} \frac{y_{21}}{y_{22}} v_1 = \left(\frac{y_{11} y_{22} - y_{12} y_{21}}{y_{22}} \right) v_1 \tag{4-94}$$

Consequently, z_{11} is

$$z_{11} = \left. \frac{v_1}{i_1} \right|_{i_2 = 0} = \frac{y_{22}}{\Delta y} \tag{4-95}$$

where

$$\Delta y = y_{11} y_{22} - y_{12} y_{21} \tag{4-96}$$

Figure 4-17 is a conversion table for the small-signal parameters. z_{11} expressed in Eq. (4-95) is indeed listed at the proper entry in the figure. The derivations of other relationships are straightforward and not discussed further.

$$z' = z/R_o. \quad y' = yR_o. \quad h_{11}' = h_{11}/R_o; \; h_{12}' = h_{12}; \; h_{21}' = h_{21}; \; h_{22}' = h_{22}R_o.$$

$$s_{11} = \frac{(z_{11}'-1)(z_{22}'+1) - z_{12}' \cdot z_{21}'}{(z_{11}'+1)(z_{22}'+1) - z_{12}' \cdot z_{21}'} \qquad y_{11}' = \frac{(1-s_{11})(1+s_{22}) + s_{12} \cdot s_{21}}{(1+s_{11})(1+s_{22}) - s_{12} \cdot s_{21}} \qquad z_{11}' = \frac{(1+s_{11})(1-s_{22}) + s_{12} \cdot s_{21}}{(1-s_{11})(1-s_{22}) - s_{12} \cdot s_{21}} \qquad h_{11}' = \frac{(1+s_{11})(1+s_{22}) - s_{12} \cdot s_{21}}{(1-s_{11})(1+s_{22}) + s_{12} \cdot s_{21}}$$

$$s_{12} = \frac{2 z_{12}'}{(z_{11}'+1)(z_{22}'+1) - z_{12}' \cdot z_{21}'} \qquad y_{12}' = \frac{-2 s_{12}}{(1+s_{11})(1+s_{22}) - s_{12} \cdot s_{21}} \qquad z_{12}' = \frac{2 s_{12}}{(1-s_{11})(1-s_{22}) - s_{12} \cdot s_{21}} \qquad h_{12}' = \frac{2 s_{12}}{(1-s_{11})(1+s_{22}) + s_{12} \cdot s_{21}}$$

$$s_{21} = \frac{2 z_{21}'}{(z_{11}'+1)(z_{22}'+1) - z_{12}' \cdot z_{21}'} \qquad y_{21}' = \frac{-2 s_{21}}{(1+s_{11})(1+s_{22}) - s_{12} \cdot s_{21}} \qquad z_{21}' = \frac{2 s_{21}}{(1-s_{11})(1-s_{22}) - s_{12} \cdot s_{21}} \qquad h_{21}' = \frac{2 s_{21}}{(1-s_{11})(1+s_{22}) + s_{12} \cdot s_{21}}$$

$$s_{22} = \frac{(z_{11}'+1)(z_{22}'-1) - z_{12}' \cdot z_{21}'}{(z_{11}'+1)(z_{22}'+1) - z_{12}' \cdot z_{21}'} \qquad y_{22}' = \frac{(1+s_{11})(1-s_{22}) + s_{12} \cdot s_{21}}{(1+s_{11})(1+s_{22}) - s_{12} \cdot s_{21}} \qquad z_{22}' = \frac{(1-s_{11})(1+s_{22}) + s_{12} \cdot s_{21}}{(1-s_{11})(1-s_{22}) - s_{12} \cdot s_{21}} \qquad h_{22}' = \frac{(1-s_{11})(1-s_{22}) - s_{12} \cdot s_{21}}{(1-s_{11})(1+s_{22}) + s_{12} \cdot s_{21}}$$

$$s_{11} = \frac{(1-y_{11}')(1+y_{22}') + y_{12}' \cdot y_{21}'}{(1+y_{11}')(1+y_{22}') - y_{12}' \cdot y_{21}'} \qquad y_{11} = \frac{z_{22}}{z_{11} z_{22} - z_{21} z_{12}} \qquad z_{11} = \frac{y_{22}}{y_{11} y_{22} - y_{21} y_{12}} \qquad h_{11} = \frac{z_{11} z_{22} - z_{21} z_{12}}{z_{22}}$$

$$s_{12} = \frac{-2 y_{12}'}{(1+y_{11}')(1+y_{22}') - y_{12}' \cdot y_{21}'} \qquad y_{12} = \frac{-z_{12}}{z_{11} z_{22} - z_{21} z_{12}} \qquad z_{12} = \frac{-y_{12}}{y_{11} y_{22} - y_{21} y_{12}} \qquad h_{12} = \frac{z_{12}}{z_{22}}$$

$$s_{21} = \frac{-2 y_{21}'}{(1+y_{11}')(1+y_{22}') - y_{12}' \cdot y_{21}'} \qquad y_{21} = \frac{-z_{21}}{z_{11} z_{22} - z_{21} z_{12}} \qquad z_{21} = \frac{-y_{21}}{y_{11} y_{22} - y_{21} y_{12}} \qquad h_{21} = -\frac{z_{21}}{z_{22}}$$

$$s_{22} = \frac{(1+y_{11}')(1-y_{22}') + y_{12}' \cdot y_{21}'}{(1+y_{11}')(1+y_{22}') - y_{12}' \cdot y_{21}'} \qquad y_{22} = \frac{z_{11}}{z_{11} z_{22} - z_{21} z_{12}} \qquad z_{22} = \frac{y_{11}}{y_{11} y_{22} - y_{21} y_{12}} \qquad h_{22} = \frac{1}{z_{22}}$$

$$s_{11} = \frac{(h_{11}'-1)(1+h_{22}') - h_{12}' \cdot h_{21}'}{(1+h_{11}')(1+h_{22}') - h_{12}' \cdot h_{21}'} \qquad y_{11} = \frac{1}{h_{11}} \qquad z_{11} = \frac{h_{11} h_{22} - h_{21} h_{12}}{h_{22}} \qquad h_{11} = \frac{1}{y_{11}}$$

$$s_{12} = \frac{2 h_{12}'}{(1+h_{11}')(1+h_{22}') - h_{12}' \cdot h_{21}'} \qquad y_{12} = -\frac{h_{12}}{h_{11}} \qquad z_{12} = \frac{h_{12}}{h_{22}} \qquad h_{12} = -\frac{y_{12}}{y_{11}}$$

$$s_{21} = \frac{-2 h_{21}'}{(1+h_{11}')(1+h_{22}') - h_{12}' \cdot h_{21}'} \qquad y_{21} = \frac{h_{21}}{h_{11}} \qquad z_{21} = -\frac{h_{21}}{h_{22}} \qquad h_{21} = \frac{y_{21}}{y_{11}}$$

$$s_{22} = \frac{(1+h_{11}')(1-h_{22}') + h_{12}' \cdot h_{21}'}{(1+h_{11}')(1+h_{22}') - h_{12}' \cdot h_{21}'} \qquad y_{22} = \frac{h_{11} h_{22} - h_{21} h_{12}}{h_{11}} \qquad z_{22} = \frac{1}{h_{22}} \qquad h_{22} = \frac{y_{11} y_{22} - y_{21} y_{12}}{y_{11}}$$

FIGURE 4-17. Conversion table between the s-, y-, z-, and h-parameters.

The y-, z-, or h-parameters, though conceptually simple, may be too difficult to be measured experimentally. For example, a direct measurement of z_{11} involves enforcing an open at the output so that $i_2 = 0$. However, from transmission line theories, an open at a certain frequency may behave as short at other frequencies. The abrupt termination also often results in oscillation, preventing the measurement from being useful. Therefore, the direct measurement of these parameters at high frequencies becomes a difficult task. Instead, the s- (scattering) parameters are typically employed to characterize the high-frequency performance of a transistor. The s-parameters are first measured and then converted to other parameters for analysis if needed.

§ 4-4 PARASITIC RESISTANCES

Though the intrinsic portion of the transistor was well characterized in the previous sections, we have yet to determine parasitic resistances/impedances that can significantly modify the high-frequency characteristics of the transistor. According to their associated terminals, these resistances/impedances are classified as the emitter resistance, base impedance, and the collector resistance. Each terminal resistance/impedance, in turn, consists of three components: that associated with the semiconductor layer, that associated with the semiconductor/metal contact, and that associated with the metal line. As shown in Fig. 4-18, the epitaxial resistances for the emitter and the collector are $R_{E(\mathrm{epi})}$ and $R_{C(\mathrm{epi})}$, respectively, and the contact resistances are R_{EE} and R_{CC}, respectively. As for the base, the epitaxial resistance of the extrinsic base layer is denoted as $R_{Bx(\mathrm{epi})}$; the subscript x emphasizes that the resistances is in the extrinsic base region. The epitaxial resistance associated with the intrinsic base region, R_{Bi}, was considered as a part of the intrinsic transistor. Rather than being purely resistive, the base contact is best characterized by an impedance, as will be shown later. Therefore, it is denoted as Z_{BB}. Figure 4-18 also contains the dimensions of the transistor geometry. Based on these dimensions, we derive the expressions for the epitaxial resistances and the contact impedances/resistances.

We find that the epitaxial resistances depend on the geometrical dimensions and the respective epitaxy resistivities. With the transistor geometries specified in Fig. 4-18, we write the epitaxial resistances for the emitter and the extrinsic base layers:

$$R_{E(\mathrm{epi})} = \rho_{E\mathrm{cap}} \frac{X_{E\mathrm{cap}}}{W_E \cdot L_E} + \rho_E \frac{X_E - X_{\mathrm{dep}}}{W_E \cdot L_E} \tag{4-97}$$

$$R_{Bx(\mathrm{epi})} = \rho_B \frac{S_{BE}}{X_B \cdot L_E} \tag{4-98}$$

where ρ's are the layer resistivites; $X_{E\mathrm{cap}}$, X_E, X_B are the cap emitter, emitter, and base thicknesses, respectively, and S_{BE} and W_B are the base-emitter contact

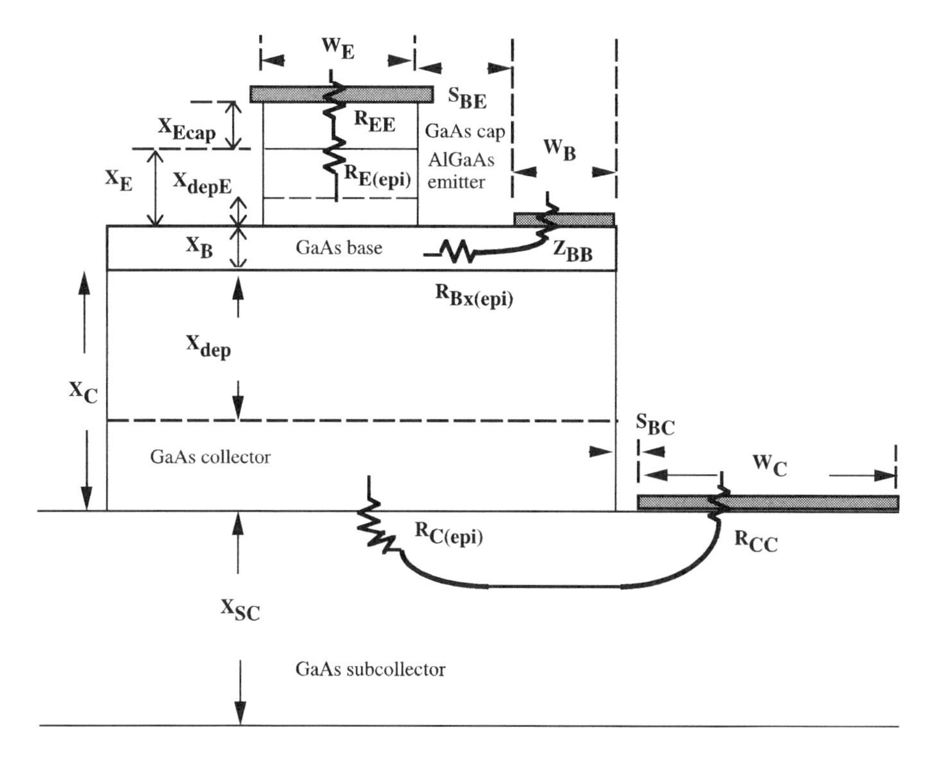

FIGURE 4-18. Locations of various parasitic resistances in a HBT.

spacing and the base contact width, respectively. The active emitter layer in HBTs is typically AlGaAs. Because it does not allow good ohmic contact, a GaAs cap layer is often grown on top of it to facilitate contact. The two terms in Eq. (4-97) reflect the fact that the total epitaxial emitter resistance consists of the resistance in the undepleted active emitter layer as well as the resistance in the emitter cap layer. In general, $R_{E(\mathrm{epi})}$ can be neglected completely because it is small compared to the emitter contact resistance. Similarly, $R_{Bx(\mathrm{epi})}$ used to be important in past technologies, but is minimized in modern technologies wherein S_{BE} is made zero by self-aligning the base–emitter contacts.

We go on to examine the epitaxial resistances associated with the collector and the subcollector layers. Figure 4-19*a* emphasizes the direction of the current flow, showing that the current spreads out from the intrinsic emitter region to the extrinsic region. If we assume that the amount of current spreading is negligible and that practically all of the collector current flows within the intrinsic region, then

$$R_{C(\mathrm{epi})} = \rho_C \frac{X_C - X_{\mathrm{dep}}}{W_E \cdot L_E} \qquad \text{(assume negligible current spreading)} \qquad (4\text{-}99)$$

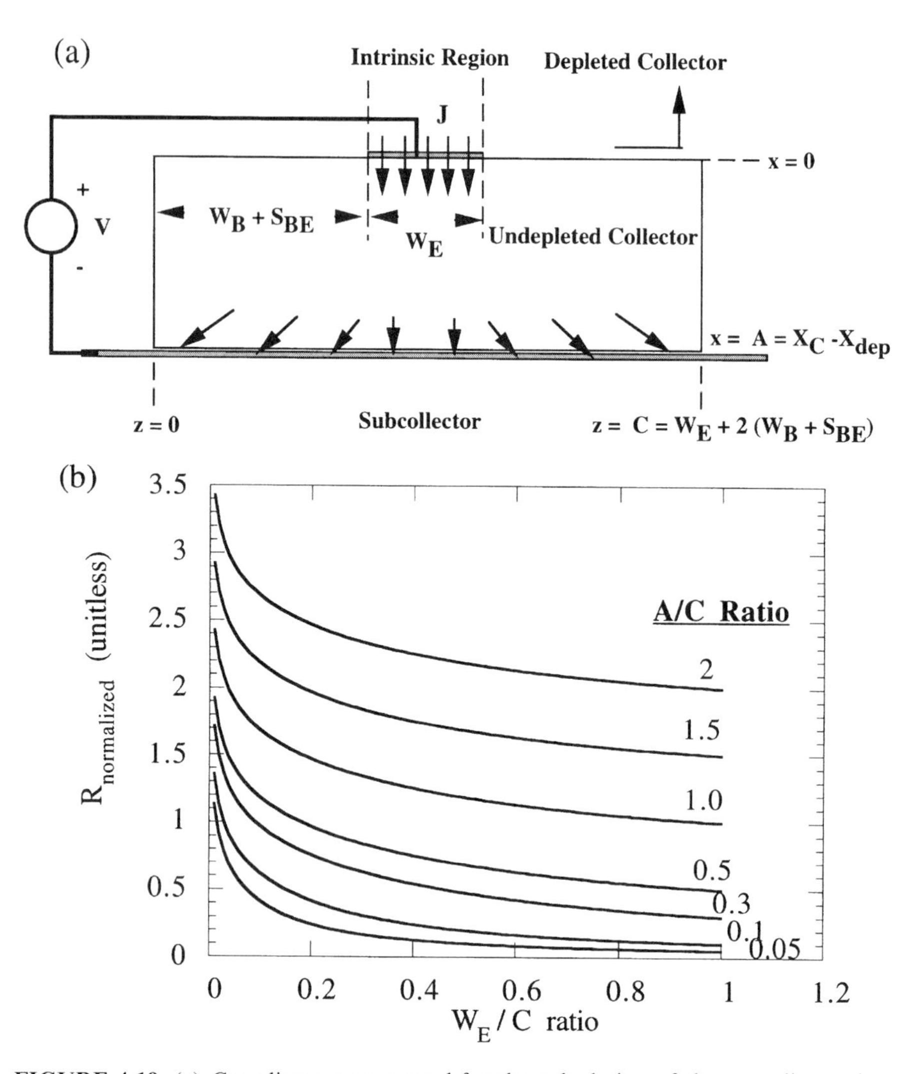

FIGURE 4-19. (*a*) Coordinate system used for the calculation of the spreading resistance between the collector and the subcollector layers. (*b*) Plot of normalized spreading resistance as a function of the ratio W_E/C for several values of the ratio A/C.

X_C and X_{dep} denote the designed collector thickness and the base–collector depletion thickness inside the collector side, respectively. In reality, however, a certain amount of current spreading is present and the $R_{C(\text{epi})}$ given is not accurate. The actual collector epitaxial resistance is a *spreading resistance* whose derivation is described in the following. To simplify the derivation, we assume that the emitter length is infinitely long. The device's cross-section under consideration is shown in Fig. 4-19*a*. At $x = 0$, the current density flowing in the intrinsic region is uniform, equal to J. Though the current density outside the intrinsic region is zero when $z = 0$, the current spreads out and J at $x = A = X_C - X_{\text{dep}}$ is nonzero in both the intrinsic and the extrinsic regions. We assume that the subcollector layer, being so heavily doped compared to the collector layer, is a metal plate and that $V(X_C - X_{\text{dep}}) = 0$. Because of the current flow is not purely one-dimensional, we do not expect the voltage developed by the resistive drops with respect to the subcollector ground plane to be identical at every location on the $x = 0$ plane. For simplicity, we define the spreading resistance as:

$$R_{C(\text{epi})} = \frac{V(x = 0, z = C/2)}{I} = \frac{V(x = 0, z = C/2)}{J \cdot W_E L_E} \tag{4-100}$$

where

$$C = W_E + 2(W_B + S_{BE}) \tag{4-101}$$

C represents the total width of the emitter mesa under consideration. To determine the spreading resistance, we need first to determine the voltage profile given that a uniform current density injects through the intrinsic region at the top. We ask, what is the governing equation of the voltage profile? From the study of electrostatics (or Maxwell's equations), we have:

$$\nabla \cdot \varepsilon = \nabla \cdot (\nabla V) = \nabla^2 V = \frac{Q}{\epsilon_s} \tag{4-102}$$

where Q is the net charge at the location at which $\nabla \cdot \varepsilon$ is evaluated. Since we continue with the quasi-neutrality assumption in these epitaxial layers, $Q = 0$. Therefore, for the particular example shown in Fig. 4-19a, we are solving the potential profile $V(x, y)$, which is governed by:

$$\frac{\partial^2 V}{\partial x^2} + \frac{\partial^2 V}{\partial z^2} = 0 \tag{4-103}$$

The appropriate boundary conditions for the problem at hand are:

$$\left. \frac{1}{\rho_C} \frac{\partial V}{\partial x} \right|_{x=0} = \begin{cases} -J & \text{if } C/2 - W_E/2 \leq z \leq C/2 + W_E/2 \\ 0 & \text{if otherwise} \end{cases}, \tag{4-104}$$

$$V|_{x = A = X_C - X_{\text{dep}}} = 0 \tag{4-105}$$

$$\left. \frac{\partial V}{\partial z} \right|_{z=0,C} = 0 \tag{4-106}$$

where A and C are marked on the figure. This is a linear second-order partial differential equation. This equation can be solved by an integral transformation technique. According to Ref. [1], we write:

$$R_{C(\text{epi})} = R_{\text{spreading}}$$

$$= \frac{\rho_C}{L_E}\left[\sum_{q=1}^{\infty} \frac{2}{q^2\pi^2}\frac{C}{W_E}\cos\left(\frac{q\pi}{2}\right) G \tanh\left(q\pi\frac{A}{C}\right) + \frac{A}{C}\right] \tag{4-107}$$

where

$$G = \sin\left[\frac{q\pi}{2}\left(1 + \frac{W_E}{C}\right)\right] - \sin\left[\frac{q\pi}{2}\left(1 - \frac{W_E}{C}\right)\right] \tag{4-108}$$

For the simplificity of presenting the spreading resistance, we define a normalized spreading resistance as:

$$R_{\text{normalized}} = R_{\text{spreading}}\frac{L_E}{\rho_C} \tag{4-109}$$

We calculate the normalized spreading resistance as a function of the ratio W_E/C for several values of the ratio A/C. The results are shown in Fig. 4-19*b*. For example, suppose that a stripe geometry has the following parameters: $A = 2\ \mu\text{m}$, $C = 2\ \mu\text{m}$, $W_E = 1\ \mu\text{m}$. The ratios W_E/C and A/C are 0.5 and 1, respectively. Locating these ratios on the figure, we find that $R_{\text{normalized}} = 1.18$. If the semiconductor has a resistivity of $\rho_C = 0.1\ \Omega\cdot\text{cm}$ and a stripe length L_E of $20\ \mu\text{m}$, then the spreading resistance from Eq. (4-109) is $R_{\text{spreading}} = 59\ \Omega$.

We note that, if the ratio of W_E/C approaches 1, then approximation of $R_{C(\text{epi})}$ of Eq. (4-107) by Eq. (4-99) is sufficient. $R_{C(\text{epi})}$ is important in that maximizing its value generally involves thickening the subcollector layer, an action that can conflict with other processing requirements. This trade-off between a low $R_{C(\text{epi})}$ and the easiness of processing is especially evident in *Pnp* HBTs since the *p*-type subcollector layer has relatively high resistivity values.

The epitaxial resistance associated with the subcollector resembles the emitter epitaxial resistance. Because there is no current spreading, the subcollector epitaxial resistance is given as:

$$R_{SC(\text{epi})} = \rho_{SC}\frac{S_{BE} + S_{BC} + W_B}{X_{SC}\cdot L_E} \tag{4-110}$$

S_{BC} denotes the base–collector contact spacing and X_{SC} is the designed thickness of the subcollector layer. A sample calculation of some of the resistances or impedances is found in §4-8.

We now describe the contact resistances/impedance associated with the three terminals of the transistor. A contact to the transistor terminal is typically made by depositing a layer of metal directly onto the semiconductor. Depending on whether a subsequent thermal or alloying treatment is applied to the wafer, the contact is classified as either a *nonalloyed* or an *alloyed* contact. Because the nonalloyed contacts undergo no subsequent heat treatment after the metal deposition, the metal and semiconductor each resides at its side of the interface. If the nonideal natures of contact formation (such as contamination) can be neglected, then a nonalloyed contact represents an ideal semiconductor–metal junction whose theoretical properties were well analyzed. These studies established that, although the contact resistance depends on the geometry of the contact, its specific contact resistance (ρ_σ) depends only on the metal–semiconductor parameters. Therefore, the specific contact resistance is an important parameter characterizing the contact, and the contact resistance is simply calculated from the specific contact resistance with some formula. The specific contact resistance, having a unit of $\Omega \cdot \text{cm}^2$, is defined as:

$$\rho_\sigma = \left.\frac{dV}{dJ}\right|_{V \to 0} \tag{4-111}$$

where V is the applied voltage across the contact and J is the resulting current density flowing through the contact.

We do not attempt to review these theoretical studies. The design rules for the metal contact to be used in practical devices are obtainable from a qualitative understanding about the contact. Figure 4-20 illustrates the band diagram of an ideal metal–semiconductor system. The metal can be thought as a very heavily doped n-type semiconductor whose Fermi level lies below the conductor band

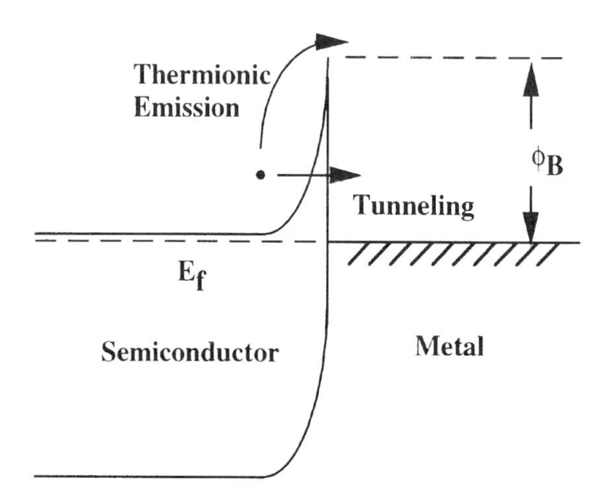

FIGURE 4-20. Band diagram of an ideal metal–semiconductor system.

edge. Because of the large number of available state below E_f, the metal can be characterized by its Fermi level alone, without the need to show its position in reference to its conduction band edge, as in the case of the semiconductor. During thermal equilibrium, the Fermi levels of the metal and of the semiconductor are aligned. Deep inside the bulk of the semiconductor, the potential difference between the Fermi level and the conductor band edge is determined by the doping. Near the surface, it is found that the E_f of the metal and the E_c of the semiconductor is fixed at a value independent of the semiconductor doping. The difference, referred to as the potential barrier (ϕ_B), remains unchanged during bias. Since ϕ_B at the interface is generally larger than $E_f - E_c$ in the bulk of the semiconductor, a depletion region near the interface exists inside the semiconductor. In the silicon material system, the magnitude of ϕ_B varies depending on the metal material used to form the contact. In GaAs, however, the barrier energy is experimentally found to be roughly the same independently of the metal. Such an observation often is referred to as the *surface pinning*, that the Fermi level near the interface is fixed because of a large number of surface states introduced by metals.

Depending on the doping level of the substrate, the charge carrier may overcome the barrier height through one of three mechanisms. When the doping level is light, the carrier may traverse the barrier only by acquiring thermal energy to pass over it. This process is called *thermionic emission*. If the semiconductor doping is very large, the depletion width associated with the barrier is so narrow that the carriers quantum mechanically tunnel through the barrier in a process called *field emission*. For intermediate doping levels, carriers cross the barrier with a combination of thermionic and field emission. This combined process is called the *thermionic-field emission*. Generally, a good ohmic contact is not possible with low semiconductor doping because, with the thermionic emission mechanism, only a small fraction of the carriers have enough thermal energy to traverse the barrier height. The contact resistance becomes tolerable when the doping is increased, so that thermionic-field emission is possible. The least resistive contact is obtained when conduction is purely field emission. In this case, because the Schottky barrier height does not appreciably impede the current flow, the contact is practically an ohmic contact. Therefore, as a rule of thumb, the semiconductor underneath the metal contact should be doped to its maximum level (the solubility of the dopant) to reduce the parasitic contact resistance.

The formula for the specific contact resistance of a nonalloyed contact depends on the exact conduction mechanism by which the carriers traverse the Schottky barrier. For the three conduction mechanisms described, the current as a function of applied voltage across the contact has each been derived, and the specific contact resistances are calculated using the definition of Eq. (4-111). An explicit formula for ρ_σ as a function of the semiconductor doping, the band density of states, the carrier effective mass, and the metal–semiconductor barrier height is known. However, we will not present the formula here. For one reason, they are complicated and a computer algorithm is necessary to calculate the values. Furthermore, because a contact is never as ideal as the theoreticians would want it to be, the theoretical calculation often does not agree well with the experimental result.

From these discussions about the nonalloyed contacts, we find that ρ_σ is an accepted figure of merit to characterize a given a contact because of the theoretical possibility of its being a constant, independent of the contact geometry. The desired contact resistance is calculated from ρ_σ once the contact geometry is known. For an alloyed contact, such as AuGe/Ni/Au on a *n*-type GaAs or Au/Zn on a *p*-type GaAs, the contact is alloyed after the deposition of the metal. A typical alloying schedule is to anneal the contact for 30 s at 440°C, at which temperature Au-GaAs eutectic is formed. (The *eutectic* of a compound AB has a lower melting point than that of either element A or element B.) This alloying process causes the contact metal (Au) to melt and penetrate into the GaAs substrate where it forms various compounds in localized areas. A theoretical analysis of such an inhomogeneous alloyed contact is difficult, and the contact is best characterized by empirical means. Extending the understanding from the nonalloyed contacts to the situation here, we characterize the alloyed contact with the specific contact resistance that is theoretically independent of the contact geometry. Typically, ρ_σ in a AuGe/Ni/Au *n*-type contact is about $10^{-6}\ \Omega \cdot \text{cm}^2$, although values approaching $10^{-7}\ \Omega \cdot \text{cm}^2$ were reported. Comparable values also are obtained in an Au/Zu *p*-type alloying contact.

The alloying contact scheme is not a popular choice to contact the base; very few publications reported alloying base contacts, partly because a good nonalloyed contact is readily obtainable from a heavily doped *p*-type base layer. Another reason relates to the fact that the base layer thickness is on the order of 1000 Å. If alloying were used, the metal spiking during the high temperature process could easily short the base with the collector underneath. The alloying contact is not desirable for the emitter, either. Although the emitter thickness is about 3000 Å, which is thick enough to prevent base–emitter shorting during processing, the metal spiking can eventually diffuse to the base after long operating hours. Therefore, instead of instant failure, the device degrades over time, causing reliability problems. The alloying contact, however, is ideal for the collector. The collector contact is made on the heavily doped subcollector layer. There is no active layer beneath it. In addition, the subcollector is on the order of 1 μm. Therefore, there is no worry that the metal spiking will effect device performance.

Once the specific contact resistance is either measured or obtained from a theoretical calculation, the contact resistance is readily calculated. Figure 4-21 illustrates the current flow patterns in the structure of a typical high-frequency HBT. The emitter current flows vertically downward from its contact. In the so-called *vertical-flow contact*, the contact resistance is given by the specific contact resistance divided by the contact area. This results follows from the definition of Eq. (4-111). Therefore,

$$R_{EE} = \frac{\rho_{\sigma E}}{A_E} = \frac{\rho_{\sigma E}}{L_E \cdot W_E} \tag{4-112}$$

where $\rho_{\sigma E}$ is the specific contact resistance of the emitter contact.

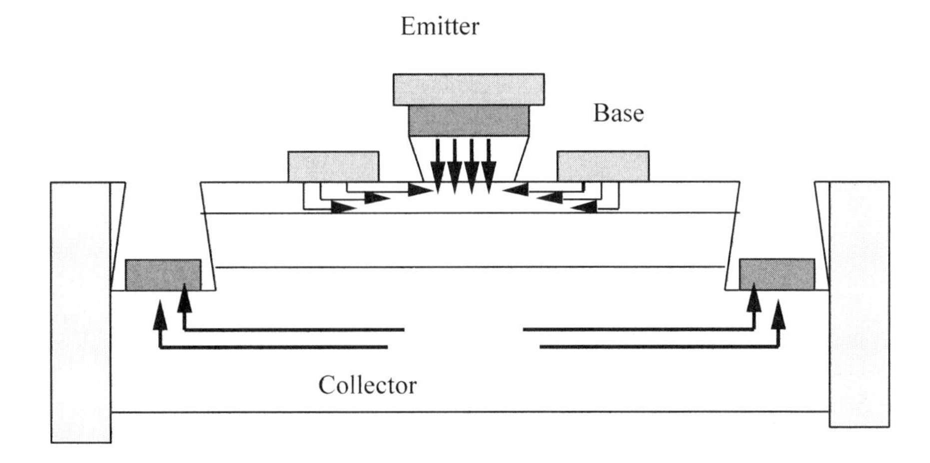

FIGURE 4-21. Current flows are horizontal for the base and collector currents. The emitter current flow is vertical. (From Liu, W. [1991]. Microwave and d.c. characterizations of Npn and Pnp HBTs. Ph.D. Dissertation, Stanford University, CA)

In the case of base and collector contacts, the associated current flow horizontally to the contacts. For these *horizontal-flow* contacts, the relationship between the contact resistance and the specific contact resistance is not as simple as depicted in Eq. (4-112). Figure 4-22 illustrates the model of a horizontal-flow contact. The dimensions labeled in this figure pertain to a base contact, but the results applies equally well to the collector contact with proper subscript changes. A capacitance is introduced in the model to account for the high-frequency behavior of the horizontal contact. This shunt capacitance is more important in the analysis of the base contact than of the collector contact because the larger base contact resistance causes a significant portion of the current to flow through the capacitive elements at high frequencies.

Because of the capacitive conduction of the base contact, the term *contact impedance* should have been used, instead of contact resistance. We hereafter use the term *base contact impedance* to emphasize this capacitive conduction contribution. In the case of the emitter contact, however, the term *emitter contact resistance* remains valid because usually the emitter contact resistance is very small and the majority of the current flows only through the resistive elements, even at high frequencies.

The base contact has dimensions of $W_B \times L_B$ as shown in Fig. 4-22. The base thickness is X_B. In this model, we define $R(\Omega/\text{cm})$, $G(\Omega^{-1}\cdot\text{cm}^{-1})$, and $C(\text{F/cm})$ as:

$$R = \frac{R_{SHB}}{L_E} \tag{4-113}$$

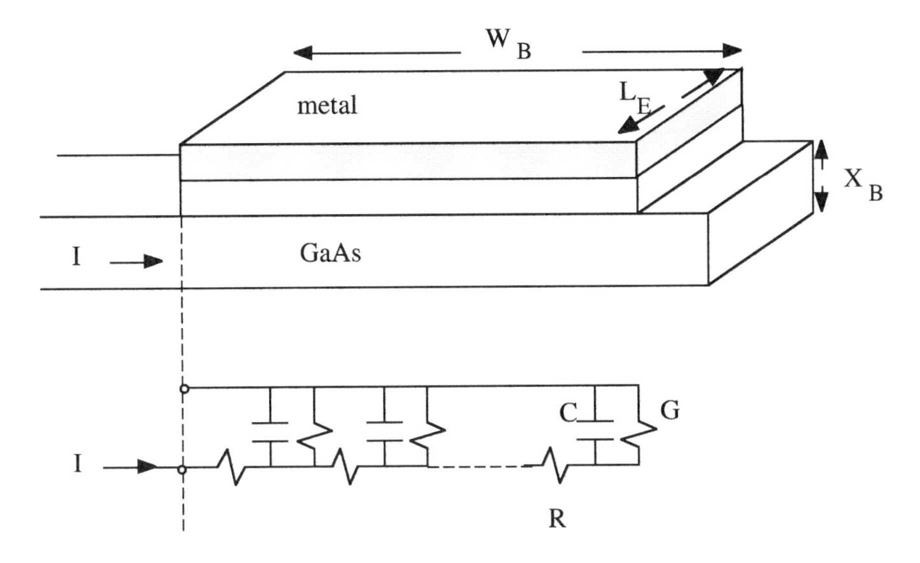

FIGURE 4-22. Base contact and its circuit representation. (From Liu, W. Microwave and d.c. characterizations of Npn and Pnp HBTs. Ph.D. Dissertation, Stanford University, CA)

$$G = \frac{L_E}{\rho_{\sigma B}} \tag{4-114}$$

$$C = C' \cdot L_E \tag{4-115}$$

R_{SHB} is the sheet resistance in the base layer, equal to $\rho_B / X_B . C'$ is the capacitance per unit area, and it can be estimated as:

$$C' = \sqrt{\frac{\epsilon_s q N_B}{2\phi_B}} \tag{4-116}$$

where N_B is the base doping, ϕ_B is the contact barrier height. It should be emphasized that the value of C' is a rough estimate.

From a transmission line analysis, the base contact impedance, Z_{BB}, can be expressed as:

$$Z_{BB} = z_{oB} \coth(\gamma_{BB} \cdot L_B) \qquad \text{(for horizontal contact)} \tag{4-117}$$

where

$$z_{oB} = \frac{\sqrt{R_{SHB} \cdot \rho_{\sigma B}}}{L_E} \frac{1}{\sqrt{1 + j\omega\rho_{\sigma B} C'}} \tag{4-118}$$

$$\gamma_{BB} = \sqrt{\frac{R_{SHB}}{\rho_{\sigma B}}} \sqrt{1 + j\omega\rho_{\sigma B} C'} \tag{4-119}$$

In defining the HBT small-signal model, it is desirable to represent the transmission-line-like base contact impedance of Eq. (4-117) with lumped circuit elements. For contacts on highly doped base material ($\sim 10^{20}$ cm^{-3}), the conduction through capacitive elements is negligible, and the base contact impedance is represented by a simple base contact resistance, R_{BB}:

$$R_{BB} = \frac{\sqrt{R_{SHB} \cdot \rho_{\sigma B}}}{L_E} \coth\left(W_B \sqrt{\frac{R_{SHB}}{\rho_{\sigma B}}} \right) \qquad (N_B \text{ large}) \qquad (4\text{-}120)$$

However, for transistors with lighter base doping, the modeling of the input transmission line impedance by lumped circuit elements becomes somewhat more complex [1].

Since the collector contact is typically an alloying contact in which the contact metal sinks into the semiconductor, the collector contact is purely resistive. The expression for the collector contact resistance (R_{CC}) is thus similar to R_{BB} of Eq. (4-120):

$$R_{CC} = \frac{\sqrt{R_{SHSC} \cdot \rho_{\sigma SC}}}{L_E} \coth\left(W_C \sqrt{\frac{R_{SHSC}}{\rho_{\sigma SC}}} \right) \qquad (4\text{-}121)$$

where R_{SHSC} is the sheet resistance of the subcollector layer and $\rho_{\sigma SC}$ is the specific contact resistance of the collector contact that is formed on the subcollector material.

§ 4-5 CUTOFF FREQUENCY (f_T)

The most important figures of merit characterizing a transistor's high-frequency performance are the cutoff frequency, f_T, and the maximum oscillation frequency, $f_{\max}$. We will discuss $f_{\max}$ in the next section. The cutoff frequency is defined as the frequency at which a magnitude of h_{21} decreased to unity. Formerly, we said that f_T is the frequency at which $|\beta_{ac}|$ becomes unity, where β_{ac} is the a.c. current gain. This was a rather casual statement. With different v_{cb}'s, β_{ac} may attain different values, leading to various values of f_T if v_{cb} were not specified. For convenience, f_T has been defined to be associated with the β_{ac} under short-circuit output condition. From the h-parameter definition Eq. (4-92) we find indeed that h_{21e} fits the required definition that:

$$h_{21e} = \left. \frac{i_2}{i_1} \right|_{v_2 = 0} = \left. \frac{i_c}{i_b} \right|_{v_{ce} = 0} \qquad (4\text{-}122)$$

The subscript e emphasizes that the transistor is in the common-emitter configuration. There is no sense to find f_T in a transistor biased in the common-base

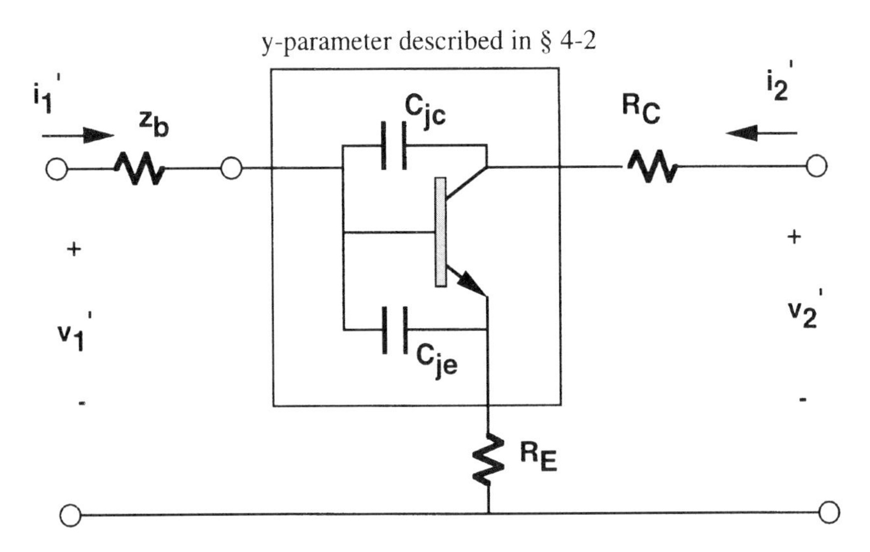

FIGURE 4-23. Circuit representation of a HBT used to calculate h_{21}. (From Liu, W. [1991]. Microwave and d.c. characterizations of Npn and Pnp HBTs. Ph.D. Dissertation, Stanford University, CA)

configuration, wherein the transistor current gain (i_c/i_e) never exceeds unity, not even under the d.c. condition.

The two-port network under consideration is shown in Fig. 4-23, which originates from the hybrid-π model discussed § 4-3. We will find the h'_{21e} corresponding to the primed two-port, which represents the realistic transistor incorporating the parasitic resistance or impedances. If the parasitic resistance/impedances are stripped away, the unprimed two-port representing the ideal intrinsic transistor appears. To clarify the terminology being used, we write out the resistances shown in Fig. 4-23:

$$R_E = R_{E(\text{epi})} + R_{EE} + R_{Em} \tag{4-123}$$

$$z_b = R_{Bi} + \frac{R_{Bx(\text{epi})}}{2} + \frac{Z_{BB}}{2} + \frac{R_{Bm}}{2} \tag{4-124}$$

$$R_C = R_{C(\text{epi})} + \frac{R_{SC(\text{epi})}}{2} + \frac{R_{CC}}{2} + \frac{R_{Cm}}{2} \tag{4-125}$$

The factors of 2 reflect the fact that there are two base and two collector contacts. We note that the R_{Bi} of the intrinsic device is lumped with the parasitic base contact impedance to form the overall z_b. This way, the y-parameters for the inner two-port marked as the "intrinsic transistor" are those given by Eq. (4-85). The epitaxial resistance ($R_{E(\text{epi})}$, R_{Bx}, $R_{C(\text{epi})}$) and contact impedances (R_{EE}, Z_{BB}, R_{CC}) were discussed in the previous sections. R_{Em}, R_{Bm}, and R_{Cm}

denote the metal resistances associated with the emitter, base, and collector interconnect metals, respectively. These resistances are generally disregarded because metals are thought to have low resistivity.

Before deriving an expression for the cutoff frequency, we first write out the expression itself and discuss the significance of each term. The cutoff frequency and the related emitter–collector transit time (τ_{ec}) are:

$$\tau_{ec} = \frac{1}{2\pi f_T} = \tau_e + \tau_b + \tau_{sc} + \tau_c \tag{4-126}$$

The first term, the emitter-charging time, is the time required to change the base potential by charging up the capacitances through the differential base–emitter junction resistance:

$$\tau_e = \frac{\eta k T}{q I_c} \cdot (C_{je} + C_{jc}) \tag{4-127}$$

C_{je} and C_{jc} denote the junction capacitances for the base–emitter and the base–collector junctions, respectively. They were given by Eqs. (4-18) and (4-67), respectively. C_{je} does not include the so-called diffusion capacitance which accounts for the stored minority carriers on both sides of the base–emitter junction under forward bias. While the base–emitter junction is forward-biased and the base–collector junction is reverse-biased, C_{jc} can be larger than C_{je} because the base–collector junction area is much larger than the base–emitter junction area.

This charging time has an inverse dependence on the collector current. This is the reason why a bipolar transistor designed for high-frequency performance operates at high collector current levels. At low current levels, this term often is the dominant component of the overall collector–emitter transit time.

τ_b is the base transit time. It is defined as the time required to discharge the excess minority carriers in the base through the collector current. It is given as:

$$\tau_b = \frac{X_B^2}{\nu \cdot D_{nB}} \tag{4-128}$$

The value of ν depends on the magnitude of the base quasi-electric field. From Eq. (3-37), we have:

$$\nu = 2 \cdot \left[\frac{2}{\kappa} \left(1 - \frac{1}{\kappa} + \frac{2}{\kappa} e^{-\kappa} \right) \right]^{-1} \tag{4-129}$$

where the field strength factor, κ, was defined in Eq. (3-35).

The space-charge transit time, τ_{sc}, is the transit time for the carriers to drift through the depletion region of the base–collector junction. It is given as:

$$\tau_{sc} = \frac{X_{\text{dep}}}{2v_{\text{sat}}} \tag{4-130}$$

where X_{dep} is the depletion thickness of the base–collector junction. The factor of 2 results from the averaging of a sinusoidal of carriers current over the period. It is assumed in this derivation that, because the electric field is large throughout the entire reverse-biased base–collector junction, the carriers travel at a constant saturation velocity, v_{sat}. For *Npn* HBTs, it is possible to design the collector layer in a way so that the field strength at the base side of the junction is small. When the electric field does not exceed a critical value, the electrons travel at the velocity determined completely by the Γ valley. Without scattering to the *L* valley, the electrons continue to travel at a velocity that is much larger than v_{sat} and velocity overshoot is said to occur. Therefore, τ_{sc} can be significantly reduced. There is no comparable scheme to reduce τ_{sc} for *Pnp* HBTs (for the same reason that the Gunn effects do not occur in *p*-type GaAs).

We reexamine the concept just mentioned. Figures 4-24 and 4-25 illustrates the epitaxial structure of a conventional HBT and a modified HBT. The notable difference is the *p*-collector layer inserted between the *n*-collector and the base in the modified structure. The band diagram of both types of transistors are also shown in their respective figures. Normally, when an electron does not acquire much kinetic energy, it resides in the Γ valley, which is characterized by an

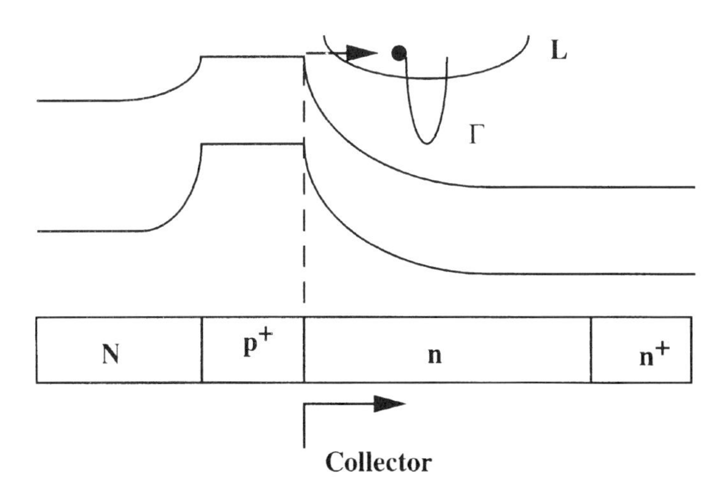

FIGURE 4-24. Band diagram of a conventional HBT whose base and collector are uniformly doped.

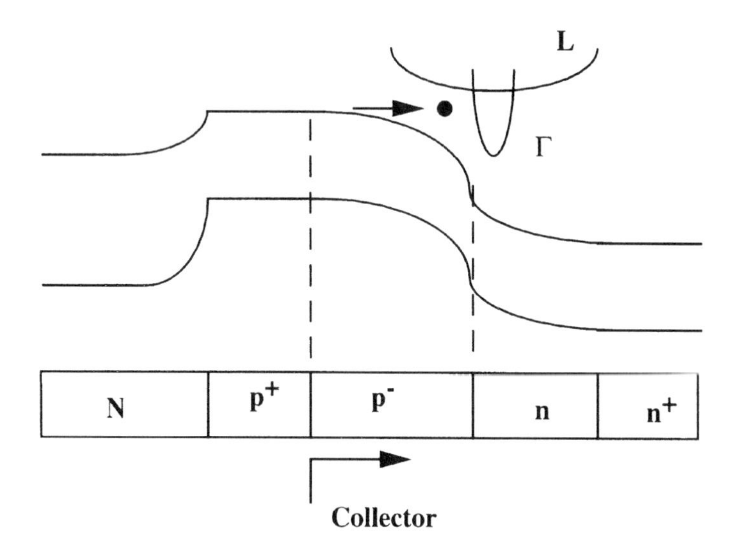

FIGURE 4-25. Band diagram of an HBT having the "p^-collector" structure. The p^- portion of the collector creates a low field region adjacent to the base.

effective mass of 0.067 m_0. When an electron leaves the base in the conventional HBT, it picks up a great amount of energy from the large field established in the base–collector junction. The electrons gain energy and move up in the Γ valley. When the energy is high enough, it may scatter to the L valley, which is characterized by a significantly larger effective mass. When a large population of electrons move through the base–collector junction, most electrons quickly gain enough energy and reside in the L valley, with only a minor fraction residing in the Γ valley. The value of saturation velocity accounts for the fact that electrons are found in both valleys and the overall effective mass is an averaged value of those of the L and Γ valleys over the associated population (or density of states) in the valleys.

By placing the p^- collector layer adjacent to the p^+ base, the electric field near the base is relatively small. The electrons do not acquire significant energy, and all of them remain in the Γ valley where its effective mass is small. Therefore, they travel at a velocity significantly higher than v_{sat}. This phenomenon is called *velocity overshoot*.

The last term, the collector charging time, τ_c, is given as:

$$\tau_c = (R_E + R_C) \cdot C_{jc} \tag{4-131}$$

The value of this charging time depends greatly on the parasitic emitter resistance. It can often dominate other terms in Eq. (4-126) at high collector current levels.

The overall transit time is a sum of the terms just listed:

$$\tau_{ec} = \frac{\eta kT}{qI_C} \cdot (C_{je} + C_{jc}) + \frac{X_B^2}{\nu \cdot D_n} + \frac{X_{dep}}{2v_{sat}} + (R_E + R_C) \cdot C_{jc} \tag{4-132}$$

To derive the transit time expression of Eq. (4-132), we need to evaluate h'_{21e} of the primed two-port shown in Fig. 4-23. We start with the common-emitter y-parameters for the intrinsic transistor. Subsequently, the modification of the lead resistances to intrinsic transistor two-port parameters is determined. From Eq. (4-83), the intrinsic common-emitter y-parameters are:

$$[y]_e = \begin{bmatrix} \sum y & -y_{cc} - y_{ec} \\ -y_{cc} - y_{ce} & y_{cc} \end{bmatrix}$$

where Σy is given by Eq. (4-84). The common-base y-parameters y_{ee}, y_{ce}, y_{ec}, and y_{cc} are given by Eq. (4-70b), and repeated here for convenience:

$$[y]_b = \begin{bmatrix} y_{ee} & y_{ec} \\ y_{ce} & y_{cc} \end{bmatrix} = \begin{bmatrix} g_e\left(1 + jm\dfrac{\omega}{\omega_0}\right) + j\omega C_{je} & 0 \\ -\alpha_{To} g_e\left(1 - j(1-m)\dfrac{\omega}{\omega_0} - j\dfrac{\omega\tau_m}{2}\right) & j\omega C_{jc} \end{bmatrix}$$

The constant m is a fitting parameter for the transistor's high-frequency characteristics. This constant ranges between 0 and 1 and is typically chosen to be 2/3 or 5/6. ω_0 is $1/\tau_b$, the reciprocal of the base transit time. ω_0 increases substantially if the transistor employs a base quasi-electric field.

The common-emitter $[z]$ matrix describing the intrinsic transistor without parasitics is obtained from the transformation of the $[y]$ matrix of Eq. (4-83). From Fig. 4-17:

$$[z]_e = \frac{1}{\Delta y}\begin{bmatrix} y_{cc} & +(y_{cc} + y_{ec}) \\ +(y_{cc} + y_{ce}) & \sum y \end{bmatrix} \tag{4-133}$$

where Σy was given in Eq. (4-84) and Δy is:

$$\Delta y = y_{ee} \cdot y_{cc} - y_{ec} \cdot y_{ce} \tag{4-134}$$

When the series resistive parasitics are considered, the $[z']$ matrix, which includes the parasitics, is easily obtained by adding elements to the original $[z]$ matrix. From Fig. 4-23, we find that:

$$v'_1 = v_1 + i_1 z_b + (i_1 + i_2) \cdot R_E = (z_{11} + z_b + R_E) \cdot i_1 + (z_{12} + R_E) \cdot i_2 \tag{4-135}$$

$$v'_2 = v_2 + i_2 R_C + (i_1 + i_2) \cdot R_E = (z_{11} + R_E) \cdot i_1 + (z_{12} + R_C + R_E) \cdot i_2 \tag{4-136}$$

Since $i_1' = i_1$ and $i_2' = i_2$, the $[z]$ matrix for the primed two-port is:

$$[z']_e = \frac{1}{\Delta y}\begin{bmatrix} y_{cc} + (R_E + z_b)\cdot\Delta y & y_{cc} + y_{ec} + R_E\cdot\Delta y \\ y_{cc} + y_{ce} + R_E\cdot\Delta y & \sum y + (R_E + R_C)\cdot\Delta y \end{bmatrix} \tag{4-137}$$

From Fig. 4-17, we find that h'_{21e} is $-(z'_{21}/z'_{22})$. Hence, the common-emitter forward current gain can be written as:

$$h'_{21e} = -\frac{y_{cc} + y_{ce} + R_E\cdot\Delta y}{\Psi} \tag{4-138}$$

which Ψ denotes:

$$\Psi = \Sigma y + (R_E + R_C)\cdot\Delta y \tag{4-139}$$

Inserting Eq. (4-70b) into Eq. (4-138), the numerator for h'_{21e} is approximately equal to $g_e\cdot\alpha_{To}$. The denominator of h'_{21e}, which is Ψ, is determined to be:

$$\Psi = g_e\cdot\left[\left(1 - \alpha_{To} - m(R_E + R_C)\cdot C_{jc}\frac{\omega^2}{\omega_0} - (R_E + R_C)\cdot C_{jc}\frac{\omega^2 C_{je}}{g_e}\right) + j\omega\tau_{ec}\right] \tag{4-140}$$

where τ_{ec} is exactly the same emitter–collector transit time as that given in Eq. (4-132). If we neglect the second-order terms, then Ψ is simplified to $g_e\cdot[(1-\alpha_{To}) + j\omega\tau_{ec}]$. Consequently, h'_{21e} can be written in a more familiar form:

$$h'_{21e} = \frac{g_e\cdot\alpha_{To}}{g_e\cdot[(1-\alpha_{To}) + j\omega\tau_{ec}]} = \frac{\alpha_{To}}{(1-\alpha_{To}) + j(\omega/\omega_T)} \tag{4-141}$$

where ω_T, equal to $1/\tau_{ec}$, is:

$$\omega_T = \left[\frac{\eta kT}{qI_C}\cdot(C_{je} + C_{jc}) + \frac{X_B^2}{\nu\cdot D_n} + \frac{X_{\text{dep}}}{2\,v_{\text{sat}}} + (R_E + R_C)\cdot C_{jc}\right]^{-1} \tag{4-142}$$

η is the collector current ideality factor. At high frequencies, as $\omega \to \omega_T$, the imaginary part of the denominator of h'_{21e} dominates. Therefore, $|h'_{21e}| \to 1$ when $\omega \to \omega_T$, agreeing with the definition of the cutoff frequency. Another useful expression of h'_{21e} is:

$$h'_{21e} = \frac{\alpha_{To}}{(1-\alpha_{To})}\frac{1}{1 + j(\omega/\omega_{3dB})} \tag{4-143}$$

where ω_{3dB} is the 3-dB rolloff radian frequency of h'_{21e}. It relates to ω_T as:

$$\omega_T = \omega_{3dB} \cdot \frac{1}{1 - \alpha_{To}} \tag{4-144}$$

Some interesting observations about Eq. (4-141) are in order. First, when the frequency is high, $j\omega/\omega_{3dB} \gg 1$. Therefore, for a 10-fold increase in frequency, $|h'_{21e}|$ decreases by 10-fold. In other words, $|h'_{21e}|$ rolls off with frequency at a slope of 20 dB/decade at high frequencies. There are occasions when the HBT's cutoff frequency exceeds 40 GHz, which is about the measurement limit of popular network analyzers. In this case, the cutoff frequency is then extrapolated from the last taken data, using the 20-dB/decade slope to find out the frequency when $|h'_{21e}|$ rolls off to unity. Second, the lack of dependence of the transit time expression on d.c. current gain suggests that the cutoff frequency of a HBT does not depend on the magnitude of the d.c. current gain. Therefore, even though the d.c. current gain of a HBT is dramatically lowered by extrinsic base surface recombination or base bulk recombination, ω_T is not affected. Another way to understand this is to realize that ω_T, defined as the frequency at which the current gain decreases to unity, is the gain-bandwidth product of the transistor. Consequently, ω_T is a constant, independent of the d.c. current gain. ω_{3dB}, in contrast, varies with the d.c. current gain according to Eq. (4-144).

Finally, we note that the common-emitter forward gain of Eq. (4-141) does not depend on the base resistance at all. If the maximization of f_T is the ultimate goal, we simply reduce the base thickness so that the base transit time is minimized. Although the base resistance increases, the increase has inconsequential effects on f_T. The resistance, however, reduces the power gain and lowers f_{max}. Therefore, a trade-off in the base layer design is unavoidable if both f_T and f_{max} are to be maximized. Often when a new device is claimed to have excellent high-frequency performance, only the f_T will be reported. This is not because somehow someone forgot to measure f_{max}, since f_T and f_{max} are determined from the same set of measured s-parameters. It is only that the device has a high base resistance, preventing the device from having a decent f_{max}.

§ 4-6 MAXIMUM OSCILLATION FREQUENCY (f_{max})

The cutoff frequency does not depend strongly on the parasitic base contact and intrinsic base impedances because the current that must charge the capacitances does not come from the base, but is injected through the emitter. The most important sources of parasitic resistance affecting the cutoff frequency are the emitter resistances that are along the main current path. As will be demonstrated shortly, the importance of the base resistance emerges as we consider the maximum oscillation frequency, which is a measure of the power gain.

The maximum oscillation frequency is the frequency at which the unilateral power gain of the transistor rolls off to unity. It is the frequency that marks the boundary between an active and a passive network. When a transistor whose reverse transmission parameter y_{12} (or z_{12}, h_{12}) is zero, it is said to be *unilateral*. That is, the output is completely isolated from the input. In practice, some amount of reverse transmission always exists. The process by which a lossless feedback network is added to the transistor such that the overall two-port is unilateral is called the *unilateralization*. The unilateral power gain of a transistor is the power gain after the transistor plus the lossless network is made unilateral. It represents the maximum power gain achievable by the transistor. A formula for the unilateral power gain was derived in terms of the z-matrix

$$U = \frac{|z_{21} - z_{12}|^2}{4[\mathrm{Re}(z_{11}) \cdot \mathrm{Re}(z_{22}) - \mathrm{Re}(z_{12}) \cdot \mathrm{Re}(z_{21})]} \tag{4-145}$$

Re denotes the function that takes the real part of its argument. It should not be confused with the emitter resistance R_E. If we wish to express U in terms of the y-parameters, we write using the conversion relationships specified in Fig. 4-17:

$$U = \frac{|(y_{12}/\Delta y) - (y_{21}/\Delta y)|^2}{4[\mathrm{Re}(y_{22}/\Delta y) \cdot \mathrm{Re}(y_{11}/\Delta y) - \mathrm{Re}(y_{21}/\Delta y) \cdot \mathrm{Re}(y_{12}/\Delta y)]} \tag{4-146}$$

where Δy was given by Eq. (4-134).

After a rather elaborate algebraic manipulation, the formula simplifies to:

$$U = \frac{|y_{21} - y_{12}|^2}{4[\mathrm{Re}(y_{11}) \cdot \mathrm{Re}(y_{22}) - \mathrm{Re}(y_{12}) \cdot \mathrm{Re}(y_{21})]} \tag{4-147}$$

From a cursory examination of Eqs. (4-145) and (4-147), we might conclude that the two equations are equivalent only if Δy is a real number such that it can be factored out and canceled from the numerator and the denominator. However, with Δy specified as in Eq. (4-134), it can be shown that Eq. (4-145) simplifies to Eq. (4-147), even though Δy is a complex quantity.

The stability factor, k, measures whether a transistor will be unconditionally stable given arbitrary passive loads. If $k > 1$, the transistor is *inherently stable*; that is, no passive terminations can cause oscillation without suitable external feedback. A device must be inherently stable for simultaneous conjugate matching to be possible and for the maximum power gain to be finite. When $k < 1$, the transistor is *potentially unstable*; that is, sustained oscillations can occur for certain passive terminations. It is found that the Rollett's stability factor has a particular functional form, independent of whether y- or z- or h-parameters are used inside the equations. Expressed in the universal equation,

$$k = \frac{2\,\mathrm{Re}(\vartheta_{11})\mathrm{Re}(\vartheta_{22}) - \mathrm{Re}(\vartheta_{12} \cdot \vartheta_{21})}{|\vartheta_{12} \cdot \vartheta_{21}|} \quad \text{(Rollett stability number)} \tag{4-148}$$

where ϑ may be any of the conventional z-, y-, or h-matrix parameters. This property is partially shared for the unilateral gain, but only for the y- and z-parameters, as shown by Eqs. (4-145) and (4-147).

We caution that the stability factor of Eq. (4-148) is not the same as the Stern stability factor. For completeness, we give Stern stability number here:

$$\text{Stern stability factor} = \frac{2\,\mathrm{Re}(y_{11})\,\mathrm{Re}(y_{22})}{\mathrm{Re}(y_{12}\cdot y_{21}) + |y_{12}\cdot y_{21}|} \tag{4-149}$$

Stern stability as given in Eq. (4-149) does not have the "universal" property enjoyed by the Rollett stability factor given in Eq. (4-148). That is, substituting y_{11}, y_{12}, y_{21}, and y_{22} of Eq. (4-149) by z_{11}, z_{12}, z_{21}, and z_{22}, respectively, results in a different value. Stern stability number, when expressed in terms of the z-parameters, has a different form from Eq. (4-149). Despite the different definitions in the Rollett and the Stern factors, when either stability factor is greater than 1, then the other also is greater than 1, so the transistor is unconditionally stable under either criterion.

Example 4-5:

Find the Rollett and Stern stability factors for the common-emitter HBT that has the following parameters: $y_{11e} = (0.5 + \mathrm{j}2)$ mS; $y_{12e} = (-0.005 - \mathrm{j}0.06)$ mS; $y_{21e} = (20 - \mathrm{j}3)$ mS; $y_{22e} = (0.02 + \mathrm{j}0.06)$ mS.

The Rollett stability is given by Eq. (4-148):

$$k_{\text{Rollet}} = \frac{2\,\mathrm{Re}(y_{11})\,\mathrm{Re}(y_{22}) - \mathrm{Re}(y_{12}\cdot y_{21})}{|y_{12}\cdot y_{21}|} = 0.246$$

$$\text{Stern stability factor} = \frac{2\mathrm{Re}(y_{11})\,\mathrm{Re}(y_{22})}{\mathrm{Re}(y_{12}\cdot y_{21}) + |y_{12}\cdot y_{21}|} = 0.02$$

Although their values differ, they are both smaller than 1, indicating that the transistor is potentially unstable.

In the previous section, we expressed the cutoff frequency as a function of device parameters. We accomplished the task by finding an expression of h'_{21e} in terms of the common-base y-parameters such as y_{ee}, y_{ec}, y_{ce}, and y_{cc}. We attempt to perform a similar task for the unilateral power gain and, therefore, f_{max}. Figure 4-23 is again used as the circuit representation. The y-parameters for the intrinsic device (unprimed two-port) are well known from the discussion in § 4-1. After some transformations, the z-parameters for the primed two-port were obtained, as found in Eq. (4-137). Plugging the primed z-parameters into Eq. (4-145), we find U to be:

$$U = \frac{|(y_{ce} - y_{ec})/\Delta y|^2}{4\{\mathrm{Re}[(\Sigma y/\Delta y) + R_E + R_C] \cdot \mathrm{Re}[(y_{cc}/\Delta y) + R_E + z_b]} \\ - \mathrm{Re}\,[(y_{cc} + y_{ce})/\Delta y + R_E] \cdot \mathrm{Re}\,[(y_{cc} + y_{ec})/\Delta y + R_E]\} \quad (4\text{-}150)$$

Δy in the terms in both the numerator and the denominator do not cancel out readily because Δy is in general a complex (as opposed to real) quantity. The derivation of f_{max} leading to a simple expression is painstaking, and we need all the approximations we can get. We simplify the common-base parameters of Eq. (4-70a) somewhat:

$$[y]_b = \begin{bmatrix} y_{ee} & y_{ec} \\ y_{ce} & y_{cc} \end{bmatrix} = \begin{bmatrix} g_e + j\omega C_\pi & 0 \\ -\alpha_{T_0} g_e e^{-j\omega\theta} & j\omega C_{jc} \end{bmatrix} \quad (4\text{-}151)$$

C_π is the sum of C_{je} and mg_e/ω_0, and θ is simply $\tau_m/2$. Noting that $y_{ec} \approx 0$, we then simplify Eq. (4-150) as:

$$U = \frac{|y_{ce}/y_{ee}y_{cc}|^2}{4\{[\mathrm{Re}(1/y_{cc}) + (1/y_{ee}) + (y_{ce}/y_{ee}y_{cc}) + R_E + R_C] \cdot \mathrm{Re}[(1/y_{ee}) + R_E + z_b]} \\ - \mathrm{Re}[(1/y_{ee}) + (y_{ce}/y_{ee}y_{cc}) + R_E] \cdot \mathrm{Re}[(1/y_{ee}) + R_E]\} \quad (4\text{-}152)$$

After canceling out terms and noticing $\mathrm{Re}(y_{cc}) = 0$, we arrive at:

$$U = \frac{|y_{ce}/(y_{ee} \cdot y_{cc})|^2}{4\mathrm{Re}(z_b)\mathrm{Re}[(1/y_{ee}) + (y_{ce}/y_{ee} \cdot y_{cc}) + R_E + R_C]} \quad (4\text{-}153)$$

Thus far the only assumption we have made is to use Eq. (4-151) for the common-base y-parameters. After this point, it is easier to carry on the algebra by first inverting both sides of the equation. Consequently:

$$\frac{1}{U} = 4\,\mathrm{Re}(z_b) \frac{\mathrm{Re}\left(\dfrac{1}{g_e + j\omega C_\pi}\right) + \mathrm{Re}\left[\dfrac{-\alpha_{T_0} g_e(1 - j\omega\theta)}{j\omega C_{jc}(g_e + j\omega C_\pi)}\right] + R_E + R_C}{\left|\dfrac{-\alpha_{T_0} g_e e^{-j\omega\theta}}{j\omega C_{jc}(g_e + j\omega C_\pi)}\right|^2}$$

$$= 4\,\mathrm{Re}(z_b) \frac{\mathrm{Re}\left(\dfrac{g_e - j\omega C_\pi}{g_e^2 + \omega^2 C_\pi^2}\right) + \mathrm{Re}\left[\dfrac{-\alpha_{To} g_e(1 - j\omega\theta)}{j\omega C_{jc} g_e - \omega^2 C_\pi C_{jc}}\right] + R_E + R_C}{\dfrac{\alpha_{To}^2 g_e^2}{\omega^2 C_{jc}^2(g_e^2 + \omega^2 C_\pi^2)}}$$

$$= 4\,\mathrm{Re}(z_b)\frac{\dfrac{g_e}{g_e^2+\omega^2 C_\pi^2}+\mathrm{Re}\left[\dfrac{-\alpha_{To}g_e(1-j\omega\theta)(-j\omega C_{jc}g_e-\omega^2 C_\pi C_{jc})}{\omega^2 C_{jc}^2 g_e^2+\omega^4 C_\pi^2 C_{jc}^2}\right]+R_E+R_C}{\dfrac{\alpha_{To}^2 g_e^2}{\omega^2 C_{jc}^2(g_e^2+\omega^2 C_\pi^2)}}$$

$$\approx 4\,\mathrm{Re}(z_b)\left[\frac{g_e\omega^2 C_{jc}^2}{\alpha_{To}^2 g_e^2}+\frac{\alpha_{To}g_e(\omega^2 C_\pi C_{jc}+\theta\omega^2 C_{jc}g_e)}{\omega^2 C_{jc}^2 g_e^2+\omega^4 C_\pi^2 C_{jc}^2}\frac{\omega^2 C_{jc}^2(g_e^2+\omega^2 C_\pi^2)}{\alpha_{To}^2 g_e^2}\right.$$
$$\left.+(R_E+R_C)\frac{\omega^2 C_{jc}^2(g_e^2+\omega^2 C_\pi^2)}{\alpha_{To}^2 g_e^2}\right]$$

$$\approx 4\,\mathrm{Re}(z_b)\left[\frac{C_{jc}\omega^2}{\alpha_{To}^2}\left(\frac{C_{jc}}{g_e}\right)+\frac{\alpha_{To}g_e(\omega^2 C_\pi C_{jc}+\omega^2 C_{jc}g_e\tau_m/2)}{\alpha_{To}^2 g_e^2}\right.$$
$$\left.+(R_E+R_C)\frac{\omega^2 C_{jc}^2 g_e^2}{\alpha_{To}^2 g_e^2}\right]$$

$$\approx 4\,\mathrm{Re}(z_b)\left[\frac{C_{jc}\omega^2}{\alpha_{To}^2}\left(\frac{C_{jc}}{g_e}\right)+\frac{C_{jc}\omega^2}{\alpha_{To}^2}\left(\frac{C_\pi}{g_e}+\frac{\tau_m}{2}\right)+C_{jc}\frac{C_{jc}\omega^2}{\alpha_{To}^2}(R_E+R_C)\right] \quad (4\text{-}154)$$

Noting that the cutoff frequency ω_T is that expressed in Eq. (4-142), we find U to be:

$$U=\frac{\alpha_{To}^2}{4\,\mathrm{Re}(z_b)C_{jc}\omega^2}\left[\frac{C_{jc}}{g_e}+\frac{C_\pi}{g_e}+\frac{\tau_m}{2}+(R_E+R_C)\cdot C_{jc}\right]^{-1}=\frac{\alpha_{To}^2\omega_T}{4\,\mathrm{Re}(z_b)C_{jc}\omega^2} \quad (4\text{-}155)$$

This is one of the most popular equations that are quoted by textbooks without much derivation. Note that the power gain in decibels (dB) is obtained by multiplying the logarithm of the power gain by a factor of 10, rather than 20 as for the current gain. For every 10-fold increase in frequency, U given by Eq. (4-155) decreases by 100-fold, which amounts to 20 dB. Therefore, the magnitude maximum power gain in dB rolls off at 20 dB/decade, which is the same rate at which $|h_{21}|$ decreases with frequency.

An important property of the unilateral gain lies in its invariance with the terminal chosen as the reference terminal. Therefore, Eq. (4-155) applies equally well to HBTs biased in the common-base configuration, although it was derived based on the common-emitter configuration. A proof of such property is as follows. Using the common-emitter y-parameters of Eq. (4-83), we write the unilateral power gain according to Eq. (4-147) as:

$$U_{(e)}=\frac{|-y_{cc}-y_{ce}+y_{cc}+y_{ec}|^2}{4[\mathrm{Re}(y_{ee}+y_{ce}+y_{ec}+y_{cc})\cdot\mathrm{Re}(y_{cc})-\mathrm{Re}(y_{cc}+y_{ce})\cdot\mathrm{Re}(y_{cc}+y_{ec})]} \quad (4\text{-}156)$$

After expanding the terms out, we find that

$$U_{(e)} = \frac{|-y_{ce} + y_{ec}|^2}{4[\mathrm{Re}(y_{ee}) \cdot \mathrm{Re}(y_{cc}) - \mathrm{Re}(y_{ce}) \cdot \mathrm{Re}(y_{ec})]} \tag{4-157}$$

which is precisely, $U_{(b)}$, the unilateral gain for the common-base configuration.

The maximum oscillation frequency is the frequency at which U equals unity. Equation (4-155) shows that the exact formula depends on the frequency dependence of z_b. Of the four components of z_b given in Eq. (4-124), the intrinsic base impedance and the contact impedance can have frequency dependences. If the base doping is high, the base contact impedance is purely resistive, with a value given in Eq. (4-120). Therefore:

$$z_b = r_b = \frac{1}{2}\frac{\sqrt{R_{SHB} \cdot \rho_{\sigma B}}}{L_E} \coth\left(W_B\sqrt{\frac{R_{SHB}}{\rho_{\sigma B}}}\right) + \frac{1}{2}R_{Bx(\mathrm{epi})} + \frac{1}{12}R_{Bi(\mathrm{tvs})} + \frac{1}{2}R_{Bm} \tag{4-158}$$

Note that a factor of $\frac{1}{2}$ in front of the contact, extrinsic base, and base metal resistances accounts for the fact that there are two base contacts. Since z_b has no frequency dependence, we simply say that $z_b = r_b$. The maximum oscillation frequency obtained from Eq. (4-155) is:

$$f_{\max} = \sqrt{\frac{f_T}{8\pi r_b C_{jc}}} \tag{4-159}$$

Example 4-6:

Estimate the maximum frequency of oscillation of a HBT whose $r_b = 20\ \Omega$. The device has a collector area of 180 $\mu\mathrm{m}^2$. The HBT operates at a current level such that the collector depletion thickness is 1 μm and the measured cutoff frequency is 20 GHz.

From Eq. (4-67):

$$C_{jc} = 180 \times 10^{-8}\,\frac{1.159 \times 10^{-12}}{1 \times 10^{-4}} = 2.1 \times 10^{-14}\ \mathrm{F}$$

Based on Eq. (4-159), the maximum oscillation frequency is:

$$f_{\max} = \sqrt{\frac{20 \times 10^9}{8\pi \cdot 20 \cdot 2.1 \times 10^{-14}}} = 43 \times 10^9\ \mathrm{Hz}$$

§ 4-7 EXAMPLE: CALCULATION OF f_T and f_{max}

Using the equations developed in the preceding sections, we present a step-by-step calculation for both the cutoff frequency and the maximum oscillation frequency. This exercise not only serves as a review of the various components of the equations, but also elucidates some of the design trade-offs. The exemplar transistor dimensions and epitaxial structure used for the calculation are shown in Fig. 4-26. The operating condition is: $V_{BE} = 1.5$ V, $V_{CB} = 4$ V, and $J_C = 10^4$ A/cm^2. The contact resistivities typically measured for a transistor of this structure are 7.4×10^{-8}, 3.3×10^{-6} and 8.4×10^{-7} $\Omega \cdot$cm^2 for the emitter, base, and collector contacts, respectively. R_{SHB} and R_{SHSC} for a transistor of this structure are generally measured to be 280 and 8.5 Ω/sq, respectively.

The first step is to calculate the values of the resistive and capacitive elements. The emitter epitaxial resistances consists of two components: the resistances in the GaAs transition layer and the undepleted AlGaAs active emitter layer.

The electron mobility in a n-type GaAs layer as a function of the doping level is given by Eq. (1-73). For a doping level of 2×10^{18} cm^{-3} in the emitter

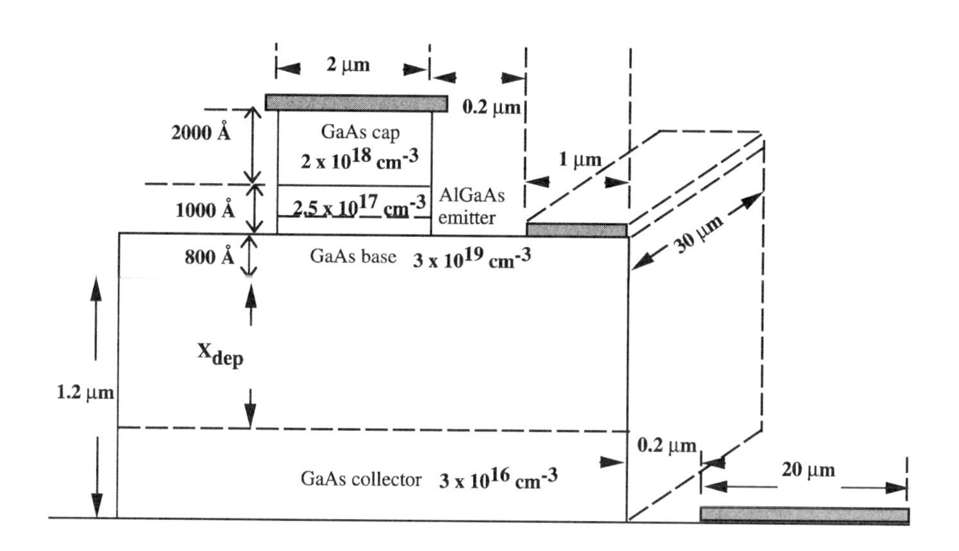

FIGURE 4-26. Transistor structure used in the exemplar calculation of § 4-7.

transition layer, μ_n is 2402 cm^2/V-s. ρ_{Ecap}, given by an equation similar to Eq. (3-53), is $1.3 \times 10^{-3}\ \Omega \cdot$cm. For the AlGaAs active emitter layer, the data of the electron mobility as a function of doping are not as comprehensive as those of GaAs. From an experimental work, the electron mobility measured in an $Al_{0.35}Ga_{0.65}As$ layer doped at 2.5×10^{17} cm^{-3} is 1000 cm^2/V-s. Therefore, ρ_E is 0.025 $\Omega \cdot$cm. The built-in potential of the AlGaAs/GaAs base-emitter heterojunction is equal to the $Al_{0.35}Ga_{0.65}As$ energy gap minus the valence band discontinuity of the base-emitter heterojunction. From Eq. (4-15):

$$\phi_{BE} = 1.4242 + 1.247 \times 0.35 - 0.55 \times 0.35 = 1.668 \text{ V}.$$

According to Eq. (4 14), the depletion thickness at a base-emitter bias of 1.5 V is:

$$X_{\text{dep}E} = \sqrt{\frac{2 \cdot 1.159 \times 10^{-12}}{1.6 \times 10^{-19} \cdot 2.5 \times 10^{17}}(1.668 - 1.5)} = 310 \text{ Å}$$

According to Eq. (4-97):

$$R_{E(\text{epi})} = 1.3 \times 10^{-3} \frac{2000 \times 10^{-8}}{2 \times 10^{-4} \cdot 30 \times 10^{-4}} + 0.025 \frac{(1000 - 310) \times 10^{-8}}{2 \times 10^{-4} \cdot 30 \times 10^{-4}} = 0.33\ \Omega$$

According to Eq. (4-112) and with a typical emitter contact resistivity of $7.4 \times 10^{-8}\ \Omega \cdot \text{cm}^2$,

$$R_{EE} = \frac{7.4 \times 10^{-8}}{2 \times 10^{-4} \cdot 30 \times 10^{-4}} = 0.123\ \Omega$$

According to Eq. (3-52), the transverse resistance associated with the intrinsic base layer is:

$$R_{Bi(tvs)} = \frac{\rho_B}{X_B}\frac{W_E}{L_E} = R_{SHB}\frac{W_E}{L_E} = 280 \cdot \frac{2 \times 10^{-4}}{30 \times 10^{-4}} = 18.67\ \Omega$$

In the calculation, we have replaced ρ_B/X_B by R_{SHB}, which is usually measured to be 280 Ω/sq for the base layer structure of Fig. 4-26. We could have calculated ρ_B if the hole mobility was known. However, since the measured data often reflect the actual quality of the base layer, we use the experimental data in place of a theoretical determination. The intrinsic base resistance in only a fraction of the transverse base resistance. According to Eq. (3-60):

$$R_{Bi} = \frac{R_{Bi(tvs)}}{12} = \frac{18.67}{12} = 1.56\ \Omega$$

Equation (4-98) gives the extrinsic base resistance:

$$R_{Bx(\text{epi})} = R_{SHB}\frac{S_{BE}}{L_E} = 280 \frac{0.2 \times 10^{-4}}{30 \times 10^{-4}} = 1.87\ \Omega$$

The base contact resistance is found from Eq. (4-120) using a typical measured contact resistivity of $3.3 \times 10^{-6}\ \Omega \cdot \mathrm{cm}^2$ for the given base layer structure,

$$R_{BB} = \frac{\sqrt{280 \cdot 3.3 \times 10^{-6}}}{30 \times 10^{-4}} \coth\left(1 \times 10^{-4} \cdot \sqrt{\frac{280}{3.3 \times 10^{-6}}}\right) = 13.9\ \Omega$$

By far the contact resistance is the largest component of the base resistance in this transistor. Because it dominates the overall base resistance, we examine its functional forms in more details. The value of the hyperbolic cotangent function of Eq. (4-120) is about 1.38 in this example. If a larger W_B is chosen (wider base contact), the value of the hyperbolic cotangent can be reduced toward its lowest value of unity. However, the benefits of reducing R_{BB} by using a wider base contact width is offset by the increase in the parasitic capacitances. Some kind of trade-off in W_B is inevitable.

According to Eq. (4-110), the resistance associated with the subcollector layer is:

$$R_{SC\,(\mathrm{epi})} = R_{SHSC} \frac{S_{BE} + S_{BC} + W_B}{L_E}$$

$$= 8.5 \frac{0.2 \times 10^{-4} + 0.2 \times 10^{-4} + 1 \times 10^{-4}}{30 \times 10^{-4}} = 0.40\ \Omega$$

Note that the measured sheet resistance of $8.5\ \Omega/\mathrm{sq}$ is used instead of ρ_{SC}/X_{SC}.

Although R_{SH} of the subcollector is measurable, it is generally difficult to measure the R_{SH} of the collector layer. We therefore resort to a theoretical calculation to calculate $R_{C\,(\mathrm{epi})}$ and R_{CC}. Using the electron mobility expression of Eq. (1-73), we find that for the $3 \times 10^{16}\ \mathrm{cm}^{-3}$ collector layer, μ_n is $5736\ \mathrm{cm}^2/\mathrm{V}$-s. The resistivity of the layer is calculated using an equation similar to Eq. (3-53), and found to be $3.63 \times 10^{-2}\ \Omega \cdot \mathrm{cm}$. Once this material parameter is known, we still need to find the depletion thickness to determine $R_{C\,(\mathrm{epi})}$. The reverse base–collector bias is given to be 4 V, and the built-in potential of the base–collector junction is obtained from Eq. (3-71):

$$\phi_{CB} = \frac{1.424}{2} + 0.0258 \cdot \ln \frac{3 \times 10^{16}}{1.79 \times 10^{6}} = 1.32\ \mathrm{V}$$

The depletion thickness is obtained from Eq. (3-78). We assume that $J_2 \gg J_C$. With $J_1 = qN_C \cdot v_{\mathrm{sat}}$, we find:

$$X_{\mathrm{dep}} \sim \sqrt{\frac{2\epsilon_s(V_{CB} + \phi_{CB})}{qN_c}} \left(1 - \frac{J_C}{J_1}\right)^{-1/2}$$

$$= \sqrt{\frac{2 \cdot 1.159 \times 10^{-12}(4 + 1.32)}{1.6 \times 10^{-19} \cdot 3 \times 10^{16}}} \left(1 - \frac{10^4}{1.6 \times 10^{-19} \cdot 3 \times 10^{16} \cdot 8 \times 10^6}\right)^{-1/2}$$

$$= 5.91 \times 10^{-5} \text{ cm}$$

So, $X_{\text{dep}} = 0.591$ μm. According to Eq. (4-99):

$$R_{C(\text{epi})} = 3.63 \times 10^{-2} \frac{1.2 \times 10^{-4} - 0.591 \times 10^{-4}}{2 \times 10^{-4} \cdot 30 \times 10^{-4}} = 3.67\ \Omega$$

The contact resistance is obtained according to Eq. (4-121):

$$R_{CC} = \frac{\sqrt{8.5 \cdot 8.4 \times 10^{-7}}}{30 \times 10^{-4}} \coth\left(20 \times 10^{-4} \cdot \sqrt{\frac{8.5}{8.4 \times 10^{-7}}}\right) = 0.89\ \Omega$$

The hyperbolic cotangent term here has a near-unity value, unlike that found in the base contact expression. This is in part due to the lower specific contact resistance in the collector contact. But more important, it is also due to the fact that the collector contact width is significantly larger than the base contact width. As mentioned, W_B directly affects the magnitude of the parasitic base–collector junction capacitance; it cannot be increased arbitrarily just for the purpose of reducing the base contact resistance. In contrast, the collector contact width does not affect any device parameters other than R_{CC}. There is the flexibility to increase the contact width to minimize the contact resistance.

The emitter, base, and collector resistances appearing in the model of Fig. 4-23 are calculated from Eqs. (4-123) to (4-125). We neglect the metal resistances associated with the three terminals. Therefore:

$$R_E = R_{E(\text{epi})} + R_{EE} = 0.33 + 0.123 = 0.453\ \Omega$$

$$r_b = R_{Bi} + \frac{R_{Bx(\text{epi})}}{2} + \frac{R_{BB}}{2} = 1.56 + \frac{1.87}{2} + \frac{13.9}{2} = 9.45\ \Omega$$

$$R_C = R_{C(\text{epi})} + \frac{R_{CC}}{2} + \frac{R_{SC(\text{epi})}}{2} = 3.67 + \frac{0.89}{2} + \frac{0.40}{2} = 4.32\ \Omega$$

The main component of the collector resistance is that associated with the undepleted collector. This signifies that the designed collector thickness is unnecessarily thick. For an operation at a base–collector bias of 4 V and a collector density of 1×10^4 A/cm^2, the designed thickness should be about 0.6 μm ($\sim X_{\text{dep}}$).

All of the resistive elements are calculated. We proceed to determine the values of the capacitances. During the calculation of $R_{E(\text{epi})}$ and $R_{C(\text{epi})}$, the depletion thicknesses in the emitter and collector were found to be:

$X_{depE} = 0.0360$ μm and $X_{dep} = 0.591$ μm. From Eq. (4-18), the base–emitter capacitance is:

$$C_{je} = 2 \times 10^{-4} \cdot 30 \times 10^{-4} \frac{1.159 \times 10^{-12}}{0.031 \times 10^{-4}} = 2.24 \times 10^{-13} \text{ F}$$

From Eq. (4-67), the base–collector capacitance is:

$$C_{jc} = 2(1 + 0.2 + 1) \times 10^{-4} \cdot 30 \times 10^{-4} \frac{1.159 \times 10^{-12}}{0.591 \times 10^{-4}} = 2.59 \times 10^{-14} \text{ F}$$

Once the values of the resistive and capacitive elements are all known, we are ready to calculate various components of the emitter–collector transit time. The minority electron diffusion coefficient in the base, D_{nB}, is given as kT/q times μ_{nB}, the minority electron mobility in the base. Its dependence on the base doping is given by Eq. (1-75):

$$\mu_{nB} = 8300 \cdot \left(1 + \frac{N_B}{3.98 \times 10^{15} + N_B/641}\right)^{-1/3}$$

For a base doping level of 3×10^{19} cm^{-3}, $\mu_{nB} = 988.62$ cm^2/V-s. From Eqs. (4-127), (4-128), (4-130), and (4-131), we find the various charging/transit times as:

$$\tau_e = \frac{kT}{qJ_c W_E L_E}(C_{je} + C_{jc}) = \frac{0.0258(2.24 \times 10^{-13} + 2.59 \times 10^{-14})}{1 \times 10^4 \cdot 2 \times 10^{-4} \cdot 30 \times 10^{-4}} = 1.08 \text{ ps}$$

$$\tau_b = \frac{X_B^2}{\nu \cdot kT/q \cdot \mu_{nB}} = \frac{(800 \times 10^{-8})^2}{2 \cdot 0.0258 \cdot 988.62} = 1.25 \text{ ps}$$

$$\tau_{sc} = \frac{X_{dep}}{2v_{sat}} = \frac{0.591 \times 10^{-4}}{2 \cdot 8 \times 10^6} = 3.69 \text{ ps}$$

$$\tau_c = (R_E + R_C) \cdot C_{jc} = (0.453 + 4.32) \cdot 2.59 \times 10^{-14} = 0.124 \text{ ps}$$

In the calculation of the base transit time, we used the fact that the base is uniform (hence $\nu = 2$). The emitter–collector transit time is the sum of the four time constants (Eq. 4-126):

$$\tau_{ec} = \tau_e + \tau_b + \tau_{sc} + \tau_c = 1.08 + 1.25 + 3.69 + 0.124 = 6.14 \text{ ps}$$

The cutoff frequency is therefore (Eq. 4-126):

$$f_T = \frac{1}{2\pi\tau_{ec}} = \frac{1}{2 \cdot 3.14 \cdot 6.14 \cdot 10^{-12}} = 26.0 \text{ GHz}$$

From this exercise, we find that τ_{sc} (the collector space-charge region transit time) accounts for the largest portion of the overall transit time. Typically in silicon bipolar junction transistors (Si BJTs), τ_b and τ_e are the dominant time constants because of the lower electron mobility and the lack of base–emitter heterojunction. Therefore, whereas the collector design is more important for the HBTs, the base and emitter designs are more critical for the Si BJTs.

The maximum oscillation frequency is calculated from Eq. (4-159):

$$f_{\max} = \sqrt{\frac{f_T}{8\pi r_b C_{jc}}} = \left(\frac{26.0 \times 10^9}{8 \cdot 3.14 \cdot 9.45 \cdot 2.59 \times 10^{-14}}\right)^{1/2} = 65.0\ \text{GHz}$$

REFERENCE

1. Liu, W. (1998). *Handbook of III-V Heterojunction Bipolar Transistors*. New York: Wiley.

PROBLEMS

1. Sometimes the circuit model of Fig. 4-27 is used. The elements R_Ξ and C_Ξ are due to the Early effects. Show that they are equal to:

$$R_\Xi = \frac{1}{\Xi(1 - \alpha_{To})g_e}$$

$$C_\Xi = \Xi C_D$$

   Also show that C_D in the equation is that of Eq. (4-45), not Eq. (4-42).

2. We are given the common-emitter parameters: $y_{11e} = (1.3 + j\,3.4)$ mS; $y_{21e} = -(-9.8 + j\,8.7)$ mS; $y_{12e} = -(0.2 + j\,0.005)$ mS; and $y_{22e} = (0.3 + j\,2.5)$ mS.
   - **a.** Find its common-base parameters.
   - **b.** Find its common-collector parameters.
   - **c.** Determine the (Rollett's) stability factors in the common-emitter and the common-base configuration.
   - **d.** Find the h-parameters in the common-emitter configuration.

3. A graded HBT has a base doping of 3×10^{19} cm^{-3} and a base thickness of 800 Å. The operating current density is 10^4 A/cm^2, at which level the collector depletion thickness is 0.5 μm. The emitter area is 2×30 μm^2 and the collector area is 2.5 times as large. Neglect the effect of the emitter capacitance.
   - **a.** Determine α_{To} and ω_o.
   - **b.** Determine g_e, C_{jc}, and τ_m.
   - **c.** At 10 GHz, find the common-emitter y-parameters.

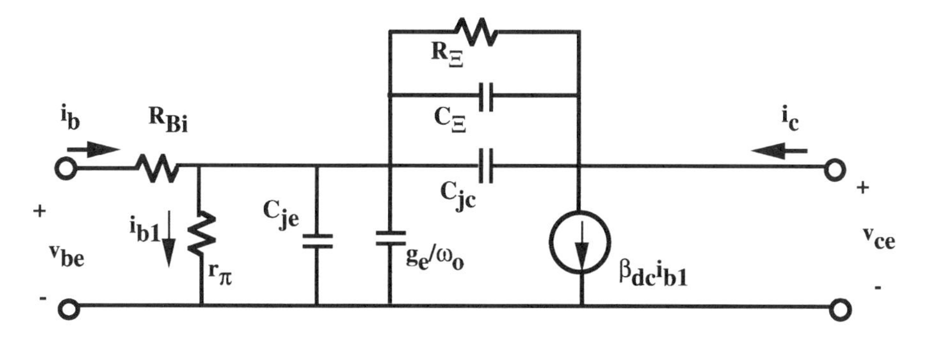

FIGURE 4-27. Circuit model accounting for the Early effect.

4. The HBT in the previous problem has $R_E \sim R_{EE} = 6\ \Omega$, $R_{BB} = 20\ \Omega$, $R_C = 1\ \Omega$.

a. What is the total base resistance? Assume that the extrinsic base resistance is negligible. Assume that two symmetrical base contacts are placed at each side of an emitter mesa. There is no current crowding.

b. Find the cutoff frequency.

c. Find the maximum oscillation frequency.

5. An *Npn* has the following material parameters for the emitter, base, and collector contacts: $\rho_{\sigma E} = \rho_{\sigma B} = \rho_{\sigma SC} = 10^{-6}\ \Omega \cdot \text{cm}^2$; $R_{SHE} = R_{SHSC} = 20\ \Omega/\text{sq}$ and $R_{SHB} = 200\ \Omega/\text{sq}$.

a. $W_E \times L_E = 2 \times 30\ \mu\text{m}^2$. Find R_{EE}.

b. $W_B = 1\ \mu\text{m}$. Find R_{BB} (assuming it is purely real).

c. $W_C = 10\ \mu\text{m}$. Find R_{CC}.

6. A $2 \times 30\ \mu\text{m}^2$ graded *Npn* HBT has a collector doping of $3 \times 10^{16}\ \text{cm}^{-3}$ and a thickness of 1.3 μm. Its base doping is $3 \times 10^{19}\ \text{cm}^{-3}$ and the thickness is 800 Å. The collector area is three times the emitter area. It is operated at $J_C - 10^4\ \text{A/cm}^2$ and $V_{CB} = 4$ V. Assume that $\phi_{CB} = 1.3$ V.

a. Find the depletion thickness in the collector at this current density.

b. Find τ_{sc} and τ_c. Assume that $R_E = 5\ \Omega$ and $R_C = 1\ \Omega$.

7. Consider the transistor of § 4-7, except that the emitter width is 1 μm instead of 2 μm. All the other parameters remain the same.

a. Determine f_T. What is the percentage difference between the cutoff frequency of the 1-μm-wide HBT and the 2-μm-wide HBT?

b. Determine f_{max}. What is the percentage difference between the maximum oscillation frequency of the 1-μm-wide HBT and the 2-μm-wide HBT?

8. Consider the transistor § 4-7, except that the base doping level is modified from 3×10^{19} cm^{-3} to 1×10^{20} cm^{-3}. All the other parameters remain the same, except that the base sheet resistance is reduced by the ratio of the doping levels.

a. Determine f_T. What is the percentage difference between the cutoff frequency of the two HBTs?

b. Determine f_{max}. What is the percentage difference between the maximum oscillation frequency of the two HBTs?

9. Do not work out quantitatively for this problem. Rank the order of the following modifications from the HBT of § 4-7, in terms of the percentage changes in f_T. Also rank the order of the impact on f_{max}. The operating current density is to be held constant.

a. Change the emitter length from 30 to 100 μm.

b. Change the base width from 1 to 3 μm.

c. Change the base-emitter spacing from 0.2 to 1 μm.

d. Change the base thickness from 800 to 1500 Å.

e. Change the collector thickness from 1.2 to 1 μm.

10. Consider the transistor of § 4-7, except that the collector doping level is modified from 3×10^{16} cm^{-3} to 7×10^{16} cm^{-3}. All the other parameters remain the same.

a. What is the value of f_T?

b. What is the value of f_{max}?

c. What is a good reason to increase the collector doping level?

CHAPTER 5

FET D.C. CHARACTERISTICS

§ 5-1 METAL-SEMICONDUCTOR JUNCTION

The metal-semiconductor junction is a crucial element of III-V field-effect transistors (FETs). When reverse-biased, the junction is used as the gate of the transistor to modulate the channel potential, and hence the channel current. We mentioned the metal-semiconductor junction twice previously, but neither in the context of field-effect transistors. In § 2-5, we were interested in finding the current-voltage characteristics of a forward-biased abrupt *p-n* heterojunction. We modified the thermionic current expression originally developed for the metal-semiconductor junction and applied the result to the *p-n* junction. The current expression is not of interest here. This is because the current flow across a reverse-biased metal-semiconductor junction, that used for field-effect transistors, is so insignificant that we may just neglect it. The second instance of mentioning the metal-semiconductor junction took place in § 4-4, where we considered the contact resistances in an HBT. We stated that, in order to minimize the resistance through an ohmic contact, we dope the semiconductor layer heavily to enhance the thermionic-field emission of carriers across the barrier height. The discussion then does not apply to this section, either. Here we are concerned with using the metal-semiconductor junction as a *Schottky contact*, a contact which is characterized by a high barrier height and a negligible amount of current flow in the reverse-bias condition.

With the application for the field-effect transistors in mind, we expand our analysis of the metal-semiconductor junction in this section. We first consider the ideal contact. The practical contact with the presence of the surface states will be described subsequently.

The first MESFET was produced by evaporating aluminum metal on *n*-GaAs to form the gate electrode. During that time, aluminum was considered more thermally stable than other common metals. Since then, the Al/GaAs contact system has been investigated and it was concluded that interdiffusion of the metal and semiconductor species occurs at a relatively low temperature of 125°C. The gallium outdiffusion constitutes a major long-term degradation mechanism. Presently, there are two popular metal schemes. One is based on the Ti/Pt/Au system, which is simple yet yields low gate resistances. The second metal choice relates to the industry move toward smaller device geometry, which has called for the self-aligned implantation to form the source and drain. In this fabrication technology, the Schottky barrier metalization must withstand the annealing temperature at which the implanted atoms are activated. Because the annealing process is around 800–950°C, a refractory metal such as tungsten is used as the gate metal.

Figure 5-1 illustrates the band diagram of a metal-semiconductor system in the charge-neutral condition. The semiconductor is taken to have a doping concentration of 10^{17} cm^{-3}. Gold is used as the gate metal in this example, which has a work function of about 4.3 eV. The *work function* of a metal is the energy it takes to move an electron to the vacuum energy level. It represents the energy separation between the metal Fermi level and the vacuum level. For semiconductors, the corresponding material parameter of interest is the electron affinity discussed in § 1-7. The *electron affinity* measures the energy separation between the bottom of the conduction band and the vacuum level. The Fermi level of the (*n*-type) semiconductor is located at a certain amount of energy below the conduction band, determined by the ratio of the doping level to the

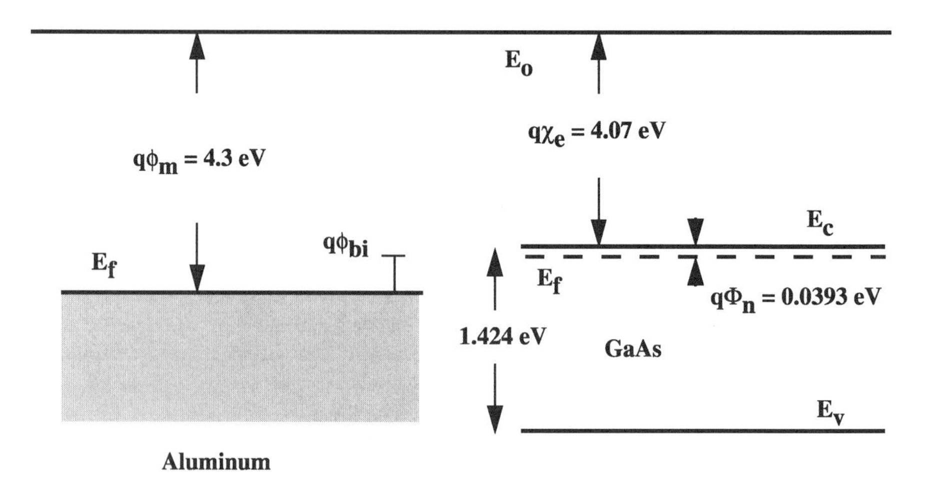

FIGURE 5-1. Band diagram of a metal-semiconductor system in the charge-neutral condition.

conduction band density of states. For the exemplar junction of Fig. 5-1, $n = N_d = 10^{17}$ cm^{-3}. Because N_c for GaAs is 4.7×10^{17} cm^{-3}, n/N_c, the y-axis value of Fig. 1-21, is equal to 0.22. We read off the corresponding x-axis value from the figure and determine that the Fermi level in the semiconductor is ~0.039 eV below the conduction band.

When two materials are brought into contact, the vacuum levels at both sides are aligned. They are in the charge-neutral condition; the thermal equilibrium is not yet established. As shown in Fig. 5-1, the electrons in the semiconductor layer have a higher chemical potential than those in the metal. There is a tendency for the electrons to move from the semiconductor, across the junction interface, and into the metal. This movement is somewhat surprising, since we tend to think that the semiconductor layer, whether p-type or n-type, has fewer electrons than the metal. It seems that the electrons in the metal should diffuse toward semiconductor, rather than the other way around. However, other factors determine the electron chemical potential, which is the ultimate quantity that dictates the movement of electrons. When two materials are distinct, as in this case, it is perfectly possible for the electrons in the semiconductor to diffuse toward the metal. Of course, in a homogenous system made of the same material, such as in a p-n junction, the electrons always diffuse from the n-type layer to the p-type layer because there are more electrons in the n-type layer. Our simplistic picture of carriers diffusing from more populated areas to less populated areas will then work.

As the electrons move from the semiconductor toward the metal, they leave behind the ionized donors in the region near the junction. This region, depleted of the mobile carriers, is called the *depletion region*, in a fashion similar to that in the p-n junction. The electrons arriving at the metal constitute negative charges, while the ionized donor charges in the semiconductor are positive charges. These charges of opposite polarity establish an electric field within the device. According to Gauss's law, the sum of these charges must be zero; otherwise, the electric field established by these charges would have extended outside the device. As the electron movement progresses, more positive charges are ionized in the semiconductor and more electrons congregate at the metal side of the interface, causing the electric field to be increasingly larger. Because this field points from the semiconductor to the metal, it tends to retard the electron movement which sets up the field in the first place. Eventually, when a certain amount of electron transfer has taken place, the electric field becomes so large that the tendency for electrons to diffuse from the semiconductor toward the metal is exactly counterbalanced by the retardation from the field. At this instant, thermal equilibrium is achieved. Because there is no further electron movement, the Fermi levels on both sides of the junction are aligned, as shown in Fig. 5-2. It is clear from the figure that the barrier height (or Schottky barrier height, $q\phi_B$) is the difference between the metal work function (ϕ_m) and the electron affinity of the semiconductor (χ_e):

$$q\phi_B = q\phi_m - q\chi_e \tag{5-1}$$

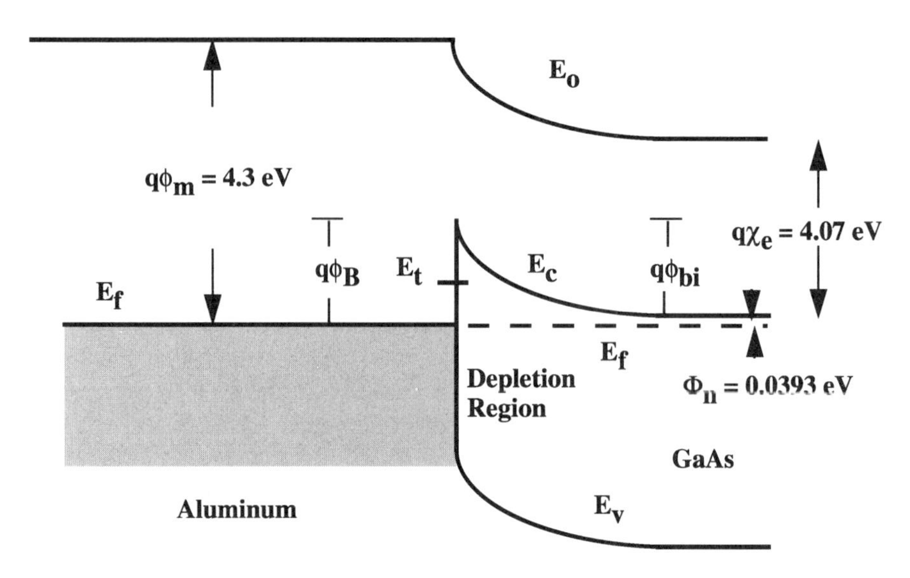

FIGURE 5-2. Band diagram of a metal-semiconductor system under the thermal equilibrium. The energy level E_t, existing in a nonideal contact, will be discussed shortly.

The built-in potential of the junction, according to § 2-3, is equal to the separation between the Fermi levels at the two materials in the charge-neutral condition. From Fig. 5-1, the built-in potential is given by:

$$\phi_{\text{bi}} = \phi_m - \chi_e - \frac{\Phi_n}{q} = \phi_B - \frac{\Phi_n}{q} \tag{5-2}$$

where Φ_n was defined in Eq. (2-15), and is shown in the figure. It is equal to 0.039 eV in this example, as calculated previously. The amount of the built-in potential is indicated in Fig. 5-2.

The above discussion pertains to the idealized metal-semiconductor contact. According to Eq. (5-1), it would be possible to change the barrier height merely by replacing the metal materials with different work functions. In practice, however, it is observed that the barrier height remains roughly the same, independent of the metal deposited on top of the GaAs. As a rule of thumb, the barrier height is about one-half to two-thirds of the energy gap of the semiconductor material.

To understand the independence of the barrier height with the metal, we reexamine the concept of the surface states first presented in § 3-4. We start by commenting that, despite 50 years of fairly intensive research, there is no consensus on the physical mechanism(s) responsible for the constant-barrier-height property. The following discussion is based on several surface science studies.

In an idealized one-dimensional bulk material with a linear array of atoms extending to infinity, the electrons are found to occupy energy states in a set of bands. Each state is characterized by a wave number ($\boldsymbol{k}$), and an electron is described by a Bloch wave function. The energy state is not *local*; we cannot identify the exact spatial position of the electron occupying a particular state. The quantum mechanical treatment is later applied to a more realistic solid, which includes a surface where the array of atoms abruptly ends. The formation of a surface involves with the breaking of covalent bonds, a process which leaves the surface bonds dangling without being properly bonded to other atoms across the junction. A calculation indicates that, in addition to the nonlocal band states, there are a band of energy states due specifically to the abrupt termination, with an energy level exactly at the middle of the energy gap. Because they are spatially located at the surface, these local states are termed the *surface states*. On the (111) surface of Si, there is one broken bond per surface atom and thus two surface states per atom. One of the states is filled and the other is empty.

Suppose that a donor-like energy state due to the dangling bonds exists at E_t, as shown in Fig. 5-2. If the Fermi level is below E_t, the surface state will have donated an electron. The surface state becomes positively charged after the donated electron diffuses to the metal/semiconductor interface. In this case, where we are considering only one energy state, the position of the Fermi level is not much perturbed by the presence of the surface state. The Fermi level is still determined primarily by the bulk properties of the semiconductor, such as the doping density and the electron affinity.

In reality, the number of surface states greatly exceeds 1, and the electron flow needed to reach equilibrium can be provided by these states. For convenience, we take silicon for example, for which we determined from Example 1-1 that the primitive unit cell containing two silicon basis atoms occupies a volume of 40 Å^3. This corresponds to an atomic density of $5.0 \times 10^{22}\,\text{cm}^{-3}$. The density of surface states is about the atomic density's two-thirds power, equal to $1.4 \times 10^{15}\,\text{cm}^{-3}$. When the number of surface states is taken into consideration, the band diagram shown in Fig. 5-3 at thermal equilibrium differs from that of Fig. 5-2. The figure shows that, independent of metal work function, the Fermi level passes through the energy level at which the number of the surface states is roughly at its maximum. A small variation of the Fermi level around E_t would involve with a great change in the electron concentration. Because the number of electrons available for conduction is greatly less than the number of surface state, the Fermi level is *pinned* near E_t, causing the metal/semiconductor barrier height (ϕ_B) to be equal to $E_g - E_t$. The Fermi level passes through an energy level E_t and stays there regardless of the external bias. Therefore, the depletion thickness is independent of the metal work function. It is instead dependent on ϕ_B, because the depletion thickness is determined by the voltage drop across the semiconductor, which is equal to $\phi_B - \Phi_n$.

The above dangling-bond-state theory is the usual model used to describe the constant barrier height observed in various metal-semiconductor junctions. As

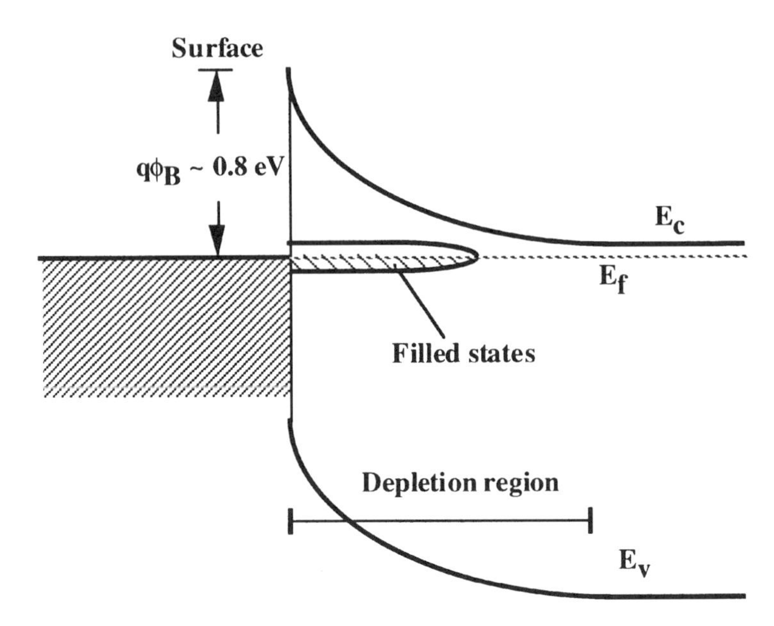

FIGURE 5-3. Band diagram of a metal-semiconductor system when the number of the surface levels is large. The Fermi level is pinned at the trap energy level.

far as explaining the Fermi level pinning is concerned, the model is sufficient. The model, however, has a shortcoming in that it assumes that the atoms at the surface retain their bulk positions. In reality, the surface atoms do not assume the same positions as the bulk atoms, but move by as much as 0.5 Å in an effort to minimize the loss of bonding at the surface. One factor is an attempt to re-form the covalent bonds between the surface atoms. This is the dominant effect for silicon. Another factor is also important for GaAs. When a covalent bond is broken during surface formation, the Ga and As atoms revert back toward the atomic configuration from the sp^3 covalent configuration of the bulk semiconductor. The surface As atom moves toward five valence electrons, achieving a surface covalent bonding configuration of $(s^2)p^3$ (the bonding is p^3 with a filled s^2 orbit). In contrast, the Ga atom, with three valence electrons, moves toward sp^2. The p^3 As configuration moves the As atom out of the surface, whereas the sp^2 bonding pulls the Ga atom back into the surface. This process is called *surface reconstruction* [1]. The net result is that the original surface states at the midgap move completely out of the bandgap region. The empty surface states lie above the conduction band and the filled states lie, below the valence band. Because the surface states are moved out of the energy gap region, they no longer determine the Fermi-level position at the surface. The Fermi level at a sufficiently perfect surface will take up the bulk position as shown schematically in Fig. 5-4. We call the states formed at the freshly cleaved

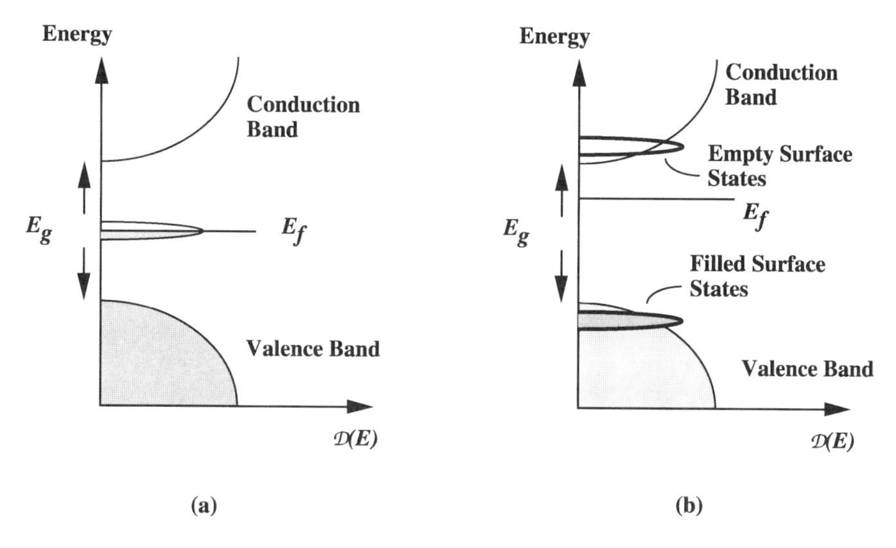

FIGURE 5-4. Fermi level and surface trap level locations: (*a*) freshly cleaved surface; (*b*) after surface reconstruction.

surface the *intrinsic surface states*. They so named to contrast with the defect-generated surface states described in the following.

Accounting for the fact that surface reconstruction takes place right after the surface is formed, the intrinsic surface states so created without any adatom or defect do not affect the electronic property, and hence are not responsible for the Fermi-level pinning. According to the *a*dvanced *u*nified *d*efect *m*odel (AUDM) [1], the pinning is induced by the surface defects instead. A series of studies experimented with depositing Au, Cs, and O overlayer species on the (110) GaAs surface. [Typical GaAs device layers are grown on (100) substrate, as mentioned in § 1-2 and § 1-3. It is believed that AUDM applies equally well to the (100) surface]. Despite the fact that these adatoms range over a wide spectrum of electronegativity, all of the resulting devices form Schottky contacts with the Fermi levels pinned at either 0.75 eV or 0.5 eV above the valence band. These energy levels correspond closely to the dual donor energy levels associated with the As_{Ga} antisites (As occupying a normally Ga lattice position). According to AUDM, the number of As atoms on the surface is critical; the more As atoms there are, the more As_{Ga} antisites are created and thus the more surface states are available for pinning.

The two energy levels introduced by the As_{Ga} antisites are of the donor type, not the acceptor type. (Donor states are neutral if they are filled and acceptor states are neutral if they are empty.) This implies that, in order to pin the surface Fermi level of *n*-GaAs at 0.75 eV, there must exist at least an acceptor level near the valence band. Otherwise, as shown in Fig. 5-5*a*, the electrons in the 0.75 and 0.5 eV levels remain occupying the states. With the electrons at E_d (introduced

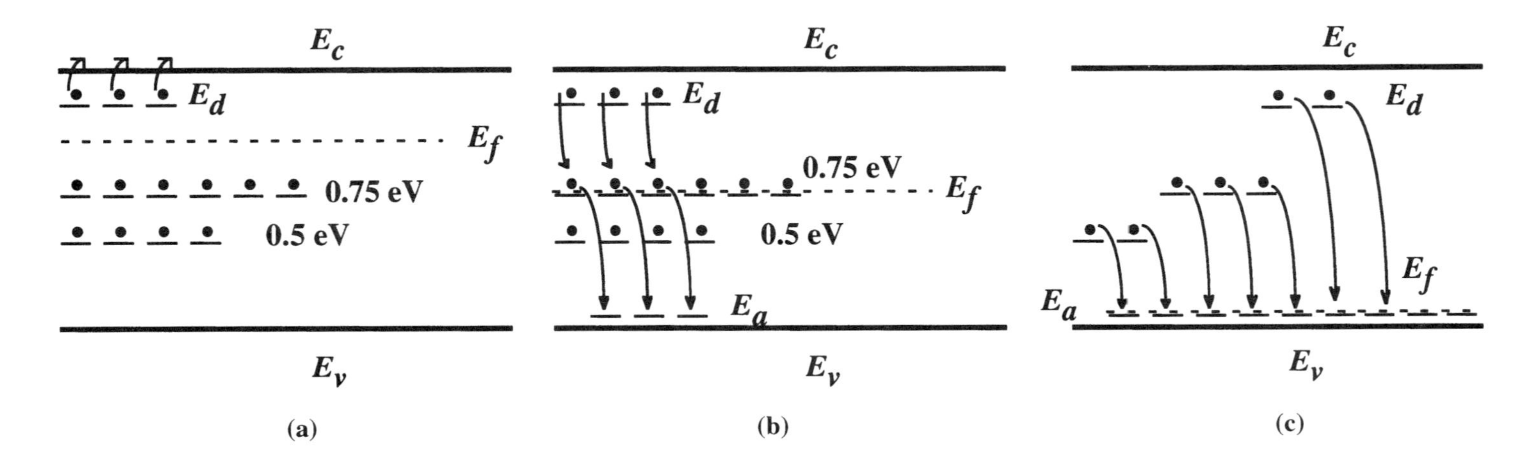

FIGURE 5-5. Band diagrams showing the existence of surface acceptor levels. (*a*) Without the acceptor level, E_f is not pinned. (*b*) The number of acceptor traps is smaller than the donor levels; E_f is pinned near the midgap. (*c*) These are excessive acceptor traps; E_f is pinned near the acceptor level.

by the dopant) gaining enough thermal energy to escape to the conduction band, E_f would be characterized entirely by the bulk doping concentration and would stay well above the 0.75 eV level. The surface states, being completely filled, are not ionized. Only when there is at least one acceptor level (Fig. 5-5*b*) will many of the electrons depart from their original surface donor levels. They leave behind a large number of empty states at 0.75 eV for the Fermi level to be pinned there. The number of acceptor states cannot be exceedingly large. Otherwise, all electrons from the bulk dopants and the surface states will fall to the acceptor level. The Fermi level would then be pinned at the acceptor level (Fig. 5-5*c*).

The origin of the acceptor level is not specified in the AUDM, though it has been hypothesized to be due to the Ga_{As} antisites (Ga atoms occupying As sites). From the basic principles of the AUDM, the following experimental results can be understood: (1) pinning of *n*-GaAs at 0.75 eV and *p*-GaAs at 0.5 eV by most metals in general; (2) pinning of *n*-GaAs by Au at 0.5 V; (3) surface pinning at lower ambient temperatures; (4) the effects of annealing on surface pinning; and finally (5) the modification to the surface with sodium sulfide treatment mentioned in § 3-4.

Let us consider a metal/semiconductor junction as shown in Fig. 5-6. The semiconductor layer has a doping level of N_d and a thickness of a. The positive terminal is taken to be at the gate metal, and the negative terminal at the semiconductor substrate. This polarity convention is adopted even when the semiconductor is doped *p*-type. For field-effect transistor applications, the metal/semiconductor junction is most often reverse-biased, a situation in which the applied voltage (V_a) is negative. When V_a is zero, the junction is at the thermal equilibrium. The band diagram in the *y*-direction was given in Fig. 5-3.

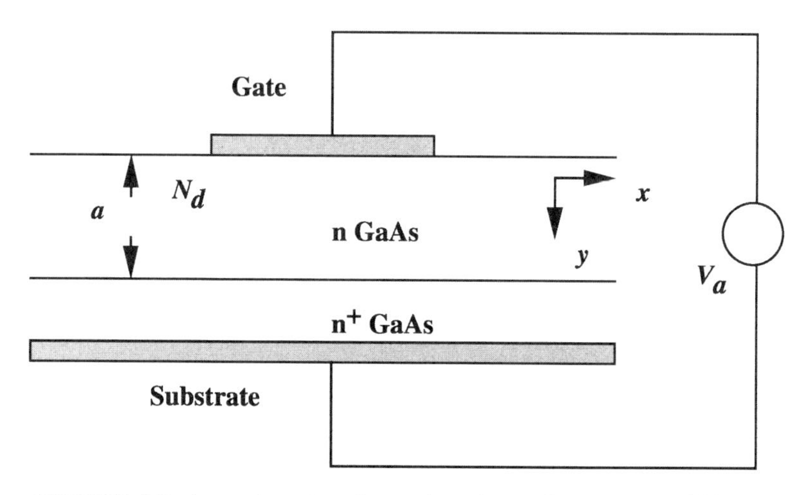

FIGURE 5-6. A metal-semiconductor junction under an external bias V_a.

We determine a relationship between the built-in voltage and the depletion thickness. Figure 5-7 illustrates the charge and electric field profiles of the metal/semiconductor junction of Fig. 5-6. These profiles are similar to those of a (p^+) base/(n^-) collector junction shown in Fig. 3-19. Inside the semiconductor region, the Poisson equation is given by:

$$\frac{d\varepsilon}{dy} = \frac{q}{\epsilon_s} N_d \tag{5-3}$$

We denote the depletion thickness inside the semiconductor layer as X_{dep}. At the end the depletion region, where $y = X_{\text{dep}}$, there is no more ionized charge and the electric field is zero. The maximum electric field which occurs at $y = 0$ is equal to the slope of the field given by Eq. (5-3) times the depletion thickness:

$$|\varepsilon(0)| = \frac{q}{\epsilon_s} N_d X_{\text{dep}} \tag{5-4}$$

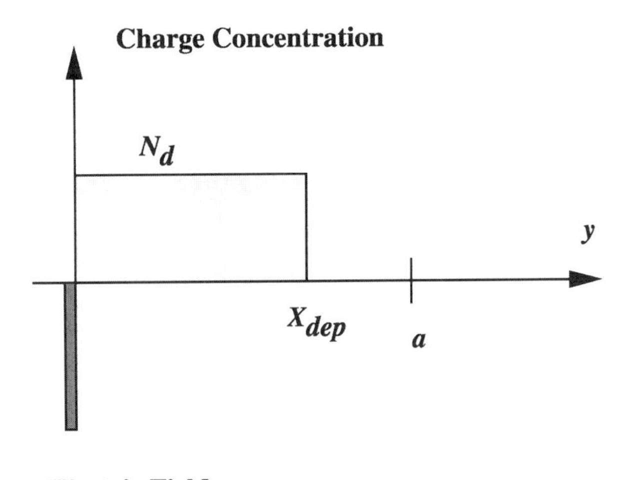

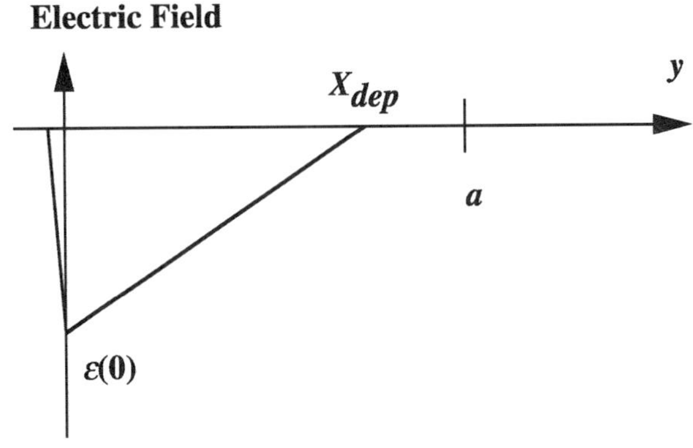

FIGURE 5-7. Charge and electric field profiles of the junction shown in Fig. 5-6.

The ionized charge in the n-type semiconductor is positive, while the charge at the gate side of the metal/semiconductor interface is negative. The direction of the electric field points from the semiconductor toward the gate. In our coordinate system, this field is negative. We enclose $\varepsilon(0)$ in absolute-value signs in Eq. (5-4) to stress that it is the magnitude which is equal to the right-hand side of the equation. The area enclosed by the electric field is equal to the built-in voltage of the metal-semiconductor junction (ϕ_{bi}). Therefore,

$$\phi_{bi} = \frac{1}{2} |\varepsilon(0)| X_{dep} \tag{5-5}$$

The depletion thickness is found by combining Eqs. (5-4) and (5-5):

$$X_{dep,0} = \sqrt{\frac{2\epsilon_s}{qN_d} \phi_{bi}} \tag{5-6}$$

We replace X_{dep} by $X_{dep,0}$ in the above expression to emphasize that the result applies to the thermal equilibrium condition.

When a negative bias is applied to the junction ($V_a < 0$), the electron quasi-Fermi level in the substrate is pulled down relative to the quasi-Fermi level in the gate metal. Due to the pinning of the Fermi level by the surface states, the barrier height of the system ($q\phi_B$) remains unchanged. Therefore, all of the applied voltage drops across the semiconductor, with an amount equal to $\phi_{bi} - V_a$. The depletion thickness must expand outward to accommodate the additional voltage drop.

To derive an expression of the depletion thickness under a reverse bias, we examine the equations leading to $X_{dep,0}$ in Eq. (5-6). Because the semiconductor layer is uniformly doped at N_d, both Eqs. (5-3) and (5-4) remain applicable (but the variable X_{dep} takes on a larger value when $V_a < 0$). However, there is now an additional voltage drop across the semiconductor depletion region; hence, we need to replace ϕ_{bi} in Eq. (5-5) by $\phi_{bi} - V_a$. Accordingly, we modify the depletion thickness given in Eq. (5-6):

$$X_{dep} = \sqrt{\frac{2\epsilon_s}{qN_d} (\phi_{bi} - V_a)} \tag{5-7}$$

The depletion thickness can reach the semiconductor layer thickness (a) when the magnitude of the reverse bias is large enough. When this happens, the semiconductor layer is said to be *fully depleted*. The potential drop in the depletion region during full depletion, $\phi_{bi} - V_a$, is defined as the *full-depletion potential*, ϕ_{00}. According to Eq. (5-7), it can be written as:

$$\phi_{00} = \frac{qN_d}{2\epsilon_s} a^2 \tag{5-8}$$

We write out explicitly the potential and electric field as functions of y, when the applied voltage is V_a. From the differential equation of Eq. (5-3) and the boundary condition that $\varepsilon(0) = 0$ at $y = X_{\text{dep}}$, we obtain:

$$\varepsilon(y) = -\frac{q}{\epsilon_s} N_d(X_{\text{dep}} - y) \tag{5-9}$$

The potential profile is equal to the negative of the integral of the electric field, according to Eq. (1-9). We arbitrarily choose the reference potential to be at $y = 0$. With $\phi_s(0) = 0$, we obtain:

$$\phi_s(y) = \frac{q}{\epsilon_s} N_d\left(X_{\text{dep}} \cdot y - \frac{y^2}{2}\right) \tag{5-10}$$

The subscript s denotes that it is a potential in the semiconductor layer. Equation (5-10), when combined with Eq. (5-7), indicates that $\phi_s = \phi_{\text{bi}} - V_a$ at $y = X_{\text{dep}}$.

Example 5-1:

For the two-terminal device of Fig. 5-6, what is the bias voltage when the active layer is fully depleted? N_d and a of the semiconductor layer are $1 \times 10^{17}\,\text{cm}^{-3}$ and 2000 Å, respectively. The barrier height of the metal/semiconductor junction is 0.9 eV.

According to the analysis leading to Fig. 5-1, Φ_n for N_d of $10^{17}\,\text{cm}^{-3}$ is 0.039 eV. Therefore, according to Eq. (5-2), the built-in voltage is $\phi_{\text{bi}} = 0.9 - 0.039 = 0.861$ V. The full-depletion potential is:

$$\phi_{00} = \frac{1.6 \times 10^{-19} \cdot 10^{17}}{2 \cdot 13.1 \times 8.85 \times 10^{-19}} (2000 \times 10^{-8})^2 = 2.76\,\text{V}$$

The bias voltage at which the active layer is fully depleted is $\phi_{\text{bi}} - \phi_{00}$, or

$$V_a = \phi_{\text{bi}} - \phi_{00} = 0.861 - 2.76 = -1.90\,\text{V}$$

A relationship based on the Poisson equation is useful in the study of FET structures. When there is a two-dimensional charge density (σ, in cm^{-2}) at a particular location x, then the electric fields at the two sides of the sheet carriers are governed by:

$$\varepsilon(x^+) - \varepsilon(x^-) = \pm\frac{q\sigma}{\epsilon_s} \tag{5-11}$$

A negative sign on the right-hand side is used if the accumulated charges are electrons. The sign is positive if the accumulated charges are holes.

We demonstrate the use of Eq. (5-11) through an example. In the semiconductor of a metal/semiconductor junction, there is space-charge density N_d between $y = 0$ and $y = X_{\text{dep}}$. Because the net charge of the device is zero, there is an equal amount of charges residing on the metal/semiconductor interface, where $y = 0$. The charge density at the interface is $q\sigma = -q \cdot N_d \cdot X_{\text{dep}}$; it is negative because the charges are made of electrons. The electric field inside the metal is zero, hence $\varepsilon(0^-) = 0$. Applying Eq. (5-11), we find $\varepsilon(0^+) = -qN_dX_{\text{dep}}/\epsilon_s$. This is consistent with the result obtained by substituting $y = 0$ in Eq. (5-9).

§ 5-2 BASIC MESFET OPERATION

The metal semiconductor *f*ield *e*ffect *t*ransistor (MESFET), shown in Fig. 5-8, is a three-terminal device consisting of the metal/semiconductor junction described in § 5-1, and two additional contacts at the two sides of the metal. The metal and the contacts are situated on top of a doped semiconductor layer, called the *channel*. The channel layer, in turn, sits on top of a semiinsulating buffer layer, whose main function is to prevent the leakage conduction between the channel and the substrate. The metal modulates the amount of depletion into the channel layer, allowing more current to flow when the amount of depletion is small and vice versa. Because the metal controls the amount of the charge flow directly, it is called the *gate*. The two metal contacts beside the gate provide the source and the collection of the charges which flow through the channel layer. They are called the *source* and the *drain* contacts. The function of the gate metal is to control the depletion of

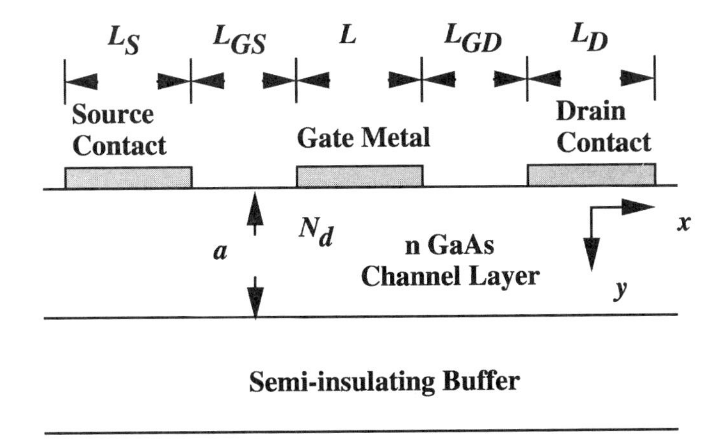

FIGURE 5-8. A MESFET structure.

the channel; it is not meant to inject charges. Therefore, the metal-semiconductor junction is most often reverse-biased. Even during the rare event when the junction is forward-biased, the amount of forward bias is significantly smaller than the built-in potential of the junction so that the exponential turn-on characteristics discussed in § 2-5 do not take place. The design of a MESFET in a circuit must take the gate polarity into consideration, ensuring that the gate is reverse-biased compared to the source and drain. A particular challenge is to maintain the reverse bias during a transient in which the terminal voltages are switching between high and low values irregularly.

The most important device design parameters of MESFETs are the gate width (W), the gate length (L), the active layer thickness (a), and the active layer doping (N_d). When the breakdown voltage is a concern, the gate-drain spacing (L_{GD}) is also important. The channel thickness would have the maximum thickness of a in the absence of any depletion from the gate bias. It has the minimum thickness of zero when the channel is fully depleted.

For simplicity, we first analyze the device operation when both the source and the drain are tied to the same potential, a configuration which is not useful in practice. The device orientation as well as the band diagrams are shown in Fig. 5-9. The band diagrams are drawn at three longitudinal (in parallel with the charge flow) locations: $x = 0$ (source), $L/2$, and L (drain). In each band diagram, the electron energy is drawn as a function of y, the depth from the gate/semiconductor interface along the transverse direction. The gate-to-source bias (V_{GS}) is negative so that the metal-semiconductor junction is reverse-biased. Because the drain and the source are at the same potential, the depletion thickness is uniform across the entire channel. The potential drop between the interface ($y = 0$) and the channel region is equal to $\phi_{bi} - V_{GS}$ for all x values, where ϕ_{bi} is the built-in potential of the metal/semiconductor junction. The transistor is essentially the two-terminal device discussed in § 5-1, with the source and drain tied together to form the substrate terminal of the two-terminal device (Fig. 5-6). There is no current flow between the source and the drain contacts.

Figure 5-10 illustrates the scenario when both the source and the drain contacts are connected to a common substrate bias, V_{SUB}. If V_{SUB} is positive, the metal-semiconductor junction becomes even more reverse-biased and the depletion region extends deeper into the channel layer. The total amount of potential drop between $y = 0$ and the channel increases from $\phi_{bi} - V_{GS}$ to $\phi_{bi} - V_{GS} + V_{SUB}$. Again, due to the lack of a potential difference between the source and the drain, the depletion region at various x locations extends to the same distance. The whole device is still symmetrical along the $x = L/2$ cross section.

The symmetry is broken whenever a finite drain-to-source bias (V_{DS}) is applied, as illustrated in Fig. 5-11. The entire V_{DS} is dropped across the channel. Therefore, in addition to the existing potential due to $\phi_{bi} - V_{GS}$, there is another

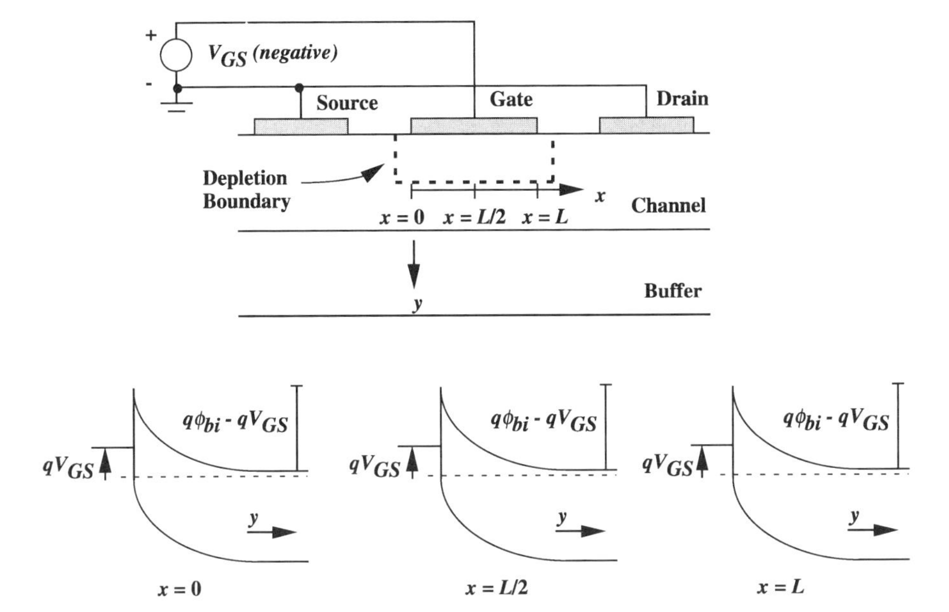

FIGURE 5-9. Band diagrams along the y direction at three different x locations in a MESFET whose drain and source are tied together to the ground. The depletion region extends to the same distance, independent of x.

potential component in the channel due to the application of V_{DS}. This additional potential is called the *channel-source potential* (or simply, *channel potential*), $V_{CS}(x)$. When $V_{DS} \neq 0$, the channel potential varies with x, rather than being a constant value as in the previous two cases. (In Fig. 5-9, $V_{CS} = 0$ for all x, and in Fig. 5-10, $V_{CS} = V_{\text{SUB}}$ for all x.) The channel potential measures the potential difference between any point x along the channel with respect to the potential at the source. It has the minimum value of zero at the source side, and increases monotonically with x until it reaches the maximum value of V_{DS} at the drain side. The channel-to-source potential can be treated as the reverse-bias (due to V_{DS}) biasing the channel and the gate metal, in addition to $\phi_{\text{bi}} - V_{GS}$. Hence, right at the source where $V_{CS} = 0$, the bias V_{DS} brings forth no effect. The only externally applied bias that appears between the metal-semiconductor junction at $x = 0$ remains V_{GS}, just as the situation depicted for the source in Fig. 5-9 (or Fig. 5-10 with $V_{\text{SUB}} = 0$). Accordingly, the depletion thickness corresponds to that which sustains a potential drop of $\phi_{\text{bi}} - V_{GS}$. This potential drop is consistent with the band diagram shown for $x = 0$ in Fig. 5-11. At $x = L$, there is an additional potential drop across the depletion region, that due to V_{DS}. In an analogous fashion similar to the right side of Fig. 5-10 (with $V_{\text{SUB}} = V_{DS}$), this V_{DS} acts

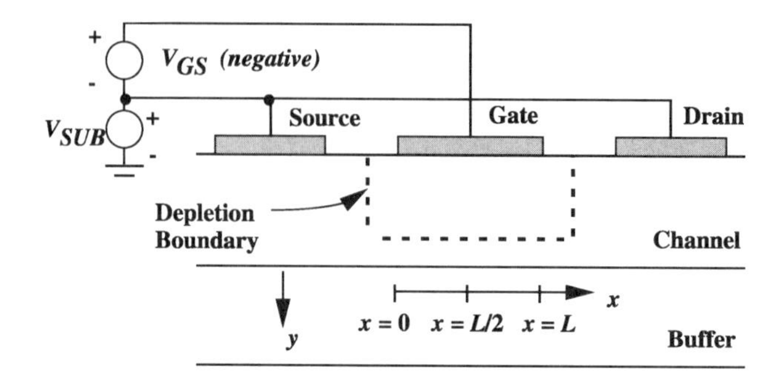

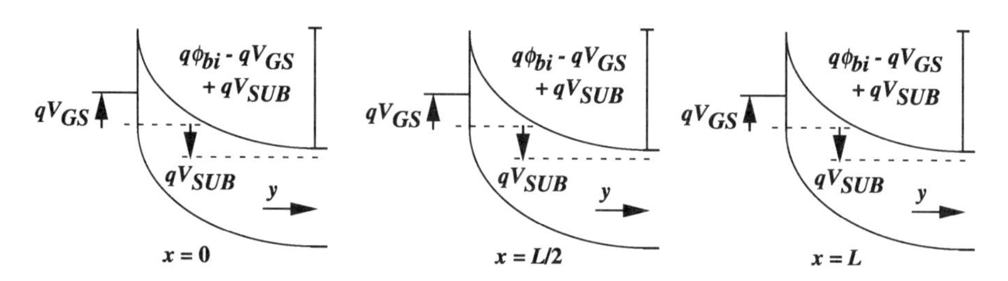

FIGURE 5-10. Same as Fig. 5-9, except that the drain and source are tied together to a certain voltage V_{SUB}. The depletion region still extends to the same distance, for all x values. However, note that the depletion thickness is larger than that of Fig. 5-9.

as an additional substrate bias which causes the depletion region to extend beyond that at the source side. The total potential drop at $x = L$ is $\phi_{bi} - V_{GS} + V_{DS}$. In the middle of the channel, $V_{CS}(x)$ is equal to some value between 0 and V_{DS}. At a given location x, the depletion region extends to a distance to sustain the voltage drop of $\phi_{bi} - V_{GS} + V_{CS}(x)$. The depletion region is therefore thinnest at the source and becomes progressively thicker along with x, reaching the maximum value at the drain. The depletion region boundary in the y-direction is a function of x, denoted $X_{dep}(x)$. For simplicity, $X_{dep}(x)$ is often drawn to vary linearly with position as shown in Fig. 5-11. It will be shown shortly that $X_{dep}(x)$, in reality, is a transcendental function of x. At any particular position x, X_{dep} cannot be expressed as a straightforward function of x. Its value is determined iteratively from a numerical technique.

A nonzero V_{DS} creates an electric field in the $-x$ direction, pointing from the drain to the source. This ε_x field is uninteresting in the depletion region. First, the depletion region is depleted of mobile carriers; therefore, no current can be established by the field. Second, the magnitude of ε_x is insignificant in comparison to the vertical field ε_y across the metal/semiconductor

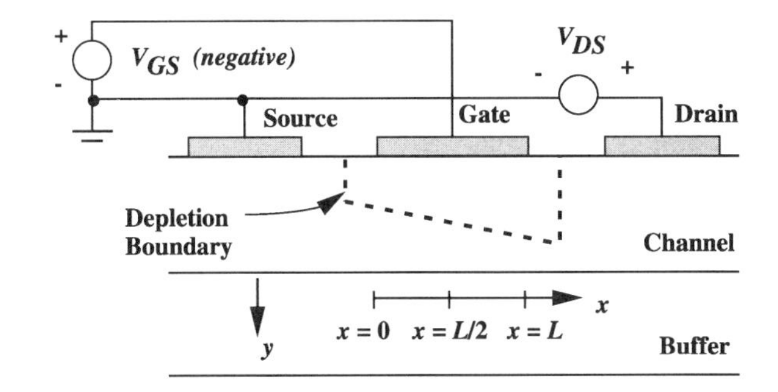

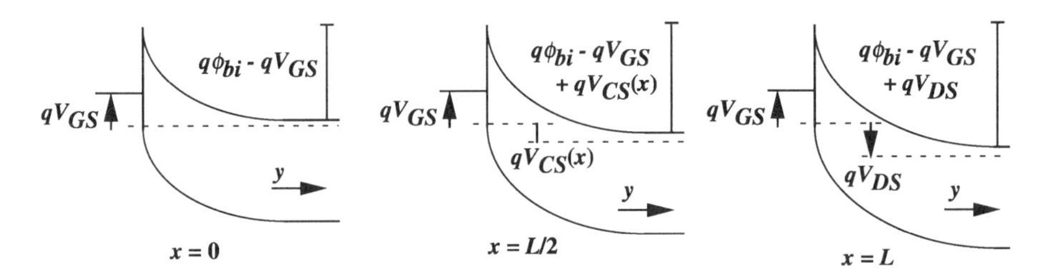

FIGURE 5-11. Band diagrams along the y direction at three different x locations in a MESFET whose drain and source are connected to different biases. The depletion region is a function of x. The channel-to-source potential, $V_{CS}(x)$, equals 0 at the source, and V_{DS} at the drain.

junction. Most of the electrical properties of the depletion region are determined by ε_y, which is governed by Eq. (5-3). In the undepleted channel region, however, ε_y field emanated from the gate has decreased to zero. The small but finite electric field in the x direction becomes the driving force of the current conduction. With the quasi-neutrality assumption, the free-electron concentration is exactly equal to the doping concentration at any given position. (In reality, the free-electron concentration is slightly different from the dopant concentration at any given location, to allow for a nonzero electric field in the x direction; see § 1-6). Since both the carriers and electric field are present in the channel, a current, made mostly of drift motion, flows from the drain to the source. (The actual electron movement is from the source to the drain.) The drain current, based on the drift component of Eq. (1-35), is given by:

$$I_D = qW\mu_n n(x)\varepsilon(x)b(x) \tag{5-12}$$

where $b(x)$ is the amount of the channel opening at x. It is equal to $a - X_{\text{dep}}(x)$. Unlike the bipolar transistors, the dominant carriers carrying the device current in field-effect transistors are majority carriers. Without a significant amount of minority carriers in the channel, the recombination current is close to nil. Hence, the current is a constant in the d.c. operation. Although $b(x)$ and $\varepsilon(x)$ differ in values from one position to the next, their product is a constant, giving rise to the same amount of current throughout the channel.

We examine the change in I_D when V_{GS} increases from some negative value to 0, which is usually the largest value of V_{GS} in MESFET operation. V_{DS} is assumed to be some small positive value. Because V_{DS} is finite, the depletion thickness is larger at the drain side than at the source side, as illustrated in Fig. 5-12*a*. When V_{GS} increases from a negative value (solid line) to zero (dotted line), there is less reverse bias across the metal/semiconductor junction, for any position x in the entire channel. As the depletion region shrinks in response to increased V_{GS}, more electron carriers become available to conduct current. If we fix V_{DS} while V_{GS} is increased, the same amount of the longitudinal electric field then results in more current flow. Therefore, increasing V_{GS} (by making V_{GS} less negative) increases the current flow.

We now examine the effects of drain-source bias on the transistor operation when V_{GS} is fixed at a constant value. An increase in V_{DS} increases the amount of reverse-bias at the drain side, resulting in the narrowing of the channel opening. This is illustrated in Fig. 5-12*b*. At the drain, where $x = L$, the current is given by $aqWb(L)\varepsilon(L)$. It is not clear whether the overall current should increase or decrease in response to the V_{DS} increase, since $b(L)$ decreases while $\varepsilon(L)$ increases. This uncertainty in the trend of current, however, can be resolved if we choose to evaluate the current at the source side. We have stated that in d.c. operation, the current flow in the channel is constant, independent of the position. Expressed mathematically, the statement is tantamount to $I_D = I_D(L) = I_D(0) = I_D(x)$. Hence, we can write $I_D = aqWb(0)\varepsilon(0)$. Because the reverse-bias potential at the source end is unperturbed by V_{DS}, $b(0)$ remains the same, prior and after the V_{DS} increase. With the channel electric field increasing, we conclude that the drain current increases in response to the V_{DS} increase.

Figure 5-13 displays the I-V characteristics of a MESFET, plotting the drain current as a function of V_{DS}, with V_{GS} as a parameter. As shown, I_D increases with V_{GS} for a fixed V_{DS}. I_D also increase with V_{DS} for a fixed V_{GS} (as long as V_{DS} is smaller than some V_{DS} value to be discussed in the following). These current behaviors are consistent with the above qualitative analysis.

As V_{DS} continues to increase at a given V_{GS}, the depletion thickness at the drain side eventually reaches the designed active-layer thickness, a. When this happens, the channel is said to be *pinched off*. The amount of V_{DS} resulting in the pinchoff condition is called the *pinchoff voltage* (V_P). The exact formula for V_P will be given in Eq. (5-20). The operating region prior to pinchoff is called the *linear region*. The name comes from the fact that, when V_{DS} is small, the channel layer is roughly a uniformly resistive layer.

Pinchoff is an operating condition in which our model can produce paradoxical results if the model limitation is not well comprehended. During

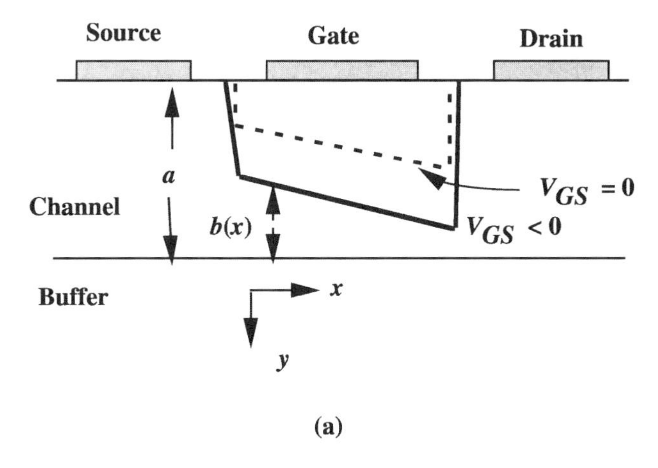

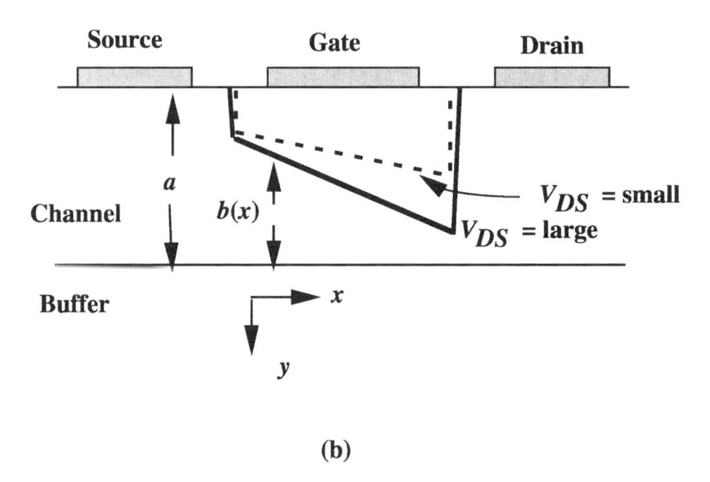

FIGURE 5-12. (*a*) The depletion boundary as a function of V_{GS} for a MESFET when V_{DS} is fixed. (*b*) The depletion boundary as a function of V_{DS} when V_{GS} is fixed.

this condition, our model states that the channel opening at $x = L$ is 0, since $X_{\text{dep}} = a$. Consequently, according to Eq. (5-12), $I_D = 0$. This is at odds with empirical results that the drain current maintains at some large value when V_{DS} increases past V_P. In reality, when V_{DS} increases toward and beyond the pinchoff, a conducting channel of a finite (albeit thin) thickness still exists at the drain end. The carrier concentration never truly reaches zero but

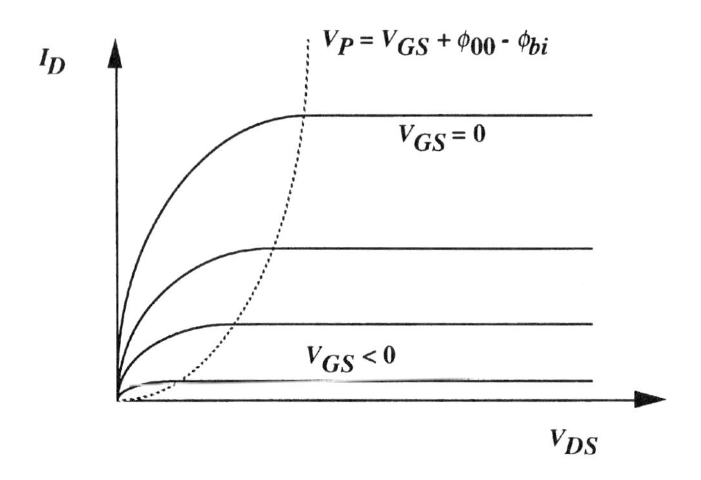

FIGURE 5-13. Schematic I-V characteristics of a long-channel MESFET.

maintains a finite value. Our simple-minded picture, in which we predicted that the depletion region at $x = L$ touches the semiinsulating substrate, is a result of our implicit assumption that the two-dimensional problem in the MESFET can be separated into a y-dimension problem and an x-dimension problem. In this approximation, we solved in the y-dimension based on the one-dimensional Poisson equation of Eq. (5-3), predicting that $b(L)$ will reach zero when a certain amount of V_{DS} is applied. Afterwards, we write the current equation on the x direction, using Eq. (5-12). In each of these two steps, the effects of the other dimension are neglected. If the current and the potential profile are solved simultaneously in both dimensions, a channel of finite thickness is shown to exist. An accurate solution to the two-dimensional analysis requires numerical techniques [2]. However, with certain assumptions, it is still possible to obtain an approximate solution from a quasi-two-dimensional analysis (to be discussed in § 5-3). For the present discussion, we continue to separate the two dimensions into two one-dimensional analyses, with the understanding that the paradox of zero carrier concentration coexisting with a finite drain current is merely a result of the simplified assumption embodied in the analysis.

The formal name for this approximate analysis is *gradual channel approximation* (GCA). The channel opening, $b(x)$, is assumed to vary slowly with x. Therefore, its value is determined from the y-dimension analysis, with Eq. (5-3). We shall revisit the GCA at the end of this section.

Right at the pinchoff condition, the channel potential increases monotonically from 0 at the source to $V_{DS} = V_P$ at the drain. What happens beyond the pinchoff, such that $V_{DS} > V_P$? If we allowed the channel potential at the drain to increase past V_P, then the depletion thickness calculated from Eq. (5-7) would exceed the designed active-layer thickness a. Because the depletion region is limited to a, we necessarily restrict the channel potential at the drain side to the maximum value of V_P in our simple model. The potential in excess of V_P must have dropped across the region between the gate and the drain contacts. As far as the current conduction is concerned, the amount of V_{DS} in excess of V_P produces no effects because the potential and electron concentrations at any position along the channel are unmodified from that when V_{DS} first reaches V_P. Therefore, the current shown in Fig. 5-13 appears as a horizontal extension of the current value attained when $V_{DS} = V_P$. The saturation of the current is a direct manifestation of channel pinchoff. This is generally observed in long-channel device. We shall see in § 5-3 that in short-channel devices, the current saturation is a result of velocity saturation, without the pinchoff of the channel charges at the drain. In order to maintain a consistent appearance of symbols, we shall call $V_{DS,\text{sat}}$ the drain-to-source bias at which the current saturation takes place. For the long-channel device under consideration, $V_{DS,\text{sat}}$ is then identical to V_P. However, in short-channel devices, the pinchoff does not occur. There is always a significant channel thickness even when the current reaches saturation due to velocity saturation. The channel thickness, though noticeable, is still smaller than that in the linear region. Hence, sometimes the term *pinchdown* is used to describe the operating condition when the current saturates as a result of velocity saturation. The pinchoff voltage (V_P) then becomes a meaningless quantity in short-channel devices. However, we may still use $V_{DS,\text{sat}}$ to denote the voltage when current saturation occurs.

We used a drain current expression (Eq. 5-12) to aid our understanding of MESFET operation. Now we develop in a rigorous manner two equations governing FET devices in general, at any position and at any instant of time (rather than just d.c.). We will later simplify the equations specifically for d.c. operation. In Chapter 6 we will reuse these equations to develop both large-signal and small-signal transistor models.

The first equation is called the *drift equation.* It is essentially Eq. (5-12), but with some differences to emphasize the spatial and time dependencies. The coordinate system adopted in this derivation is shown in Fig. 5-14. At any position x and any time t, the instantaneous channel current is:

$$i_{\text{CH}}(x, t) = qWN_d\mu_n b(x, t)\varepsilon(x, t) \tag{5-13}$$

More generally, N_d in the above equation should be replaced by $n(x, t)$, the free-electron concentration. The substitution of n by N_d is justified with the quasi-neutrality approximation discussed in § 1-6.

We briefly discuss our notation for current and voltage. A d.c. quantity, such as I_{CH}, is expressed with all letters capitalized. A small-signal quantity is expressed in small letters, such as i_{ch}. Their sum represents the instantaneous total value, and is given by a small letter with capitalized subscripts. Therefore, i_{CH} in Eq. (5-13) is an instantaneous value, hinting that we are concerned with a transient problem. If we were focusing only on d.c., we would have

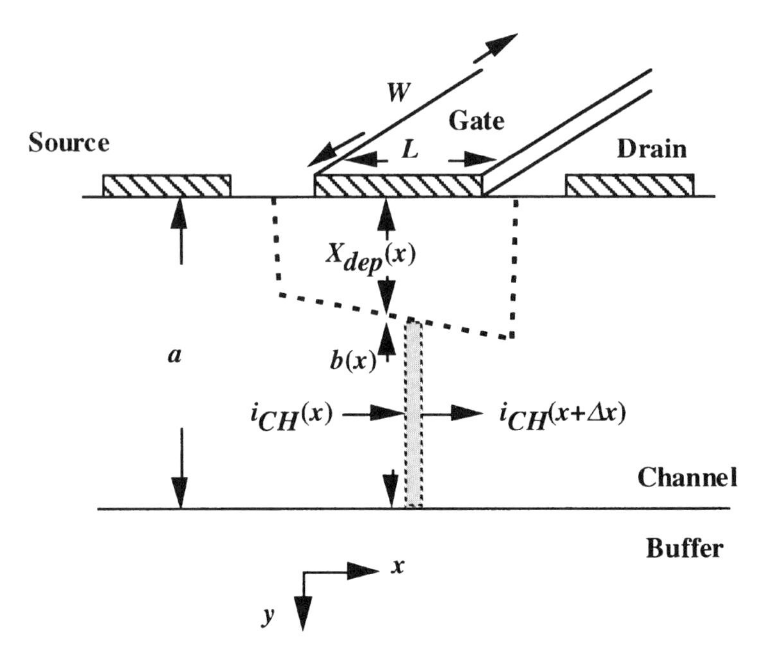

FIGURE 5-14. Device structure showing the involved parameters used in the derivation of the governing equations of MESFET.

used I_{CH} instead. This notation convention has also been used in the discussion of HBTs.

The electric field is the spatial derivative of the potential variation. In accordance with Eq. (1-9), we write:

$$\varepsilon(x, t) = -\frac{\partial \phi_s(x, t)}{\partial x} \tag{5-14}$$

where

$$\phi_s(x, t) = \phi_{bi} - v_{GS}(t) + v_{CS}(x, t) \tag{5-15}$$

$\phi_s(x, t)$ measures the potential drop from the metal-semiconductor interface to the quasi-neutral channel region at a given position x and a given time t. As shown in the band diagrams of Fig. 5-11, ϕ_s at any position consists of two components. One is the potential drop through the gate, equal to $\phi_{bi} - v_{GS}$, which is the same for all x values. The second component, the channel-to-source potential (v_{CS}), is the reverse bias between a position x in the channel and the source. It is brought about by a nonzero v_{DS}, having a value of zero at the source side and gradually increasing to v_{DS} at the drain. In a manner similar to

Eq. (5-7), we write the depletion thickness of the MESFET as:

$$X_{\text{dep}} = \sqrt{\frac{2\epsilon_s}{qN_d}\phi_s} = a\sqrt{\frac{\phi_s}{\phi_{00}}} \tag{5-16}$$

In an n-type MESFET (the channel layer is doped n-type), the applied drain-source bias $v_{DS}(t)$ is positive. $v_{CS}(x, t)$ increases from zero at $x = 0$ to $v_{DS}(t)$ at $x = L$. Therefore:

$$\phi_s(0, t) = \phi_{\text{bi}} - v_{GS}(t) \tag{5-17}$$

$$\phi_s(L, t) = \phi_{\text{bi}} - v_{GS}(t) + v_{DS}(t) \qquad \text{(linear region only; no saturation)} \tag{5-18}$$

Strictly speaking, $\phi_s(L, t)$ given by Eq. (5-17) is valid only when the transistor is in the linear operating region, in which the depletion thickness at the drain side is less than the channel thickness, a. As v_{DS} increases past the saturation voltage $v_{DS,\text{sat}}$, the channel on the drain side becomes fully depleted. The relationship between a and $v_{DS,\text{sat}}$ is obtained from Eq. (5-7):

$$a = \sqrt{\frac{2\epsilon_s}{qN_d}(\phi_{\text{bi}} - v_{GS} + v_{DS,\text{sat}})} \tag{5-19}$$

Alternatively, we can write:

$$v_{DS,\text{sat}} = \phi_{00} - \phi_{\text{bi}} + v_{GS} \tag{5-20}$$

As mentioned previously, in a long-channel FET without velocity saturation, $V_{DS,\text{sat}}$ is identical to the pinchoff voltage V_P.

Incorporating the saturation region of operation as well as the linear region, we rewrite Eq. (5-18) as:

$$\phi_s(L, t) = \begin{cases} \phi_{\text{bi}} - v_{GS}(t) + v_{DS}(t) & \text{if } v_{DS}(t) \leqslant v_{DS,\text{sat}}(t) \\ \phi_{\text{bi}} - v_{GS}(t) + v_{DS,\text{sat}}(t) & \text{if } v_{DS}(t) > v_{DS,\text{sat}}(t) \end{cases} \tag{5-21}$$

Because the surface potential increases in value as x increases from 0 to L, Eq. (5-14) indicates that the channel electric field is negative. The negative sign of the electric field is consistent with the field direction that, with a positive v_{DS}, the field points from the drain to the source.

We need to be careful about the sign of the current flow. Because the electrons flow from the source to the drain in response to a positive v_{DS}, the channel current is negative, consistent with the sign presented in Eq. (5-13), in which ε is a negative quantity. The channel current is basically the drain current at $x = L$, which is essentially the source current at $x = 0$. However, adopting the convention that the device current is defined positive when it enters a terminal, we therefore write:

$$i_D(t) = -i_{CH}(x, t)|_{x=L} \qquad i_S(t) = +i_{CH}(x, t)|_{x=0} \tag{5-22}$$

Replacing $b(x, t)$ of Eq. (5-13) by $a - X_{\text{dep}}(x, t)$, we then have:

$$i_{CH}(x, t) = qWN_d\mu_n a\left(1 - \sqrt{\frac{\phi_s(x, t)}{\phi_{00}}}\right)\left(-\frac{\partial\phi_s(x, t)}{\partial x}\right) \tag{5-23}$$

This marks the first governing equation, the drift equation. The second equation is the *continuity equation.* Consider the shaded region of Fig. 5-14. In an incremental time Δt, the charge entering the shaded region is $i_{CH}(x)\cdot\Delta t$, while the charge leaving the region is $i_{CH}(x + \Delta x)\cdot\Delta t$. Let us suppose that the charge per unit area at a given channel position x and at a given time t is $q'_{CH}(x, t)$. The prime suggests that q'_{CH} is a per-unit area quantity. In terms of the variables introduced previously, $q'_{CH}(x, t)$ is:

$$q'_{CH}(x, t) = qN_d[a - b(x, t)] \tag{5-24}$$

Because there is no carrier lost due to recombination, the charge conservation at the shaded region demands that:

$$[q'_{CH}(x, t + \Delta t) - q'_{CH}(x, t)]\cdot\Delta x\cdot W = i_{CH}(x, t)\cdot\Delta t - i_{CH}(x + \Delta x, t)\cdot\Delta t \tag{5-25}$$

Rearranging the terms, we obtain:

$$W\frac{q'_{CH}(x, t + \Delta t) - q'_{CH}(x, t)}{\Delta t} = \frac{i_{CH}(x, t) - i_{CH}(x + \Delta x, t)}{\Delta x} \tag{5-26}$$

Or, equivalently in the differential form, Eq. (5-26) is expressed as:

$$\frac{\partial i_{CH}(x, t)}{\partial x} = -W\frac{\partial q'_{CH}(x, t)}{\partial t} \tag{5-27}$$

Substituting Eq. (5-24) into the above equation, we write the second governing equation of the MESFET as:

$$\frac{\partial i_{CH}(x)}{\partial x} = aqWN_d\frac{\partial}{\partial t}\sqrt{\frac{\phi_s(x, t)}{\phi_{00}}} \tag{5-28}$$

In d.c. operation, all the time dependencies drop out and the two governing equations reduce to:

$$I_{CH}(x) = qWN_d\mu_n a\left[1 - \sqrt{\frac{\phi_{s,dc}(x)}{\phi_{00}}}\right]\left[-\frac{\partial\phi_{s,dc}(x)}{\partial x}\right] \tag{5-29}$$

$$\frac{\partial I_{CH}(x)}{\partial x} = 0 \tag{5-30}$$

Equation (5-30) states that the channel current at any position x along the channel is a constant. We stress that the independence of the channel current with position, while being intuitive, is correct only under d.c. operation. In general, such as during a transient, the channel current varies with the channel position because the time derivative of the surface potential may not be zero.

In the following, we attempt to use the above two d.c. equations to establish an expression for the drain current and the potential variation in the channel. According to Eq. (5-30), the d.c. channel current is a constant, independent of position. Taking x to be L (at the drain), we see from Eq. (5-22) that this constant channel current is equal to the negative of the d.c. drain current:

$$I_{CH}(x) = -I_D \qquad \text{(independent of } x\text{)} \tag{5-31}$$

Substituting Eq. (5-31) into Eq. (5-29) and subsequently integrating both sides from $x' = x$ to $x' = L$, we find:

$$\int_x^L I_D\,dx' = \int_x^L qWN_d\mu_n a\left[1 - \sqrt{\frac{\phi_{s,dc}(x')}{\phi_{00}}}\right]\left[\frac{\partial\phi_{s,dc}(x')}{\partial x'}\right]dx' \tag{5-32}$$

At this point, it is convenient to introduce some normalizing variables to simplify the integration. We define:

$$u(x) = \frac{\phi_{s,dc}(x)}{\phi_{00}} \qquad s = \frac{\phi_{s,dc}(0)}{\phi_{00}} \qquad d = \frac{\phi_{s,dc}(L)}{\phi_{00}} \tag{5-33}$$

where, according to Eqs. (5-17) and (5-21):

$$\phi_{s,dc}(0) = \phi_{bi} - V_{GS} \tag{5-34}$$

$$\phi_{s,dc}(L) = \begin{cases} \phi_{bi} - V_{GS} + V_{DS} & \text{if } V_{DS} \leq V_{DS,sat} \\ \phi_{bi} - V_{GS} + V_{DS,sat} & \text{if } V_{DS} > V_{DS,sat} \end{cases} \tag{5-35}$$

Noting that $b(x) = a - X_{\text{dep}}(x)$, where X_{dep} was given in Eq. (5-16), we give a useful relationship for $u(x)$:

$$u(x) = \left[1 - \frac{b(x)}{a}\right]^2 \tag{5-36}$$

Because the d.c. drain current is a constant independent of x, the integration of Eq. (5-32) is straightforward:

$$I_D = \frac{1}{L - x}\int_{\phi_{s,\text{dc}}(x)}^{\phi_{s,\text{dc}}(L)} qWN_d\mu_n a\left(1 - \sqrt{\frac{\phi_{s,\text{dc}}}{\phi_{00}}}\right) d\phi_{s,\text{dc}} \tag{5-37}$$

The integration is further simplified with the reduced variables given in Eq. (5-33):

$$I_D = \frac{1}{L - x}\int_u^d qaWN_d\mu_n\phi_{00}(1 - \sqrt{u})\,du \tag{5-38}$$

A straightforward integration yields:

$$I_D = \frac{W}{L - x} aqN_d\mu_n\phi_{00}\left[d - u(x) - \frac{2}{3}d^{3/2} + \frac{2}{3}u(x)^{3/2}\right] \tag{5-39}$$

When the above equation is evaluated at $x = 0$, where $u(x) = s$, we obtain a d.c. current expression for MESFETs:

$$I_D = I_{\max}\left(d - s - \frac{2}{3}d^{3/2} + \frac{2}{3}s^{3/2}\right) \tag{5-40}$$

where $I_{\max}$ is:

$$I_{\max} = \frac{qaWN_d\mu_n\phi_{00}}{L} \tag{5-41}$$

There is a physical significance to $I_{\max}$. In an ideal situation where $s = 0$ (V_{GS} is positive enough to offset the built-in voltage ϕ_{bi} so that there is no depletion in the source side) and $d = 1$ (pinchoff), then $I_{\max}$ represents the maximum operating current. In practice, s is never zero. The maximum device current is always at a value smaller than $I_{\max}$.

In order to find how the d.c. surface potential varies as a function of x, we divide Eq. (5-40) by Eq. (5-39). Rearranging the terms, we find $u(x)$ to satisfy:

$$u(x) - s - \frac{2}{3}u(x)^{3/2} + \frac{2}{3}s^{3/2} = \frac{x}{L}\left(d - s - \frac{2}{3}d^{3/2} + \frac{2}{3}s^{3/2}\right) \tag{5-42}$$

This is a transcendental equation. Although $u(x)$ cannot be determined in a straightforward manner, it can be calculated with any root-finding numerical routine once d and s are specified (see Example 5-2).

Once $u(x)$ is determined from Eq. (5-42), the d.c. surface potential ($\phi_{s,\mathrm{dc}}$) is known. The channel thickness and the electric field can also be established. In terms of $u(x)$, we have:

$$b(x) = a[1 - \sqrt{u(x)}] \tag{5-43}$$

$$\varepsilon(x) = -\phi_{00}\frac{du(x)}{dx} = -\frac{\phi_{00}}{L}\frac{d - s - \frac{2}{3}d^{3/2} + \frac{2}{3}s^{3/2}}{1 - u(x)^{1/2}} \tag{5-44}$$

Example 5-2:

A MESFET has the following parameters: $\phi_B = 0.9$ eV; $a = 2000$ Å; $N_d = 1 \times 10^{17}$ cm^{-3}; $V_{GS} = 0$ V; $V_{DS} = 3$ V. Find the surface potential, electric field, and the depletion thickness as a function of position.

From Example 5-1, the depletion potential, ϕ_{00}, is determined to be 2.76 V. The drain-to-source saturation voltage is found from Eq. (5-20):

$$V_{DS,\mathrm{sat}} = 2.76 - 0.861 + 0 = 1.899 \text{ V}$$

Since the drain operating voltage exceeds $V_{DS,\mathrm{sat}}$, the transistor is in saturation and $\phi_s(L)$ is determined by the saturation voltage rather than the actual drain bias:

$$\phi_s(L) = 0.861 - 0 + 1.899 = 2.76 \text{ V}$$

$$\phi_s(0) = 0.861 - 0 = 0.861 \text{ V}$$

Therefore, $d = 1$ and $s = 0.312$. We use a numerical routine to determine $u(x)$ from Eq. (5-42). Once $u(x)$ is known, $\phi_s(x)$ is obtained by $u(x) \cdot \phi_{00}$, $\varepsilon(x)$ from Eq. (5-44), and $X_{\mathrm{dep}}(x)$ from Eq. (5-16). The results are plotted in Fig. 5-15. Note that the depletion thickness does not increase linearly with position, although in the schematic drawing of Fig. 5-14, for example, we draw a straight line to mark the depletion boundary.

Example 5-3:

Find the drain current of the transistor of Example 5-2. The transistor has a gate dimension $L \times W = 0.5$ μm × 500 μm. Assume the mobility to be 4000 cm²/V-s.

$I_{\max}$ calculated from Eq. (5-41) is:

$$I_{\max} = \frac{1.6 \times 10^{-19} \cdot 2000 \times 10^{-8} \cdot 500 \times 10^{-4} \cdot 1 \times 10^{17} \cdot 4000 \cdot 2.76}{0.5 \times 10^{-4}} = 3.53 \text{ A}$$

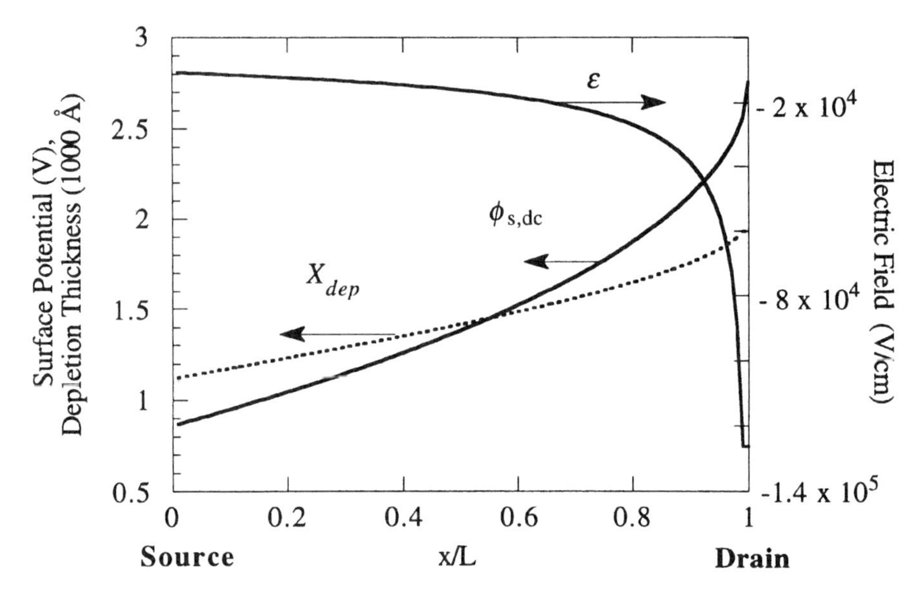

FIGURE 5-15. The surface potential, electric field, and depletion thickness as a function of x in the MESFET examined in Example 5-2. The depletion thickness extends from about 1100 Å at the source to 2000 Å at the drain.

With $d = 1$ and $s = 0.312$, the drain current is calculated from Eq. (5-40) to be:

$$I_D = 3.53 \times \left(1 - 0.312 - \frac{2}{3}1^{3/2} + \frac{2}{3}0.312^{3/2}\right) = 0.485 \text{ A}$$

The gradual channel approximation has the obvious drawback that a finite drain current exists even though the channel near the drain side is said to be depleted of carriers beyond pinchoff. Before presenting an analysis to improve the gradual channel approximation, we review the governing equations of the device. We limit ourselves to the d.c. operation to simplify our discussion:

$$\nabla^2 V = -\frac{q}{\epsilon_s}(N_d - n) \tag{5-45}$$

$$J_n = qD_n \nabla n + q\mu_n n \nabla V \approx q\mu_n n \nabla V \tag{5-46}$$

$$\nabla \cdot J_n = 0 \tag{5-47}$$

These equations apply to the entire device region and should be solved simultaneously. With the GCA, we approximated the two-dimensional problem into a series of one-dimensional problems, solving only one particular equation at a time. More specifically, we solved the following equations:

$$\frac{d^2\phi_{s,\mathrm{dc}}(y)}{dy^2} = -\frac{q}{\epsilon_s} N_d \qquad \text{for } b \le y \le a$$

$$\phi_{s,\mathrm{dc}}(y=0) = 0; \qquad \left.\frac{d\phi_{s,\mathrm{dc}}}{dy}\right|_{y=b} = 0 \tag{5-48}$$

$$I_D = aqWN_d\mu_n\left(1 - \sqrt{\frac{\phi_{s,\mathrm{dc}}(x)}{\phi_{00}}}\right)\frac{d\phi_{s,\mathrm{dc}}(x)}{dx} \qquad \text{for } 0 \le x \le L, 0 \le y < b$$

$$\phi_{s,\mathrm{dc}}(x=L) = \phi_{\mathrm{bi}} - V_{GS} + V_{DS};\ \phi_{s,\mathrm{dc}}(x=0) = \phi_{\mathrm{bi}} - V_{GS} \tag{5-49}$$

$$\frac{dI_D}{dx} = 0 \tag{5-50}$$

In writing down Eq. (5-48) instead of Eq. (5-45), we replace the charge concentration $N_d - n$ by N_d (besides the trivial replacement of the potential symbol of V by $\phi_{s,\mathrm{dc}}$). This is a fair assumption since we restrict the validity of the equation to the space-charge region underneath the gate. A more error-prone simplification lies in assuming:

$$\frac{\partial^2\phi_{s,\mathrm{dc}}}{\partial x^2} = 0 \qquad \text{for } b \le y \le a \tag{5-51}$$

Strictly, this approximation is correct only when $V_{DS} = 0$, such that the variation in the x direction is absent (Fig. 5-9). When the current is finite, $\partial\phi_{s,\mathrm{dc}}/\partial x \neq 0$ at the boundary between the space-charge region and the channel region. Hence, in general, the $\partial^2\phi_{s,\mathrm{dc}}/\partial x^2$ does not vanish. The approximation can be valid if the channel depletion thickness $X_{\mathrm{dep}}(x)$ varies gradually with x, such that $\partial^2\phi_{s,\mathrm{dc}}/\partial y^2 \gg \partial^2\phi_{s,\mathrm{dc}}/\partial x^2$. If $X_{\mathrm{dep}}(x)$ instead varies rapidly with x, then the charge at position $x + \Delta x$ can influence the potential distribution right at x (Fig. 15-14). As a rule of thumb, the GCA is said to be valid when

$$\frac{db}{dx} < \frac{b}{a} \tag{5-52}$$

The position corresponding to where the GCA ceases to be valid is found in Problem 5-3. It is given by:

$$x = L - \frac{a}{2} \tag{5-53}$$

That is, at a distance $a/2$ from the drain, the gradual channel approximation becomes incorrect, according to the criterion specified in Eq. (5-52). This points out that, if a device's active-layer thickness a is comparable to the channel length L, the gradual channel approximation may fail in as much as half of the channel region. In this case, we would not expect the drain current equation derived in Eq. (5-40) to be accurate.

MESFETs are almost exclusively made in the GaAs material system. Although Si MESFETs have been demonstrated, the device fabrication is more complicated than Si MOSFETs due to the great care required to prevent native oxide formation at the metal/Si interface. Furthermore, Si MESFETs are generally outperformed by Si MOSFETs and bipolar transistors. MESFETs made of other III-V compounds, such as InP or InGaAs, suffer from a low Schottky barrier height. The leakage current as a result of the thermionic field emission between the gate and the channel limits the performance of these devices.

§ 5-3 VELOCITY SATURATION IN THE MESFET

An implicit assumption of the previous discussion is that the electron velocity is equal to the product of the mobility and the longitudinal electric field. In reality, the electron velocity cannot increase indefinitely as the electric field increases. Figure 1-37 illustrates the electron drift velocity versus the electric field, showing that the velocity increases linearly with ε only when the field is smaller than the critical electric field (ε_{crit}) of about 10^3 V/cm. Depending on the doping level in the channel, velocity either increases monotonically toward the saturation velocity (v_{sat}), or increases toward a peak velocity (v_{peak}) at first before dwindling back to v_{sat}. In either case, the linear relationship between the drift velocity and the field breaks down when the field is high.

It is found that the assumption of constant mobility works well in long-channel FETs. When the channel length is short, the field in the channel easily surpasses ε_{crit} during device operation. Therefore, we generally refer the analysis, of § 5-2 as a long-channel MESFET analysis. In contrast, we shall refer the following analysis, which considers velocity saturation, as a short-channel analysis.

We consider only the situation when the electron velocity increases monotonically toward the maximum value of v_{sat} (rather than increasing to v_{peak} first, and then decreasing toward v_{sat}). Under a prescribed V_{GS} and a small V_{DS}, the depletion boundary is shown for point A in Fig. 5-16. If V_{DS} is so small that $\varepsilon < \varepsilon_{crit}$, the velocity saturation has not occurred. Just as in Fig. 5-15, the depletion region varies smoothly within the channel, reaching the maximum value right at the drain. At the moment V_{DS} increases to a value that $\varepsilon(L) = \varepsilon_{crit}$ (Fig. 5-16, point B), velocity saturation takes place right at the drain. We shall call this drain bias the saturation voltage, $V_{DS,sat}$. When the voltage is increased further, the point at which $\varepsilon = \varepsilon_{crit}$ shifts inward toward the source. The field at the drain now exceeds ε_{crit}. In this situation,

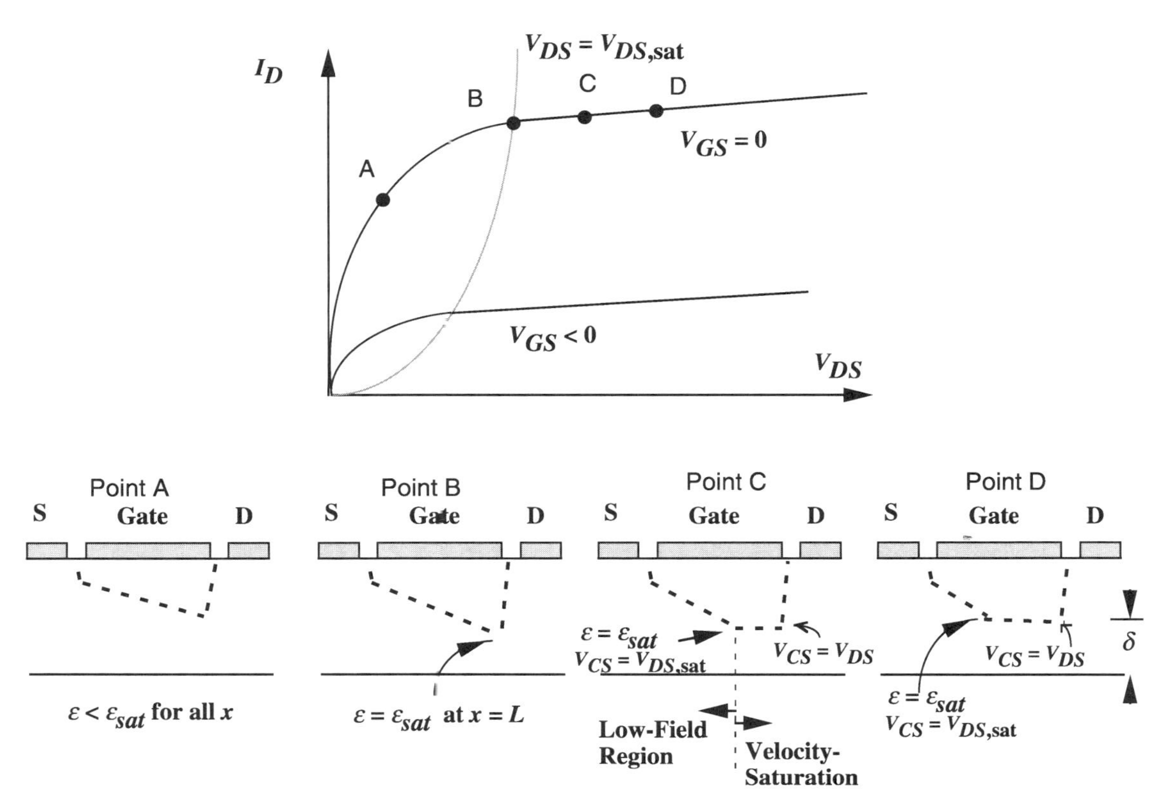

FIGURE 5-16. Drain current as a function of V_{DS} and the depletion boundaries at various bias points (A, B, C, and D).

it is convenient to replace the drain current equation of Eq. (5-12) with the following:

$$I_D = qW \cdot n(x) \cdot b(x) \cdot v_{\text{drift}}(x) \tag{5-54}$$

As shown in Fig. 5-16, point C, we can classify the channel into two subregions, the low-field region adjacent to the source, where $v_{\text{drift}}(x) < v_{\text{sat}}$, and the velocity-saturation region adjacent to the drain, where $v_{\text{drift}}(x) = v_{\text{sat}}$. In the low-field region, $v_{\text{drift}}(x)$ is still given by $\mu_n \cdot \varepsilon$, increasing monotonically with x because ε increases with x. The channel opening, in contrast, decreases with x because of the increased depletion. These variations [in $v_{\text{drift}}(x)$ and $b(x)$] are such that their product always maintains the same value along the channel, so that I_D is a constant. Upon reaching the boundary of the saturation region, $v_{\text{drift}}(x) = v_{\text{sat}}$ for the remainder of the channel length. In order to maintain a constant drain current, $b(x)$ in the velocity-saturation region must also become a constant, in accordance with Eq. (5-54). The fixed amount of channel opening is shown for point C in Fig. 5-16.

When V_{DS} increases even further, the point at which ε reaches $\varepsilon_{\text{crit}}$ then occurs at an even smaller x value at point D of Fig. 5-16 than at point C. The boundary between the low-field and the saturation regions continues to move toward the source. In other words, the extent of the low-field region becomes shorter, while the velocity-saturation region becomes larger. In the discussion of long-channel MESFET operation in § 5-2, we stated that the drain current continues to increase with V_{DS} until V_{DS} reaches the pinchoff voltage V_p. Afterward, the additional increment in V_{DS} is dropped between the pinchoff point and the drain contact. The drain current remains constant, unaffected by V_{DS}. In the present discussion, where we find that the low-field region gradually turns into the velocity-saturation region after a certain $V_{DS} = V_{DS,\text{sat}}$, it is not clear whether the drain current becomes insensitive to V_{DS} after $V_{DS} > V_{DS,\text{sat}}$. We shall establish later that, because the extent of the low-field region becomes less as V_{DS} increases, I_D continues to increase after V_{DS} exceeds $V_{DS,\text{sat}}$. However, the rate of I_D increase is much smaller than the rate before V_{DS} reaches $V_{DS,\text{sat}}$. Hence, we shall say that the current saturates when $V_{DS} = V_{DS,\text{sat}}$, although in reality the current does not saturates at a constant value as in the case of long-channel devices.

An analytical study of the channel current in the velocity-saturated transistors is difficult. There are generally two methods to treat this problem. In one, the mobility is treated as a field-dependent quantity, for example:

$$\mu_n = \frac{\mu_{n,0}}{1 + \varepsilon/\varepsilon_{\text{sat}}} \tag{5-55}$$

where $\mu_{n,0}$ is the low-field mobility and ε_{sat}, the saturation field, is a constant used to obtain a best fit to the empirical data. The other method, to be used in the following, is to treat the mobility as a constant until the drift velocity is equal to the saturation velocity. It is set to be v_{sat} when the field reaches the saturation field (ε_{sat}):

$$\varepsilon_{sat} = \frac{v_{sat}}{\mu_n} \tag{5-56}$$

The drift velocity as a function of electric field is displayed in Fig. 5-17. If the channel doping is in the neighborhood of $10^{17}\,\text{cm}^{-3}$, the mobility calculated from Eq. (1-73) is about $4650\,\text{cm}^2/\text{V-s}$. If the saturation velocity is taken to be 10^7 cm/s, then the saturation electric field is fixed at 2.15×10^3 V/cm. This value is somewhat smaller than the measured value of $\varepsilon_{crit} = 3\text{–}4 \times 10^3$ V/cm observed for these doping levels. We use the symbol ε_{crit} (the critical electric field) to mean the field at which the measured drift velocity reaches its saturation value. It is a material parameter. ε_{sat}, in contrast, is a parameter which is defined strictly by Eq. (5-56), and does not have as much physical significance as ε_{crit}.

We divide the channel into two regions, according to the carrier drift velocity. In region I, the velocity is smaller than v_{sat}, and the mobility is a constant. In region II, the carrier velocity saturates at v_{sat}. Because the mobility is constant in region I, the governing equations are nearly identical to those developed previously. For example, the drain current is similar to Eq. (5-29), written as (with $I_D = -I_{CH}$):

$$I_{D,\text{I}} = qWN_d\mu_n a\left[1 - \sqrt{\frac{\phi_{s,\text{dc}}(x')}{\phi_{00}}}\right]\left[\frac{\partial \phi_{s,\text{dc}}(x')}{\partial x'}\right] \tag{5-57}$$

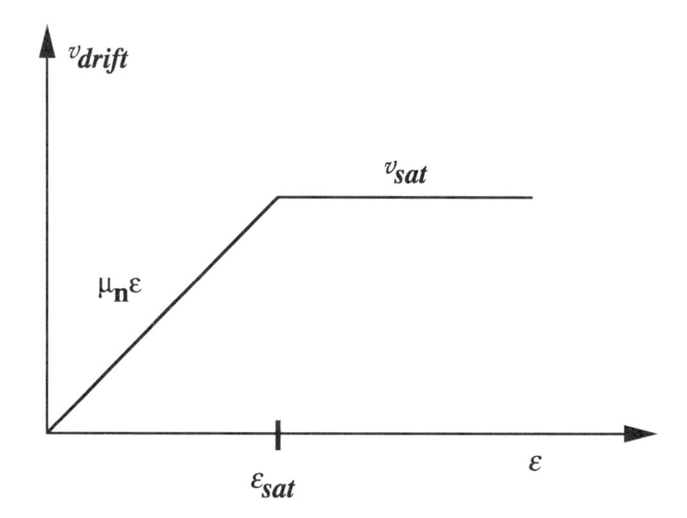

FIGURE 5-17. Drift velocity as a function of electric field used to study the d.c. I-V characteristics of transistors exhibiting velocity saturation.

As another example, the equation governing $b(x)$ is still that given in Eq. (5-36). However, because region I does not extend all the way to $x = L$ (drain), we need to modify the variables given in Eq. (5-33):

$$u(x) = \frac{\phi_{s,\mathrm{dc}}(x)}{\phi_{00}} \qquad s = \frac{\phi_{s,\mathrm{dc}}(0)}{\phi_{00}} \qquad d = \frac{\phi_{s,\mathrm{dc}}(L)}{\phi_{00}} \qquad p = \frac{\phi_{s,\mathrm{dc}}(L_\mathrm{I})}{\phi_{00}} \tag{5-58}$$

where the surface potentials are:

$$\phi_{s,\mathrm{dc}}(0) = \phi_\mathrm{bi} - V_{GS} \tag{5-59}$$

$$\phi_{s,\mathrm{dc}}(L) = \phi_\mathrm{bi} - V_{GS} + V_{DS} \tag{5-60}$$

$$\phi_{s,\mathrm{dc}}(L_\mathrm{I}) = \phi_\mathrm{bi} - V_{GS} + V_{CS}(L_\mathrm{I}) \tag{5-61}$$

L_I is the length of region I. It is thus far an unknown. $V_{CS}(L_\mathrm{I})$ is the channel-to-source potential at $x = L_\mathrm{I}$, the boundary between region I and region II.

Similar to the process used in Eq. (5-32), we integrate Eq. (5-57) from $x = 0$ to $x = L_\mathrm{I}$. The drain current is found to be:

$$\begin{aligned} I_{D,\mathrm{I}} &= \frac{qWN_d\mu_n a\phi_{00}}{L_\mathrm{I}} \int_s^p [1 - \sqrt{u(x)}]\, du \\ &= \frac{qWN_d\mu_n a\phi_{00}}{L_\mathrm{I}} \left(p - s - \frac{2}{3}p^{3/2} + \frac{2}{3}s^{3/2}\right) \end{aligned} \tag{5-62}$$

As mentioned in the discussion of point D of Fig. 5-16, as V_{DS} increases past $V_{DS,\mathrm{sat}}$, L_I decreases. Hence, I_D increases as V_{DS} increases, rather than truly saturating at a constant value. Current saturation to a constant value occurs only in long-channel FETs at $V_{DS} > V_p$.

We shift our focus to region II. Here, the electrons are assumed to travel at a constant speed of v_sat. In order to maintain constant I_D, the channel opening $b(x)$ must be constant in this region. Consequently, $X_\mathrm{dep}(x)$ is constant, and $X_\mathrm{dep}(x) = X_\mathrm{dep}(L_\mathrm{I})$ for all x values between L and L_I. We shall denote this constant depletion thickness as $X_{\mathrm{dep,II}}$, which is given by:

$$X_{\mathrm{dep,II}} = \sqrt{\frac{2\epsilon_s}{qN_d}\left[\phi_\mathrm{bi} - V_{GS} + V_{CS}(L_\mathrm{I})\right]} \qquad \text{for } L_\mathrm{I} \le x \le L \tag{5-63}$$

Let us denote the channel thickness in region II as δ (see point D of Fig. 5-16), which is equal to $a - X_{\mathrm{dep,II}}$. Using the definition of ϕ_{00} in Eq. (5-8) and the normalized variables given in Eq. (5-58), we have:

$$\delta = a - a\sqrt{\frac{\phi_\mathrm{bi} - V_{GS} + V_{CS}(L_\mathrm{I})}{\phi_{00}}} = a(1 - p^{1/2}) \qquad \text{for } L_\mathrm{I} \le x \le L \tag{5-64}$$

The drain current in region II is therefore:

$$I_{D,\mathrm{II}} = qN_d W_a (1 - p^{1/2})\, v_{sat} \equiv I'_{max}(1 - p^{1/2}) \tag{5-65}$$

where the maximum current, I'_{max}, is:

$$I'_{max} = qWN_d a v_{sat} \tag{5-66}$$

In this particular model, I'_{max} can be alternatively written as $aqWN_d\mu_n \cdot \varepsilon_{sat}$. I'_{max} represents the ideal maximum current obtainable from the device when $p = 0$. In practice, p is never zero, so the drain current is always somewhat less than I'_{max}. We add a prime in the symbol to differentiate it from I_{max} used in Eq. (5-41), which is the maximum possible current in a long-channel MESFET without velocity saturation.

Since we are considering d.c. operation, $I_{D,\mathrm{I}} = I_{D,\mathrm{II}}$. Equating Eq. (5-65) to Eq. (5-62), we find a relationship for L_I:

$$L_\mathrm{I} = \frac{\mu_n \phi_{00}}{v_{sat}(1 - p)} \left(p - s - \frac{2}{3} p^{3/2} + \frac{2}{3} s^{3/2} \right) \tag{5-67}$$

Example 5-4:

A short-channel MESFET with $a = 2000$ Å, $N_d = 1 \times 10^{17}\mathrm{cm}^{-3}$, has a width $W = 500$ μm and a length $L = 0.5$ μm. Its low-field mobility is 4000 cm^2/V-s, and at high field the electron velocity saturates at 1×10^7cm/s. When $V_{GS} = -0.2$ V, the drain current is 0.053 A at a particular V_{DS}. Estimate the length of the region in which the carriers travel at the saturation velocity.

I'_{max} according to Eq. (5-66) is:

$$I'_{max} = 1.6 \times 10^{-19} \cdot 500 \times 10^{-4} \cdot 1 \times 10^{17} \cdot 2000 \times 10^{-8} \cdot 1 \times 10^7 = 0.16\,\mathrm{A}$$

With $I_D = 0.053$ A and $I'_{max} = 0.16$ A, the normalized potential p can be found from Eq. (5-65) to be 0.45. The transistor has the same epitaxial structure as that of Example 5-1, from which we find the full-depletion potentials to be $\phi_{00} = 2.76$ V and the built-in potential to be $\phi_{bi} = 0.861$ V. Therefore, s, according to Eq. (5-58), is:

$$s = \frac{\phi_{bi} - V_{GS}}{\phi_{00}} = \frac{0.861 + 0.2}{2.76} = 0.384$$

The distance L_I is given by Eq. (5-67):

$$L_\mathrm{I} = \frac{4000 \cdot 2.76}{10^7 \cdot (1 - 0.45)} \left(0.45 - 0.384 - \frac{2}{3} 0.45^{3/2} + \frac{2}{3} 0.384^{3/2} \right) = 4.7 \times 10^{-5}\,\mathrm{cm}$$

Therefore, $L_\mathrm{II} = L - L_\mathrm{I} = 0.5$ μm $- 0.47$ μm $= 0.03$ μm.

In § 5-2, where the velocity saturation was not taken into consideration, we were able to find the d.c. drain current as soon as V_{GS} and V_{DS} were specified. That is, once s and d were known from Eqs. (5-33)–(5-35), the drain current was immediately found from Eq. (5-40). When the velocity saturation is taken into account, a direct determination of the drain current from the bias voltages is more difficult. With the analysis developed so far, we cannot ascertain the drain current even though s and d are known. This is because the current equations given above require knowledge of p, which relates to the boundary condition at the interface of regions I an II. If p is known, then I_D is straightforwardly calculated from Eq. (5-65).

In order to predict the drain current from V_{GS} and V_{DS} (which give rise to s and d), we need somehow to relate p to d, a task which is equivalent to relating $V_{CS}(L_{\mathrm{I}})$ to V_{DS}. Because the difference of these two voltages is the voltage drop in region II, we focus on the Poisson equation governing that region. From the analysis about the validity of the GCA in § 5-2, we know that the longitudinal field cannot be neglected in comparison to the transverse field in the velocity-saturated region. According to Eq. (5-45), we therefore have (we shall omit the subscript dc for ϕ_s in the rest of the section to reduce complexity in notation):

$$\frac{\partial^2 \phi_s(x, y)}{\partial x^2} + \frac{\partial^2 \phi_s(x, y)}{\partial y^2} = -\frac{q}{\epsilon_s} N_d \qquad \text{for } L_{\mathrm{I}} \le x \le L;\ 0 < y < (a - \delta) \approx a \tag{5-68}$$

δ, the channel thickness in region II, is assumed to be much smaller than the designed active device thickness a. To ensure continuous transitions of the potential and the electric field between regions I and II, we invoke the following boundary conditions:

$$\phi_s(x, 0) = 0 \tag{5-69}$$

$$\phi_s(L_{\mathrm{I}}, y) = \frac{q}{\epsilon_s} N_d \left(X_{\mathrm{dep,II}}\, y - \frac{y^2}{2} \right) \tag{5-70}$$

$$\left.\frac{\partial \phi_s(x, y)}{\partial y}\right|_{y=a} = 0 \tag{5-71}$$

$$\left.\frac{\partial \phi_s(x, y)}{\partial x}\right|_{x=L_{\mathrm{I}},\, y=a} = \mathcal{E}_{\mathrm{sat}} \tag{5-72}$$

The first boundary condition states that the gate (at $y = 0$) is at a constant potential, independent of x. We arbitrarily choose ϕ_s to be zero there, to be consistent with the choice used throughout this chapter. The second boundary condition, embodied in Eq. (5-70), fulfills the requirement that the potential be continuous across the boundary between regions I and II. The potential profile given on the right-hand side of the equation is that obtained in region I at

$x = L_{\mathrm{I}}$, which was given by Eq. (5-10). The vertical electric field in the channel in region II is taken to be zero. The field is entirely in the longitudinal direction. Hence, Eq. (5-71) specifies that the derivative in the y direction is zero at $y = a - \delta \approx a$. The last boundary condition concerns the continuity of the electric field across the boundary between regions I and II. We stated in our model that region II is the region where the electron velocity reaches v_{sat} or, equivalently, where the longitudinal electric field reaches ε_{sat}. Although the electric field increases past ε_{sat} as x increases above L_{I}, the velocity remains at v_{sat} in accordance with Fig. 5-16. Right at $x = L_{\mathrm{I}}$ and in the channel (so that $y = a - \delta \approx a$), the electric field is $-\varepsilon_{sat}$. The electric field points from the drain to the source, so it is negative (ε_{sat}, however, is a positive quantity as defined in Eq. 5-65). Because the electric field is the negative of the differential of ϕ_s with respect to position, Eq. (5-72) results.

Equation (5-68) is a linear partial differential equation. The solution technique, described step by step in Problem 6, leads eventually to an approximate relationship between the applied V_{DS} and $V_{CS}(L_{\mathrm{I}})$. It is given by:

$$V_{DS} = \frac{2a}{\pi} \varepsilon_{sat} \sinh \frac{\pi L_{\mathrm{II}}}{2a} + \phi_{oo}(p - s) \tag{5-73}$$

where L_{II}, the length of region II, is $L - L_{\mathrm{I}}$. Equations (5-73) contains two unknowns: L_{II}, and p. When combined with Eq. (5-67), we then have a system of two equations with two unknowns. Therefore, as soon as the bias voltages V_{DS} and V_{GS} are specified, L_{II} and p can be determined. Other variables, such as L_{I} and I_D, can be subsequently established as well.

§ 5-4 NONUNIFORM DOPING PROFILES

We have discussed two MESFET models. The first one, in § 5-2, assumes the mobility to be constant and that the velocity never reaches the saturation velocity. The second model, discussed in § 5-3, accounts for velocity saturation by introducing the concept of two regions. The first region is characterized by a constant mobility, and the second, by a constant velocity of v_{sat}. When the mobility is high and the channel length is short (so that the field is high), it is often fair to make the assumption that the electrons drift at v_{sat} throughout the entire channel. This assumption helps us to highlight some design strategies to increase the linearity of MESFET.

We have considered MESFETs whose active channel layer is doped uniformly at N_d. In this section, we consider transistors whose doping density varies in the transverse (y) direction. It helps to review a lesson learned from the two-region model. In region II where the velocity saturates, the channel potential $\phi_s(x)$ continues to increase from $x = L_{\mathrm{I}}$ toward $x = L$. The potential was obtained from solving a two-dimensional quasi-Poisson equation. Although the potential varies with x, the channel thickness (thin but finite) remains constant so that the channel current is a constant. The amount of channel thickness available for the carriers to conduct current is equal to that at the $x = L_{\mathrm{I}}$, as indicated by Eq. (5-64). Equivalently, the amount of the depletion thickness in region II is given by Eq. (5-63), independent of the channel potential variation.

The above concept is employed in this section, which considers transistor geometry such that region II prevails in the entire channel. We imagine that the length of region I approaches zero, as we assume that the electrons travel at v_{sat} right after they exit the source and enter the channel. Extending the aforementioned lesson, we state that although the channel potential varies between $x = 0$ and $x = L$, the channel thickness is the same so that the channel current is independent of x. Likewise, the depletion thickness is uniform across the channel, not extending to a larger value toward the drain as in the case of constant mobility. Further, the depletion thickness is determined at the source (which is basically the boundary between region I and region II as L_{I} approaches zero). In other words, X_{dep}, which is a constant along the channel, is a function of V_{GS} only and does not depend on V_{DS}. Writing out the Poisson equation in one-dimension for a nonuniformly doped channel layer, we have:

$$\frac{d\phi_{s,\mathrm{dc}}(y)}{dy} = -\varepsilon(y) \qquad \text{or} \qquad \frac{d^2\phi_{s,\mathrm{dc}}(y)}{dy^2} = -\frac{q}{\epsilon_s} N_d(y) \tag{5-74}$$

For a given applied gate-to-source voltage V_{GS}, the total amount of voltage dropping across the depletion region is equal $\phi_{\mathrm{bi}} - V_{GS}$. The boundary conditions for Eq. (5-74) are hence $\phi_{s,\mathrm{dc}}(0) = 0$, $\phi_{s,\mathrm{dc}}(X_{\mathrm{dep}}) = \phi_{\mathrm{bi}} - V_{GS}$, and $d\phi_{s,\mathrm{dc}}/dy = 0$ at $y = X_{\mathrm{dep}}$. Integrating the differential equation twice, we identify the following equation for the potential drop:

$$\phi_{\mathrm{bi}} - V_{GS} = \frac{q}{\epsilon_s} \int_0^{X_{\mathrm{dep}}} \int_y^{X_{\mathrm{dep}}} N_d(y')\,dy'\,dy \tag{5-75}$$

In a constant-doping case where $N_d(y) \equiv N_d$, Eq. (5-75) yields a familiar expression:

$$\phi_{\mathrm{bi}} - V_{GS} = \frac{q}{2\epsilon_s} N_d X_{\mathrm{dep}}^2 \qquad (\text{constant } N_d) \tag{5-76}$$

The depletion potential (ϕ_{00}) is the value of $\phi_{\mathrm{bi}} - V_{GS}$ such that the designed channel becomes fully depleted (i.e., X_{dep} reaches the value a). In this constant-doping case, ϕ_{00}, according to Eq. (5-76), is then equal to that given in Eq. (5-8). For the more general case in which the channel doping is not uniform, the expression of ϕ_{00} will be given shortly.

The drain current is equal to:

$$I_D = qWv_{\text{sat}} \int_{X_{\text{dep}}}^{a} N_d(y)\,dy \tag{5-77}$$

In the case of constant channel doping, Eq. (5-77) yields:

$$I_D = qWv_{\text{sat}}(a - X_{\text{dep}})\,N_d = I'_{\text{max}} \times \left[1 - \left(\frac{\phi_{\text{bi}} - V_{GS}}{\phi_{00}}\right)^{1/2}\right] \quad \text{(constant } N_d\text{)} \tag{5-78}$$

where I'_{max}, the maximum current (when no portion of the channel is depleted—clearly an unachievable scenario), is identical to that of Eq. (5-66). More generally, for MESFETs with arbitrary doping profiles, I'_{max} is given by:

$$I'_{\text{max}} = qWv_{\text{sat}} \int_{0}^{a} N_d(y)\,dy \tag{5-79}$$

The current as a function of V_{GS} is shown in Fig. 5-18 for the case of constant doping, power doping, and step doping profiles. They are obtained by applying Eqs. (5-75), (5-77) and (5-79). The expressions for the depletion profile and drain current are also listed.

For many applications, a transistor is desired to be *linear*. In a linear transistor, the output sinusoidal is always at the same frequency as the input sinusoidal, without the generation of higher harmonics. For FETs, the input is the gate-to-source voltage and the output is the drain current. Therefore, we desire a transistor whose drain current increases linearly with the gate voltage. A good measure of the device linearity is the *mutual transconductance* (g_m), defined as:

$$g_m = \left.\frac{\partial I_D}{\partial V_{GS}}\right|_{V_{DS} = \text{const.}} \tag{5-80}$$

In accordance with the desire to have a linear transistor, g_m should be designed to be a constant, preferably independent of the gate bias. In the case of constant channel doping, the drain current was expressed in Eq. (5-78). The mutual transconductance is:

$$g_m = \frac{qWN_d v_{\text{sat}} a}{2}\left[\phi_{00}(\phi_{\text{bi}} - V_{GS})\right]^{-1/2} \quad \text{(constant } N_d\text{)} \tag{5-81}$$

There is a inverse square-root dependence on V_{GS}, indicating that the transistor is not entirely linear.

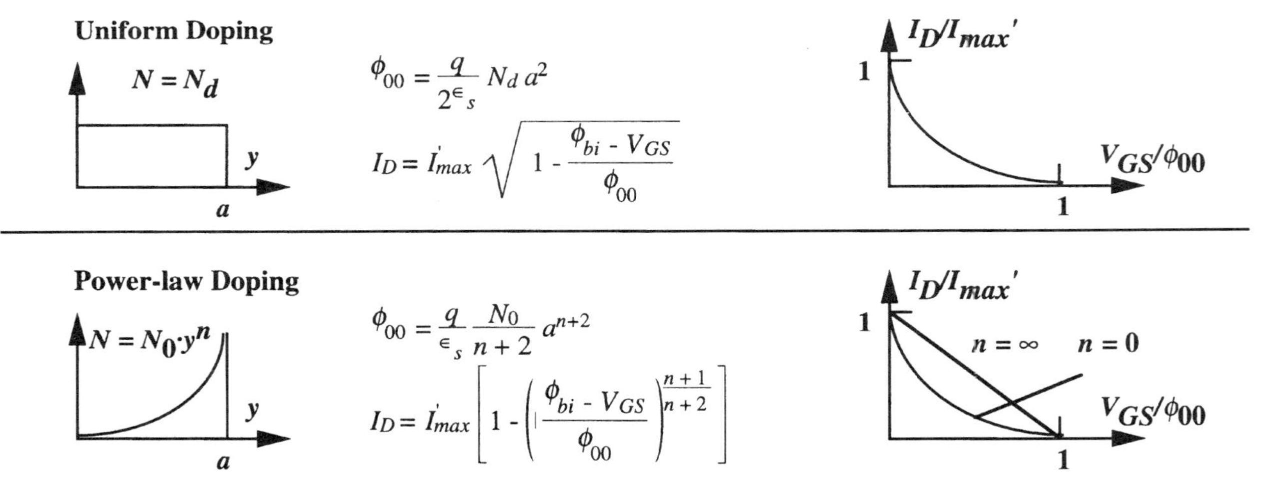

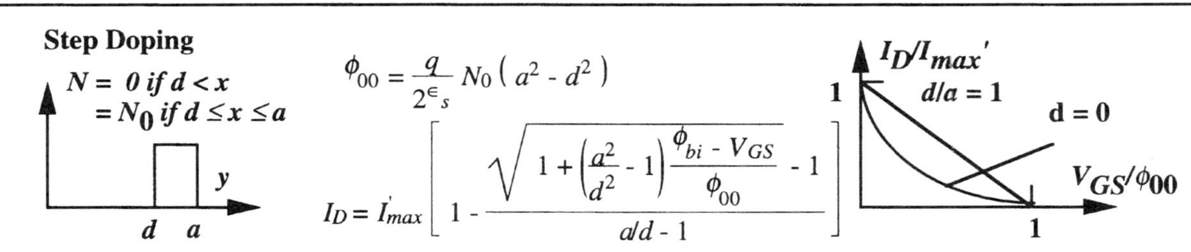

FIGURE 5-18. Device parameters for velocity-saturated MESFETs with various doping profiles. (From R. Williams and D. Shaw, Graded channel FET's: improved linearity and noise figure, *IEEE Trans. Electron. Dev.* **25,** 600–605, 1978. © IEEE, reprinted with permission. The figure notations are modified to conform to the symbols used in this book.)

The above derivation to establish whether g_m is a strong function of V_{GS} works well for the constant-doping case. For arbitrary doping profiles, the derivation becomes more difficult. It is useful to estimate the linearity of the transistor with the following approach. Since the total enclosed voltage under an electric field profile is $\phi_{bi} - V_{GS}$, we write:

$$\phi_{bi} - V_{GS} = -\int_0^{X_{dep}} \varepsilon(y)\,dy = -y\varepsilon(y)\Big|_0^{X_{dep}} + \int_{\varepsilon(0)}^{\varepsilon(X_{dep})} y\,d\varepsilon \qquad (5\text{-}82)$$

The integration by parts used in the above evaluation involves setting $u = \varepsilon(y)$ and $dv = dy$, so that $du = d\varepsilon(y)$ and $v = y$. Equation (5-82) is further simplified by noting that $\varepsilon(X_{dep}) = 0$, so:

$$\phi_{bi} - V_{GS} = \int_{\varepsilon(0)}^{\varepsilon(X_{dep})} y\frac{d\varepsilon}{dy}\,dy = \frac{q}{\epsilon_s}\int_0^{X_{dep}} yN_d(y)\,dy \qquad (5\text{-}83)$$

The above formulation also gives a simple formula for ϕ_{00} with arbitrary doping profiles, which is the potential at which X_{dep} reaches the maximum channel thickness a:

$$\phi_{00} = \frac{q}{\epsilon_s}\int_0^a yN_d(y)\,dy \qquad (5\text{-}84)$$

Our goal is to determine g_m. According to Eq. (5-83):

$$\frac{dV_{GS}}{dX_{dep}} = -\frac{q}{\epsilon_s} X_{dep} N_d(X_{dep}) \qquad (5\text{-}85)$$

In addition, according to Eq. (5-77):

$$\frac{dI_D}{dX_{dep}} = -qWv_{sat} N_d(X_{dep}) \qquad (5\text{-}86)$$

Therefore, g_m is:

$$g_m(V_{GS}) = \frac{dI_D}{dX_{dep}} \div \frac{dV_{GS}}{dX_{dep}} = \frac{Wv_{sat}\epsilon_s}{X_{dep}(V_{GS})} \qquad (5\text{-}87)$$

Equation (5-87) points out that, in order to acquire a constant g_m, the doping profile should be such that the depletion thickness does not vary much with the applied V_{GS}. If X_{dep} is a strong function of V_{GS}, g_m likewise varies with V_{GS}. To minimize the depletion thickness' dependence on V_{GS}, dX_{dep}/dV_{GS} should be small. Alternatively, according to Eq. (5-85), $X_{dep}N_d(X_{dep})$ should be large. This is achieved by having a δ-doped MESFET structure, or δ-FET, all of whose

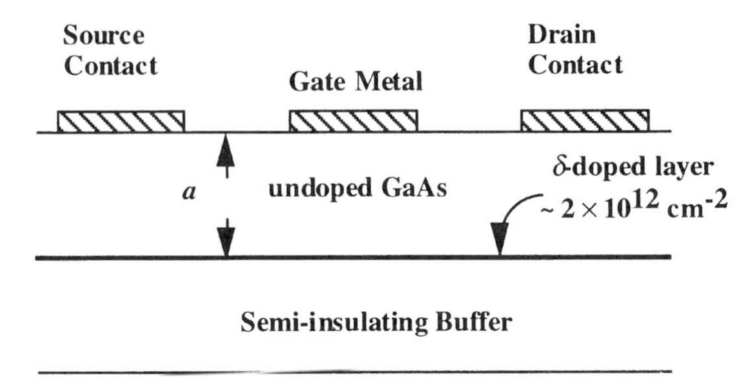

FIGURE 5-19. A delta-FET (δ-FET) structure.

available dopant charges are located at the channel/buffer interface, as shown in Fig. 5-19. The drain current varies linearly with V_{GS}, resulting in a high degree of linearity.

Example 5-5:

Find g_m's of two short-channel MESFETs when $V_{GS} = 0$ V. One has a conventional channel layer of a constant $N_d = 1 \times 10^{17}$ cm^{-3} and $a = 2000$ Å. The other is a δ-FET with a delta doping density of $\sigma = N_d a = 2 \times 10^{12}$ cm^{-2}, which is located at a (fairly long) distance $d = 1000$ Å from the gate. The gate width is $W = 1$ mm and $v_{\text{sat}} = 10^7$ cm/s.

From Example 5-1, for the constant-doping FET, $\phi_{00} = 2.76$ V and $\phi_{\text{bi}} = 0.861$ V. From Eq. (5-81), g_m for the conventional device is:

$$g_m = \frac{1.6 \times 10^{-19} \cdot 0.1 \cdot 10^{17} \cdot 10^{7} \cdot 2000 \times 10^{-8}}{2} [2.76(0.861 - 0)]^{-1/2} = 0.104\ \Omega^{-1}$$

For the δ-FET, ϕ_{00} is determined from Eq. (5-84) to be:

$$\phi_{00} = \frac{q\sigma}{\epsilon_s} d = \frac{1.6 \times 10^{-19} \cdot 2 \times 10^{12}}{1.159 \times 10^{-12}} 1000 \times 10^{-8} = 2.76 \text{ V}$$

This ϕ_{00} is the same as that of the conventional FET. This will allow comparison between the g_m's of the two devices more meaningful. From Eq. (5-87), g_m of the δ-FET is calculated to be:

$$g_m = \frac{W v_{\text{sat}} \epsilon_s}{d} = \frac{0.1 \times 10^{7} \cdot 1.159 \times 10^{-12}}{1000 \times 10^{-8}} = 0.116\ \Omega^{-1}$$

§ 5-5 MODULATION DOPING

In a perfect crystal whose lattice presented a periodic potential to the electrons, the electrons would not interchange energy with the lattice. There would be no scattering. However, at any temperature above absolute zero, the lattice atoms vibrate, disturbing the perfect periodicity of the otherwise stationary lattice. The breaking of the periodicity of the lattice potential enables the electrons to exchange energy with the lattice. Because the lattice vibration is characterized by phonons, the resultant scattering process is called *phonon scattering*. In a material whose electron movement is limited by phonon scattering, the mobility decreases with increasing temperature.

As mentioned in § 1-7, there is another mechanism affecting the electron mobility: the impurity scattering. When there is a significant amount of impurity (dopant) atoms, the local lattice potential is distorted and scattering with electrons occurs. The resultant scattering process is referred to as *impurity scattering*. Unlike phonon scattering, whose magnitude is determined primarily by the lattice temperature, the amount of impurity scattering is controllable. We can eliminate impurity scattering by removing the impurity atoms. Certainly, there are other concerns which preclude us from doing so. For example, in the heavily doped base layer of HBTs, the mobilities of the majority holes and the minority electrons are determined primarily by the impurity scattering. If we improved the carrier mobilities by lowering the base doping, we would suffer from increased base resistance and worsened high-frequency performance. In this instance, the drawbacks associated with the impurity reduction more than offset the benefits.

How about the field-effect transistors? In the channel layer of a MESFET doped to $\sim 10^{17}$ cm^{-3}, the impurity scattering is fairly important. The desire to increase the mobility, and thereby the current, leads eventually to the concept of *modulation doping*. This is a doping method in which the conduction carriers are provided by the dopants, yet the dopant location is maintained at a distance from the conduction channel. In this manner, the amount of scattering for the conduction carriers is minimized and the conduction current is maximized. Figure 5-20 illustrates the epitaxial structure used for modulation doping. (The figure also shows the transistor structure of a high electron mobility transistor, to be discussed in the next section.) The structure consists of an n^--GaAs or intrinsic GaAs layer at the bottom, a thin undoped AlGaAs layer of thickness δ, and then a N-AlGaAs layer. Without any external bias, the bands bend in a manner such that the Fermi levels are aligned throughout the entire structure. This naturally forms a depletion region in the N-AlGaAs layer, as well as an accumulation region in the lightly doped GaAs layer. The accumulation region, right near the AlGaAs/GaAs interface, is filled with electrons originating from the doped AlGaAs layer. If somehow an electric field is established in the direction parallel to the dashed line in Fig. 5-20, then the electrons in the conduction channel will move in response to the field. Because this electron movement will occur in the lightly doped GaAs layer, and because the electrons

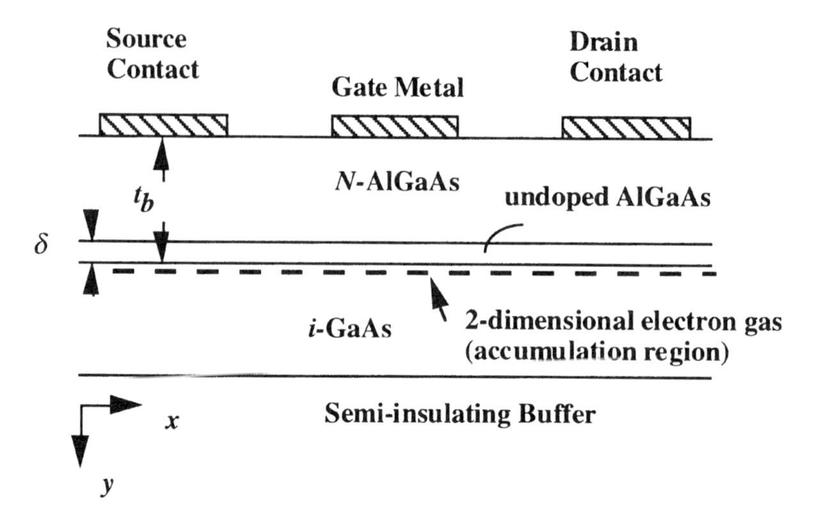

FIGURE 5-20. Epitaxial structure used for modulation doping in the AlGaAs/GaAs material system. An undoped layer of thickness δ is inserted between an *N*-AlGaAs layer and a lightly doped (or undoped) GaAs layer. It also shows the transistor structure of heterojunction field-effect transistor.

are located far away from the nearest dopants (with a separation of at least δ), the amount of impurity scattering will be minimal. The electrons will travel with a high mobility, limited only by the phonon scattering induced by the ambient temperature.

To simplify the discussion about modulation doping, we shall consider the AlGaAs/GaAs heterojunction of Fig. 5-20 with the assumption that the AlGaAs layer is infinitely thick. The band diagram of the AlGaAs/GaAs heterojunction under thermal equilibrium is shown in Fig. 5-21. The electrons in the conduction (accumulation) region are restricted to motion only in the two-dimensional plane perpendicular to the paper. The electrons are not allowed to move in the *y* direction because of the energy barriers at the both sides of the potential well. Hence, the electrons accumulated in the well are referred to as the *two-dimensional electron gas*. At room temperature, the de Broglie wavelength of a thermal electron is approximately 260 Å (it is even longer at lower temperatures). Because the distance between the two sides of the potential well is generally shorter than the de Broglie wavelength, the electrons cannot have an arbitrary amount of momentum in the direction perpendicular to the interface. The momentum in this direction is quantized. This gives rise to a steplike density-of-state function, different from that of a three-dimensional solid. In addition to the quantization of momentum along the *y* direction (Fig. 5-21), the electrons can have only certain eigenfunctions, and associated with them, certain eigen energies. These are the discrete energy levels with which the electrons can reside in the potential well. Each collection of continuum states associated with one particular energy level is called a *subband*. The quantization effect

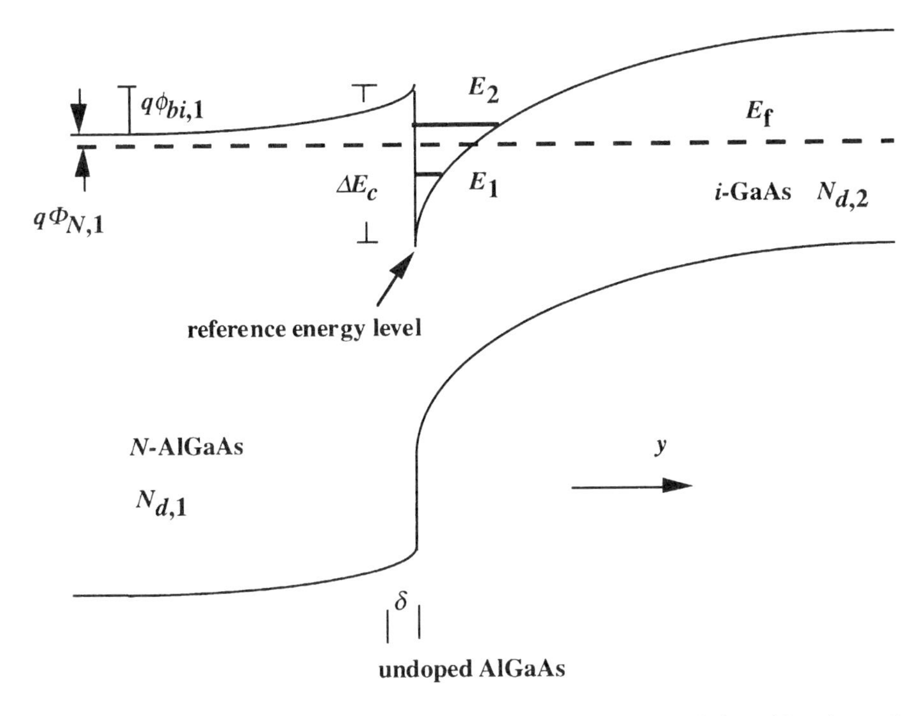

FIGURE 5-21. Band diagram of the modulation doping structure of Fig. 5-20 at thermal equilibrium.

is observable in silicon inversion layers, but it is more pronounced in the AlGaAs/GaAs structure because of the smaller electron effective mass of GaAs, which translates to a longer deBroglie wavelength.

In § 1-5, where we related the carrier concentrations to the Fermi level, we made an implicit assumption that the solid is three-dimensional. When the carriers are bound to move in only two dimensions, the relationships need to be modified. We briefly review how we obtain the equations for the three-dimensional case given in § 1-5. Subsequently, we derive a relationship between the electron density and the Fermi level in a two-dimensional system.

When the region of interest is three-dimensional, the electron concentration can be written as:

$$n = \int_0^\infty \mathscr{D}(E) f(E)\, dE \tag{5-88}$$

where $\mathscr{D}(E)$ is the density-of-state function and $f(E)$ is the probability function for finding an electron at a particular energy E. Independent of the dimensionality, $f(E)$, given by the Fermi-Dirac statistics, is:

$$f(E) = \frac{1}{1 + \exp\left[(E_f - E_c)/kT\right]} \tag{5-89}$$

The density-of-state function, in contrast, depends critically on the dimensionality of the sample. In a three-dimensional sample, it is a continuous function, having the following dependence on the electron energy E:

$$\mathscr{D}(E) = \begin{cases} 4\pi\left(\dfrac{2}{h^2}\right)^{3/2} m_{de}^{*3/2}\sqrt{E - E_c} & \text{if } E \geq E_c \\ 0 & \text{if } E < E_c \end{cases} \tag{5-90}$$

The unit of $\mathscr{D}(E)$ in the three-dimensional sample is $\text{cm}^{-3} \cdot \text{J}^{-1}$. With $\mathscr{D}(E)$ and $f(E)$ given above, the electron density in a three-dimensional sample is equal to:

$$n = \int_{E_c}^{\infty} 4\pi\left(\frac{2}{h^2}\right)^{3/2} m_{de}^{*3/2}\sqrt{E - E_c}\, \frac{1}{1 + \exp\left[(E_f - E_c)/kT\right]}\, dE \tag{5-91}$$

After letting $x = (E - E_c)/kT$, and substituting the three-dimensionl density of states (N_c) from Eq. (1-23), we transform the above integral to:

$$\frac{n}{N_c} = \frac{2}{\sqrt{\pi}} \int_0^{\infty} \frac{\sqrt{x}}{1 + \exp\left[x - (E_f - E_c)/kT\right]}\, dx \tag{5-92}$$

This is exactly the expression given in Eq. (1-28). As is, the integral cannot be evaluated analytically, although the Joyce-Dixon approximation given in Eq. (1-29) can be used to approximate the integral. In the case that $E_c - E_f \gg 1$, the exponential function in the denominator greatly exceeds unity. This corresponds to the Maxwell-Boltzmann statistics. When the unity is dropped from the denominator, the integration becomes straightforward, and Eq. (1-21) results.

In the two-dimensional case, the electron sheet density is given by:

$$n_s = \int_0^{\infty} \mathscr{D}(E) f(E)\, dE \tag{5-93}$$

This resembles Eq. (5-88), except that n in Eq. (5-88) is as the unit of cm^{-3} whereas n_s here has the unit cm^{-2}, consistent with the dimensionality of each case. For convenience, all of the energies given in the following are referenced to the bottom of the triangular well. For example, E_1 means the energy difference between the first subband and the energy at the bottom of the well.

The density-of-state function in a two-dimensional sample depends on the semiconductor material, as illustrated in Fig. 5-22. Part *a* of the figure applies to GaAs (or $\text{Al}_\xi\text{Ga}_{1-\xi}\text{As}$ with $\xi \leq 0.45$) under consideration, whose minimum energy valley, the Γ valley, is spherical with respect to the momentum. The energy valley is characterized by a single effective mass, m_Γ^*, given in Eq. (1-79a). Fig. 5-22*b* is pertinent to Si, whose minimum energy valley is the X valley. When expressed in the momentum space, the constant energy surface in the X valley is

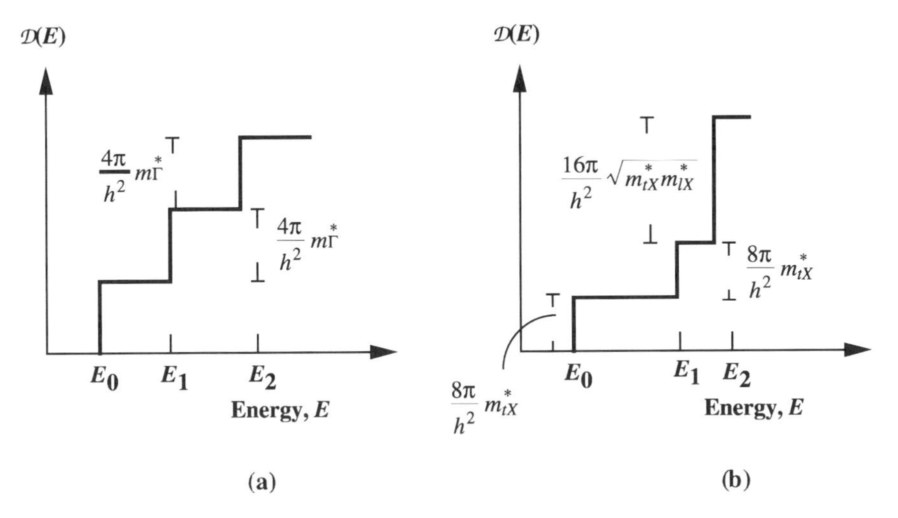

FIGURE 5-22. Density-of-state functions in two-dimensional samples: (*a*) GaAs; (*b*) silicon. (From S. Tiwari, *Compound Semiconductor Device Physics*, 1992. © Academic Press, reprinted with permission. The figure notations are modified to conform to the symbols used in this book.)

ellipsoidal, characterized by a transverse effective mass (m_{tX}^*) and a longitudinal effective mass (m_{lX}^*). In terms of mathematical expression, the density-of-state function for GaAs is given by (for $E < E_3$, that is, considering only two lowest subbands):

$$\mathscr{D}(E) = \begin{cases} 0 & \text{if } E_1 \geq E \\ \dfrac{4\pi}{h^2} m_\Gamma^* & \text{if } E_2 \geq E \geq E_1. \\ 2 \times \dfrac{4\pi}{h^2} m_\Gamma^* & \text{if } E_3 \geq E \geq E_2 \end{cases} \tag{5-94}$$

Often in the literature about two-dimensional electron gas, a constant D is defined as:

$$D = \frac{4\pi}{h^2} m_\Gamma^* \tag{5-95}$$

Note the difference between $\mathscr{D}(E)$ and D. The former is referred to as the density-of-state function, having a dependence on the electron energy. The latter is purely a constant, as given in Eq. (5-95). They both have the unit $\text{cm}^{-2}\cdot\text{J}^{-1}$ in the two-dimensional solid under consideration. ($D = 1.743 \times 10^{32}\ \text{cm}^{-2}\cdot\text{J}^{-1} = 2.79 \times 10^{13}\ \text{cm}^{-2}\cdot\text{eV}^{-1}$ for GaAs.)

The probability function finding an electron at a given energy level remains that of Eq. (5-89). Therefore, when we consider only two subbands, the electron density is:

$$n_s = D\int_{E_1}^{E_2} \frac{1}{1+\exp[(E-E_f)/kT]}\,dE + 2D\int_{E_2}^{\infty} \frac{1}{1+\exp[(E-E_f)/kT]}\,dE \tag{5-96}$$

The following integration identity is helpful in evaluating the outcome of Eq. (5-96):

$$\int \frac{1}{1+\exp(x)}\,dx = \int \frac{\exp(-x)}{\exp(-x)+1}\,dx = \ln[1+\exp(-x)] \tag{5-97}$$

Hence, n_s can be expressed as a function of Fermi energy:

$$n_s = DkT\ln\left[1+\exp\left(\frac{E_f-E_1}{kT}\right)\right] + DkT\ln\left[1+\exp\left(\frac{E_f-E_2}{kT}\right)\right] \tag{5-98}$$

At low temperatures, the above equation reduces to:

$$n_s = \begin{cases} D(E_f-E_1) & \text{if only the first subband is occupied} \\ D(E_2-E_1)+2D(E_f-E_2) & \text{if both subbands are occupied} \end{cases} \tag{5-99}$$

In general, at or near room temperature, the electron sheet concentration is given by Eq. (5-98) and both bands are occupied by a certain number of electrons.

There are four unknowns in Eq. (5-98): the electron sheet concentration (n_s), the Fermi level (E_f), and the two subband energies (E_1 and E_2). We seek to eliminate two of the variables by establishing a relationship between the subband energies and n_s. From the band diagram (Fig. 5-21), we see that the potential well in the GaAs layer can be approximated as a triangular well, with the slope of the conduction band edge being roughly equal to the electric field at the interface. The subband energies of a triangular well are well known, acquired from solving the Schrodinger equation:

$$E_n \approx \left(\frac{h^2}{8\pi^2 m_e^*}\right)^{1/3}\left(\frac{3\pi}{2}q\varepsilon_{i,2}\right)^{2/3}\left(n-\frac{1}{4}\right)^{2/3} \qquad n = 1, 2, \ldots \tag{5-100}$$

where $\varepsilon_{i,2}$ is the magnitude of the AlGaAs/GaAs interface electric field at region 2 (GaAs). Since the charge is negative in region 2 and positive in region 1, the electric field is positive in our coordinate system. We consider only the first two eigenstates. The higher subbands are generally discarded because the probability function decreases rapidly with increasing E, and because the well ceases to be exactly triangular at higher energy levels. Substituting the material parameters into Eq. (5-100), we find that the subband energy levels are given by:

$$E_1(\text{eV}) \approx 4.14 \times 10^{-5}(\varepsilon_{i,2})^{2/3} \qquad (\varepsilon_{1,2} \text{ in V/cm}) \tag{5-101}$$

$$E_2(\text{eV}) \approx 7.31 \times 10^{-5}(\varepsilon_{i,2})^{2/3} \qquad (\varepsilon_{1,2} \text{ in V/cm}) \tag{5-102}$$

The interface electric field in GaAs can be related to the electron sheet density there. We write the Poisson equation pertaining to region 2 as:

$$\frac{d\varepsilon_{i,2}}{dx} = -\frac{q}{\epsilon_s}[n(x) - N_{d,2}] \tag{5-103}$$

where $N_{d,2}$ is the concentration of the lightly doped n-type GaAs layer. We shall neglect the difference in the relative permeability between the AlGaAs (layer 1) and the GaAs (layer 2) regions. Therefore, ϵ_s is used in the above equation, without the need of differentiate between $\epsilon_{s,1}$ and $\epsilon_{s,2}$. The free-electron concentration is related to the electron sheet concentration according to:

$$n_s = \int_0^{X_{\text{dep},2}} n(x)\, dx \tag{5-104}$$

where $X_{\text{dep},2}$ is the depletion thickness in region 2, i.e., the GaAs layer. At $X_{\text{dep},2}$, the electric field $\varepsilon_{i,2}$ decreases to zero. Integrating Eq. (5-103) from $x = 0$ (the AlGaAs/GaAs interface) to $X_{\text{dep},2}$, we obtain a relationship between $\varepsilon_{i,2}$ and n_s:

$$\varepsilon_{i,2} = \frac{qn_s}{\epsilon_s} - \frac{qN_d X_{\text{dep},2}}{\epsilon_s} \approx \frac{qn_s}{\epsilon_s} \tag{5-105}$$

In the above equation, the term involving N_d is dropped because the doping is small compared to the accumulated electron concentration.

Once the relationship between $\varepsilon_{i,2}$ and n_s is established, we rewrite Eqs. (5-101) and (5-102) as:

$$E_1(\text{eV}) \approx 1.11 \times 10^{-9}(n_s)^{2/3} \qquad (n_s \text{ in cm}^{-2}) \tag{5-106}$$

$$E_2(\text{eV}) \approx 1.95 \times 10^{-9}(n_s)^{2/3} \qquad (n_s \text{ in cm}^{-2}) \tag{5-107}$$

These two energy levels in the triangular quantum well are drawn as two horizontal lines, as shown in Fig. 5-21.

After we substitute Eqs. (5-106) and (5-107) into Eq. (5-98), we then have an equation in which n_s and E_f as the only variables. Such an equation will be used in the analysis of the HFET, to be described in the next section. If it is desirable to determine these two unknowns simultaneously, the Poisson equation can be invoked, as outlined in Problem 8.

Example 5-6:

In a modulation doping structure whose band diagram is shown in Fig. 5-21, the accumulated electron sheet density is $n_s = 2 \times 10^{12}\ \text{cm}^{-2}$. If the Fermi level is known to be 0.24418 eV above the bottom tip of the triangular well, what are the electron sheet densities in the two subbands?

From Eqs. (5-106) and (5-107), the two subband energy levels are:

$$E_1 = 1.11 \times 10^{-9}(2 \times 10^{12})^{2/3} = 0.1762\ \text{eV}$$

$$E_2 = 1.95 \times 10^{-9}(2 \times 10^{12})^{2/3} = 0.3095\ \text{eV}$$

The electron concentration in the first subband is the first term in Eq. (5-98). Similarly, the second subband electron concentration corresponds to the second term. Therefore, with $D = 2.79 \times 10^{13}\ \text{cm}^{-2} \cdot \text{eV}^{-1}$ for GaAs, we have:

$$n_{s,1} = 2.79 \times 10^{13} \cdot 0.0258 \cdot \ln\left[1 + \exp\left(\frac{0.24418 - 0.1762}{0.0258}\right)\right]$$

$$= 1.946 \times 10^{12}\ \text{cm}^{-2}$$

$$n_{s,2} = 2.79 \times 10^{13} \cdot 0.0258 \cdot \ln\left[1 + \exp\left(\frac{0.24418 - 0.3095}{0.0258}\right)\right]$$

$$= 0.055 \times 10^{12}\ \text{cm}^{-2}$$

The electron population of the second subband is much less than that of the first subband. This confirms that neglecting the electron population in even higher subbands is valid.

§ 5-6 THE HETEROJUNCTION FET (HFET)

Modulation doping is a doping scheme whereby the conduction carriers are separated from the donor impurities which supply the electrons. The MESFET device structure can be modified to incorporate the modulation doping concept. For example, in order to modulate the amount of carriers in the channel of the modulation doping structure (the bound accumulation region in GaAs), a Schottky gate is placed on top of the AlGaAs layer. To establish an electric field along the channel direction to induce the carriers to drift, the drain and source contacts are placed symmetrically about the gate. Just like the MESFET, the metals forming the drain and the source contacts are driven inward through an alloying process, to ensure that contacts are made to the channel inside the bulk of the epitaxial layer structure. The resulting FET structure, which utilizes modulation doping to increase the carrier mobility in the channel, is called a *h*igh *e*lectron *m*obility *t*ransistor (HEMT).

A number of names have been used to describe HEMTs. All of them focus on a particular physical aspect of the device. The name modulation-doped FET (MODFET) emphasizes the fact that modulation doping is used to provide the charges in the channel. Because modulation doping is sometimes referred to as selective doping, the term selectively doped FET (SDFET) is also used. Finally, the name *t*wo-dimensional *e*lectron *g*as FET (TEGFET) is used to emphasize the reduced dimensionality of the electron momentum in the accumulated region.

Sometimes a thin strained InGaAs layer is grown at where the 2-dimensional electron gas is located. The InGaAs layer is *pseudomorphic*, i.e., thin enough so the lattice mismatch between InGaAs and GaAs is absorbed by the strain in the InGaAs layer. This type of transistor, pseudomorphic HEMT (pHEMT), is found to have improved performance due to superior electron transport properties in InGaAs than GaAs.

HEMT falls into a broader category of *h*eterojunction *f*ield-*e*ffect *t*ransistor (HFET), which can be used to describe FETs with undoped AlGaAs and doped GaAs channel. Since the equations governing their electrical characteristics are identical, and since HFET is a more generic name, we shall refer HEMT as a kind of HFET in the following development.

Just like the MESFET, the HFET is best studied by first examining a two-terminal structure as shown in Fig. 5-23. The band diagram in the y direction is shown in Fig. 5-24. The charge diagram and the corresponding electric field profile are shown in Fig. 5-25. $\phi_{G,1}$ is used to denote the potential drop in the AlGaAs layer in this gated structure. $\phi_{G,1}$ can be obtained from a visual inspection of the band diagram:

$$q\phi_{G,1} + \Delta E_c = q\phi_B - qV_{GB} + E_f \tag{5-108}$$

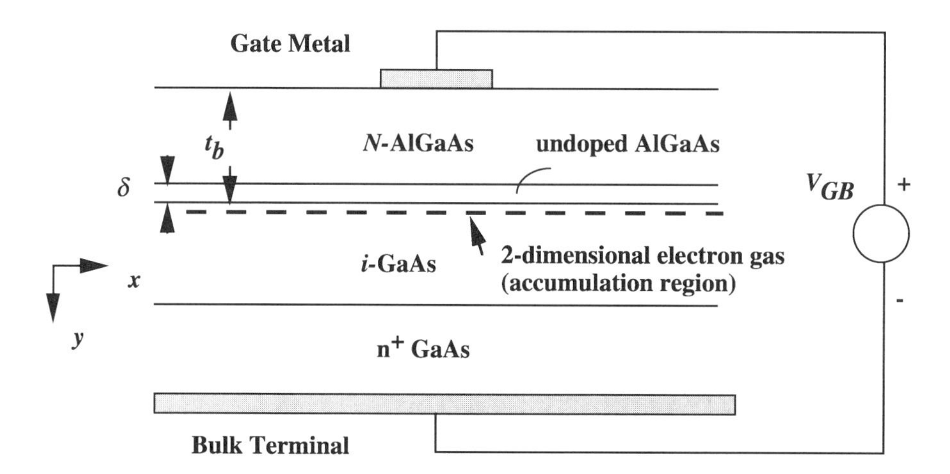

FIGURE 5-23. A two-terminal diode made of the AlGaAs/GaAs modulation doping structure shown in Fig. 5-20. The diode is biased at V_{GB}.

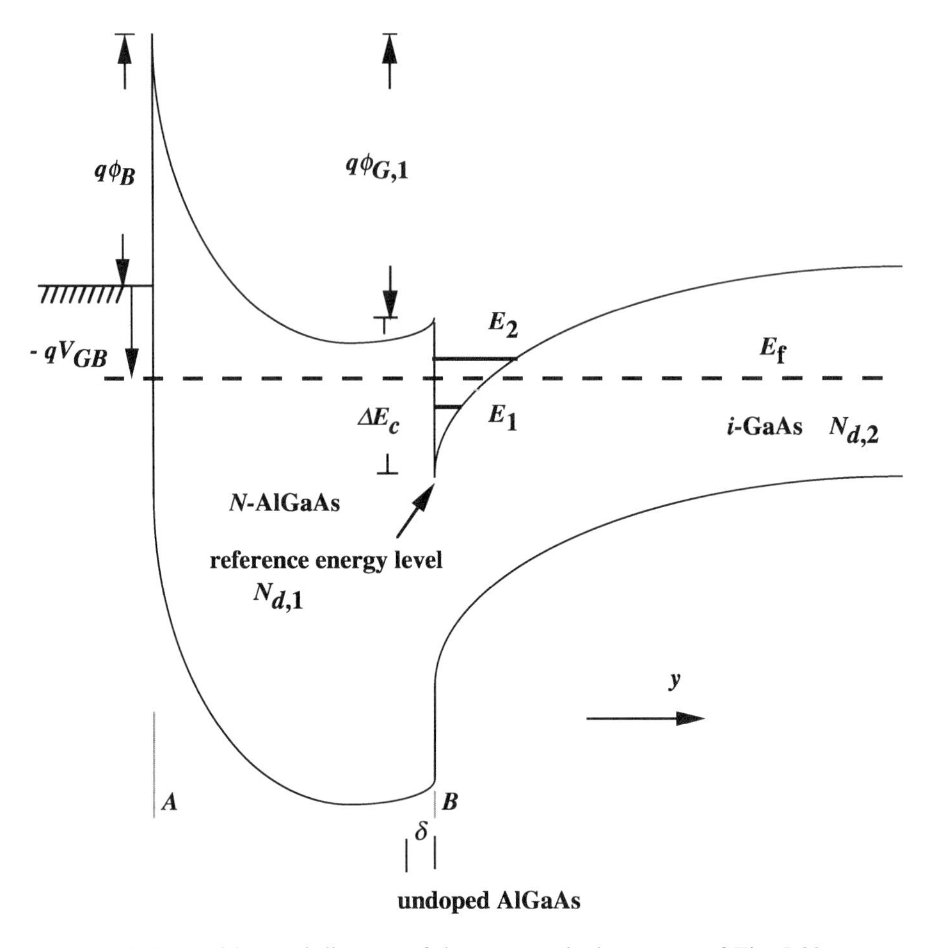

FIGURE 5-24. Band diagram of the two-terminal structure of Fig. 5-23.

where V_{GB} is the applied bias between the gate and the bulk. The band diagram plots out the electron energy, which is equal to $-q$ times the potential. Hence, if we label point A at the gate/AlGaAs interface and point B at the AlGaAs/GaAs interface (Fig. 5-24), then $q\phi_{G,1}$ is given by:

$$q\phi_{G,1} = E_A - E_B = -qV_A + qV_B = -q\int_A^B \varepsilon(x)\,dx \tag{5-109}$$

E_A in the above expression means the electron energy at point A. We can carry out the integral graphically, with the help of Fig. 5-25:

$$\phi_{G,1} = -\left[\frac{\varepsilon_i\delta}{2} + \frac{\varepsilon_i t_b}{2} - \frac{qN_{d,1}}{2\epsilon_s}(t_b - \delta)^2 + \frac{qN_{d,1}}{2\epsilon_s}h(t_b - \delta)\right] \tag{5-110}$$

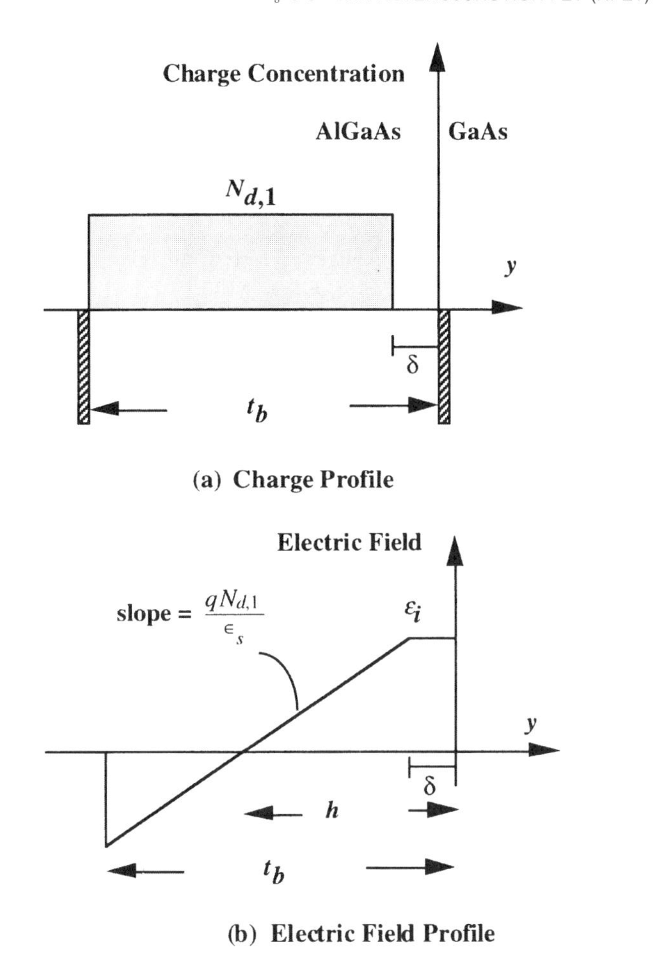

FIGURE 5-25. Charge diagram and electric field profile of the two-terminal structure of Fig. 5-23.

where ε_i is the electric field at the AlGaAs/GaAs interface. We use the symbol t_b to denote the thickness of the AlGaAs barrier layer. The distance h marks the location where the electric field changes sign. From an inspection of the electric field diagram, h is related to ε_i by:

$$\varepsilon_i = \frac{q}{\epsilon_s} N_{d,1} h \tag{5-111}$$

Substituting the value of h from Eq. (5-111) into Eq. (5-110), we obtain:

$$\epsilon_s \varepsilon_i = \frac{\epsilon_s}{t_b} (\phi'_{00} - \phi_{G,1}) \tag{5-112}$$

where

$$\phi'_{00} = \frac{qN_{d,1}}{2\epsilon_s}(t_b - \delta)^2 \tag{5-113}$$

The voltage ϕ'_{00} is the voltage required to deplete the entire AlGaAs layer with doping layer thickness $t_b - \delta$ and doping density $N_{d,1}$. We add a prime to the symbol so that it will not be confused with ϕ_{00} in the case of MESFET, as given in Eq. (5-8). According to Eq. (5-11), $\epsilon_s \varepsilon_i$ is equal to $q \cdot n_s$, the electron charge sheet density per unit area:

$$qn_s = \epsilon_s \varepsilon_i = \frac{\epsilon_s}{t_b}\left(\phi'_{00} + \frac{\Delta E_c - E_f}{q} + V_{GB} - \phi_B\right) \tag{5-114}$$

Of the components enclosed in parentheses, ϕ'_{00}, ΔE_c, and ϕ_B are constant once a device geometry and epitaxial-layer structure are fixed. V_{GB} is a forcing variable, whose changing values modify the amount of carrier sheet concentration. The Fermi level, in contrast, is a variable which is dependent on n_s. Therefore, even though at a certain bias condition V_{GB} is specified, n_s still cannot be calculated straightforwardly from Eq. (5-114). We need to know exactly how E_f varies with n_s. Their relationship was in fact established some time ago, as given by Eq. (5-98). With E_1 and E_2 given in Eqs. (5-101) and (5-102), the dependence of E_f on n_s can be evaluated numerically. We plot in Fig. 5-26 the relationship between E_f and n_s. At a typical value of n_s exceeding $5 \times 10^{11}\ \text{cm}^{-2}$, Fig. 5-26 shows that E_f can be approximated as:

$$E_f(n_s) = E_{f0} + a \cdot n_s \tag{5-115}$$

where $a = 0.09427 \times 10^{-12}\ \text{eV} \cdot \text{cm}^2$ and $E_{f0} \approx 0.0518\ \text{eV}$ at 300 K. Substituting this finding into Eq. (5-114), we arrive at:

$$qn_s = \frac{\epsilon_s}{t_b + \Delta t_b}\left(\phi'_{00} + \frac{\Delta E_c - E_{f0}}{q} + V_{GB} - \phi_B\right) \tag{5-116}$$

where

$$\Delta t_b = \frac{\epsilon_s}{q^2} a = \frac{\epsilon_s}{q^2}\frac{dE_f}{dn_s} \tag{5-117}$$

The value of Δt_b is calculated to be about 68 Å. (Note that the unit conversion for the last equation can be confusing. It is good to know that 1 Joule = 1 Coulomb × 1 Volt.) The exact value of Δt_b depends on the approximation made to fit the n_s-versus-E_f curve of Fig. 5-26. Sometimes a value of 80 Å is used for Δt_b when a somewhat different approximation is made. Regardless of the exact value, Δt_b is significant compared to the AlGaAs thickness t_b in practical devices, which is on the order of 300 Å.

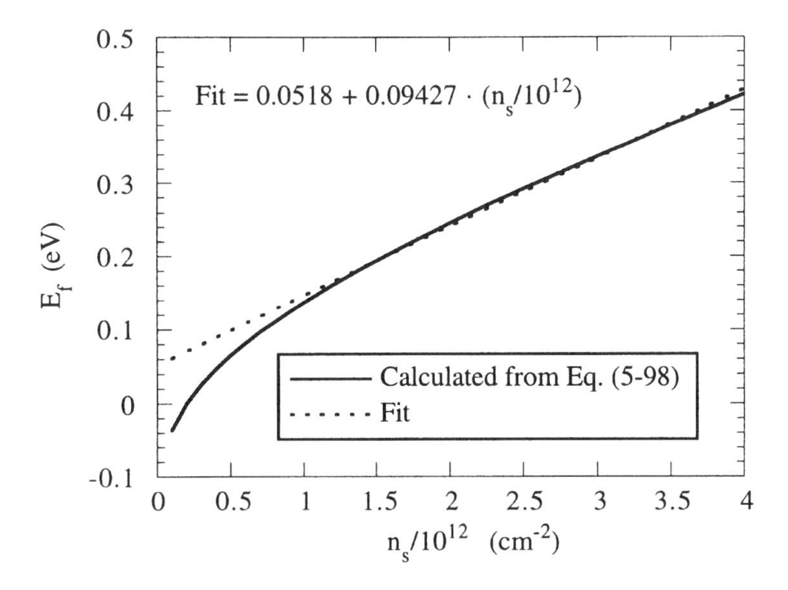

FIGURE 5-26. The Fermi level as a function of the surface charge density as specified in Eq. (5-98) for a modulation-doped structure.

Δt_b has its origin from the fact that the Fermi level is a function of the gate bias, and hence the charge sheet density. In the derivation leading to Eq. (5-117), we have consistently assumed that the accumulation of electrons on the GaAs side of a heterojunction resides right on the interface (see Fig. 5-25*a*). Sometimes, erroneously, Δt_b is solely attributed to the fact that the centroid of the electron wave functions in the quantum well is not right at the interface, but some distance away. When this fact is taken into consideration, Δt_b may in fact increase beyond 68 Å, by an amount equal to the distance between the interface and the centroid of the electron wave functions.

The Δt_b value of 68 Å is typical of a standard HFET whose quantum well is approximately triangular in shape. In a modified HFET structure as shown in Fig. 5-27, Δt_b can be made smaller, thus improving device performance. This structure, called a double heterojunction FET (or D-HFET), employs two heterojunctions instead of a single heterojunction. The quantum well in this case resembles a rectangular potential well. The reason for the smaller Δt_b in this structure is elucidated in Problem 9.

Using the terminology adopted in the discussion of the MESFET, we write the channel charge per unit area as q'_{CH}, which is equal to qn_s in our present notation. More correctly, we should equate qn_s, to the *negative* of q'_{CH}. The accumulated charge in the channel electron, is of negative charge. Hence, q'_{CH}, is a negative quantity. However, in the previous calculation, qn_s has been

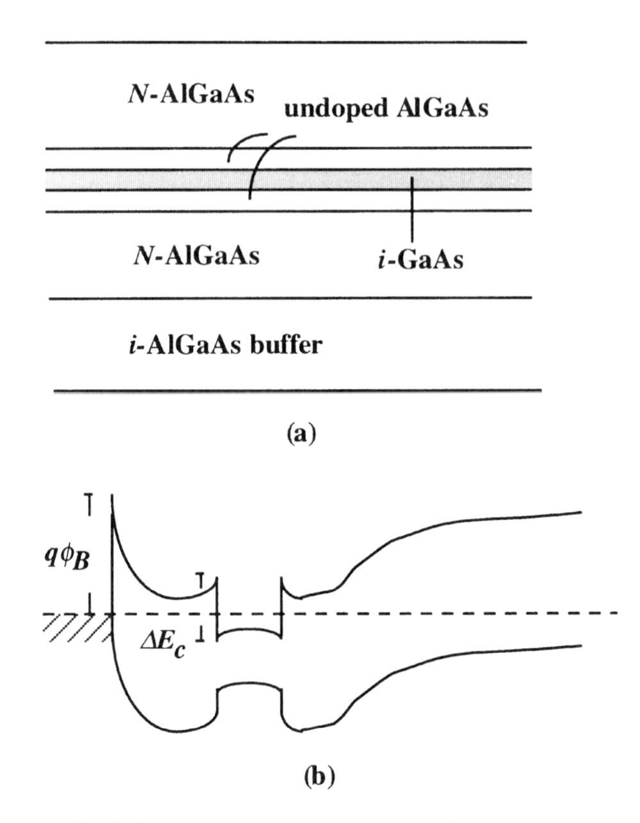

FIGURE 5-27. Epitaxial structure and band diagram of a double heterojunction field-effect transistor.

used to mean the magnitude of such a charge density. Therefore, Eq. (5-116) is expressed as

$$q'_{\mathrm{CH}} = -\frac{\epsilon_s}{t_b + \Delta t_b}\left(V_{GB} - V_T\right) \tag{5-118}$$

where the threshold voltage is given by

$$V_T = \phi_B + \frac{E_{f0}}{q} - \phi'_{00} - \frac{\Delta E_c}{q} \tag{5-119}$$

Example 5-7:

Find the threshold voltage of a HEMT whose $Al_{0.35}Ga_{0.65}As$ barrier layer has a thickness of 330 Å. The undoped portion of the AlGaAs layer adjacent to the channel is 30 Å. The doped portion adjacent to the gate is doped at 1×10^{18} cm^{-3}. Assume that the barrier height between the gate metal and the AlGaAs barrier layer is 1.0 eV.

The various parameter values given in the description are: $t_b = 300$ Å; $\delta = 30$ Å; $N_{d,1} = 5 \times 10^{17}\ \text{cm}^{-3}$; and $\phi_B = 1$ eV. According to the discussion about Eq. (5-115), $E_{f0} = 0.0518$ eV. From Eq. (1-90), ΔE_c of an $Al_{0.35}Ga_{0.65}As/GaAs$ heterojunction is 0.244 eV. According to Eq. (5-113):

$$\phi'_{00} = \frac{1.6 \times 10^{-19} \cdot 1 \times 10^{18}}{2 \cdot 1.159 \times 10^{-12}} (330 \times 10^{-8} - 30 \times 10^{-8})^2 = 0.62\ \text{V}$$

The threshold voltage is found from Eq. (5-119):

$$V_T = 1.0 + 0.0518 - 0.62 - 0.244 = 0.188\ \text{V}$$

By the fact that $V_T > 0$, this is an enhancement-mode transistor.

There is a restriction to the validity of Eq. (5-118). In reality, the charge density in the channel region has a minimum value of zero. It does not decrease to an arbitrary negative number when V_{GB} is negative, as predicted by Eq. (5-118). One simple correction to Eq. (5-118) is to restrict the validity of the equation to $V_{GB} \geq V_T$. We arbitrarily set q'_{CH} to zero when $V_{GB} < V_T$. In this manner, q'_{CH} maintains a zero when $V_{GB} < V_T$, and abruptly increases above zero right at $V_{GB} = V_T$. We caution that, in actual device operation, q'_{CH} increases gradually from zero and eventually agrees completely with Eq. (5-118) at *strong inversion* ($V_{GB} \gg V_T$). The charge sheet density as a function of V_{GB} is shown in Fig. 5-28.

The above analysis is based on a two-terminal device formed with a gate and a substrate. The device is symmetrical about a midplane through the center of the gate, and the channel charge density q'_{CH} is constant in the entire channel. In the practical HFET shown in Fig. 5-20, there is a potential drop in the direction parallel to the channel, causing q'_{CH} to be a function of position x. The band diagrams at three positions along the channel are illustrated in Fig. 5-29. Similar to the MESFET analysis, we denote the channel-to-source potential (or simply, the channel potential) resulting from the applied V_{DS} as $V_{CS}(x)$. Further, the gate voltage is now made with respect to the source terminal, so we replace V_{GB} by V_{GS}. The channel charge density is then given by, in an analogous fashion to Eq. (5-118),

$$q'_{\text{CH}}(x) = -\frac{\epsilon_s}{t_b + \Delta t_b} [V_{GS} - V_T - V_{CS}(x)] \tag{5-120}$$

The proportional constant $\epsilon_s/(t_b + \Delta t_b)$ is common formula for parallel-plate-capacitance. In order to relate the HFET equations to the well-developed MOSFET equations, we shall define a per-area gate "oxide" capacitance as:

$$C'_{\text{ox}} \equiv \frac{\epsilon_s}{t_b + \Delta t_b} \tag{5-121}$$

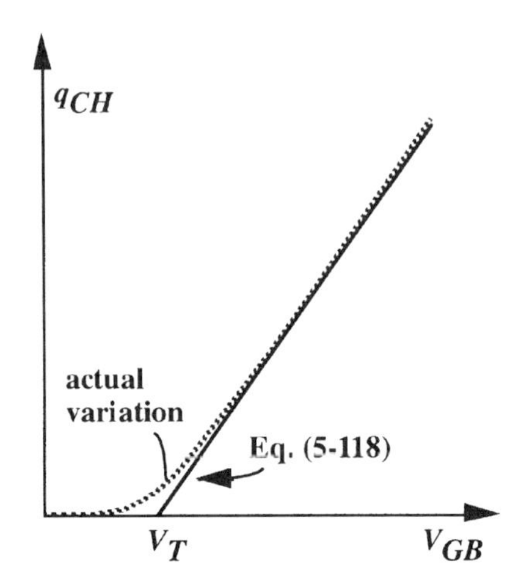

FIGURE 5-28. Channel charge sheet density as a function of the applied bias for the two-terminal structure of Fig. 5-23.

Therefore, the channel charge sheet density is expressed as:

$$q'_{\mathrm{CH}}(x,t) = -C'_{\mathrm{ox}}[v_{GS}(t) - V_T - v_{CS}(x,t)] \tag{5-122}$$

We added the time dependence to generalize the equation so that it becomes applicable in a transient situations. v_{GS} is a function of time because the external bias often varies with time. The channel potential is a function of both time and position.

We mentioned in the study of MESFET that two equations govern field-effect transistors: the drift equation and the continuity equation. The latter is based on the fundamental charge-conservation principle. Its form, given in Eq. (5-27), remains applicable for the HFET, although the exact expression of q'_{CH} needs to be changed from that of the MESFET [Eq. (5-24)] to that of the HFET [Eq. (5-122)]. We have, for HFETs:

$$\frac{\partial i_{\mathrm{CH}}(x,t)}{\partial x} = WC'_{\mathrm{ox}}\frac{\partial[v_{GS}(t) - V_T - v_{CS}(x,t)]}{\partial t} \tag{5-123}$$

The drift equation is also in principle similar to the MESFET's, in that the channel current is proportional to the cross-sectional area of the conduction, the charge density, the mobility, and the electric field. Due to the differences in the geometries involved, Eq. (5-13), applicable to MESFET, is modified to the

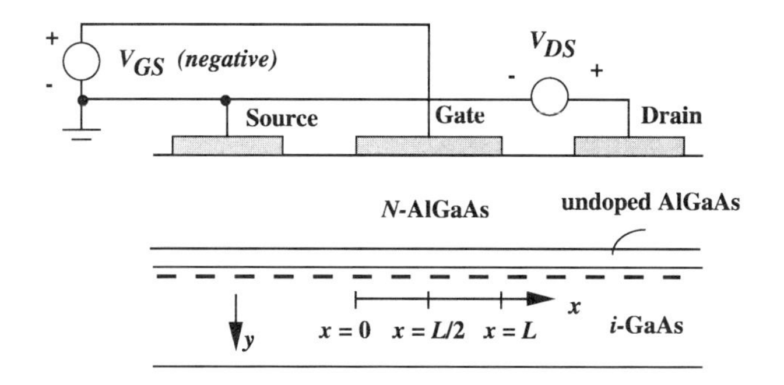

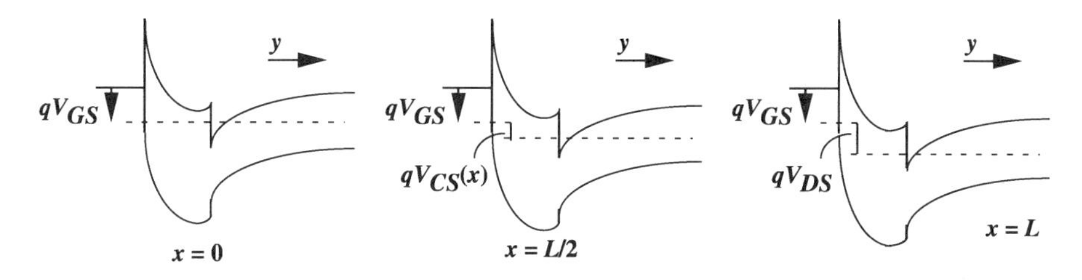

FIGURE 5-29. Band diagrams at three different locations along the channel of a HFET.

following for HFET:

$$i_{\mathrm{CH}}(x, t) = W \mu_n q'_{\mathrm{CH}}(x, t) \frac{\partial v_{CS}(x, t)}{\partial x} \tag{5-124}$$

We note that q'_{CH} is a negative quantity in HFETs, (since electrons accumulated in the channel have negative charge.) Hence, $i_{\mathrm{CH}}(x, t)$ is a negative number, consistent with the fact that the electrons, originating from the source and going toward the drain in the $+x$ direction, constitute a negative channel current. For readers who are familiar with the equation $I = qA\mu_n n\varepsilon$ (A = area), Eq. (5-124) can be rationalized as the following. First, q times the electron density (cm^{-3}) can be taken to be q'_{CH} (C/cm^2) divided by Λ, where Λ represents the depth of the sheet of accumulated electrons at a particular location x. The area of current conduction is $W \times \Lambda$ and the electric field is $-dV_{CS}/dx$. The product of these terms lead to the channel current expression given in Eq. (5-124).

Substituting q'_{CH} of Eq. (5-122) into Eq. (5-124), we obtain the final form of the drift equation in HFET:

$$i_{\mathrm{CH}}(x, t) = -WC'_{\mathrm{ox}}\mu_n[v_{GS}(t) - V_T - v_{CS}(x, t)] \frac{\partial v_{CS}(x, t)}{\partial x} \tag{5-125}$$

We are interested in the d.c. analysis in this section. The continuity equation of Eq. (5-123) and the drift equation of Eq. (5-125) are reduced to the following:

$$\frac{dI_{\mathrm{CH}}(x)}{dx} = 0 \tag{5-126}$$

$$I_{\mathrm{CH}}(x) = -WC'_{\mathrm{ox}}\mu_n[V_{GS} - V_T - V_{CS}(x)]\frac{dV_{CS}(x)}{dx} \tag{5-127}$$

The continuity equation at d.c. indicates that $I_{\mathrm{CH}}(x)$ is constant, independent of x. This is consistent with our understanding that, in the absence of any recombination, the current flow is continuous. Whatever current flows in one end flows out in the other. In fact, if we choose $x = L$ at the drain, we see from Eq. (5-22) that this constant channel current is equal to the negative of the drain current. Hence, we have $I_D = -I_{\mathrm{CH}}$, as stated for MESFETs in Eq. (5-31). Applying this equality to Eq. (5-127), we find:

$$\int_0^L I_D\,dx = \int_{V_{CS}(0)}^{V_{CS}(L)} WC'_{\mathrm{ox}}\mu_n[V_{GS} - V_T - V_{CS}(x)]\,dV_{CS}(x) \tag{5-128}$$

In the following analysis, we consider the long-channel HFETs whose channel mobility is treated as a constant. The short-channel HFETs with velocity saturation will be analyzed in the subsequent section. To carry out the integration in Eq. (5-128), we shall assume temporarily that we are working in the linear region such that current saturation due to channel pinchoff at the drain does not occur. The I-V characteristics after pinchoff will be dealt with shortly. In the linear operating region, the boundary conditions are $V_{CS}(L) = V_{DS}$ and $V_{CS}(0) = 0$. Hence, Eq. (5-128) leads to:

$$I_D = \frac{WC'_{\mathrm{ox}}\mu_n}{L}\left[(V_{GS} - V_T)V_{DS} - \frac{V_{DS}^2}{2}\right] \tag{5-129}$$

Equation (5-129) is plotted schematically in Fig. 5-30, with I_D shown as a function of V_{DS}. Because this is a parabolic equation, the current initially increases, reaching a highest value denoted by $I_{D,\mathrm{sat}}$, and then decreases toward zero. The value of V_{DS} corresponding to the attainment of $I_{D,\mathrm{sat}}$ is denoted as $V_{DS,\mathrm{sat}}$, the saturation voltage. (It can also be called the pinchoff voltage V_P in this long-channel FET.) The saturation voltage can be obtained by taking the derivative of I_D with respect to V_{DS} and setting the result to zero. We find that:

$$V_{DS,\mathrm{sat}} = V_{GS} - V_T \tag{5-130}$$

(Note that the saturation voltage in a MOSFET has an additional multiplication constant which accounts for the transistor's bulk charge.) At this saturation voltage, q'_{CH} calculated from Eq. (5-122) is identically zero at the drain (pinchoff).

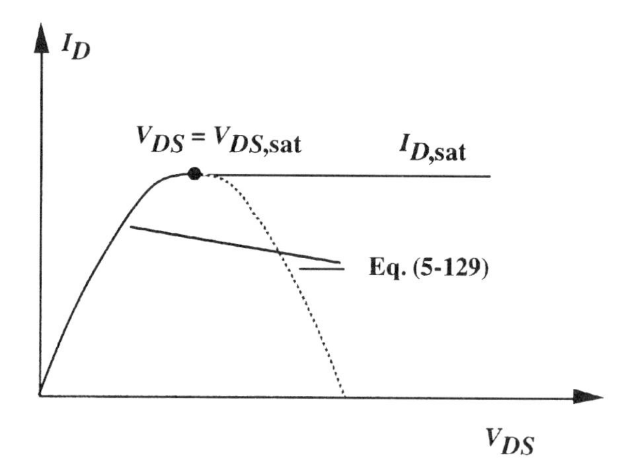

FIGURE 5-30. Actual I-V characteristics and those predicted by Eq. (5-129) for a HFET.

While we have no problem with using Eq. (5-129) to calculate drain current at V_{DS} much smaller than $V_{DS,\mathrm{sat}}$, continuously using the same equation at V_{DS} arbitrarily close to $V_{DS,\mathrm{sat}}$ then assumes that Eq. (5-122) remains applicable even as q'_{CH} approaches zero. This assumption is clearly incorrect, as mentioned in the discussion about Eq. (5-118) and shown in Fig. 5-28. However, rather than resorting to numerical analysis to ascertain the charge concentration when q'_{CH} approaches zero, we continue to use Eq. (5-122) and accept the drain current calculated from Eq. (5-129) for V_{DS} values all the way up to $V_{DS,\mathrm{sat}}$.

Since $q'_{\mathrm{CH}}(L) = 0$ at $V_{DS} = V_{DS,\mathrm{sat}}$, we say that the channel is *pinched off* at the drain end, just as in the long-channel MESFETs. With the charges at the drain being zero, the carriers would have to travel at infinite velocity in order to maintain a nonzero drain current. This is clearly unphysical. However, we understand that this conclusion originates from the fact that we are extending the validity of Eq. (5-122) all the way to where $q'_{\mathrm{CH}}(L)$ is identically zero. Physically, the channel at the drain does not pinch off completely. Instead, there is a finite thickness of accumulation of charges and q'_{CH} at $x = L$ is nonzero. The drift velocity is high, but nonetheless finite, so a constant current is maintained throughout the channel. When $V_{DS} > V_{DS,\mathrm{sat}}$, a narrow region is taken to exist between the tip of the pinchoff region and the drain contact. In this narrow region, q'_{CH} is small, with carriers traveling at high speed. Because the amount of accumulation thickness is small, this region can be viewed as a depletion region (similar to region II of the MESFET analysis in Eq. 5-68, where $a - \delta$ is approximated as a). The accumulation layer between the source and the tip of the pinchoff supports only a voltage amount of $V_{DS,\mathrm{sat}}$. Any excess voltage, $V_{DS} - V_{DS,\mathrm{sat}}$, is dropped across this depletion region. As V_{DS} increases, the length of this narrow depletion region increases somewhat, but is taken to be still fairly small in comparison to the channel length L. Therefore, the distance between the source and the pinchoff point remains close to L. Because the voltage at the pinchoff point remains $V_{DS,\mathrm{sat}}$, the current is still $I_{D,\mathrm{sat}}$. Therefore, a complete model of the drain current is given by:

$$I_D = \begin{cases} \dfrac{WC'_{\mathrm{ox}}\mu_n}{L}\left[(V_{GS} - V_T)V_{DS} - \dfrac{V_{DS}^2}{2}\right] & \text{for } V_{DS} < V_{DS,\mathrm{sat}} \\ \dfrac{WC'_{\mathrm{ox}}\mu_n}{L}\dfrac{(V_{GS} - V_T)^2}{2} = I_{D,\mathrm{sat}} & \text{for } V_{DS} \geq V_{DS,\mathrm{sat}}. \end{cases} \tag{5-131}$$

$I_{D,\mathrm{sat}}$ is obtained by substitution $V_{DS,\mathrm{sat}}$ of Eq. (5-130) into V_{DS} of Eq. (5-129).

For HFETs, it is convenient to define the *saturation index* (α) as:

$$\alpha = \begin{cases} 1 - \dfrac{v_{DS}}{v_{DS,\mathrm{sat}}} & \text{for } v_{DS} < v_{DS,\mathrm{sat}} \\ 0 & \text{for } v_{DS} \geq v_{DS,\mathrm{sat}}. \end{cases} \qquad v_{DS,\mathrm{sat}} = v_{GS} - V_T \tag{5-132}$$

Since we are concerned with d.c. operation for now, we can replace v_{DS} by V_{DS}, $v_{DS,\mathrm{sat}}$ by $V_{DS,\mathrm{sat}}$ and v_{GS} by V_{GS} in the above expression.

The saturation index differs from the current gain transfer ratio of §4-8, also denoted as α. There should be no confusion about exactly what α refers to, since one is used in HBTs, and the other in HFETs.

With the definition of the saturation index, the drain current of Eq. (5-131) is expressed in a compact manner:

$$I_D = I_{D,\mathrm{sat}}(1 - \alpha^2) = \frac{WC'_{\mathrm{ox}}\mu_n}{L}\frac{(V_{GS} - V_T)^2}{2}(1 - \alpha^2) \tag{5-133}$$

Example 5-8:

The HFET described in Example 5-7 has a gate width of 500 μm and a gate length of 0.25 μm. The mobility is 6500 $\mathrm{cm^2/V\text{-}s}$. Find the drain current when $V_{GS} = 0.5$ V and $V_{DS} = 2$ V.

The HFET has a V_T of 0.188 V. The gate capacitance per area is found from Eq. (5-121):

$$C'_{\mathrm{ox}} = \frac{1.159 \times 10^{-12}}{330 \times 10^{-8} + 68 \times 10^{-8}} = 2.91 \times 10^{-7}\ \mathrm{F/cm^2}$$

According to Eq. (5-130), $V_{DS,\text{sat}} = 0.5 - 0.188 = 0.312$ V. Since $V_{DS} > V_{DS,\text{sat}}$, the transistor is in saturation, and α given by Eq. (5-132) is 0. Therefore, I_D calculated from Eq. (5-133) is:

$$I_D = \frac{500 \times 10^{-4} \cdot 2.91 \times 10^{-7} \cdot 6500}{0.25 \times 10^{-4}} \frac{(0.5 - 0.188)^2}{2}(1 - 0) = 0.184 \text{ A}$$

Although we have expressed the drain current in terms of the external bias voltages, we shall carry out an alternative derivation. This derivation is less straightforward, as it involves an intermediate variable, the channel potential, which is denoted by $u_{\text{CH}}(x, t)$ in general, and by $U_{\text{CH}}(x)$ in d.c. operation. The channel potential will prove useful in the derivation of the small-signal y-parameters of the HFETs, as demonstrated in Appendix B. It is defined as:

$$u_{\text{CH}}(x, t) \equiv v_{GS}(t) - V_T - v_{CS}(x, t) \tag{5-134}$$

The continuity and the drift equations, in terms of the channel potential, are given by:

$$\frac{\partial i_{\text{CH}}(x, t)}{\partial x} = WC'_{\text{ox}} \frac{\partial u_{\text{CH}}(x, t)}{\partial t} \tag{5-135}$$

$$i_{\text{CH}}(x, t) - WC'_{\text{ox}} \mu_0 u_{\text{CH}} \frac{\partial u_{\text{CH}}(x, t)}{\partial x} \tag{5-136}$$

There are many ways to express the basic concepts embodied in the above two equations. For example, sometimes x is defined from the drain toward the source; then the equations may have negative signs. Let us verify qualitatively that the signs of the above equations make sense. For Eq. (5-135), suppose that $\partial u_{\text{CH}}/\partial t$ is positive. That means the number of the channel electron charges increases with time. If $\partial i_{\text{CH}}/\partial x$ is positive, then there are more charges flowing out a particular region than coming in. There are thus fewer positive charges or, equivalently, more negative charges as time passes. These two statements are consistent with one another. Hence, the terms $\partial u_{\text{CH}}/\partial t$ and $\partial i_{\text{CH}}/\partial x$ have the same sign. For Eq. (5-136), we know that i_{CH} is a negative quantity in our coordinate system (see § 5-2). This is consistent with the fact that u_{CH} is positive and that $\partial u_{\text{CH}}/\partial x$ is negative.

In d.c. operation, we have:

$$U_{\text{CH}}(x) = V_{GS} - V_T - V_{CS}(x) \tag{5-137}$$

with the two boundary conditions given by:

$$U_{\mathrm{CH}}(0) = V_{GS} - V_T$$

$$U_{\mathrm{CH}}(L) = \begin{cases} V_{GS} - V_T - V_{DS} & \text{for } V_{DS} < V_{DS,\mathrm{sat}} \\ V_{GS} - V_T - V_{DS,\mathrm{sat}} & \text{for } V_{DS} < V_{DS,\mathrm{sat}} \end{cases} = \alpha U_{\mathrm{CH}}(0) \tag{5-138}$$

where α was given in Eq. (5-132).

Noting that the drain current is the negative of the channel current, we can rewrite the drain current in terms of $U_{\mathrm{CH}}(x)$:

$$I_D = -WC'_{\mathrm{ox}}\mu_n U_{\mathrm{CH}}(x)\frac{dU_{\mathrm{CH}}(x)}{dx} \tag{5-139}$$

We carry out an integral similar to Eq. (5-128), except that the integration of I_D is now taken to be from $x' = x$ to $x' = L$. Utilizing the property that the channel current is a constant independent of position, we find that:

$$I_D = \frac{WC'_{\mathrm{ox}}\mu_n}{L - x}\frac{U_{\mathrm{CH}}^2(x) - U_{\mathrm{CH}}^2(L)}{2} \tag{5-140}$$

When $x = 0$ is substituted into the above equation, we then obtain the drain current in terms of the two boundary conditions:

$$I_D = \frac{WC'_{\mathrm{ox}}\mu_n}{L}\frac{U_{\mathrm{CH}}^2(0) - U_{\mathrm{CH}}^2(L)}{2} \tag{5-141}$$

Applying the boundary conditions given in Eq. (5-138) results in the same current expression as that of Eq. (5-131).

An interesting result is obtained by dividing the two drain current equations given in Eqs. (5-140) and (5-141). After rearranging the terms, we find the channel potential's variation as a function of x:

$$U_{\mathrm{CH}}(x) = U_{\mathrm{CH}}(0)\sqrt{1 - \frac{x}{L}(1 - \alpha^2)} \tag{5-142}$$

This leads readily to an expression for the channel-to-source potential:

$$V_{CS}(x) = (V_{GS} - V_T)\left[1 - \sqrt{1 - \frac{x}{L}(1 - \alpha^2)}\right] \tag{5-143}$$

Therefore, the channel-to-source potential is an analytical function of x. Once a particular value of x is chosen, $V_{CS}(x)$ can be easily calculated. This is not the case for MESFET. As shown in Eq. (5-42), the relationship between the channel-to-source potential $[\phi_s(x) = \phi_{00} \cdot u(x)]$ and the position x is governed by a transcendental equation. Numerical techniques must be used to determine such a potential.

§ 5-7 VELOCITY SATURATION IN THE HFET

We used the two-region model to analyze the velocity saturation behavior in a MESFET. In one region, the mobility is assumed to be constant, and in another the velocity is assumed to be at v_{sat}. An alternative approach is to use a field-dependent mobility of Eq. (5-55), without assuming *a priori* the existence of two separate regions. In the interest of not repeating the analysis done for the MESFET, we adopt the latter approach to analyze the velocity saturation behavior of HFET. To simplify the algebra somewhat, we find it useful to set ε_{sat} arbitrarily to a value such that $\mu_0 \cdot \varepsilon_{sat}/2 = v_{sat}$. It is possible to use Eq. (5-55) exclusively for all magnitudes of field, so that the drift velocity equals the saturation velocity only when ε approaches infinity. However, the algebra becomes more tedious, as treated in Problem 10. Therefore, we adopt a drift velocity model given by:

$$v_{drift} = \begin{cases} \mu_{n,0}|\varepsilon| \Big/ \left(1 + \dfrac{|\varepsilon|}{\varepsilon_{sat}}\right) & \text{if } \varepsilon < \varepsilon_{sat} \\ v_{sat} & \text{if } \varepsilon \geq \varepsilon_{sat}. \end{cases} \tag{5-144}$$

Absolute signs are attached to the electric field since it is negative in our coordinate system, with $x = 0$ being the source and $x = L$ being the drain.

For HFETs with velocity saturation, $\mu_n\varepsilon$ (or $\mu_n \cdot dV_{CS}(x)/dx$) in Eq. (5-127) should be replaced by v_{drift} of Eq. (5-144). Hence, the drain current is given by (in the linear region where $\varepsilon < \varepsilon_{sat}$):

$$I_D = W\mu_0 C'_{ox}[V_{GS} - V_T - V_{GS}(x)] \times \frac{\mu_{n,0}|\varepsilon|}{1 + |\varepsilon|/\varepsilon_{sat}} \tag{5-145}$$

When expressed in terms of $U_{CH}(x)$ defined in Eq. (5-137), $|\varepsilon| = -\varepsilon = -dU_{CH}/dx$. Hence, the drain current can be rewritten as:

$$I_D = W\mu_0 C'_{ox}\left\{\frac{-U_{CH}(x)\,[(d/dx)U_{CH}]}{1 - (1/\varepsilon_{sat})\,[(d/dx)U_{CH}]}\right\} \tag{5-146}$$

The d.c. drain current is a constant. The differential equation of $U_{CH}(x)$ given in Eq. (5-146) can be solved, with $U_{CH}(x)$ expressed in a transcendental form:

$$\frac{W\mu_0 C'_{ox}}{2}\frac{U_{CH}^2(x)}{I_D} - \frac{U_{CH}(x)}{\varepsilon_{sat}} + x = k \tag{5-147}$$

where k is an integration constant.

There is a subtlety which is often overlooked. Despite its appearance, Eq. (5-146) applies only from $x = 0$ to L_I, the location in the channel where velocity first reaches v_{sat}. At any $x > L_I$, Eq. (5-146) no longer applied. The same statements can be said about Eq. (5-147), which derives from Eq. (5-146). To demonstrate these points, we assume for a moment that these two equations would work for all regions of x under all bias conditions. The boundary conditions for the transistors are:

$$U_{CH}(0) = V_{GS} - V_T \quad U_{CH}(L) = \begin{cases} V_{GS} - V_T - V_{DS} & \text{if } V_{GS} - V_T > V_{DS} \\ 0 & \text{if } otherwise. \end{cases} \tag{5-148}$$

With these two boundary conditions, we find I_D and k as:

$$I_D = \frac{W\mu_0 C'_{ox}}{2L} \times \frac{[U_{CH}^2(0) - U_{CH}^2(L)]}{1 + [U_{CH}(0) - U_{CH}(L)]/(\varepsilon_{sat}L)} \tag{5-149}$$

$$k = \frac{\varepsilon_{sat}LU_{CH}^2(0) - U_{CH}(L)\,U_{CH}^2(0) + U_{CH}(0)\,U_{CH}^2(L)}{\varepsilon_{sat}[U_{CH}^2(0) - U_{CH}^2(L)]} \tag{5-150}$$

Substituting these two newly found values back into Eq. (5-147) and employing the identity that $U_{CH}(L) = \alpha \cdot U_{CH}(0)$, we find that $U_{CH}(x)$ satisfies this transcendental equation:

$$\frac{U_{CH}^2(x)}{U_{CH}^2(0)}\frac{1 - \alpha + \varepsilon_{sat}L/U_{CH}(0)}{1 - \alpha^2} - \frac{U_{CH}(x)}{U_{CH}(0)} + \frac{\varepsilon_{sat}x}{U_{CH}(0)} = \frac{\varepsilon_{sat}L/U_{CH}(0) - \alpha + \alpha^2}{1 - \alpha^2} \tag{5-151}$$

Example 5-9:

We demonstrate that Eq. (5-151) applies only before velocity saturation takes place. Suppose that $\varepsilon_{sat} = 10^3$ V/cm, $L = 10\,\mu$m, and $V_{GS} - V_T = 1$ V. Find the channel potential profile $U_{CH}(x)$ for two drain biases: $V_{DS} = 0.5$ V and 1 V.

$U_{CH}(0) = V_{GS} - V_T = 1$ V, $U_{CH}(L) = 0.5$ V, and $\alpha = 0.5$. Substituting these values into Eq. (5-151), we find that the channel potential is governed by the following equation: $U_{CH}^2(x) - U_{CH}(x) + x = 1$. The resulting $U_{CH}(x)$ as a function of x is shown in Fig. 5-31. The channel potential decreases somewhat

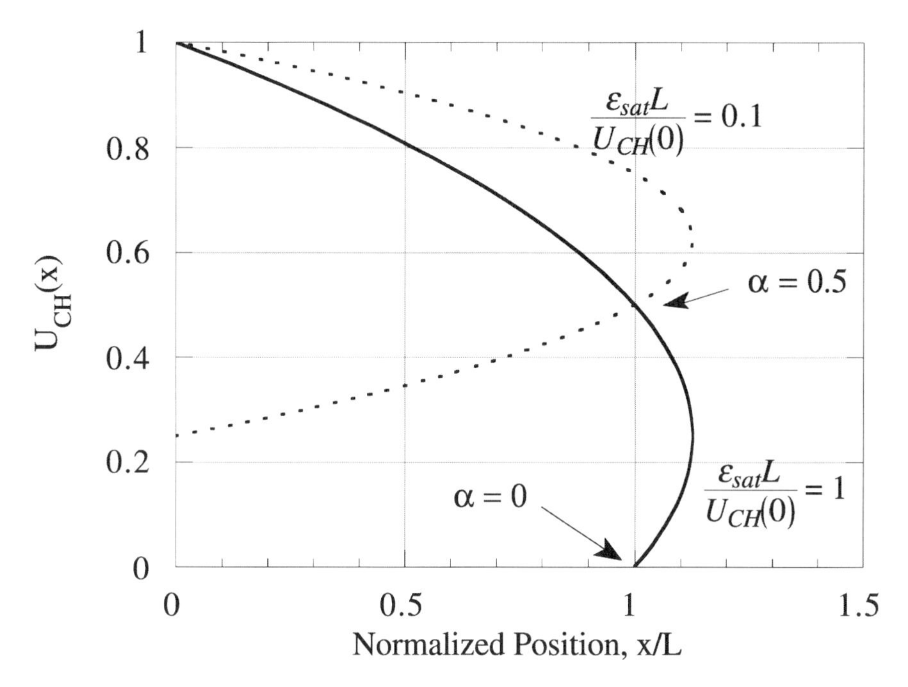

FIGURE 5-31. Channel potential as a function of normalized position under different conditions. $V_{GS} - V_T = 1$ V.

with x, indicating that the magnitude of the accumulated sheet charge density, $C'_{ox} \cdot U_{CH}(x)$, decreases with position.

Suppose that instead the drain bias is 1 V, so that $U_{CH}(L) = 0$. Substituting $\alpha = 0.5$ into Eq. (5-151), we find that again $U_{CH}^2(x) - U_{CH}(x) + x = 1$. While Eq. (5-151) predicts the correct potential profile in the previous case, it is clearly incorrect here. From Fig. 5-31, we see that at $x = L$, $U_{CH}(L)$ can be either 0 or 0.5. The problem, of course, lies in the fact that the bottom half of the parabola becomes part of the solution. If only the upper half of the parabola is used to match the boundary conditions at $x = 0$ and $x = L$, then Eq. (5-151) works fine. The reason it breaks down at higher V_{DS} is that, when the saturation region is formed, the number of accumulated charges is no longer a straightforward function of the channel potential. That is, Eq. (5-122) is no longer correct. That equation applies only in region I, where the channel potential affects the amount of charge accumulation directly. Since Eq. (5-122) is incorrect in region II, we cannot expect Eq. (5-151) to work in the entire channel.

We are interested in knowing the condition at which Eqs. (5-151) ceases to be valid. We know that the equation is correct as long as $\varepsilon < \varepsilon_{sat}$. It becomes problematic only when $U_{CH}(L)$ is at a value such that $dU_{CH}(x)/dx = -\varepsilon_{sat}$. We therefore take the derivative of Eq. (5-151) with respect to x, evaluate the derivative at $x = L$, and seek the condition such that the derivative is equal to $-\varepsilon_{sat}$. We find that the condition is met if $U_{CH}(L)$ satisfies the following:

$$U_{CH}(L) = U_{CH}(0)\left[1 + \frac{\varepsilon_{sat}L}{U_{CH}(0)}\right]^{-1} \tag{5-152}$$

For the example discussed above, where $\varepsilon_{sat}L/U_{CH}(0) = 1$, we find $U_{CH}(L)$ to be 0.5 V. When $V_{DS} = 0.5$ V, $U_{CH}(L)$ calculated from Eq. (5-152) is 0.5 V. Since $U_{CH}(L) \geq$ that calculated from Eq. (5-152), we expect Eq. (5-151) to hold. When $V_{DS} = 1$ V, $U_{CH}(L)$ calculated from Eq. (5-148) is 0 V. Since this is smaller than 0.5 V, the saturation region has been formed and Eq. (5-151) fails.

If ε_{sat} is still taken to be 10^3 V/cm but L is shortened to 1 μm, the critical $U_{CH}(L)$ value determined from Eq. (5-152) is 0.91 V. This means that the moment V_{DS} exceeds 0.09 V, a saturation region is formed. This $V_{DS,sat}$ value of 0.09 V for the $L = 1$ μm device is much smaller than the 0.5 V value for the $L = 10$ μm device discussed previously. In Fig. 5-31, we also plotted the channel potential for the $L = 1$ μm device, for which $\varepsilon_{sat}L/U_{CH}(0) = 0.1$. With $\alpha = 0.5$, we know that Eq. (5-151) is correct when $\varepsilon_{sat}L/U_{CH}(0) = 1$ for the $L = 10$ μm device. However, when L is shortened to 1 μm, the upper curve of the figure indicates that the solution of $U_{CH}(x)$ reaches 0.5 only at the bottom portion of the parabola. This is an indication that at $\alpha = 0.5$, the 1 μm device has entered saturation.

A slightly different condition from that specified in Eq. (5-152) should be used if we assume Eq. (5-65) to work for all fields, reaching v_{sat} only when ε approaches infinity. This fact is shown in Problem 10.

At $x < L_I$, Eq. (5-146) is still applicable. Noting that the d.c. current is a constant independent of x, we write:

$$\int_0^{L_I} I_D\left(1 - \frac{1}{\varepsilon_{sat}}\frac{d}{dx}U_{CH}\right)dx = -W\mu_0 C'_{ox}\int_{U_{CH}(0)}^{U_{CH}(L_I)} U_{CH}(x)\,dU_{CH} \tag{5-153}$$

This leads to an equation similar, but not identical, to Eq. (5-149):

$$I_{D,I} = \frac{W\mu_0\varepsilon_{sat}C'_{ox}}{2} \times \frac{[U^2_{CH}(0) - U^2_{CH}(L_I)]}{L_I\varepsilon_{sat} + [U_{CH}(0) - U_{CH}(L_I)]} \tag{5-154}$$

We add the subscript I to the symbol I_D, to emphasize that the drain current is obtained from consideration of region I only. In region II, the drain current is the drift current formed when all carriers travel at a constant velocity equal to v_{sat}. Since the amount of carrier inversion is $C'_{ox} \cdot U_{CH}(L_I)$, we have:

$$I_{D,II} = WC'_{ox}v_{sat} \times U_{CH}(L_I) \tag{5-155}$$

Equating Eqs. (5-154) and (5-155) gives a relationship between L_I and $U_{CH}(L_I)$. Because $\mu_0 \cdot \mathcal{E}_{sat}/2 = v_{sat}$, $U_{CH}(L_I)$ is given by:

$$U_{CH}(L_I) = \frac{U_{CH}^2(0)}{U_{CH}(0) + \mathcal{E}_{sat} L_I} \tag{5-156}$$

In the event that the drain current is known (such as when it is measured), then $U_{CH}(L_I)$ is known from Eq. (5-155). Consequently, L_I can be established from Eq. (5-156). In a situation when we are given only a set of bias conditions, such as V_{DS} and V_{GS}, then the equations derived thus far do not allow us to find the drain current. We have two variables, $U_{CH}(L_I)$ and L_I, but just one equation relating them, namely, Eq. (5-156). We need another equation. This second equation can be obtained in a similar fashion as the MESFET in saturation, where we solved the quasi-two-dimensional Poisson equation in region II. The quasi-two-dimensional Poisson equation for MESFETs, Eq. (5-68) does not apply exactly to HFETs because it does not account for the undoped layer (with thickness equal to δ) in the AlGaAs barrier layer. However, if we neglect this difference (and if we accept the approximation made in solving the quasi-Poisson equation rather than the actual Poisson equation), then we can apply the analysis of § 5-3 to the problem at hand. The final result is then similar to Eq. (5-73):

$$V_{DS} = \frac{2a}{\pi} \mathcal{E}_{sat} \sinh \frac{\pi L_{II}}{2a} + V_{CS}(L_I) \tag{5-157}$$

We rewrite Eq. (5-156) in terms of the channel-to-source potential. Substituting the definition given in Eq. (5-137), we obtain:

$$V_{CS}(L_I) = \frac{(V_{GS} - V_T)\,\mathcal{E}_{sat} L_I}{(V_{GS} - V_T) + \mathcal{E}_{sat} L_I} \tag{5-158}$$

Since $L_I + L_{II} = L$, Eqs. (5-157) and (5-158) can be combined to solve for L_I, L_{II}, $V_{CS}(L_I)$ simultaneously. The drain current is therefore determined once V_{GS} and V_{DS} are given.

The application of quasi-Poisson solution to HFETs was done in one study of the velocity saturation [3]. It was commented that such a model at times can result in a large drain conductance. An alternative approach to analyze the velocity saturation region of the channel is to solve a one-dimensional Poisson equation [4]. This approach is discussed in Problem 11.

§ 5-8 AVALANCHE BREAKDOWN

The basic governing equations about the impact ionization in field-effect transistors are the same as those in HBTs. However, a quantitative study of the breakdown voltage in FET structures is considerably more difficult than in bipolar structures. The bipolar transistor has a vertical device structure. Because the base layer is on top of the collector layer, the field inside the reverse-biased base-collector junction is predominantly one-dimensional. The field is accurately determined from the solution of a one-dimensional Poisson equation. The avalanche multiplication (M) can then be determined from the ionization integral of Eq. (3-107). When a given collector-emitter bias results in a certain filed profile which causes M to be $1/(\alpha_T \cdot \Gamma)$, then avalanche breakdown is said to occur and the corresponding V_{CE} is the breakdown voltage. In bipolar structures, it is relatively straightforward to tabulate the breakdown voltage as a function of the collector doping and thickness.

In contrast, the electric field in FETs is two-dimensional, due to the fact that the FET is a horizontal structure, with the drain/source contacts placed perpendicular to the channel current flow. It is difficult to establish the electric field profile with much accuracy. A two-dimensional device simulator is almost always necessary for an accurate breakdown analysis. The simulations demonstrate that two main factors affect the FET breakdown (besides the channel doping and thickness): the surface pinning potential and the geometric shape of the transistor.

We first analyze the effects of the surface pinning potential on the breakdown voltage. The MESFET structure [5] shown in Fig. 5-8 has the following dimensions: $L = L_D = L_S = L_{GD} = L_{GS} = 1\ \mu m$, $a = 0.2\ \mu m$, and $N_d = 10^{17}\ cm^{-3}$. Figure 5-32 illustrates the equipotential lines in the device with three different pinning potentials, $\phi_{sb} = 0, 0.65$ and 1.0 V. The subscript sb in ϕ_{sb} signifies that the pinning potential measures the potential difference between the surface and the bulk of a given layer. A ϕ_{sb} of 0 means that there is no surface trap level. Without any surface Fermi-level pinning, the Fermi level with respect to the conduction band is the same at the surface and at the bulk. When $\phi_{sb} \neq 0$, some amount of surface traps are present, with increasing number of traps associated with increasing values of ϕ_{sb}. With Fermi-level pinning, the relative differences of the Fermi-level at the surface and at the bulk is an amount equal to ϕ_{sb}.

In one extreme, when $\phi_{sb} = 0$ V (Fig. 5-32*a*), the equipotentials show a high gradient near the gate edge, indicating a high electric field there. The high electric field brings about high impact ionization. The gate edge is therefore the location of avalanche breakdown. This conclusion, however, results from a simulation which neglects the surface pinning effects of the exposed GaAs layer. As shown in Fig. 5-32*b* for the case $\phi_{sb} = 0.65$ V, the equipotential lines are relatively more uniform, not exhibiting a particularly high gradient at any location. In the extreme that ϕ_{sb} approaches 1 V, Fig. 5-32*c* shows that the equipotentials tend to crowd near the drain contact, suggesting the presence of a large electric field there. In this extreme, breakdown occurs near the drain contact rather than the gate.

We give a semiquantitative analysis of the above phenomenon, that the surface potential greatly influences the breakdown voltage of FETs. The

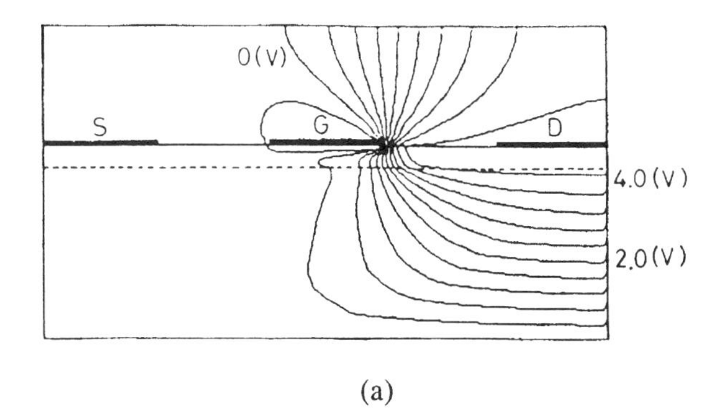

(a)

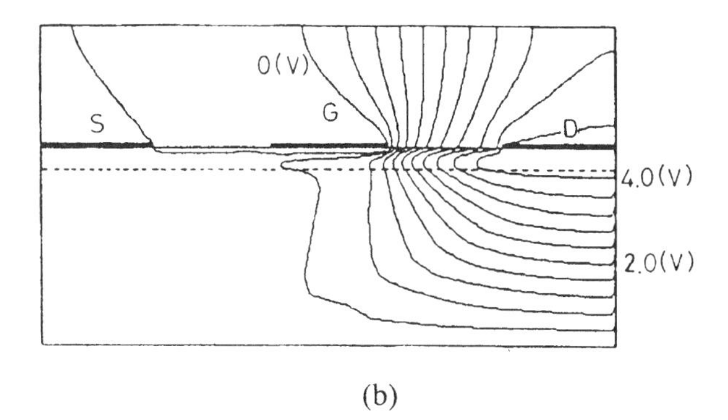

(b)

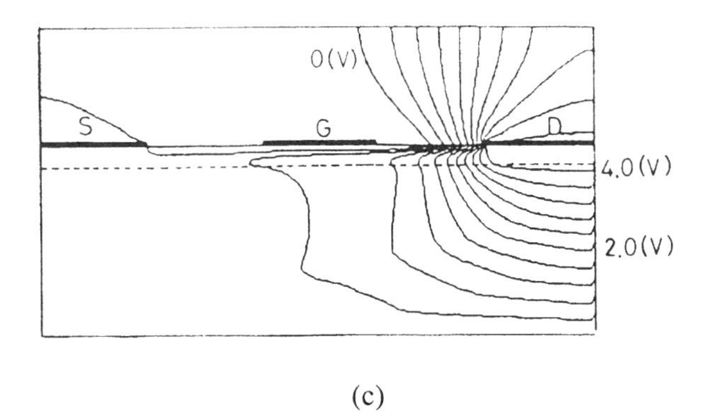

(c)

FIGURE 5-32. Equipotential lines in MESFET with three different pinning potentials; $\phi_{sb} = 0$, 0.65, and 1.0 V. (From H. Mizuta, K. Yamaguchi, and S. Takahashi, Surface potential effect on gate-drain avalanche breakdown in GaAs MESFETs, *IEEE Trans. Electron. Dev.* **34,** 2027–2033, 1987, © IEEE, reprinted with permission.)

analysis is not exact because we consider only one dimension, along the gate-drain contact at the top surface. However, this analysis renders important physical insight to the FET breakdown voltage, predicting well the breakdown voltage's variation with various FET geometries.

We examine the field and potential profiles at the top surface ($y = 0$) between the gate-drain contacts, along the x direction. To start, we assume that there is no surface trap states on the exposed active layer. Hence $\phi_{sb} = 0$. Let N_d be the doping density of the active layer and the depletion thickness along the x direction be X_{dep}. In order to maintain the overall device charge neutrality, an equal amount of negative charges exists on the metal/semiconductor interface. The transistor structure and the charge profile are shown in Fig. 5-33. According to Eq. (5-11), this charge density results an positive electric field equal to $X_{\text{dep}} \cdot N_d/\epsilon_s$ right at the semiconductor side of the interface, since the electric filed inside the gate metal is zero. The Poisson equation and the boundary conditions governing the electrostatics of the region are therefore:

$$\frac{d^2\phi_s(x)}{dx^2} = -\frac{qN_d}{\epsilon_s} \qquad \text{for } x \leq X_{\text{dep}} \tag{5-159}$$

$$\phi_s(0) = 0 \qquad \phi_s(X_{\text{dep}}) = \phi_{\text{bi}} + V_{DG} \tag{5-160}$$

$$\left.\frac{d\phi_s}{dx}\right|_{x=0} = -\frac{qN_d X_{\text{dep}}}{\epsilon_s} \qquad \left.\frac{d\phi_s}{dx}\right|_{x=X_{\text{dep}}} = 0 \tag{5-161}$$

The solution to the second-order linear differential equation is

$$\varepsilon = -\frac{d\phi_s}{dx} = \frac{qN_d}{\epsilon_s}(x - X_{\text{dep}}) \qquad \phi_s = \frac{qN_d}{\epsilon_s}\left(X_{\text{dep}} \cdot x - \frac{x^2}{2}\right) \tag{5-162}$$

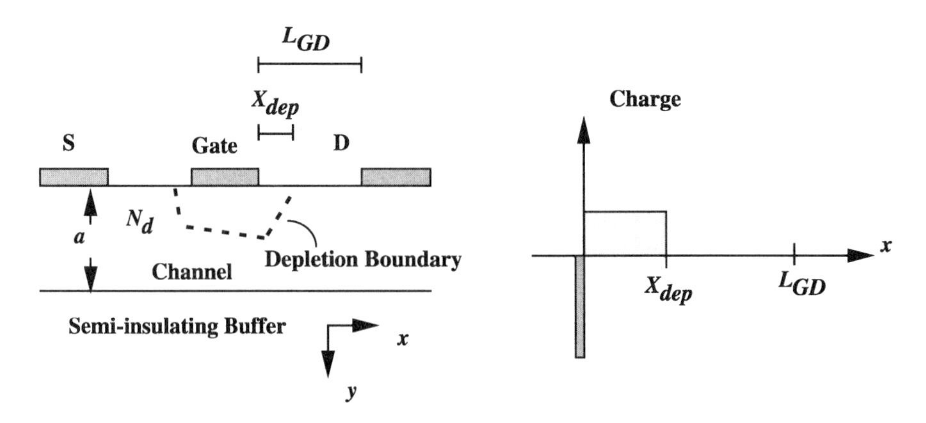

FIGURE 5-33. Depletion boundary and charge profile at $y = 0$ of an idealized MESFET without surface Fermi-level pinning ($\phi_{sb} = 0$).

where X_{dep} is given by

$$X_{\text{dep}} = \sqrt{\frac{2\epsilon_s}{qN_d}(\phi_{\text{bi}} + V_{DG})} \tag{5-163}$$

V_{DG} is the drain bias relative to the gate bias. The solution given in Eq. (5-162) is applicable for $x \leq X_{\text{dep}}$. At $x > X_{\text{dep}}$, $\varepsilon = 0$ and $\phi_{sb} = \phi_{\text{bi}} + V_{DG}$.

The calculation agrees qualitatively with the device simulator's results, which indicate that, in the absence of surface traps, the breakdown always occurs near the gate electrode. The electric field is shown to have the largest magnitude there, while being nearly zero at the drain side. This is reflected in Fig. 5-32*a*, which shows the constant potential lines to be crowding near the gate.

The effects of nonzero surface potential are now analyzed. As mentioned previously, the surface Fermi-level pinning results from the fact that a large amount of traps exists on the surface. We assume that the electrons captured by the surface states distribute uniformly on the surface between the gate and the drain electrodes, with a density equal to N_s (in cm^{-3}). In accordance with Ref. [5], we simulate these electron's effects by placing a thin acceptor sheet charge on the surface, with a density equal to N_s. In this way, we do not have to deal with the surface trap dynamics; the situation simplifies into a straightforward electrostatic problem. Figure 5-34 illustrates the transistor geometry and the charge profile under consideration. The amount of the negative gate charge is taken to exactly compensate the positive ionized charge in the channel layer. Hence, the gate charge sheet density (in cm^{-2}) is equal to $N_d \cdot X'_{\text{dep}}$, where X'_{dep} is in the presence of a nonzero N_s. X_{dep} without the prime represents the depletion thickness when $N_s = 0$. We caution that X'_{dep} is not the depletion thickness in its true sense. With N_s, the entire surface is depleted, as marked in Fig. 5-34. X'_{dep} simply denotes the

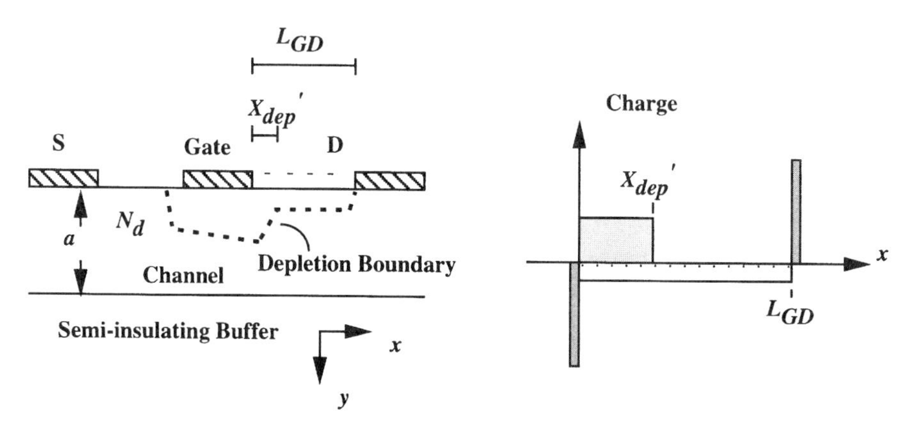

FIGURE 5-34. Depletion boundary and charge profile at $y = 0$ of a realistic MESFET with surface Fermi-level pinning ($\phi_{sb} \neq 0$).

distance which, when multiplied by N_d, is equal to the gate charge density. Because of the requirement of overall charge neutrality, there must be positive charges at the drain contact to compensate for the surface acceptor charges N_s. The positive charges cannot reside inside the metal, so they form a charge sheet, residing on the semiconductor/drain interface, with a density equal to $N_s \cdot L_{GD}$. The Poisson equation and the boundary conditions governing the electrostatics of the region are therefore:

$$\frac{d^2\phi_s(x)}{dx^2} = \begin{cases} -\frac{q}{\epsilon_s}(N_d - N_s) & 0 \le x \le X'_{\text{dep}} \\ \frac{q}{\epsilon_s} N_s & X'_{\text{dep}} \le x \le L_{GD}. \end{cases} \tag{5-164}$$

$$\phi_s(0) = 0 \qquad \phi_s(L_{GD}) = \phi_{\text{bi}} + V_{DG} \tag{5-165}$$

$$\left.\frac{d\phi_s}{dx}\right|_{x=0} = -\frac{qN_d X'_{\text{dep}}}{\epsilon_s} \qquad \left.\frac{d\phi_s}{dx}\right|_{x=L_{GD}} = -\frac{qN_s L_{GD}}{\epsilon_s} \tag{5-166}$$

The last boundary conditions, relating to the electric fields at the gate/semi-conductor and the drain/semiconductor interfaces, are obtained from Eq. (5-11).

We require both the potential and the electric field to be continuous at $x = X'_{\text{dep}}$. The solution to Eq. (5-164) is:

$$\phi_s(x) = \begin{cases} -\frac{q(N_d - N_s)}{2\epsilon_s} x^2 + \frac{qN_d}{\epsilon_s} X'_{\text{dep}} x & 0 \le x \le X'_{\text{dep}} \\ \frac{qN_s}{2\epsilon_s} x^2 - \frac{qN_s}{2\epsilon_s} L_{GD}^2 + \phi_{\text{bi}} + V_{DG} & X'_{\text{dep}} \le x \le L_{GD} \end{cases} \tag{5-167}$$

$$\varepsilon(x) = \begin{cases} \frac{q(N_d - N_s)}{\epsilon_s} x - \frac{qN_d}{\epsilon_s} X'_{\text{dep}} & 0 \le x \le X'_{\text{dep}} \\ -\frac{qN_s}{\epsilon_s} x & X'_{\text{dep}} \le x \le L_{GD} \end{cases} \tag{5-168}$$

where X'_{dep} is equal to

$$X'_{\text{dep}} = \sqrt{\frac{2\epsilon_s}{qN_d}(\phi_{\text{bi}} + V_{DG}) - \frac{N_s}{N_d} L_{GD}^2} \tag{5-169}$$

The electric fields and surface potentials, with and without the consideration of the surface charge N_s, are shown in Figs. 5-35. Basically, the trapped electrons at the surface (modeled by fixed acceptor charge with a density of N_s) decrease the amount of total positive charge within the depletion region. This reduces the

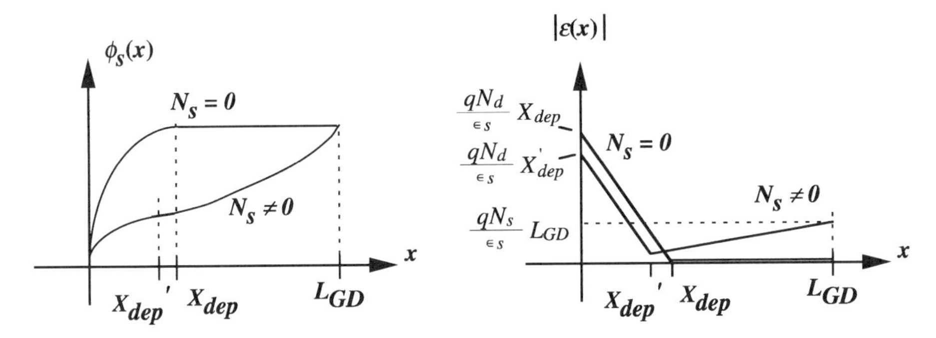

FIGURE 5-35. Electric fields and surface potentials, with and without the consideration of the surface Fermi-level pinning. (From H. Mizuta, K. Yamaguchi, and S. Takahashi, Surface potential effect on gate-drain avalanche breakdown in GaAs MESFETs, *IEEE Trans. Electron. Dev.* **34,** 2027–2033, 1987 © IEEE, reprinted with permission. The figure notations are modified to conform to the symbols used in this book.)

magnitude of the electric field at the gate terminal. The overall potential drop is preserved by having the electric field at the drain side increase above zero. Because the impact ionization rate is exponentially dependent on the field magnitude, the lowering of the field at the gate side significantly reduces the amount of impact ionization.

As the amount of N_s increases (corresponding to a higher value of surface potential, ϕ_{sb}), we expect the field to continuously decrease at the gate side, and to increase at the drain side. The impact ionization gradually shifts from the gate electrode to the drain electrode. It is perceived that, in the extreme condition, the breakdown occurs at the drain terminal, rather than at the gate. This discussion is reflected in Figs. 5-32*b* and 5-32*c*, showing that the maximum field strength moves toward the drain as the surface potential increases.

Previously, when we neglected N_s so that $\phi_{sb} = 0$, the potential and electric field contours were not much changed as the gate-drain spacing (L_{GD}) was modified. Thus, the breakdown voltage was independent of L_{GD}. This conclusion directly contradicts the empirical observation that, as L_{GD} increases, the breakdown voltage increases. It turns out that this contradiction is resolved when we consider the surface pinning effects by including a finite N_s in the analysis. According to Eq. (5-168), when $N_s \neq 0$, the electric field at the gate interface is:

$$\varepsilon(0) = \frac{qN_d}{\epsilon_s}\sqrt{\frac{2\epsilon_s}{qN_d}(\phi_{\mathrm{bi}} + V_{DG}) - \frac{N_s}{N_d}L_{GD}^2} \tag{5-170}$$

As L_{GD} increases, $\varepsilon(0)$ decreases, leading to a smaller magnitude of impact ionization for a given drain-gate bias. The gate-drain breakdown voltage thus increases.

Another phenomenon understandable with $\phi_{sb} \neq 0$ but inexplicable with $\phi_{sb} = 0$ is that the breakdown voltage is found empirically to increase when a double-recess structure is used. The double-recess structure is shown in the last portion of Fig. 7-10, and the more basic single-recess structure is shown in Fig. 6-31. When $\phi_{sb} \neq 0$, there is some voltage drop across the region where the second recess edge is made (the circled region in Fig. 7-10). The recess edge in the double-recess structure provides a sharp corner where the electric field is concentrated. While the electric field profile is greatly modified with the second recess edge, the total potential drop between the gate and the drain remains the same. With the peak electric field strength taking place at the second recess edge, the electric field elsewhere must decrease. The reduction in the electric field at the gate side increases the breakdown voltage. In the analysis with $\phi_{sb} = 0$, there is no potential drop near the second recess edge. The entire potential drop occurs near the gate edge. Consequently, the breakdown still occurs near the gate edge and the improvement with the double-recess structure could not be predicted.

We have just shown that the breakdown voltage increases when the surface of the conducting channel between the gate and drain contacts is depleted. The surface depletion is caused by the electrons captured by the surface states. The surface depletion thickness (along the x direction, not the y direction) remains relatively constant between the drain-gate contacts. It is then widened near the gate electrode when the gate depletion region joins the depletion region of the metal/semiconductor junction. We see that, in conventional MESFETs, the amount of surface depletion is not controllable, since they are fixed by material parameters. The desire to control the magnitude of the breakdown voltage leads to the so-called *overlapping-gate* structure [6]. The overlapping-gate structure has the advantage that the overlapping portion of the gate creates an additional vertical electric field on the conduction channel beneath. This electric field extends the depletion thickness, an action which is equivalent to increasing the surface trap density right near the gate. As the depletion region is thickened near the gate, the electric field decreases at the gate edge. Consequently, the breakdown voltage increases.

We have just discussed the dependence of the breakdown voltage on the transistor structure. In the following, we examine the behavior of the breakdown voltage as the gate bias is varied. Figure 5-36 shows schematically the gate current as a function of V_{GS} when V_{DS} is fixed at two certain constant values. When V_{DS} is small, the device operates in the linear region independent of the gate-source bias. Without a high-field region developed between the gate and the drain, the amount of impact ionization is small. The gate is well isolated from the channel, and the gate current is practically zero.

When V_{DS} is large, the transistor operates in saturation. Near the drain, where the channel either pinches off (with near-zero channel charge) or pinches down (due to velocity saturation), the field between the gate-drain terminals is large. Competing physical mechanisms result in the so-called *bell shape* of the gate current shown in Fig. 5-36. Initially, when V_{DS} is large but V_{GS} is fairly negative,

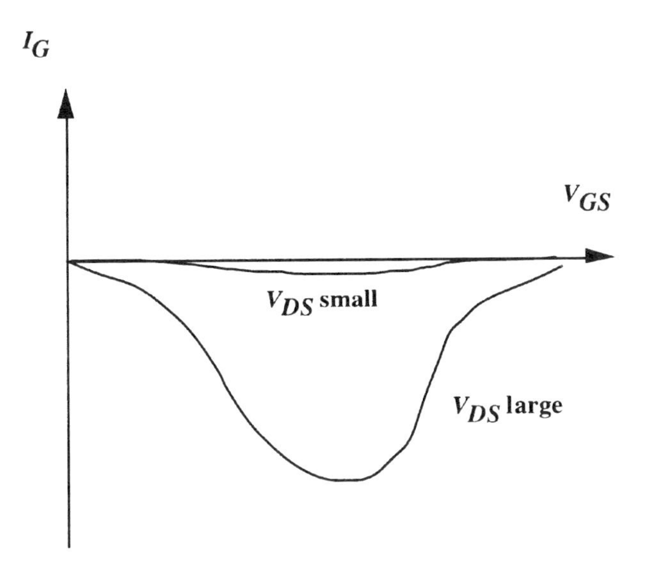

FIGURE 5-36. Gate current as a function of V_{GS} when V_{DS} is fixed at two certain constant values. The gate current is a result of impact ionization.

the channel is not turned on. There is only some finite leakage current flowing to the gate. As the gate bias increases, the amount of current flowing in the channel increases. This results in more events of impact ionization. The leakage gate current is composed of the holes which are created in ionization and which are attracted by the negative gate bias. Because the holes flow into the gate, this gate current is negative in accordance with the convention that a current flowing into a device's terminal is considered positive. The amount of holes entering the gate depends on two factors. One factor is the availability of initiating conducting carriers to start the impact ionization. If there is more channel current (hence more electron carriers), then the likelihood of impact ionization increases. The second factor relates to the magnitude of the electric field between the gate-drain contacts. If the electric field is large, there will be more impact ionization.

When V_{GS} starts small (very negative) and is allowed to increase, the fact that the gate current becomes increasingly more negative reflects that there are more carriers to initiate the impact ionization. However, this trend eventually stops. After V_{GS} increases to a certain value while V_{DS} remains fixed, the voltage drop in the saturated region decreases. This decreases the electric field and the amount of impact ionization decreases. In the extreme that V_{GS} increases to a large value such that $V_{GS} - V_{DS} > V_T$, then the transistor enters linear region and the gate current drops back to zero.

Depending on which of the two factors is the dominant effect, the breakdown voltage will either be like Fig. 5-37*a*, in which the breakdown voltage increases

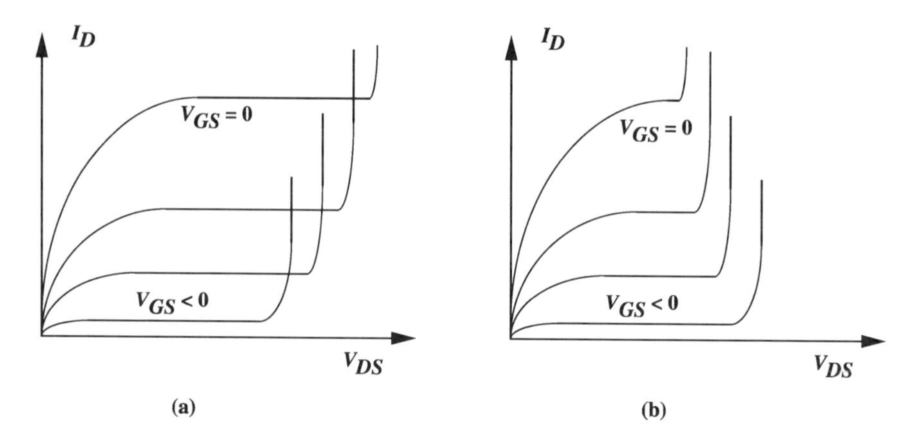

FIGURE 5-37. Typical I-V characteristics observed in: (*a*) silicon JFETs; (*b*) GaAs MESFETs.

with V_{GS}, or like Fig. 5-37*b*, in which the breakdown voltage decreases with V_{GS}. Fig. 5-37*a* represents the typical I-V characteristics of a silicon junction field-effect transistor (JFET), which is in principle the same as the MESFET except that the gate depletion is that of a *p-n* depletion rather than a metal/semiconductor depletion. In a typical JFET, apparently the amount of impact ionization depends more critically on the magnitude of the electric field, rather than the availability of carriers. Therefore, when V_{GS} increases, the breakdown voltage increases roughly by the amount that V_{GS} increases, so that the breakdown electric field is about the same before and after the V_{GS} increment. In a MESFET, however, the increase of channel current associated with an increase in V_{GS} is significant (likely due to higher g_m in III-V MESFETs than in Si JFETs). With more charges to initiate impact ionization, the field at which significant impact ionization occurs decreases in magnitude. Therefore, the breakdown voltage decreases as V_{GS} increases.

The breakdown voltage in MESFETs has been expressed as a function of the doping level (N_d) and the thickness of the channel layer (a) [7]:

$$BV_{DS} = \frac{4.4 \times 10^{13}}{a \cdot N_d} \qquad (BV_{DS} \text{ in V, } a \text{ in cm, and } N_d \text{ in cm}^{-3}) \tag{5-171}$$

This formula is semiempirical. The actual breakdown of a device is likely to depend on various factors, such as the surface treatment of the surface between the gate and drain contacts and the final etched profile of the recessed gate. Equation (5-171) serves only as a crude estimate of the breakdown voltage.

All of the above discussion about the physics of the breakdown voltage applies to both MESFETs and HFETs, although MESFETs have been used as the device of demonstration. Some additional phenomena, though they could potentially occur in all FETs, are more pronounced in HFETs. One of these phenomena is the *breakdown walkout* [8]. Breakdown walkout refers to the permanent increase in the breakdown voltage after the device is stressed with high bias voltage and high current. Although the breakdown voltage increase is considered a plus for high-power applications, breakdown walkout is generally a phenomenon to be avoided. As the breakdown voltage increases, the parasitic drain resistance also increases. Hence, the drain current decreases, resulting in worsened device characteristics. Worst of all, breakdown walkout is not a controllable process. The device continues to degrade as the stressing continues, causing reliability concern.

The breakdown walkout is observed in unpassivated and passivated HFETs. The word *passivation* as used in FETs connotes a slightly different meaning than that used in the HBT literature. In the context of HBTs, passivation predominantly means the use of a AlGaAs layer as the passivation layer, passivating the AlGaAs/GaAs interface so that the surface recombination velocity at the extrinsic base surface is small, on the order of 10^3 cm/s. In FETs, the word passivation is used to contrast an unpassivated device having a GaAs layer exposed to the air. Hence, by placing a nitride or oxide layer on top of the otherwise exposed GaAs, the FET is said to be passivated. Despite the "passivation" by a dielectric, the dielectric/GaAs interface remains characterized by a high surface recombination velocity, approaching 10^6 cm/s.

For unpassivated HFETs, the physical mechanism behind the walkout is likely related to the oxidation of the AlGaAs barrier layer during stress [9]. The oxidation process is similar to anodic oxidation, where the primary oxidizing component of water comes from the humidity in the air. As the transistor is biased with high voltage and high current, the avalanche-generated holes carries the necessary current to complete the oxidation loop. Once the exposed surface is oxidized, the effective doping becomes less than that designed. The electric field between the gate and the drain then becomes more uniform and the breakdown voltage increases.

For passivated HFETs, breakdown walkout is likely a result of an increased amount of trapped charges in the dielectric/semiconductor interface or inside the dielectric. That is, as the stress continues, more and more negative charges are built up at the surface. From the semiquantitative analysis discussed earlier, these trapped charges cause the peak electric field to shift from the gate toward the drain, reducing the peak value in the process. As more charges are trapped, the breakdown voltage increases. However, the trapped charge increases the depletion region, bringing forth a larger drain resistance.

A phenomenon peculiar to InGaAs-based HFET (such as InAlAs/InGaAs) is the kink effect in the I-V characteristics. The I-V characteristics exhibit the usual saturation characteristics at a certain $V_{DS,\mathrm{sat}}$. Beyond $V_{DS,\mathrm{sat}}$, however, there is a sudden rise in the drain current, leading to a higher drain conductance and poor linearity. A direct correlation between the impact ionization in the InGaAs channel and kink behavior has been experimentally established [10]. However, the exact mechanism of how the impact ionization can cause the kink has not been unequivocally identified. Since this kink behavior is also found in SOI (silicon-on-insulator) devices, and since both the SOI and HFETs lack the bulk terminal, it is conceivable that the mechanism well known in SOI devices is responsible for the HFETs under consideration. In this SOI-based theory, the electron-hole pairs are generated due to the impact ionization between the gate and drain electrodes. It is assumed that the impact ionized electrons are much smaller in number than the conducting electrons of the channel. Hence, as the generated electrons flow through the drain contact, not much effect is produced. The holes, in contrast, deserve more attention. In the previous discussion about the bell-type curve of the gate current, the impact-ionized holes were thought to all leave through the gate. It is conceivable that some of the holes actually escape to the substrate. In the absence of bulk contact, these holes cannot be channeled out of the device. The only way for holes to be removed is through recombination, which is a slow process. When the rate of hole generation exceeds the recombination rate, more and more holes accumulate. These holes act like a positive-charge parasitic back gate, leading to an increase in the electron current flow in the channel by an amount which is large compared to the impact ionization current itself. This increase in the electron current is thought to be responsible for the observed kink in the I-V characteristics.

REFERENCES

1. Spicer, W., Liliental-Weber, Z., Weber, E., Newman, N., Kendelewicz, T., Cao, R., McCants, C., Mahowald, P., Miyano, K., and Lindau, I. (1988). "The advanced unified defect model for Schottky barrier formation." *J. Vacuum Sci. Technol.* **B6,** 1245–1251.
2. For a review, see Yang, E. S. (1975). "Current saturation mechanisms in junction field-effect transistors." In Marton, L., ed., *Advances in Electronics and Electron Physics.* New York: Academic Press, **38,** 247–265.
3. Rohdin, H., and Roblin, P. (1986). "A MODFET dc model with improved pinchoff and saturation characteristics." *IEEE Trans. Electron Dev.* **33,** 664– 672.
4. Roblin, P., Kang, S., and Morkoc, H. (1990). "Analytic solution of the velocity-saturated MOSFET/MODFET wave equation and its application to the prediction of the microwave characteristics of MODFET's. "*IEEE Trans. Electron Dev.* **37,** 1608–1621.
5. Mizuta, H., Yamaguchi, K., and Takahashi, S. (1987). "Surface potential effect on gate-drain avalanche breakdown in GaAs MESFETs." *IEEE Trans. Electron Dev.* **34,** 2027–2033.
6. Chen, C. (1996). "Breakdown of overlapping-gate GaAs MESFET's." *IEEE Trans. Electron Dev.* **43,** 535–542.

7. Frenley, W. (1981). "Power-limiting breakdown effects in GaAs MESFET's." *IEEE Trans. Electron Dev.* **28,** 962–970.

8. Menozzi, R., Cova, P., Canali, C., and Fantini, F. (1996). "Breakdown walkout in pseudomorphic HEMT's." *IEEE Trans. Electron Dev.* **43,** 543–546.

9. Chao, P., Shur, M., Kao, M., and Lee, B. (1992). "Breakdown walkout in AlGaAs/GaAs HEMT's," *IEEE Trans. Electron Dev.* **39,** 738–740.

10. Somerville, M., del Alamo, J., and Hoke, W. (1996). "Direct correlation between impact ionization and the kink effect in InAlAs/InGaAs HEMT's," *IEEE Electron Device Lett.* **17,** 473–475.

PROBLEMS

1. Consider a long-channel MESFET.

a. What is the difference between the depletion potential ϕ_{00} and the pinchoff channel potential V_P?

b. The value of d found in Eq. (5-40) has a maximum value of 1. In order for the MESFET to conduct current, V_{GS} must be a value such that $s < 1$. Let us define V_T as the V_{GS} value corresponding to $s = 1$. Then, V_T can be treated as a threshold voltage, such as that the transistor turns on when $V_{GS} > V_T$ and cuts off when $V_{GS} < V_T$. Find V_T.

c. Sometimes this threshold voltage is called the pinchoff voltage. However, in this book the pinchoff voltage refers to V_P. What are the values of V_T and V_P for the MESFET of Example 5-2 when $V_{GS} = 0$ V?

d. What are the values of V_T and V_P for the MESFET of Example 5-2 when $V_{GS} - -0.5$ V?

2. A long-channel ($L = 1\ \mu$m) MESFET's active layer has a doping level of $2 \times 10^{17}\ \text{cm}^{-3}$ and a thickness of 1500 Å. Assume the metal/semiconductor barrier height for the gate region is 0.9 V. The gate width is 200 μm and the electron mobility is 3000 cm^2/V-s.

a. Find the drain current when $V_{GS} = 0$ V and $V_{DS} = 0.5$ V.

b. Find the drain current when $V_{GS} = 0$ V and $V_{DS} = 3$ V.

3. We examine the gradual channel approximation used in a long-channel MESFET. The GCA is generally said to be valid when $db(x)/dx < b/a$, where b is the channel opening at a particular location x. We are interested in knowing at what position along the channel the GCA breaks down.

a. show that

$$db = \frac{a}{2(1 - b/a)} \frac{1}{\phi_{00}} d\phi_{s,\text{dc}}$$

b. We just obtained $db/d\phi_s$; however, we are interested in the quantity db/dx. The equation relating $d\phi_s$ to dx is obtained by combining Eqs. (5-29) and (5-31). Show that

$$\frac{db}{dx} = \frac{I_D}{I_0}\left[\frac{2b}{a}\left(1 - \frac{b}{a}\right)\right]^{-1} \qquad \text{where } I_0 = qW\mu_n N_d\,\phi_{00}$$

c. Show that for a MESFET,

$$x = \frac{aI_0}{I_D}\left[u(x) - s - \frac{2}{3}u(x)^{3/2} + \frac{2}{3}s^{3/2}\right]$$

d. Show that the following relationship exists for a transistor operating in saturation:

$$\frac{aI_0}{I_D}\left(s - \frac{2}{3}s^{3/2}\right) = \frac{1}{3}\frac{aI_0}{I_D} - L$$

e. Suppose that MESFET operates in saturation. Find the location x such that the GCA breaks down, from the criterion that $db(x)/dx < b/a$.

4. The delta-FET shown in Fig. 5-19 has a δ-doping of $\sigma = 3 \times 10^{12}\ \text{cm}^{-2}$ at a distance $d = 300$ Å away from the gate. Assume the electrons travels at a constant velocity equal to $v_{sat} = 10^7$ cm/s throughout the channel. The transistor width is 200 μm.

a. Find the transconductance when the transistor is on.

b. Draw the band diagram between $y = 0$ and $y = d$ (see y direction in Fig. 5-19). Assume that the Fermi level in a heavily doped n-type material to lie on the edge of the conduction band.

c. What is built-in potential, ϕ_{bi}, given that the barrier height is ϕ_B?

d. Find the full-depletion potential and threshold voltage. Express in terms of d, σ, and ϕ_B.

5. A modified δ-FET has two delta-doping layers instead of just one. The first doping layer is at a distance d_1 away from the gate surface, and is doped at a sheet density of σ_1. The second layer is at a distance d_2 away, with a density of σ_2. The first layer is closer to the gate, so that $d_2 > d_1$. Find the expressions for g_m and V_T.

6. We intend to derive Eq. (5-73). Suppose that $\phi_{s,1}(x, y)$ and $\phi_{s,2}(x, y)$ satisfy

$$\frac{\partial^2\phi_{s,1}(x,y)}{\partial x^2} + \frac{\partial^2\phi_{s,1}(x,y)}{\partial y^2} = 0 \qquad \text{for } L_I \le x \le L;\ 0 < y < (a - \delta) \approx a$$

$$\phi_{s,1}(x,0) = 0;\ \phi_{s,1}(L_I, y) = 0;\ \left.\frac{\partial\phi_{s,1}(x,y)}{\partial y}\right|_{y=a} = 0;\ \left.\frac{\partial\phi_{s,1}(x,y)}{\partial x}\right|_{x=L_I,\,y=a} = \mathcal{E}_{sat}$$

and

$$\frac{\partial^2 \phi_{s,2}(x,y)}{\partial x^2} + \frac{\partial^2 \phi_{s,2}(x,y)}{\partial y^2} = -\frac{q}{\epsilon_s} N_d \qquad \text{for } L_I \le x \le L;\ 0 < y < (a-\delta) \approx a$$

$$\phi_{s,2}(x,0) = 0;\ \phi_{s,2}(L_{\mathrm{I}}, y) = \frac{qN_d}{\epsilon_s}\left(X_{\mathrm{dep,II}}\, y - \frac{y^2}{2}\right);$$

$$\left.\frac{\partial \phi_{s,2}(x,y)}{\partial y}\right|_{y=a} = 0;\ \left.\frac{\partial \phi_{s,2}(x,y)}{\partial x}\right|_{x=L_{\mathrm{I}},\, y=a} = 0$$

a. Show through linear superposition that $\phi_s(x,y)$ of Eq. (5-68) is equal to $\phi_{s,1}(x,y) + \phi_{s,2}(x,y)$. Show the equivalence in the boundary conditions in Eqs. (5-69)–(5-72), besides the differential equation.

b. Let us concentrate on $\phi_{s,1}(x,y)$ first. Show the following $\phi_{s,1}(x,y)$ solution satisfies the differential equation specified in part **a**:

$$\phi_{s,1}(x,y) = \sum_{n=1}^{\infty} A_n \sinh \beta_n (x - L_{\mathrm{I}}) \sin \nu_n y$$

In order for the $\phi_{s,1}(x,y)$ solution to satisfy the boundary conditions, what are the coefficients ν_n and β_n in terms of a?

c. Find a governing equation for A_n, determined from the last boundary condition.

d. Assume $A_n = 0$ for all n's > 1. What is the reason for making this assumption, and what is the final equation for $\phi_{s,1}(x,y)$?

e. We now focus on $\phi_{s,2}(x,y)$. Show that it is exactly equal to the right hand side of the second boundary condition, that is:

$$\phi_{s,2}(x,y) = \frac{q}{\epsilon_s} N_d \left(X_{\mathrm{dep,II}}\, y - \frac{y^2}{2}\right)$$

f. With both $\phi_{s,1}(x,y)$ and $\phi_{s,2}(x,y)$ determined, show that V_{DS} is given by Eq. (5-73).

7. Derive the ϕ_{00} and I_D expressions given in Fig. 5-18 for the power-law-doped MESFET.

8. We consider an AlGaAs/GaAs heterojunction whose band diagram was shown in Fig. 5-21. Note that the AlGaAs layer is not gated, and the AlGaAs layer is assumed to be infinitely thick. As mentioned in § 5-5, after we substitute Eqs. (5-106) and (5-107) into Eq. (5-98), we then have an equation with n_s and E_f as the only variables. If it is desirable to determine these two unknowns simultaneously, the Poisson equation can be invoked. Assume that $\epsilon_{s,1} = \epsilon_{s,1} = \epsilon_s$.

a. Draw the charge and electric field profiles of the heterojunction.

b. Let $\varepsilon_{i,1}$ be the interface electric field in region 1, $X_{\text{dep},1}$ be the depletion thickness in region 1, and $\phi_{\text{bi},1}$ be the potential drop across $X_{\text{dep},1}$. Show that:

$$\epsilon_s \varepsilon_{i,1} = \sqrt{(qN_{d,1}\delta)^2 + 2qN_{d,1}\epsilon_s\phi_{bi,1}} - qN_{d,1}\delta$$

c. One equation relating n_s and E_f was identified to be Eq. (5-98). Based on the result of part **b**, what is other equation relating them?

d. With two equations and two unknowns, n_s and E_f can be determined simultaneously. Describe (but do not implement) a computer algorithm to determine n_s and E_f simultaneously.

9. We consider the double heterojunction FET shown in Fig. 5-27. The quantum well resembles a square well more than the triangular well of a typical HFET. For simplicity, we shall approximate the well as a perfect square well, with the bottom of the well assumed to be perfectly flat, and the potential barriers to be infinitely high. From elementary quantum mechanics, the energy eigenvalues for a particle in a infinite square well with a well separation t_w are:

$$E_n = \frac{n^2 h^2}{8m_{\text{I}}^* t_w^2} \qquad n = 1, 2, \cdots$$

a. For a $t_w = 100$ Å, find E_n for $n = 1$, and 2. Notice that E_n is independent of n_s, unlike E_n's of the triangular well given in Eqs. (5-106) and (5-107).

b. To simplify the analysis, we shall assume that the electrons occupy only the first subband. This is not a bad assumption, according to Example 5-6, which finds $n_{s,1} \gg n_{s,2}$. With this assumption, express a relationship between n_s and E_f. (Hint: it is similar to Eq. 5-98.)

c. Apply Eq. (5-117) to find an expression for Δt_b in the D-HEFT.

d. Show that Δt_b is calculated to be 26 Å in a D-HEFT when n_s is large. This is a smaller value than the 68 Å calculated for the conventional HFET. Discuss its implication for the device performance of a D-HFET.

10. When we determined the I-V characteristics of a velocity-saturated HFET in § 5-7, we assumed the velocity obys the relationship given in Eq. (5-144), that the velocity reaches v_{sat} when $\varepsilon \geq \varepsilon_{\text{sat}}$. Suppose that we now assume the velocity is given exclusively by $v_{\text{drift}} = \mu_{\text{no}}|\varepsilon|/(1 + |\varepsilon|/\varepsilon_{\text{sat}})$. That is, the velocity reaches v_{sat} only when the magnitude of the electric field reaches infinity. When the transistor is in the linear region, Eqs. (5-146) to (5-151) still apply.

a. Find the condition at which Eqs. (5-151) is no longer correct. Express the result in terms of α, $U_{\text{CH}}(0)$, and $\varepsilon_{\text{sat}}L$.

b. What is the value of α when $\varepsilon_{sat}L/U_{CH}(0) = 1$? Would Eq. (5-151) hold when $U_{CH}(L) = 0.3$ V? How about $U_{CH}(L) = 0.25$ V?

c. Let us consider a shorter-channel device such that $\varepsilon_{sat}L/U_{CH}(0) = 1$. What is the critical value of $U_{CH}(L)$?

11. We continue the analysis of the HFET of Problem 10, in which the velocity is given exclusively by $v_{drift} = \mu_{no}|\varepsilon|/(1 + |\varepsilon|/\varepsilon_{sat})$. That is, the velocity reaches v_{sat} only when the magnitude of the electric field reaches infinity.

a. Supposing that the transistor is in the linear region, find the drain current in terms of V_{GS} and V_{DS}.

b. Suppose that the transistor operates in the saturation region and the channel can be divided into a GCA and a velocity saturation region. The boundary separating the two regions is located at $x = L_I$. Derive drain current expressions $I_{D,I}$ and $I_{D,II}$ for the two regions. Express the results in terms of $U_{CH}(0)$ and $U_{CH}(L_I)$, L_I and ε_{sat}.

c. Equating $I_{D,I}$ and $I_{D,II}$, determine $U_{CH}(L_I)$.

d. What is the V_{DS} value when the transistor starts to enter the current saturation region?

e. There is another way to derive the V_{DS} value of part **d.** Start with the current equation determined in part **a.** Take the derivative of I_D with respect to V_{DS}, then find the V_{DS} such that the derivative is zero.

f. Let us call the V_{DS} value found in parts **d** and **e** $V_{DS,sat}$, the saturation voltage. Likewise, let us call the corresponding drain current $I_{D,sat}$. When $V_{DS} > V_{DS,sat}$, the transistor enters the saturation operation and the channel can be divided into two distinct regions. The length of the GCA region is L_I and the length of the velocity saturation region is $L_{II} = L - L_I$. Assume that the effective channel thickness remains constant in region II, equal to δ. Set up a one-dimensional Poisson equation to find $\Delta L \equiv L_{II}$ in terms of I_D, V_{DS}, and $V_{DS,sat}$. To simplify the algebra, assume that the electric field at $x = L_I$ is equal to ε_{sat}.

g. The drain current expression given in Eq. (5-154) is valid in the linear region, whether current saturation has taken place or not. Therefore, $I_{D,I}$ given in Eq. (5-154) can be used as an expression for the drain current I_D, which is equal throughout the channel position. Express the drain current in terms of V_{GS} and $V_{DS,sat}$.

h. Let $I_{D,sat}$ be the drain current when the current saturation starts to occur, taking place right at the drain ($x = L$). Find an expression for $I_{D,sat}$ based on the result of part **g**.

i. Use I_D from part **g** and $I_{D,sat}$ from part **h**, determine ΔL in terms of I_D, $I_{D,sat}$, and $V_{DS,sat}$.

j. Equate ΔL from part **i** and part **f**. Determine a relationship between I_D and the operating voltage V_{DS}.

12. For a HFET operated at low temperatures, Eq. (5-99) is applicable. Assuming that the two lowest subbands are occupied, find E_f in terms of n_s. What is the value of E_f when $n_s = 2 \times 10^{12}\ \text{cm}^{-2}$?

13. Consider a HEMT whose $Al_{0.4}Ga_{0.6}As$ barrier layer has a thickness of 280 Å. The undoped portion of the AlGaAs layer adjacent to the channel is 30 Å. The doped portion adjacent to the gate is doped at $1 \times 10^{18}\ \text{cm}^{-3}$. Assume that the barrier height between the gate metal and the AlGaAs barrier layer is 1.0 eV. The gate width is 500 μm and the gate length is 0.25 μm. The mobility is $6500\ \text{cm}^2/\text{V-s}$.

a. Find the threshold voltage.

b. Find the drain current when $V_{GS} = 0.5$ V and $V_{DS} = 2$ V.

14. True or False:

a. If we neglect recombination in the channel, then the source current is always identical to the drain current (except for sign difference), because whatever goes in must go out.

b. When the electron in the channel travels at v_{sat}, the current does not increases with further increase in V_{DS}.

c. $U_{\text{CH}}(x) = C'_{\text{ox}}(V_{GS} - V_T - V_{CS})$ does not work the moment electrons travel at v_{sat}.

d. When V_{GS} is fixed constant yet V_{DS} increases, the impact ionization always increases.

15. Consider an n-type MESFET with the following structure: active-layer thickness of 0.12 μm and doping of $10^{17}\ \text{cm}^{-3}$; source/drain contact length = 1.0 μm; gate to source/drain spacing = 1.35 μm; and gate metal work function = 4.77 eV. Use a device simulator such as PISCES to perform the following.

a. For gate lengths of 0.3, 1.0, and 1.5 μm, plot I_D-versus-V_{DS} curves for $V_{GS} = 0$ V. Save the structure at $V_{DS} = 3.0$ V and make a 2D plot of electron density. What does it imply about the current saturation mechanism?

b. For the case of gate length of 0.3 μm, plot I_D-versus-V_{GS} curves (linear and log) for $V_{DS} = 3.0$ V. Change the gate to source/drain spacing from 1.35 μm to 0.3 μm. Replot the I_D-versus-V_{GS} curves. Compare the results.

c. Consider a recessed gate structure: gate etch depth is 0.05 μm; gate length is 0.3 μm; and gate-to-source/drain spacing is 1.35 μm. Plot I_D-versus-V_{GS} curves (linear and log) for $V_{DS} = 3.0$ V. Compare the results with those for part **b**.

Below is an example input deck for a device simulator:

```
Title GaAs MESFET
# Define the mesh
mesh space.mult=1.0
#
x.mesh loc=0.00 spac=0.3
x.mesh loc=2.3 spac=0.02
x.mesh loc=2.7 spac=0.02
x.mesh loc=5.0 spac=0.3
y.mesh loc=0.00 spac=0.01
y.mesh loc=0.04 spac=0.03
y.mesh loc=0.12 spac=0.02
# Region specification: Defined as two regions for
different properties
region  num=1 GaAs x.min=0 x.max=5.0 y.min=0 y.max=0.12
# Electrode specification
elec  num=1  name=source  x.min=0.0  y.min=0.0  x.max=1.0
y.max=0.0
elec  num=2 name=drain x.min=4.0 y.min=0.0 x.max=5.0 y.max=0.0
elec  num=3 name=gate x.min=2.35 length=0.3
# Doping specification
doping region=1 uniform conc=1.0e17 n.type
# Set models, material and contact parameters
contact num=3 work=4.77
models region=1 print conmob fldmob srh optr
material material=GaAs taun0=1.0e-9 taup0=1.0e-9
# Solution – use gummel newton for initial then switch to full
newton
method gummel newton trap
output con.band val.band efield
solve init
save outfile=mesfet2.str
log outf=mesfet2d.log
solve vdrain=0.025 vstep=0.025 vfinal=0.1 name=drain
method newton trap
solve vdrain=0.2 vstep=0.1 vfinal=0.4 name=drain
solve vdrain=0.5 vstep=0.5 vfinal=3 name=drain
save outf=mesfet2.out
log outf=mesfet2g.log master
solve vgate=0 vstep=0.1 vfinal=0.5 name=gate
solve vgate=0.6 vstep=0.5 vfinal=1.0 name=gate
solve vgate=0 vstep=–0.1 vfinal=–0.5 name=gate
solve vgate=–0.6 vstep=–0.2 vfinal=–1.0 name=gate
quit
```

16. An AlGaAs/GaAs HEMT structure has the following parameters: AlGaAs doped layer thickness = 150 Å; n-type doping density = $4 \times 10^{18}\,\text{cm}^{-3}$; Al mole fraction = 0.3; gate length = 0.5 μm; AlGaAs surface state density = $10^{12}\,\text{cm}^{-2}$; and gate metal work function = 4.73 eV.
    a. Plot the I_D-versus-V_{DS} curve at $V_{GS} = 0$ and the I_D-versus-V_{GS} curve at $V_{DS} = 3$ V. What is the threshold voltage for the device?
    b. Change the Al mole fraction in the doped layer to 0.1 and repeat part **a**. Compare the results and explain. Feel free to include any other plots in your explanation (band diagram, carrier concentration, etc.).
    c. What is the effect of the AlGaAs surface states on the I-V characteristics? Change the surface state density to 0 and $3 \times 10^{12}\,\text{cm}^{-2}$ and repeat part **a**.
    d. In reality, a thin layer of undoped AlGaAs (spacer) is placed between the doped AlGaAs and the undoped GaAs layers. Run a simulation with a 50 Å spacer and compare the results with those of part **a**. Now set the temperature to 77 K and run the simulation again. Explain your results.

Below is an example input deck for a device simulator:

```
Title  HEMT simulation
# SECTION 1: Mesh input
mesh  space.mult=1.0
#
x.m    l=0.0  spac=0.03
x.m    l=0.05  spac=0.02
x.m    l=0.70  spac=0.02
x.m    l=0.75  spac=0.03
#
y.m    l=-0.02  spac=0.01
y.m    l=0.0  spac=0.005
y.m    l=0.015  spac=0.001
y.m    l=0.02  spac=0.001
y.m    l=0.05  spac=0.002
y.m    l=0.1  spac=0.01
#
# SECTION 2: Structure Specification
#
region  num=1 material=GaAs y.min=0.015
region  num=2 material=AlGaAs y.max=0.015 x.
composition=0.3
region  num=3 oxide y.min=-0.02 y.max=0
#
elec  num=1  name=source  x.min=0.0  x.max=0.0  y.min=0.0
y.max=0.035
```

```
elec num=2 name=gate x.min=0.1 x.max=0.6 y.min=0.0
y.max=0.0
elec num=3 name=drain x.min=0.75 x.max=0.75 y.min=0.0
y.max=0.035
#
doping uniform y.min=0 y.max=0.015 n.type conc=4.e18
doping uniform x.min=0.0 x.max=0.05 y.min=0.015
y.max=0.035
n.type conc=1.e18
doping uniform x.min=0.70 x.max=0.75 y.min=0.015
y.max=0.035
n.type conc=1.e18
#
interface x.min=0 x.max=0.75 y.min=-0.01 y.max=0.005
qf=-1.e12
#
# SECTION 3: Material Models
#
material taun0=1.e-9 taup0=1.e-9
material material=AlGaAs mun=2000 mup=350
#
model fldmob srh fermi
model material=GaAs conmob
#
contact name=gate workfun=4.73
#
# SECTION 4: Id-Vd calculation
#
method gummel newton itlim=20 trap maxtrap=6 carriers=1
elect
output con.band val.band efield
solve v#
save outf=hemt2.str
log outf=hemt2d.log master
solve vdrain=0.05
solve vdrain=0.10
solve vdrain=0.125
solve vdrain=0.15
solve vdrain=0.20
solve vdrain=0.30
method newton trap itlim = 35 maxtrap=6 carriers=1
elect
solve vdrain=0.50 vstep=0.25 name=drain vfinal=1
solve vdrain=1.5 vstep=0.5 name=drain vfinal=3
save outf=hemt2d.str
```

```
# SECTION 5: Id-Vg calculation
log off
method gummel newton trap itlim=35 maxtrap=6 carriers=1
elect
solve vgate=0 vstep=–0.225 name=gate vfinal=–0.9
log outf=hemt2g.log master
method newton trap itlim=35 maxtrap=6 carriers=1 elect
solve vgate=–0.8 vstep=0.1 name=gate vfinal=0.6
save outf=hemt2g.str
quit
```

CHAPTER 6

FET HIGH-FREQUENCY PROPERTIES

§ 6-1 QUASI-STATIC OPERATION IN THE HFET

The previous chapter concerned d.c. operation, in which the terminal voltages are held constant with time. This is clearly not the desired mode of operation. In a practical circuit, the bias voltages of a FET are always some functions of time, and we are interested in finding the time dependencies of the resulting drain, gate, and source currents. This chapter examines the transient and sinusoidal operation of both MESFETs and HFETs. For simplicity, we shall analyze the long-channel FETs exclusively, without considering the effect of velocity saturation. As will be demonstrated, some of the long-channel analyses, carried out in Appendix A for MESFETs and in Appendix B for HFETs, are already algebraically intensive. The task of analyzing the effects of velocity saturation in a.c. operation will be much more demanding.

We study HFETs first and delay the discussion of MESFETs until the next section, mainly because it is easier to trace the charge movement in HFETs. The conducting layer in a MESFET is a charge-neutral layer. With a net charge of zero there, the notion of a current due to the channel charge movement can sometimes be confusing. In contrast, the channel charges in a HFET come from the accumulated electrons. It is easy to picture the formation of the current due to the movement of these electrons. There is also another benefit in considering HFETs first. In the course of analysis we shall invoke the expression of the channel potential. It turns out that the channel potential in a HFET is an analytical function of position [Eq. (5-142)], whereas in MESFETs, it must be determined numerically from a transcendental equation [Eq. (5-42)].

The equations governing the instantaneous behaviour of HFETs are the continuity equation, Eq. (5-135), and the drift equation, Eq. (5-136). These two equations can be solved to give the channel potential, $u_{\text{CH}}(x, t)$, and consequently, the channel current, $i_{\text{CH}}(x, t)$. The instantaneous drain and source currents can then be determined, according to Eq. (5-22). To obtain the gate current, however, we need to make a distinction between the drain/source currents and the gate current. The drain current, for example, is composed of charges which flow out of the drain contact, through the drain voltage source, through the source contact, and then into the channel. There is a continuous loop of current path for the electrons to travel. The charges constituting the gate current, in contrast, flow into the gate metal and stay there. There is not a loop path (discounting the displacement current); the electrons cannot proceed further after entering the gate metal, because the metal-semiconductor barrier height $q\phi_B$ is sufficient to block the electrons from entering the semiconductor. As the gate voltage increases with time, more charges flow into and accumulate at the gate metal. Conversely, as the gate voltage decreases with time, the number of the charges in the gate decreases at the charges flow out of the gate. If we neglect any possible leakage current between the gate and the semiconductor, we can write:

$$i_G(t) = \frac{d}{dt} q_G(t) \tag{6-1}$$

In this definition, the gate current is positive when the number of positive charges in the gate increases. That is, the gate current is taken to be positive when it flows *into* the gate.

Equation (6-1) begs the question of what, exactly, the gate charge, q_G, is. We know from Gauss's law that the net charge in a device is zero; otherwise, there would be electric field outside of the device. (It is possible to have field established inside the device, however, mainly because the location of the positive charges does not coincide with that of the negative charges.) With $q_G(t) + q_{\text{CH}}(t) = 0$, we formulate the gate charge to be:

$$q_G(t) = -\, q_{\text{CH}}(t) = -\, W \int_0^L q'_{\text{CH}}(x, t)\, dx \tag{6-2}$$

According to Eqs. (5-122) and (5-134), we establish q_G to be:

$$q_G(t) = WC'_{\text{ox}} \int_0^L u_{\text{CH}}(x, t)\, dx \tag{6-3}$$

Therefore, $q_G(t)$ is known once the channel potential $u_{\text{CH}}(x, t)$ is known, and consequently, $i_G(t)$ can be determined. It turns out that in the process of solving Eqs. (5-135) and (5-136) to determine the source and drain currents, we would have already obtained $u_{\text{CH}}(x, t)$. Therefore, the determination of $i_G(t)$ does not involve more than that required to determine the drain and source currents.

In general, solving Eqs. (5-135) and (5-136) simultaneously is difficult. Together they form a second-order partial differential equation given by

$$\mu_n \frac{\partial}{\partial x}\left[u_{\text{CH}} \frac{\partial}{\partial x} u_{\text{CH}}(x, t)\right] = \frac{\partial}{\partial t} u_{\text{CH}}(x, t) \tag{6-4}$$

$u_{\text{CH}}(x, t)$ is subject to the initial and boundary conditions:

$$u_{\text{CH}}(x, 0) = 0 \tag{6-5}$$

$$u_{\text{CH}}(0, t) = v_{GS}(t) - V_T \qquad u_{\text{CH}}(L, t) = \alpha \cdot u_{\text{CH}}(0, t) \tag{6-6}$$

Equation (6-5) assumes that initially there is no charge in the channel (the transistor is off), so that u_{CH} at $t = 0$ is zero. The boundary conditions are similar to those expressed in Eq. (5-138). The saturation index, α, was given in Eq. (5-132). Because Eq. (6-4) is nonlinear, there is no analytical solution general enough to accommodate arbitrary time-dependent boundary conditions. Numerical techniques used to solve the differential equation by brute force are necessary to obtain an accurate transient solution for $u_{\text{CH}}(x, t)$. Once the channel potential is found, the drain and source currents are determined from (see Eqs. 5-22 and 5-136):

$$i_D(t) = -WC'_{\text{ox}}\mu_0 u_{\text{CH}} \left.\frac{\partial u_{\text{CH}}(x, t)}{\partial x}\right|_{x=L} \qquad i_S(t) = WC'_{\text{ox}}\mu_0 u_{\text{CH}} \left.\frac{\partial u_{\text{CH}}(x, t)}{\partial x}\right|_{x=0} \tag{6-7}$$

While the exact transient solution can be obtained only by numerical methods, there is a powerful technique which yields a fairly good transient solution without solving Eq. (6-4) directly. This solution, called the *quasi-static solution*, is obtained by assuming the transistor operation to consist of a series of static events. (In contrast, the solution obtained by solving the nonlinear differential equation, i.e., the true solution, is called the *non-static solution*).

We illustrate the meaning of the quasi-static analysis through an example. A ramp of gated bias is applied to a HFET, as shown in Fig. 6-1. To simplify discussion, we assume that the drain bias is maintained at some high value such that $\alpha = 0$ during the entire transient. The source is grounded. The gate bias is initially at a value equal to V_T, such that there is no charge accumulation in the channel. Because the entire channel layer is depleted, the transistor is initially at the cutoff mode, without any current flow. Let us consider the transistor operation at $t = t_1$, at which instant $v_{GS}(t) = v_{GS,1}$. Although only at the instant $t = t_1$ does $v_{GS}(t)$ attain the value $v_{GS,1}$, let us suppose that $v_{GS}(t)$ was equal to $v_{GS,1}$ for a long time prior to reaching the instant $t = t_1$. In this thought experiment, the device attained d.c. operation state by the time $t = t_1$ was

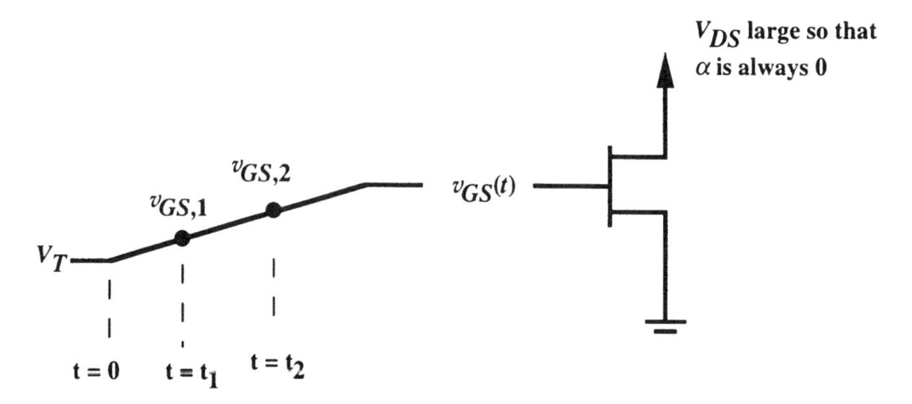

FIGURE 6-1. A HFET subjected to a ramp increase at the input. The output voltage is assumed to be large so that the transistor always stays in saturation.

reached, and $v_{GS}(t)$ was $v_{GS,1}$. The channel potential under such a scenario would be that of the d.c. solution given in Eq. (5-142):

$$u_{\text{CH}}(x, t_1) = u_{\text{CH}}(0, t_1)\sqrt{1 - \frac{x}{L}\left[1 - \frac{u_{\text{CH}}^2(L, t_1)}{u_{\text{CH}}^2(0, t_1)}\right]} \tag{6-8}$$

Now consider $t = t_2$, at which time $v_{GS}(t) = v_{GS,2}$. Again, although only at the instant $t = t_2$ does $v_{GS}(t)$ reach the value $v_{GS,2}$, we imagine that $v_{GS}(t)$ had been equal to $v_{GS,2}$ for a long time prior to reaching to the instant $t = t_2$. Basically, we assume the device had been in d.c. operation long before the time $t = t_2$, and that $v_{GS}(t)$ had been $v_{GS,2}$. The channel potential at t_2 would then be governed by the following equation:

$$u_{\text{CH}}(x, t_2) = u_{\text{CH}}(0, t_2)\sqrt{1 - \frac{x}{L}\left[1 - \frac{u_{\text{CH}}^2(L, t_2)}{u_{\text{CH}}^2(0, t_2)}\right]} \tag{6-9}$$

This procedure is repeated at other times. In each instant, $u_{\text{CH}}(x, t)$ is obtained by assuming that its steady-state distribution was obtained right at the instant t of concern. That is, $u_{\text{CH}}(x, t)$ at any moment obeys this equation:

$$u_{\text{CH,QS}}(x, t) = u_{\text{CH}}(0, t)\sqrt{1 - \frac{x}{L}\left[1 - \frac{u_{\text{CH}}^2(L, t)}{u_{\text{CH}}^2(0, t)}\right]} \tag{6-10}$$

We added the subscript QS to $u_{\text{CH}}(x, t)$ to emphasize that the solution so obtained represents the quasi-static solution. We caution that, despite the similarity in forms, $u_{\text{CH,QS}}(x, t)$ is not identical to $u_{\text{CH}}(x)$ of Eq. (5-142), which represents the d.c. channel potential. There is a time dependence in $u_{\text{CH,QS}}(x, t)$, mainly through the time variations of $v_{GS}(t)$ and $v_{DS}(t)$. If we drop the time dependencies, then u_{CH} and $u_{\text{CH,QS}}$ appear the same in form.

In the quasi-static approximation, the instantaneous channel potential given by Eq. (6-10) is solely a function of the instantaneous values of the applied v_{GS} and v_{DS}. The charge responds instantaneously to obey the above equation the moment v_{GS} or v_{DS} changes to a different value.

The gate current obtained under the quasi-static approximation, according to Eq. (6-1), is:

$$i_{G,\mathrm{QS}}(t) = \frac{dq_{G,\mathrm{QS}}(t)}{dt} \qquad q_{G,\mathrm{QS}}(t) = +WC'_{\mathrm{ox}} \int_0^L u_{\mathrm{CH,QS}}(x,t)\,dx \tag{6-11}$$

Carrying out this quasi-static analysis one more step, we express the drain and source currents as:

$$\begin{aligned} i_D(t) &= -W\mu_n C'_{\mathrm{ox}} u_{\mathrm{CH,QS}}(x,t) \left.\frac{\partial u_{\mathrm{CH,QS}}(x,t)}{\partial x}\right|_{x=L} \\ i_S(t) &= W\mu_n C'_{\mathrm{ox}} u_{\mathrm{CH,QS}}(x,t) \left.\frac{\partial u_{\mathrm{CH,QS}}(x,t)}{\partial x}\right|_{x=0} \end{aligned} \tag{6-12}$$

According to the quasi-static approximation, the channel potential reaches its steady-state value instantaneously, at a value corresponding to the bias voltages at the given time. This scenario is illustrated in Fig. 6-2. At $t = t_1$, $u_{\mathrm{CH,QS}}(0, t_1)$, the channel potential at the source edge is equal to $v_{GS}(t_1) - V_T$. $u_{\mathrm{CH,QS}}(L, t_1)$ is zero because the drain bias is assumed to be high, exceeding $v_{DS,\mathrm{sat}}(\alpha = 0)$. The

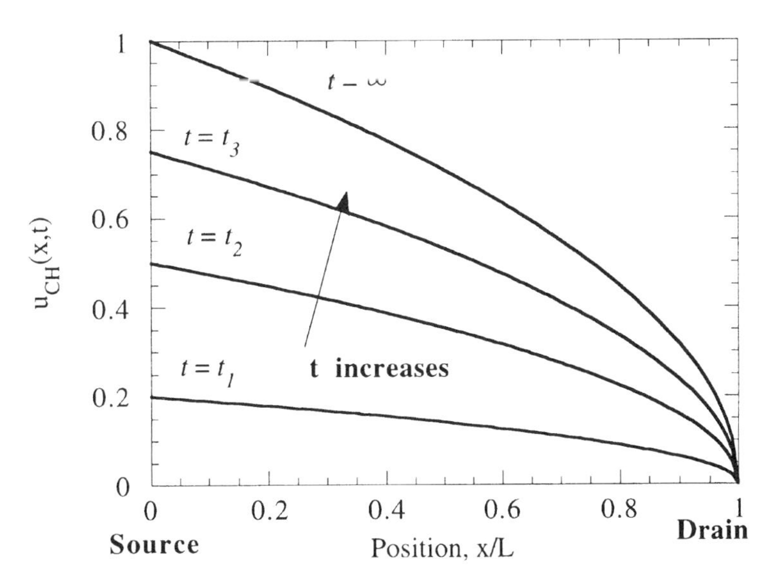

FIGURE 6-2. Channel potential profile predicted in a quasi-static analysis for the HFET of Fig. 6-1.

quasi-static channel potential basically follows the d.c. solution given by Eq. (5-142). As t increases to t_2 and $v_{GS}(t)$ increases to $v_{GS,2}$, the channel potential at the source increases, and the overall channel potential profile remains to that given by Eq. (5-142). As input voltage continues to increase over time, the channel potential at $x = 0$ increases. The curves then shift upward as shown in the figure. In the quasi-static approximation, the channel potential is then a series of square-root-like curves of Eq. (5-142) which would be truly obtained only in the steady state. Once the approximation is made, the drain current given by Eq. (6-12) has finite values, even at a time t very close to 0. In reality, as discussed in the following, the drain current remains zero for some time during the initial part of the transient. The time duration at which i_D stays zero is called the *intrinsic channel transit time* (τ_{tr}).

At $t = 0$, when the input bias is equal to V_T, the transistor is just about to turn on. Since $v_{GS}(0) = V_T$, $u_{CH,QS}$ is equal to 0 at the source. There is no accumulated charge at the interface of the AlGaAs/GaAs heterojunction. At $t > 0$, such that v_{GS} becomes greater than V_T, the electron sheet concentration at $x = 0$, $n_s(0, t)$ increases to a certain value. Propelled by the electric field, these electrons at the source start to move toward the drain. The moment these charges move, the source current at the source contact is nonzero. However, the charges have not reached the drain. Hence, the drain current remains zero initially. Since it takes a certain time for the charge to traverse through the channel, we cannot expect the steady-state profiles of Fig. 6-2 to be set up instantaneously. With numerical techniques, we solved Eq. (6-4) for a particular case where $\alpha = 0$ and where $v_{GS}(t)$ increases linearly with time. The charge profile $[n_s(x, t)]$, plotted in Fig. 6-3, reveals the existence of a charge front which progresses from the source to the drain as time elapses. There is no drain current before the charge front reaches the drain (before $t = t_3$). Although Fig. 6-3 applies to the case where $v_{GS}(t)$ increases linearly with time, the central point that drain current is initially zero remains unaffected for other functions of $v_{GS}(t)$.

We briefly examine Fig. 6-3 to show that the figure makes sense. First, with $\alpha = 0$, the boundary condition specified in Eq. (6-6) is indeed satisfied. We see that throughout the entire transient, $n_s(L, t) = 0$. At the source side, $n_s(0, t)$ increases linearly with time, consistent with the other boundary condition specified in Eq. (6-6). The source current, obtained from Eq. (6-7), is finite because $\partial u_{CH}/\partial x$ is finite at $x = 0$. When t is small, the drain current is zero because $\partial u_{CH}/\partial x = 0$ at $x = L$. I_D rises above zero only after the charge front arrives at the drain, an event which gives a nonzero $\partial u_{CH}/\partial x$ at $x = L$.

Comparing the quasi-static curves of Fig. 6-2 and the non-quasi-static curves of Fig. 6-3, we see that only one curve of Fig. 6-2 is accurate. It is the $t = \infty$ curve. It is correct because at $t = \infty$, we truly obtain the steady-state solution, consistent with the assumption used in the quasi-static analysis.

Equation (6-3) is an exact equation. However, the moment we substitute for the instantaneous $u_{CH}(x, t)$ the quasi-static $u_{CH,QS}(x, t)$, the gate current calculated from Eq. (6-11) is no longer exact. It becomes a quasi-static quantity. To obtain a non-quasi-static i_G (an i_G without the quasi-static assumption, i.e., the

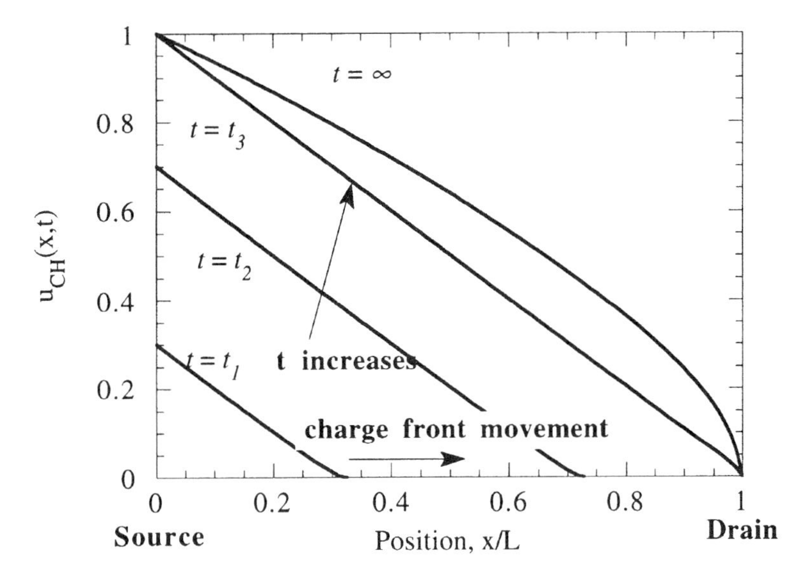

FIGURE 6-3. Channel potential profile calculated in a non-quasi-static analysis for the HFET of Fig. 6-1. The charge front is the tip location of the charges moving from the source to the drain.

exact i_G), we have to establish a non-quasi-static $u_{CH}(x, t)$ before applying in Eq. (6-3). The difficulty associated with this approach resides in the first step, that the detailed time evolution of the channel potential profile, $u_{CH}(x, t)$, needs to be determined from the partial differential equation. This almost always requires a numerical solution, although several approaches compatible with compact models can sometimes be used. Similarly, the drain current equation as specified in Eq. (6-7) is itself correct. However, supplying it with an approximate charge profile from the quasi-static assumption (as in Eq. 6-12) causes the resultant drain current to be an approximate value.

In the following, we shall be concerned exclusively with the quasi-static approximation. Although it is just an approximation, it is remarkably accurate for many practical situations, which take place at a time scale longer than the intrinsic channel transit time.

We evaluate the quasi-static gate charge in a HFET. Substituting $U_{CH}(x)$ of Eq. (5-142) into $u_{CH,QS}(x, t)$ of Eq. (6-11), we find:

$$
\begin{aligned}
Q_G(t) &= +WC'_{ox}\int_0^L U(0)\sqrt{1-\frac{x}{L}(1-\alpha^2)}\,dx \\
&= +WLC'_{ox}(v_{GS}-V_T)\frac{2}{3}\frac{1+\alpha+\alpha^2}{1+\alpha}
\end{aligned}
\qquad (6\text{-}13)
$$

We replace the notation of $q_{G,QS}(t)$ by Q_G. In our usual convention, Q_G would have meant the d.c. value of the gate charge (just like V_{GS} is the d.c. value while v_{GS} is the time-varying value). We make the replacement to be consistent with the convention, and also to simplify the look of the symbol. The time dependence of Q_G comes from the time dependence of the bias voltages, which make α appearing in Eq. (6-13) to be a function of time. Essentially:

$$Q_G(t) = Q_G(v_{GS}(t), v_{DS}(t)) \tag{6-14}$$

If $v_{GS}(t)$ and $v_{DS}(t)$ are constant with respect to time, then $Q_G(t)$ loses its time dependence.

The capacitance measures the amount of change in the charge with respect to the change in the terminal voltage. In a three-terminal FET device (as opposed to MOSFET, which has the fourth terminal at the bulk), there are generally four linearly independent capacitances of interest. They are defined as (Q_D means the drain charge, to be determined shortly):

$$C_{gg} = \frac{\partial Q_G}{\partial v_G} \qquad C_{gd} = -\frac{\partial Q_G}{\partial v_D} \qquad C_{dg} = -\frac{\partial Q_D}{\partial v_G} \qquad C_{dd} = \frac{\partial Q_D}{\partial v_D} \tag{6-15a}$$

There are five other capacitances which are capable of characterizing the device. They are nonetheless inferable from the four capacitances given in Eq. (6-14). For completeness, we write out these additional capacitances (Q_S means the source charge, to be determined shortly):

$$C_{gs} = -\frac{\partial Q_G}{\partial v_S} \quad C_{ds} = -\frac{\partial Q_D}{\partial v_S} \quad C_{sg} = -\frac{\partial Q_S}{\partial v_G} \quad C_{sd} = -\frac{\partial Q_S}{\partial v_D} \quad C_{ss} = \frac{\partial Q_S}{\partial v_S} \tag{6-15b}$$

A minus sign is inserted into the definition of C_{xy} when $x \neq y$. The physical reason for doing so will become clear in § 6-3. The addition of the minus sign makes most of the capacitances positive quantities.

With the gate charge expression of Eq. (6-13), we can determine C_{gg} and C_{gd}. It is important to realize that α, given by Eq (5-132), is a function of both v_{GS} and v_{DS}. Further, C_{gg} is the derivative of Q_G with respect to v_G rather that v_{GS}, and C_{gd} is the derivative of Q_G with respect to v_D rather than v_{DS}. After we replace v_{GS} by $v_G - v_S$ and v_{DS} by $v_D - v_S$ in Eq. (6-13), we take the derivatives of the gate charge and find:

$$C_{gg} = WLC'_{ox} \times \frac{2}{3}\frac{1 + 4\alpha + \alpha^2}{(1 + \alpha)^2} \tag{6-16}$$

$$C_{gd} = WLC'_{ox} \times \frac{2}{3}\frac{2\alpha + \alpha^2}{(1 + \alpha)^2} \tag{6-17}$$

To find C_{dg} and C_{dd}, we need to establish the drain charge as a function of time. We commented before about the difference between the gate current and the drain current, that the charges flowing into the gate do not travel on a continuous loop. But if we were to blindly omit this difference and write an i_D equation similar to the i_G's in Eq. (6-11), then the drain current would be:

$$i_D(t) = \frac{dQ_D(t)}{dt} \qquad \text{(not useful)} \tag{6-18}$$

There are some problems with this kind of definition. Foremost, this formulation is awkward in a d.c. analysis. In d.c., $i_d(t) = I_D$, so $Q_D(t)$ as defined in Eq. (6-18) would be required to increase linearly with time, even though the external bias voltages v_{GS} and v_{DS} are constant with time. This is in violation of the concept embodied in the quasi-static gate charge, whose time dependence originates from the bias voltages as discussed in Eq. (6-14). An alternative definition of the drain charge to circumvent the problem is to define $Q_D(t)$ as:

$$i_D(t) = I_D + \frac{dQ_D(t)}{dt} \tag{6-19}$$

In a non-quasi-static analysis, the definition of the drain charge in Eq. (6-19) will not be useful, either. $Q_D(t)$ will turn out to be an intractable expression, depending on the time evolution of the boundary conditions. However, the above formulation becomes meaningful in a quasi-static approximation. To see that, we apply the *first-moment technique* [1], wherein we multiply both sides of Eq. (6-4) by the space variable x, and then integrate both sides from $x = 0$ to L. That is,

$$\mu_n \int_0^L x \frac{\partial}{\partial x}\left[u_{\mathrm{CH}} \frac{\partial u_{\mathrm{CH}}}{\partial x}\right] dx = \int_0^L \frac{\partial}{\partial t} x u_{\mathrm{CH}}(x, t)\, dx \tag{6-20}$$

The left integral can be evaluated with integration by parts, that the integral of $u{\cdot}dv$ is equal to $u{\cdot}v$ minus the integral of $v{\cdot}du$. Let $u = x$ so that $du = dx$, and dv be $\partial/\partial x$ of the term in square brackets; then v is exactly equal to that enclosed in square brackets. We have:

$$\begin{aligned} \int_0^L x \frac{\partial}{\partial x}\left[u_{\mathrm{CH}} \frac{\partial u_{\mathrm{CH}}}{\partial x}\right] dx &= x u_{\mathrm{CH}} \frac{\partial u_{\mathrm{CH}}}{\partial x}\bigg|_0^L - \int_0^L u_{\mathrm{CH}} \frac{\partial u_{\mathrm{CH}}}{\partial x}\, dx \\ &= L u_{\mathrm{CH}} \frac{\partial u_{\mathrm{CH}}}{\partial x}\bigg|_{x=L} - \int_{u_{\mathrm{CH}}(0)}^{u_{\mathrm{CH}}(L)} u_{\mathrm{CH}}\, du_{\mathrm{CH}} \end{aligned} \tag{6-21}$$

Therefore, Eq. (6-20) can be rewritten as:

$$\mu_n L u_{CH}(x,t) \frac{\partial u_{CH}}{\partial x}\bigg|_{x=L} = \mu_n \int_{u_{CH}(0)}^{u_{CH}(L)} u_{CH}(x,t)\, du_{CH} + \int_0^L \frac{\partial}{\partial t} x u_{CH}\, dx \quad (6\text{-}22)$$

Thus far the derivation has been exact; no approximation was made. However, in order to proceed further, we shall invoke the quasi-static approximation, replacing $u_{CH}(x, t)$ by $U_{CH}(x)$ [or, more precisely, $u_{CH,QS}(x, t)$] in the right-hand-side terms:

$$\mu_n L u_{CH}(x,t) \frac{\partial u_{CH}}{\partial x}\bigg|_{x=L} = \mu_n \int_{U_{CH}(0)}^{U_{CH}(L)} U_{CH}\, dU_{CH} + \frac{\partial}{\partial t} \int_0^L x U_{CH}(x)\, dx \quad (6\text{-}23)$$

The first term is proportional to the instantaneous drain current in accordance with Eqs. (5-22) and (5-136), whereas the second term is proportional to the d.c. drain current in accordance with Eq. (5-128) after the replacement of $V_{GS} - V_T - V_{CS}(x)$ by $U_{CH}(x)$. Therefore:

$$-\frac{i_D(t)L}{WC'_{ox}} = -\frac{I_D L}{WC'_{ox}} + \frac{\partial}{\partial t}\int_0^L x U_{CH}\, dx \quad (6\text{-}24)$$

The above equation is simplified to:

$$i_D(t) = I_D - \frac{\partial}{\partial t}\frac{WC'_{ox}}{L}\int_0^L x U_{CH}\, dx \quad (6\text{-}25)$$

Comparing Eq. (6-25) to our definition of the quasi-static drain charge in Eq. (6-19), we find that the drain charge is given by:

$$Q_D = -\frac{WC'_{ox}}{L}\int_0^L x U_{CH}\, dx = -\frac{WC'_{ox}}{L} U_{CH}(0) \int_0^L x \sqrt{1 - \frac{x}{L}(1-\alpha^2)}\, dx$$

$$= -WLC'_{ox}(V_{GS} - V_T)\frac{6\alpha^3 + 12\alpha^2 + 8\alpha + 4}{15(1+\alpha)^2} \quad (6\text{-}26)$$

The drain charge is negative in HFETs, consistent with the fact that the electrons which accumulate in the channel are of negative charge.

The capacitances associated with the drain charge, defined in Eq. (6-14), are given by:

$$C_{dg} = WLC'_{ox} \times \frac{2}{15}\frac{2 + 14\alpha + 11\alpha^2 + 3\alpha^3}{(1+\alpha)^3} \quad (6\text{-}27)$$

$$C_{dd} = WLC'_{ox} \times \frac{2}{15}\frac{8\alpha + 9\alpha^2 + 3\alpha^3}{(1+\alpha)^3} \quad (6\text{-}28)$$

The source charge is defined in a similar manner as the drain charge. Since the d.c. current flows through the source is negative of I_D, we have:

$$i_S(t) = -I_D + \frac{dQ_S(t)}{dt} \tag{6-29}$$

$$\begin{aligned} Q_S &= -\frac{WC'_{ox}}{L}\int_0^L (L-x)\, U_{CH}\, dx \\ &= -WLC'_{ox}(V_{GS}-V_T)\frac{4\alpha^3+8\alpha^2+12\alpha+6}{15(1+\alpha)^2} \end{aligned} \tag{6-30}$$

The integrand in the source charge expression is $(L-x)\cdot U_{CH}(x)$, rather than $x \cdot U_{CH}(x)$ as in the drain charge. The derivation of the source charge is described in Problem 1.

Adding Eqs. (6-29) and (6-19), we find the sum of the source and drain currents to be:

$$i_S(t) + i_D(t) = \frac{dQ_S(t)}{dt} + \frac{dQ_D(t)}{dt} = -\frac{dQ_G(t)}{dt} \tag{6-31}$$

This result makes sense because the sum of the device currents is zero. The above equation basically states that $i_S(t) + i_D(t) = -i_G(t)$.

The plots of the charges as function of biases are shown in Fig. 6-4 with V_{GS} as a variable and in Fig. 6-5 for varying V_{DS}. When V_{DS} is small, the transistor is in the linear region. Q_S and Q_D, as shown in Fig. 6-4, do not differ much. When the transistor is in saturation, $\alpha = 0$, and the ratio of Q_D to Q_S is 2 to 3. Therefore, the quasi-static charges are sometimes referred to as the 40/60 partition scheme (i.e., Q_D = 40% of the sum of Q_D and Q_S). It is important to realize that, $Q_D : Q_S$ is equal to 40 : 60 only when $\alpha = 0$! In the linear region, the integrals of the drain and source charges produce a ratio different from 40 : 60. Furthermore, the 40/60 partition scheme is not an arbitrary partition of charges. It has been derived based entirely on device equations. The moment we make the quasi-static assumption, the 40/60 partition is *exactly* correct.

The partition of the channel charge into the drain and source charges is usually not straightforward. We employed the first-moment technique to greatly simplify the derivation of such charges. Sometimes, when the first-moment technique is not used, the so-called 0/100 partition scheme is applied to derive Q_D and Q_S. The 0/100 partition scheme *arbitrarily* assigns all the channel charge in saturation to the source charge, leaving the drain charge equal to zero. The logic is that, since the channel is pinched off at the drain, the drain charge is zero. In reality, the fact the channel is pinched off merely suggests that $q'_{CH}(x = L) = 0$. It does not necessarily lead to $Q_D = 0$. After all, Q_D is defined such that Eq. (6-19) relating the instantaneous and d.c. currents is valid. Therefore, it is difficult to establish the value of Q_D based on the value of q'_{CH} at a particular location.

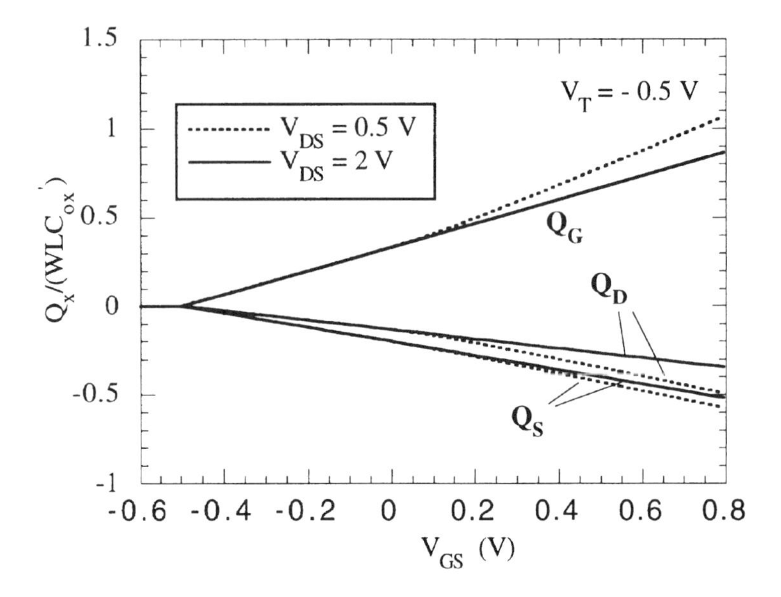

FIGURE 6-4. Gate, drain, and source charges of a HFET as a function of V_{GS}.

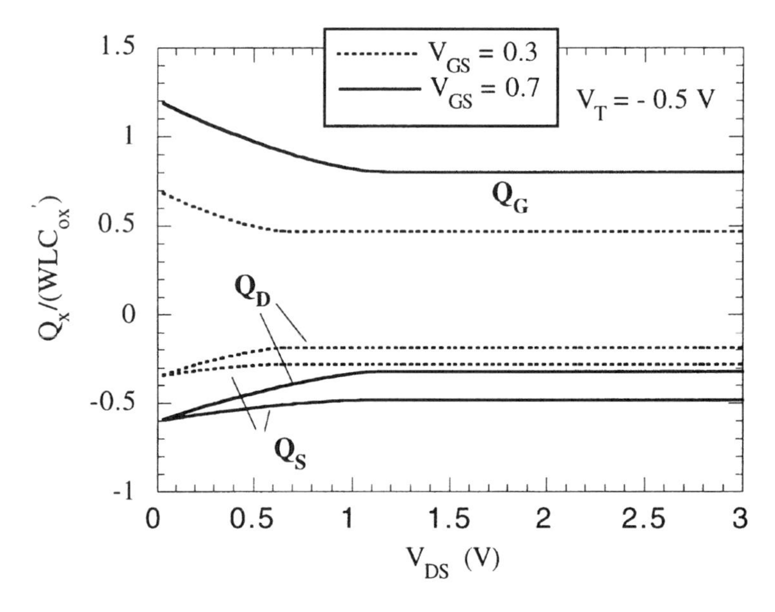

FIGURE 6-5. Gate, drain, and source charges of a HFET as a function of V_{DS}.

Despite that the 0/100 partition is physically unsound and leads to wrong delay time in a transient analysis, it is popular in some simulation tools. This is because at times the 40/60 partition scheme can produce a numerical problem in the initial part of a transient simulation during which the quasi-static analysis is not correct. However, in a small-signal analysis, the 40/60 partition should always be used.

The capacitances given in Eqs. (6-16), (6-17), (6-27), and (6-28) are plotted as a function of V_{GS} in Fig. 6-6, and as a function of V_{DS} in Fig. 6-7. It is crucial to realize that C_{dg} is not the same as C_{gd}. They would be equal if the gate and drain are the two terminals of a parallel-plate capacitor. We attempt to explain why physically these two capacitances should differ. For this purpose, it is instructive to examine the saturation region in which the channel charge is pinched off at the drain end (see Fig. 6-7). The source is tied to the ground so that $V_G = V_{GS}$ and $V_D = V_{DS}$. By definition, C_{dg} is proportional to the change in the drain charge when the gate bias is changed. Or, stated in another way, C_{dg} is related to the amount of drain current variation as the gate bias is modified. When V_G increases while the source is grounded, we expect a substantial increase in the channel charge. The additional charges are supplied, in part, through the drain. We therefore expect C_{dg} to be a large value. The situation for C_{gd} in saturation, however, is drastically different. C_{gd} is equal to the variation of the gate charge as V_D is modified. During saturation, the channel is pinched off at the drain, regardless of the exact value of V_{DS}. As long as V_{DS} exceeds the saturation voltage value, $V_{DS,\text{sat}}$, the channel charge is not modified. The channel potential right at the pinchoff point is related to $V_{DS,\text{sat}}$, but is unaffected by V_{DS}. Therefore, during saturation, $q'_{\text{CH}}(x, t)$ remains the same, independent of the

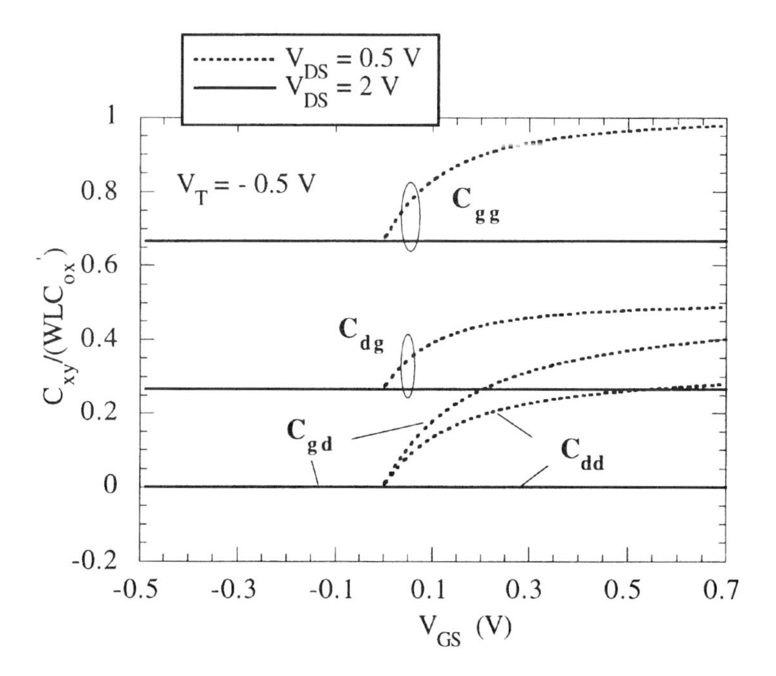

FIGURE 6-6. C_{gg}, C_{gd}, C_{dg}, and C_{dd} of a HFET as a function of V_{GS}.

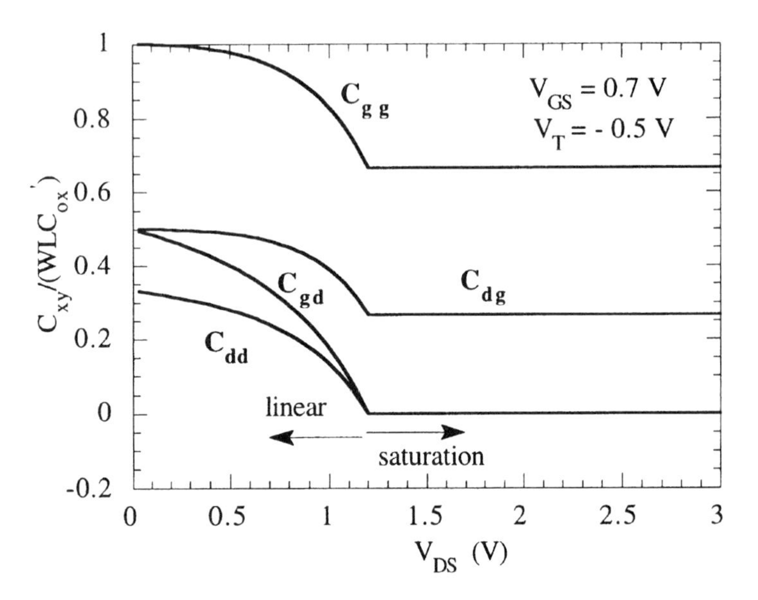

FIGURE 6-7. C_{gg}, C_{gd}, C_{dg}, and C_{dd} of a HFET as a function of V_{DS}.

V_{DS} variation. Consequently, the gate charge remains the same, despite a change in V_{DS}. This insensitivity of the gate charge with respect to V_D leads to a zero C_{gd} in saturation.

Example 6-1:

A HFET has a threshold voltage of 0.188 V. Its AlGaAs barrier layer has a thickness of 330 Å, in which the undoped portion is 30 Å. The device has a gate width of 500 μm and a gate length of 0.25 μm. Find its C_{gg}, C_{gd}, C_{dg}, and C_{dd} when $V_{GS} = 0.5$ V and $V_{DS} = 0.2$ V.

The gate capacitance per area is found from Eq. (5-121):

$$C'_{ox} = \frac{1.159 \times 10^{-12}}{330 \times 10^{-8} + 68 \times 10^{-8}} = 2.91 \times 10^{-7} \text{ F/cm}^2$$

According to Eq. (5-132),

$$V_{DS,\text{sat}} = 0.5 - 0.118 = 0.382 \text{ V}$$

and the saturation index α is equal to

$$\alpha = 1 - \frac{0.2}{0.382} = 0.476$$

Substituting $\alpha = 0.476$ into Eqs. (6-16), (6-17), (6-27), and (6-28), we find:

$$C_{gg} = 500 \times 0.25 \times 10^{-8} \cdot 2.91 \times 10^{-7} \times \frac{2}{3} \frac{1 + 4 \cdot 0.476 + 0.476^2}{(1 + 0.476)^2} = 5.23 \times 10^{-13}\ \text{F}$$

$$C_{gd} = 500 \times 0.25 \times 10^{-8} \cdot 2.91 \times 10^{-7} \times \frac{2}{3} \frac{2 \cdot 0.476 + 0 \cdot 476^2}{(1 + 0.476)^2} = 1.31 \times 10^{-13}\ \text{F}$$

$$C_{dg} = 500 \times 0.25 \times 10^{-8} \cdot 2.91 \times 10^{-7}$$
$$\times \frac{2}{15} \frac{2 + 14 \cdot 0.476 + 11 \cdot 0.476^2 + 3 \cdot 0.476^3}{(1 + 0.476)^3} = 1.73 \times 10^{-13}\ \text{F}$$

$$C_{dd} = 500 \times 0.25 \times 10^{-8} \cdot 2.91 \times 10^{-7}$$
$$\times \frac{2}{15} \frac{8 \cdot 0.476 + 9 \cdot 0.476^2 + 3 \cdot 0.476^3}{(1 + 0.476)^2} = 1.69 \times 10^{-14}\ \text{F}$$

The quasi-static analysis is accurate if the time scale under consideration is much longer than the intrinsic transistor transit time. The transit time, however, is not easily calculated since it is determined from a non-quasi-static analysis, by solving the nonlinear differential equation of Eq. (6-4). Fortunately, we may still estimate the transit time as

$$\tau_{\text{tr, QS}} = \frac{|Q_{\text{CH}}|}{I_D} \tag{6-32}$$

We add the subscript QS to emphasize that this is a quasi-static approximation of the true transit time. Q_{CH} represents the total channel charge in the quasi-static approximation. (In a more strict sense, Q_{CH} in our notation represents d.c. channel charge, just as I_D represents the d.c. current. We should have used the notation $Q_{\text{CH,QS}}$. However, just as we write Q_D in place of $q_{D,\text{QS}}$, we make the above replacement.) Q_{CH} for an HFET is given by, following the derivation of Eqs. (6-2) and (6-3):

$$Q_{\text{CH}} = -WC'_{\text{ox}} \int_0^L U_{\text{CH}}(x)\, dx \tag{6-33}$$

An absolute sign is used for Q_{CH} in Eq. (6-32) because it is a negative quantity in an HFET. $\tau_{\text{tr,QS}}$ measures the amount of time that the channel charge is removed from the channel by the d.c. drain current. In an HFET, $|Q_{\text{CH}}|$ is identical to Q_G, the gate charge, which was given in Eq. (6-13). With the d.c. drain current given by Eq. (5-133), we find the transit time (applicable to both the linear and saturation regions) to be:

$$\tau_{\text{tr, QS}} = \frac{L^2}{\mu_n(V_{GS} - V_T)} \times \left[\frac{4}{3} \frac{1 + \alpha + \alpha^2}{(1 + \alpha)(1 - \alpha^2)}\right] \tag{6-34}$$

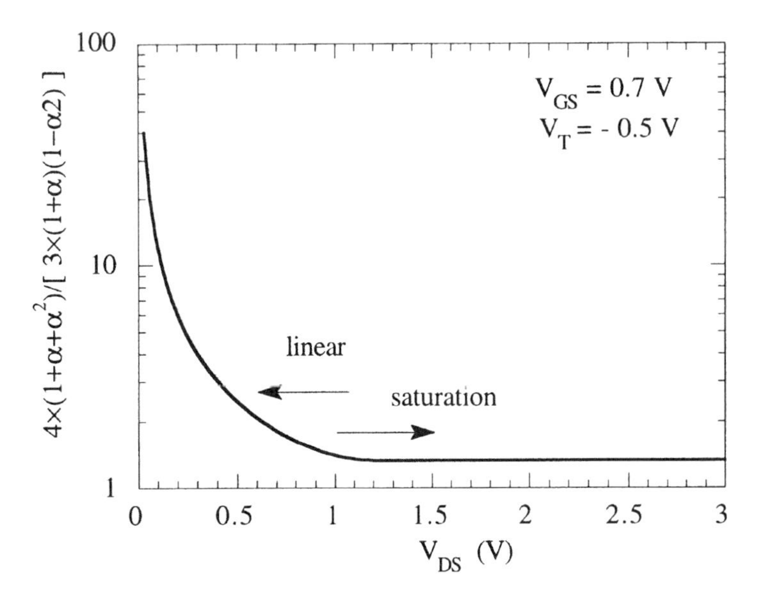

FIGURE 6-8. Quasi-static transit time of a HFET [square-brackets term of Eq. (6-34)] as a function of V_{DS}.

The function in square brackets of the above equation is plotted in Fig. 6-8. The figure demonstrates that the transit time attains the smallest value when the transistor is operated in the saturation region.

§ 6-2 QUASI-STATIC OPERATION IN THE MESFET

In the analysis of charges in MESFETs, it is important to realize the following. The channel region where the actual current conduction takes place does not have space charge. It is a charge-neutral region. Besides the gate metal, the only place where charge resides is the depletion region underneath the gate. Although it is the depletion region that contains space charges, these charges are referred to as channel charges. Therefore, the channel layer of a MESFET is quasi-neutral, yet its channel charge is nonzero.

The total charge in the channel is equal to the integral of the depleted charges per unit area between $x = 0$ and L:

$$Q_{\mathrm{CH}} = \int_0^L qN_d Wa \sqrt{\frac{\phi_{s,\,\mathrm{dc}}(x)}{\phi_{00}}}\, dx \tag{6-35}$$

It is a positive quantity because the n-type channel layer, when depleted of mobile electrons, gives rises to positive ionized charges. Because the net charge of the overall device is zero, the gate charge is exactly equal to the negative of the channel charge:

$$Q_G = -\int_0^L qN_dWa\sqrt{\frac{\phi_{s,\,dc}(x)}{\phi_{00}}}\,dx \tag{6-36}$$

It is easier to perform the integration in the potential domain. From Eqs. (5-29) and (5-31), we express dx as

$$dx = \frac{qW\mu_nN_da}{I_D}\left(1-\sqrt{\frac{\phi_{s,\,dc}}{\phi_{00}}}\right)d\phi_{s,\,dc} \tag{6-37}$$

With proper change of variables, we transform the Q_G integral as

$$Q_G = -\frac{(qN_dWa)^2\mu_n}{I_D}\phi_{00}\int_s^d\sqrt{u}(1-\sqrt{u})\,du \tag{6-38}$$

Carrying out the integration and replacing I_D from Eqs. (5-40) and (5-41), we obtain the final equation for Q_G:

$$Q_G = -qWLN_da\frac{\frac{2}{3}d^{3/2}-\frac{2}{3}s^{3/2}-\frac{1}{2}d^2+\frac{1}{2}s^2}{d-s-\frac{2}{3}d^{3/2}+\frac{2}{3}s^{3/2}} \tag{6-39}$$

As mentioned previously, the gate charge in an n-channel MESFET is negative. The gate charge in an n-channel HFET, in contrast, is positive.

The derivation of Q_G for a MESFET is more complicated than for an HFET. In an HFET, the channel potential $U_{\text{CH}}(x)$ is a simple function of x, allowing the integration of Eq. (6-11) to be carried out straightforwardly, On the contrary, the channel potential in a MESFET, $\phi_{s,\,dc}(x)$, is expressible only in transcendental equation, as shown in Eqs. (5-42) and (5-33). The evaluation of Q_G then relies on playing tricks with the drain current expression, as demonstrated in the process of writing down Eqs. (6-37) and (6-38).

According to Eqs. (5-33)–(5-35), we write d and s as

$$d = \frac{1}{\phi_{00}}\Big(\phi_{\text{bi}}-(V_G-V_S)+(V_D-V_S)\Big) \qquad s = \frac{1}{\phi_{00}}\Big[\phi_{bi}-(V_G-V_S)\Big] \tag{6-40}$$

The above expressions allow us to write the following derivatives:

$$\frac{\partial d}{\partial V_G} = \frac{\partial s}{\partial V_G} = -\frac{1}{\phi_{00}} \qquad \frac{\partial d}{\partial V_D} = \frac{1}{\phi_{00}} \qquad \frac{\partial s}{\partial V_D} = 0 \tag{6-41}$$

Taking the derivative of Q_G with respect to V_G, we find C_{gg} and C_{gd} to be:

$$C_{gg} = \frac{\partial Q_G}{\partial V_G} = \frac{qWLN_d a}{\phi_{00}} \left[\frac{d^{1/2}(1 - d^{1/2}) - s^{1/2}(1 - s^{1/2})}{d - s - \frac{2}{3}d^{3/2} + \frac{2}{3}s^{3/2}} + \frac{(d^{1/2} - s^{1/2})(\frac{2}{3}d^{3/2} - \frac{2}{3}s^{3/2} - \frac{1}{2}d^2 + \frac{1}{2}s^2)}{(d - s - \frac{2}{3}d^{3/2} + \frac{2}{3}s^{3/2})^2} \right] \tag{6-42}$$

$$C_{gd} = -\frac{\partial Q_G}{\partial V_D} = \frac{qWLN_d a}{\phi_{00}} \left[\frac{d^{1/2}(1 - d^{1/2})}{d - s - \frac{2}{3}d^{3/2} + \frac{2}{3}s^{3/2}} - \frac{(1 - d^{1/2})(\frac{2}{3}d^{3/2} - \frac{2}{3}s^{3/2} - \frac{1}{2}d^2 + \frac{1}{2}s^2)}{(d - s - \frac{2}{3}d^{3/2} + \frac{2}{3}s^{3/2})^2} \right] \tag{6-43}$$

The channel charge given in Eq. (6-35) consists of the ionized charges inside the depletion region. As mentioned, the integration does not include the channel region because the channel layer is quasi-neutral, not contributing to any net charge to the channel charge. The partial derivative of the channel charge with respect to time is equal to the sum of the drain and the source currents, as in Eq. (6-31). However, we are interested in the individual current component specifically. For example, we are interested in finding the drain current, and we would like to know exactly what portion of the channel charge belongs to the drain. Simply assigning half of the channel charge to the drain and the remaining half to the source is at best a guess, not supported by any physical principles. It turns out that the drain charge expression, like the gate charge, is a complicated function of V_{GS} as well as V_{DS} (or, alternatively, s and d). Depending on the exact bias conditions, the drain charge is a different proportion of the total channel charge. We apply the first-moment technique described in § 6-1 to find Q_D, this time with the MESFET device in mind. Following the procedure outlined in § 6-1 (see also Problem 3), we find that the drain charge in MESFET is given by:

$$Q_D = \frac{aqWN_d}{L} \int_0^L x \sqrt{\frac{\phi_{s,\text{dc}}(x)}{\phi_{00}}}\, dx \tag{6-44}$$

The evaluation of Q_D involves the use of Eq. (6-37), which gives dx in terms of $\phi_{s,\text{dc}}$. We further integrate the same equation to express x in terms of $\phi_{s,\text{dc}}$:

$$\int_0^x dx' = \frac{qW\mu_n N_d a}{I_D} \int_{\phi_{s,\text{dc}}(0)}^{\phi_{s,\text{dc}}(x)} \left(1 - \sqrt{\frac{\phi_{s,\text{dc}}}{\phi_{00}}}\right) d\phi_{s,\text{dc}} \tag{6-45}$$

This leads to an expression of x in terms of the channel potential:

$$x = \frac{qW\mu_n N_d a}{I_D}\left[\phi_{s,\mathrm{dc}}(x) - \phi_{s,\mathrm{dc}}(0) - \frac{2}{3}\frac{\phi_{s,\mathrm{dc}}^{3/2}(x)}{\sqrt{\phi_{00}}} + \frac{2}{3}\frac{\phi_{s,\mathrm{dc}}^{3/2}(0)}{\sqrt{\phi_{00}}}\right] \tag{6-46}$$

Substituting dx from Eq. (6-37) and x from Eq. (6-46) into Eq. (6-44), we find that Q_D is equal to:

$$Q_D = qaN_d WL \times \frac{\frac{2}{5}d^{5/2} + \frac{4}{9}d^{3/2}s^{3/2} - \frac{2}{3}d^{3/2}s - \frac{1}{3}d^2 s^{3/2} - \frac{5}{9}d^3 + \frac{4}{21}d^{7/2} + \frac{1}{2}d^2 s + \frac{4}{15}s^{5/2} - \frac{7}{18}s^3 + \frac{1}{7}s^{7/2}}{(d - s - \frac{2}{3}d^{3/2} + \frac{2}{3}s^{3/2})^2} \tag{6-47}$$

The capacitance is obtained by taking the derivative of the charge. Using Eq. (6-41), we find:

$$C_{dd} = \frac{\partial Q_D}{\partial V_D} = \frac{aqN_d WL}{\phi_{00}} \times \left[\frac{f_1(d, s)}{(d - s - \frac{2}{3}d^{3/2} + \frac{2}{3}s^{3/2})^2} - \frac{2(1 - d^{1/2})f_2(d, s)}{(d - s - \frac{2}{3}d^{3/2} + \frac{2}{3}s^{3/2})^3}\right] \tag{6-48}$$

where

$$f_1(d, s) = d^{3/2} + \frac{2}{3}d^{1/2}s^{3/2} - d^{1/2}s - \frac{2}{3}ds^{3/2} - \frac{5}{3}d^2 + \frac{2}{3}d^{5/2} + ds \tag{6-49}$$

$$f_2(d, s) = \frac{2}{5}d^{5/2} + \frac{4}{9}d^{3/2}s^{3/2} - \frac{2}{3}d^{3/2}s - \frac{1}{3}d^2 s^{3/2} - \frac{5}{9}d^3 + \frac{4}{21}d^{7/2} + \frac{1}{2}d^2 s + \frac{4}{15}s^{5/2} - \frac{7}{18}s^3 + \frac{1}{7}s^{7/2} \tag{6-50}$$

Similarly,

$$C_{dg} = -\frac{\partial Q_D}{\partial V_G} = \frac{aqN_d WL}{\phi_{00}} \times \left[\frac{f_3(d, s)}{(d - s - \frac{2}{3}d^{3/2} + \frac{2}{3}s^{3/2})^2} + \frac{2(d^{1/2} - s^{1/2})f_2(d, s)}{(d - s - \frac{2}{3}d^{3/2} + \frac{2}{3}s^{3/2})^3}\right] \tag{6-51}$$

where

$$f_3(d, s) = \frac{1}{3}d^{3/2} + \frac{2}{3}d^{1/2}s^{3/2} + \frac{2}{3}d^{3/2}s^{1/2} - sd^{1/2} - \frac{1}{2}d^2s^{1/2} - \frac{2}{3}ds^{3/2} - \frac{7}{6}d^2 + \frac{2}{3}d^{5/2} + ds + \frac{2}{3}s^{3/2} - \frac{7}{6}s^2 + \frac{1}{2}s^{5/2} \quad (6\text{-}52)$$

Once we know Q_G and Q_D, the source charge Q_S can be found from the fact that the sum of the terminal charges is zero. So $Q_S = -Q_G - Q_D$.

Figure 6-9 illustrates Q_G, Q_D, and Q_S vary as functions of V_{GS}. Their variations with V_{DS} are shown in Fig. 6-10. Figure 6-11 plots C_{gg}, C_{gd}, C_{dg}, and C_{dd}'s V_{GS} dependencies. Their dependencies on V_{DS} are shown in Fig. 6-12.

It is important not to regard these capacitances as parallel-plate capacitances. C_{dg} is not equal to C_{gd} in a three-terminal transistor such as a MESFET. (Likewise, $C_{sg} \neq C_{gs}$, etc.) These two capacitances are especially different in the saturation region, as demonstrated in Fig. 6-7 for the HFET and in Fig. 6-12 for the MESFET. Unfortunately, sometimes in the MESFET literature the two capacitances are erroneously equated. Without using the first-moment technique outlined above, it is difficult to derive the drain charge Q_D, although Q_G can still be obtained in a rather straightforward manner [Eq. (6-38)]. A common mistake is to derive Q_G, and claim its derivative with respect to V_D to be C_{dg}. In fact, this results in C_{gd}. The way to find C_{dg} is to establish the drain charge expression, and then take its derivative with respect to V_G.

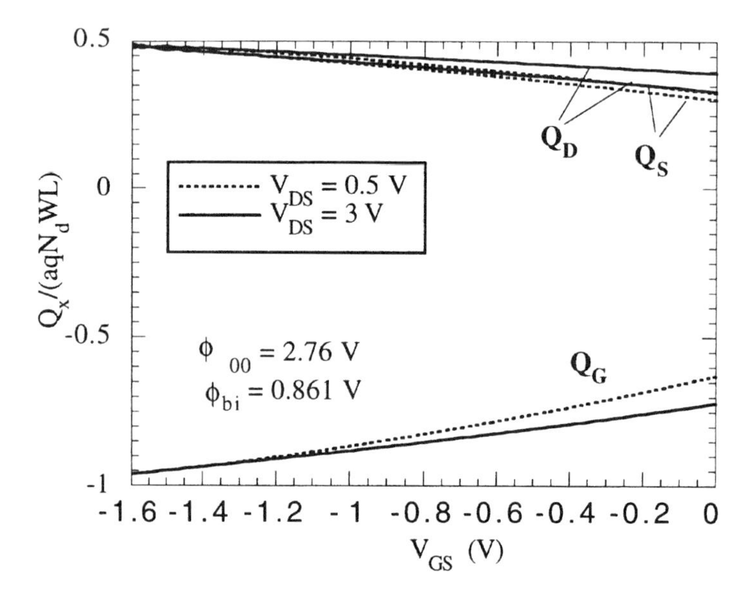

FIGURE 6-9. Gate, drain, and source charges of a MESFET as a function of V_{GS}.

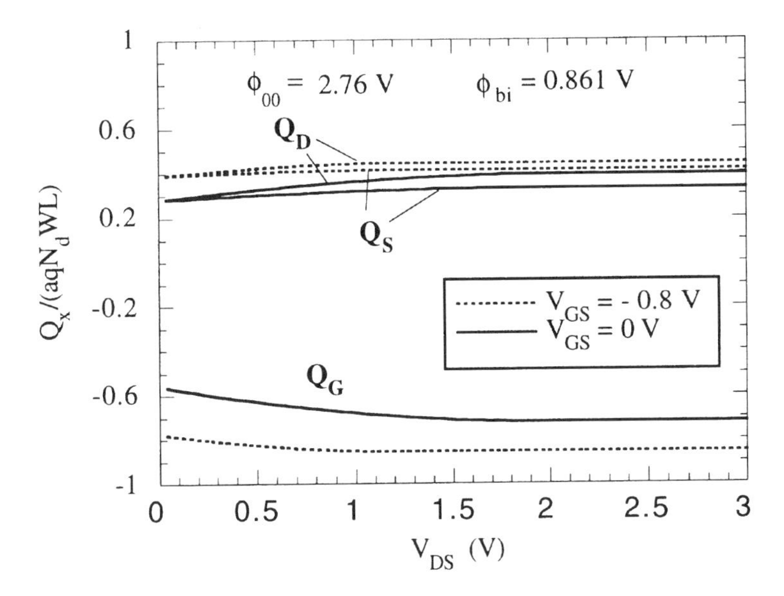

FIGURE 6-10. Gate, drain, and source charges of a MESFET as a function of V_{DS}.

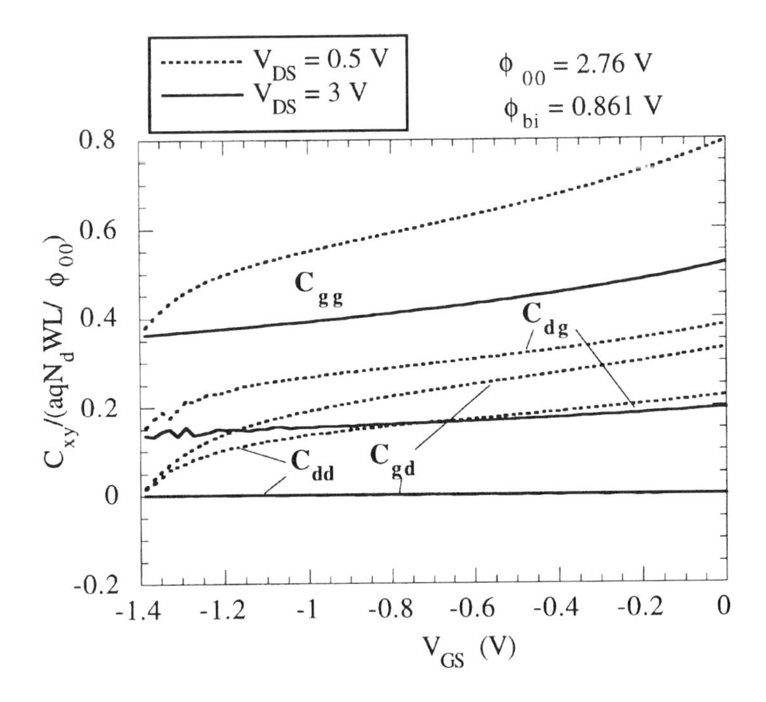

FIGURE 6-11. C_{gg}, C_{gd}, C_{dg}, and C_{dd} of a MESFET as a function of V_{GS}.

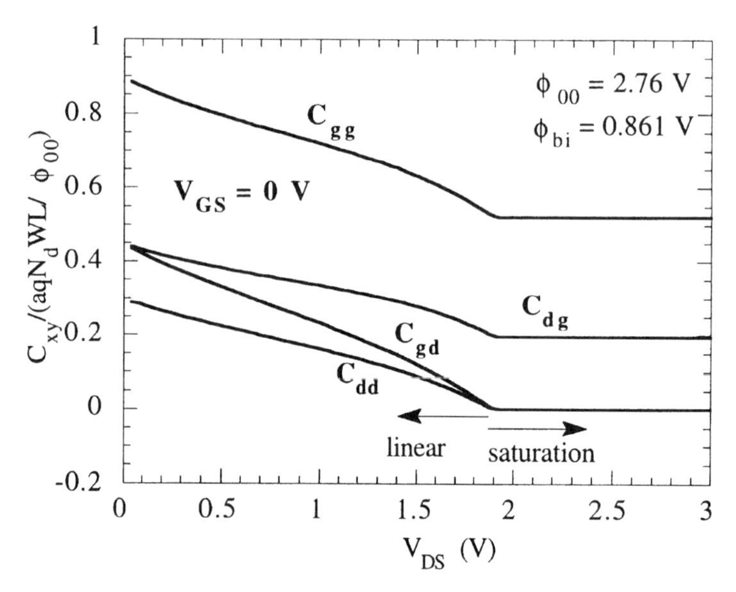

FIGURE 6-12. C_{gg}, C_{gd}, C_{dg}, and C_{dd} of a MESFET as a function of V_{DS}.

The quasi-static analysis of HFET finds that the partition of the channel charge into the drain and source charges follows the 40/60 scheme. The drain charge is 40% of the total channel charge during saturation, and the source charge is 60%. (In the linear region, the charge partition no longer follows the 40/60 ratio, however.) The same quasi-static analysis, based on the first-moment technique, leads to the Q_D of Eq. (6-44) for the MESFET. However, in a MESFET during saturation, the percentage of Q_D relative to Q_{CH} (or $-Q_G$) depends on the value of s. The percentage varies when s varies between its extreme values of 0 and 1. Because the ratio varies with the bias condition at the source, we may not simply call the partitioning scheme in a MESFET a 40/60 scheme, for example.

Example 6-2:

Find the drain and source charges for the MESFET described in Example 5-2. The transistor has a $W = 500$ μm and a L of 0.5 μm.

According to Example 5-2, the MESFET's channel layer has a thickness $a = 2000$ Å and a doping $N_d = 10^{17}$ cm^{-3}. The parameters d and s were found to be 1 and 0.312, respectively. According to Eq. (6-47):

$$Q_D = 1.6 \times 10^{-19} \cdot 2000 \times 10^{-8} \cdot 10^{17} \cdot 500 \times 0.5 \times 10^{-8}$$
$$\times \frac{\frac{2}{5} \cdot 1^{5/2} + \frac{4}{9} \cdot 1^{3/2} \cdot 0.312^{3/2} - \frac{2}{3} \cdot 1^{3/2} \cdot 0.312 + \cdots + \frac{1}{7} \cdot 0.312^{7/2}}{(1 - 0.312 - \frac{2}{3} \cdot 1^{3/2} + \frac{2}{3} \cdot 0.312^{3/2})^2}$$
$$= 3.31 \times 10^{-13} \text{ C}$$

The gate charge is calculated from Eq. (6-39):

$$Q_G = -1.6\times 10^{-19}\cdot 2000\times 10^{-8}\cdot 10^{17}\cdot 500\times 0.5\times 10^{-8}$$
$$\times \frac{\frac{2}{3}\cdot 1^{3/2} - \frac{2}{3}\cdot 0.312^{3/2} - \frac{2}{3}\cdot 1^2 + \frac{2}{3}\cdot 0.312^2}{1 - 0.312 - \frac{2}{3}\cdot 1^{3/2} + \frac{2}{3}\cdot 0.312^{3/2}} = -5.77\times 10^{-13}\ \text{C}$$

Q_S is equal to $-Q_G - Q_D$. So $Q_S = 2.46\times 10^{-13}$ coulomb. The ratio of Q_D/Q_S is 54/46.

We determine the quasi-static transit time for the MESFET, based on the definition given in Eq. (6-32). With the gate charge given by Eq. (6-39) and the d.c. drain current given in Eq. (5-40), we find that:

$$\tau_{\text{tr, QS}} = \frac{L^2}{\mu_n \phi_{00}} \times \left[\frac{\frac{2}{3}d^{3/2} - \frac{2}{3}s^{3/2} - \frac{1}{2}d^2 + \frac{1}{2}s^2}{(d - s - \frac{2}{3}d^{3/2} + \frac{2}{3}s^{3/2})^2}\right] \tag{6-53}$$

The term inside the square brackets is plotted as a function of d, for two different values of s in Fig. 6-13. A high value of s means that the applied V_{GS} is fairly negative. Hence, a smaller value of s means that the MESFET is more heavily turn-on and the conduction current is higher than that achieved with a smaller value of s. The shorter transit time associated with such a situation demonstrates the need to drive the d.c. current to its maximum value to achieve its maximum high-speed performance. In addition, the variation of the transit time with

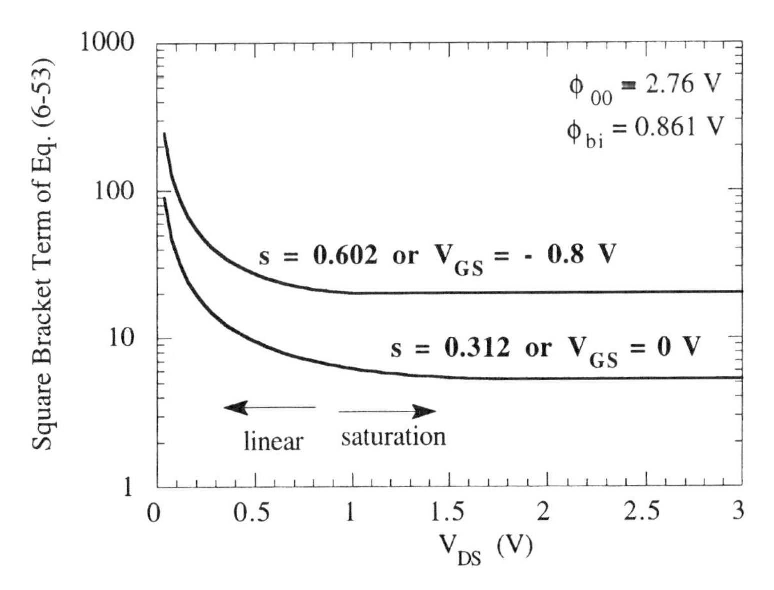

FIGURE 6-13. Quasi-static transit time of a MESFET [square-bracket term of Eq. (6-53)] as a function of V_{DS}.

respect to d shows that the MESFET should be biased in the saturation region to minimize the transistor transit time.

We summarize the expression of the terminal currents in FETs, for both MESFETs and HFETs:

$$i_G(t) = \frac{dQ_G(v_G, v_D, v_S)}{dt} = \frac{\partial Q_G}{\partial v_G} \cdot \frac{dv_G}{dt} + \frac{\partial Q_G}{\partial v_D} \cdot \frac{dv_D}{dt} + \frac{\partial Q_G}{\partial v_S} \cdot \frac{dv_S}{dt} \tag{6-54}$$

$$i_D(t) = I_D + \frac{dQ_D(v_G, v_D, v_S)}{dt} = I_D + \frac{\partial Q_D}{\partial v_G} \cdot \frac{dv_G}{dt} + \frac{\partial Q_D}{\partial v_D} \cdot \frac{dv_D}{dt} + \frac{\partial Q_D}{\partial v_S} \cdot \frac{dv_S}{dt} \tag{6-55}$$

Most often, FETs are biased in the common-source configuration, in which $dv_S/dt = 0$, $dv_G/dt = dv_{GS}/dt$, and $dv_D/dt = dv_{DS}/dt$. Therefore:

$$i_G(t) = C_{gg}(v_{GS}, v_{DS}) \cdot \frac{dv_{GS}}{dt} - C_{gd}(v_{GS}, v_{DS}) \cdot \frac{dv_{DS}}{dt} \tag{6-56}$$

$$i_D(t) = I_D(v_{GS}, v_{DS}) - C_{dg}(v_{GS}, v_{DS}) \cdot \frac{dv_{GS}}{dt} + C_{dd}(v_{GS}, v_{DS}) \cdot \frac{dv_{DS}}{dt} \tag{6-57}$$

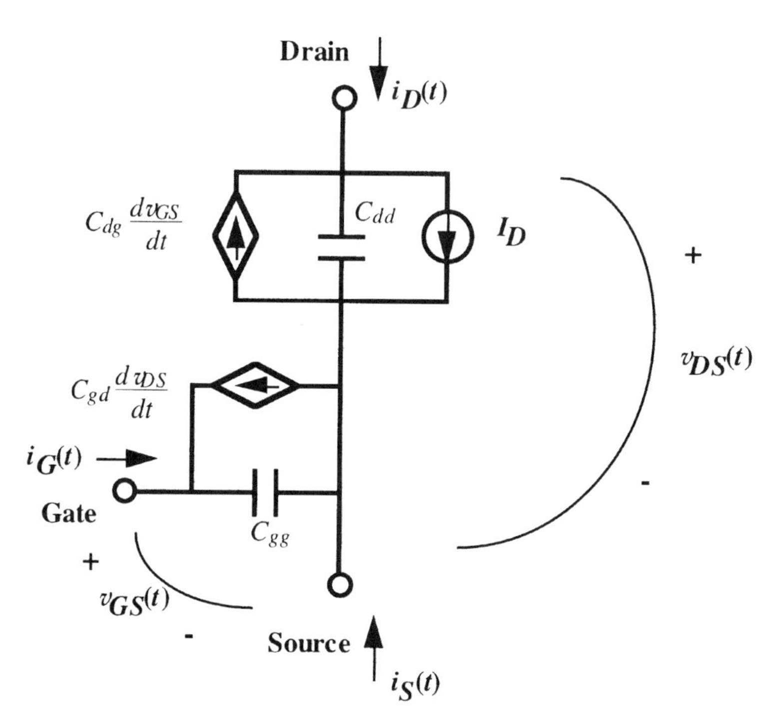

FIGURE 6-14. A large-signal quasi-static equivalent circuit for a common-source FET.

The source current can be obtained from Kirchoff's current law:

$$i_S(t) = -i_D(t) - i_G(t) \tag{6-58}$$

We have consistently followed the convention that the current flowing *into* a terminal is counted as positive.

Equations (6-56) and (6-57) are the equations governing the large-signal transients in MESFETs and HFETs, assuming that the quasi-static approximation is valid. They can be represented by an equivalent circuit shown in Fig. 6-14. In a more general case, the drift equation and the continuity equation should be solved together. The latter approach, basically the so-called non-quasi-static analysis, is necessary when the time scale involved is comparable to the intrinsic transistor delay time given in Eq. (6-53) for the MESFETs and Eq. (6-34) for the HFETs.

§ 6-3 INTRINSIC y-PARAMETERS

In digital circuits where the transistor terminal voltages toggle between two extreme values, the transistor characteristics change drastically during the transient. However, in some analog circuits, the external bias changes represent only a small perturbation to the prior operating condition. In these small-signal situations, Eqs. (6-56) and (6-57) can be simplified. One type of small-signal operation is of particular interest, in which the perturbation is sinusoidal at a frequency f. If the frequency is small, then the d/dt terms can be neglected. Equation (6-57) can be written as

$$i_D = I_D(V_{GS} + \Delta v_{gs}, V_{DS} + \Delta v_{ds}) \approx I_D(V_{GS}, V_{DS}) + \frac{\partial I_D}{\partial V_{GS}} \Delta v_{gs} + \frac{\partial I_D}{\partial V_{DS}} \Delta v_{ds} \tag{6-59}$$

The drain current increases due to the perturbations in V_{GS} and V_{DS}. The *mutual transconductance* measures the amount of current increase due to the increment in the gate bias. It was first mentioned in § 5-4 and is rewritten here for convenience:

$$g_m = \left.\frac{\partial I_D}{\partial V_{GS}}\right|_{V_{DS}=\text{const.}} \tag{6-60}$$

The *drain transconductance* measures the amount of drain current increase due to the increment in the drain bias. It is defined as

$$g_d = \left.\frac{\partial I_D}{\partial V_{DS}}\right|_{V_{GS}=\text{const.}} \tag{6-61}$$

For long-channel MESFETs in which the electrons travel at a constant mobility, the d.c. drain current expression was obtained in Eq. (5-40):

$$I_D = \frac{W}{L} aqN_d\mu_n\phi_{00}\left[d - s - \frac{2}{3}d^{3/2} + \frac{2}{3}s^{3/2}\right] \quad (6\text{-}62)$$

According to Eq. (6-41), we find g_m and g_d to be:

$$g_m = \frac{W}{L} aq\mu_n N_d[\sqrt{d} - \sqrt{s}] \qquad \text{(MESFET)} \quad (6\text{-}63)$$

$$g_d = \frac{W}{L} aq\mu_n N_d[1 - \sqrt{d}] \qquad \text{(MESFET)} \quad (6\text{-}64)$$

When the transistor enters saturation, $d = 1$. Hence $g_d = 0$. In practice, some finite value of g_d is measured.

For HFET with constant mobility, the drain current was given by Eq. (5-133), which is repeated below for convenience:

$$I_D = \frac{WC'_{ox}\mu_n}{L}\frac{(V_{GS} - V_T)^2}{2}(1 - \alpha^2) \quad (6\text{-}65)$$

The parameter α is a function of both V_{DS} and V_{GS}, given in Eq. (5-132). Knowing that $\partial\alpha/\partial V_{GS} = (1 - \alpha)/(V_{GS} - V_T)$, and that $\partial\alpha/\partial V_{DS} = -1/(V_{GS} - V_T)$, we write:

$$g_m = \frac{WC'_{ox}\mu_n}{L}(V_{GS} - V_T) \times (1 - \alpha) \qquad \text{(HFET)} \quad (6\text{-}66)$$

$$g_d = \frac{WC'_{ox}\mu_n}{L}(V_{GS} - V_T) \times \alpha \qquad \text{(HFET)} \quad (6\text{-}67)$$

We intend to simplify the large-signal equivalent circuit to one that is appropriate for the small-signal operation under consideration. We capture the essence of Eq. (6-59), rewriting $i_D(v_{GS}, v_{DS}) - I_D(V_{GS}, V_{DS})$ as

$$\Delta I_D = g_m\Delta V_{GS} + g_d\Delta V_{DS} \quad (6\text{-}68)$$

Further, since there is no d.c. current component in the gate (and we are neglecting transistor capacitances due to low frequency), we have:

$$\Delta I_G = 0 \quad (6\text{-}69)$$

Therefore, a low-frequency small-signal model is that given in Fig. 6-15. This small-signal model is meant to track only the small-signal currents in the

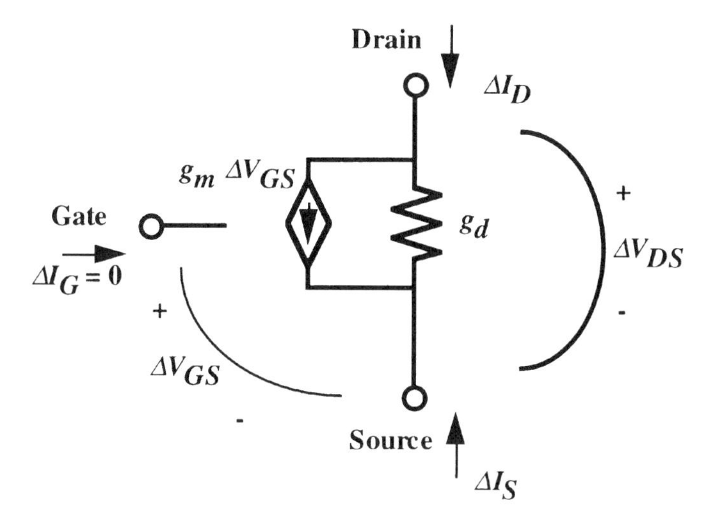

FIGURE 6-15. A small-signal, low-frequency equivalent circuit for a common-source FET.

terminals when the small-signal voltages are applied. It does not include the d.c. currents as in the large-signal equivalent circuit of Fig. 6-14.

When the small-signal voltages applied to a field-effect transistor vary rapidly with time, there are capacitive currents, as indicated in Eqs. (6-54) and (6-55), in addition to the currents indicated in Fig. 6-15. In a three-terminal FET structure, there are a total of nine capacitances, although only four are linearly independent. Each of the nine capacitances can include a parasitic component, in addition to the component associated with the intrinsic device. Here, we consider only the capacitances intrinsic to the transistor. Parasitic capacitances such as the fringing capacitances will be discussed in § 6-6.

Let us consider the intrinsic portion of a common-source MESFET shown in Fig. 6-16. (The figure applies to MESFETs. For HFETs, the charge signs are reversed.) As shown in Fig. 6-16*a*, the d.c. bias voltages V_G, V_D, and V_S are applied across the gate, drain, and source terminals, respectively. Suppose that we place an incremental voltage across the gate as shown in Fig. 6-16*b*; there will obviously be a corresponding change in the gate charge. Because the total charge of the overall device must sum to zero, an incremental amount of gate charge induces the same amount of negative charge in the channel. Some of the negative charge appears in the drain charge term, and the rest, in the source charge term. If ΔV_G is positive, then ΔQ_G is positive while ΔQ_D and ΔQ_S are negative. There are two sign conventions in defining the capacitances. We shall adopt the convention in which C_{xy} is defined as $\delta_{xy} \times \partial Q_x/\partial V_y$, where δ_{xy} is equal to 1 if $x = y$, and -1 if otherwise. That is:

$$C_{gg} = \frac{\Delta Q_G}{\Delta V_G} \qquad C_{dg} = -\frac{\Delta Q_D}{\Delta V_G} \qquad C_{sg} = -\frac{\Delta Q_S}{\Delta V_G} \tag{6-70}$$

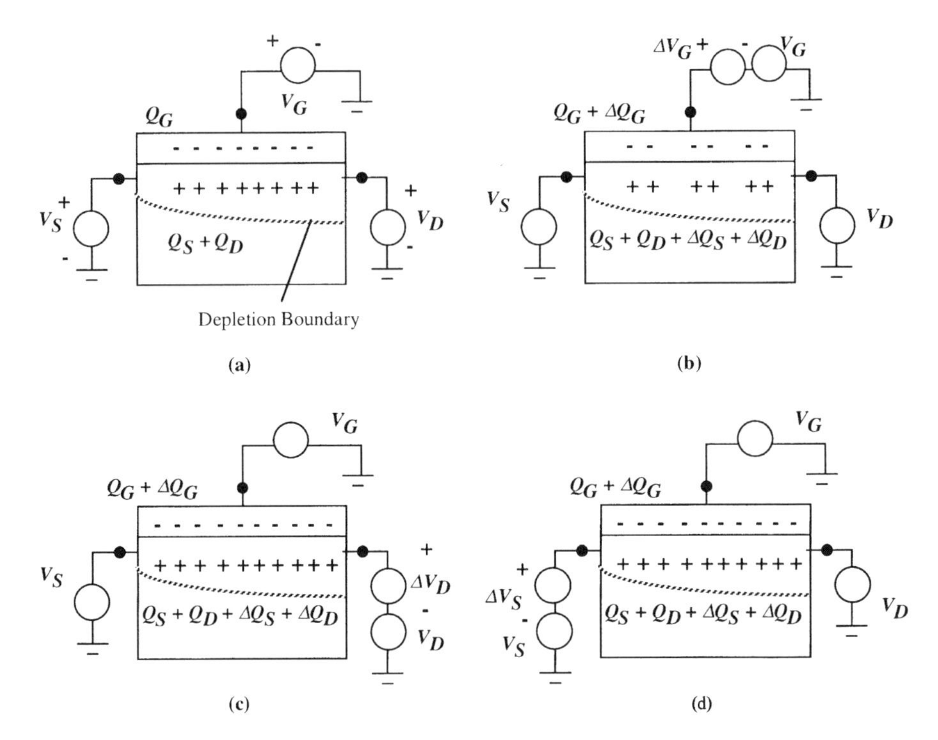

FIGURE 6-16. Determination of the capacitances in an intrinsic MESFET. (For an HFET, the charge signs are reversed.) (*a*) Transistor is biased with certain d.c. biases. (*b*) A perturbation is applied at the gate. (*c*) A perturbation is applied at the drain. (*d*) A perturbation is applied at the source.

In this convention, C_{gg}, C_{dg}, and C_{sg} are all positive quantities, in agreement with our intuitive understanding of a capacitor.

When we place a positive incremental voltage ΔV_D at the drain (Fig. 6-16*c*), there will be a positive increment of the drain charge. The source charge and the gate charge also adjust themselves in such a way that the total charge inside the device is zero. In this case, although we are confident that the incremental gate charge is negative, it is not clear what the sign of the incremental source charge is. Such a sign can be determined after we solve the device equations governing the charge transport. Following the aforementioned convention, we define the drain-related capacitances as

$$C_{gd} = -\frac{\Delta Q_G}{\Delta V_D} \qquad C_{dd} = \frac{\Delta Q_D}{\Delta V_D} \qquad C_{sd} = -\frac{\Delta Q_S}{\Delta V_D} \tag{6-71}$$

Likewise, when we place an incremental voltage ΔV_S at the source (Fig. 6-16*d*), we can determine the capacitances from the amount of charge variations:

$$C_{gs} = -\frac{\Delta Q_G}{\Delta V_S} \qquad C_{ds} = -\frac{\Delta Q_D}{\Delta V_S} \qquad C_{ss} = \frac{\Delta Q_S}{\Delta V_S} \tag{6-72}$$

Let us revisit Fig. 6-16*b*, in which a ΔV_G is applied. As mentioned, the sum of the induced charges is zero. With $\Delta Q_G + \Delta Q_D + \Delta Q_S = 0$, we establish a relationship among the gate-related capacitances:

$$C_{gg} - C_{dg} - C_{sg} = 0 \tag{6-73}$$

The same rationale is applied to Figs. 6-16*c* and 6-16*d*, resulting in the following relationships:

$$-C_{gd} + C_{dd} - C_{sd} = 0 \tag{6-74}$$

$$-C_{gs} - C_{ds} + C_{ss} = 0 \tag{6-75}$$

Further, let us consider the situation when three voltage sources of equal magnitude are connected to the three terminals, namely, $\Delta V_G = \Delta V_D = \Delta V_S = \Delta V$, as shown in Fig. 6-17. Because effectively the bias condition is the same before and after the application, $\Delta Q_G = \Delta Q_D = \Delta Q_S = 0$. Because all three

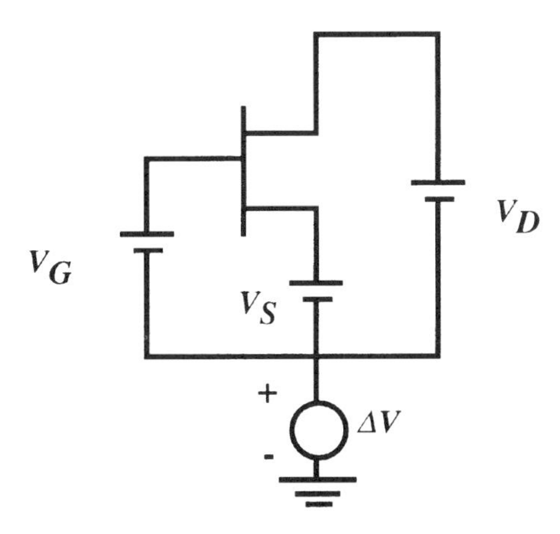

FIGURE 6-17. A perturbation of ΔV is simultaneously applied to the gate, drain, and source terminals of a FET.

terminals are subjected to the same incremental voltage, the total incremental charge ΔQ is equal to:

$$\Delta Q_G = C_{gg}\Delta V_G - C_{gd}\Delta V_D - C_{gs}\Delta V_S \tag{6-76}$$

$$\Delta Q_D = -\ C_{dg}\Delta V_G + C_{dd}\Delta V_D - C_{ds}\Delta V_S \tag{6-77}$$

$$\Delta Q_S = -\ C_{sg}\Delta V_G - C_{sd}\Delta V_D + C_{ss}\Delta V_S \tag{6-78}$$

Since we have said that $\Delta Q_G = \Delta Q_D = \Delta Q_S = 0$ and that $\Delta V_G = \Delta V_D = \Delta V_S = \Delta V$, we conclude that:

$$C_{gg} - C_{gd} - C_{gs} = 0 \tag{6-79}$$

$$-\ C_{dg} + C_{dd} - C_{ds} = 0 \tag{6-80}$$

$$-\ C_{sg} - C_{sd} + C_{ss} = 0 \tag{6-81}$$

These three relationships, together with those given by Eqs. (6-73) to (6-75), allow all of the nine capacitances to be known once a linearly independent set of four capacitances is known.

We applied the definitions of Eqs. (6-70) and (6-71) to find C_{gg}, C_{gd}, C_{dg}, and C_{dd} of MESFETs and HFETs in the previous two sections. They were given in Eqs. (6-16), (6-17), (6-27), and (6-28) for HFETs, and in (6-42), (6-43), (6-48), and (6-51) for MESFETs. For completeness, we express the five remaining capacitances for each device. For HFETs, we write:

$$C_{gs} = WLC'_{ox}\left[\frac{2}{3}\frac{2\alpha + 1}{(1+\alpha)^2}\right] \tag{6-82}$$

$$C_{ds} = WLC'_{ox}\left[-\frac{4}{15}\frac{\alpha^2 + 3\alpha + 1}{(1+\alpha)^3}\right] \tag{6-83}$$

$$C_{sg} = WLC'_{ox}\left[\frac{1}{15}\frac{4\alpha^3 + 28\alpha^2 + 22\alpha + 6}{(1+\alpha)^3}\right] \tag{6-84}$$

$$C_{sd} = WLC'_{ox}\left[-\frac{4}{15}\frac{\alpha^3 + 3\alpha^2 + \alpha}{(1+\alpha)^3}\right] \tag{6-85}$$

$$C_{ss} = WLC'_{ox}\left[\frac{1}{15}\frac{16\alpha^2 + 18\alpha + 6}{(1+\alpha)^3}\right] \tag{6-86}$$

The saturation index α was defined in Eq. (5-132). For MESFETs, the expressions are more complicated:

$$C_{gs} = \frac{qWLN_d a}{\phi_{00}} \left[-\frac{s^{1/2}(1 - s^{1/2})}{d - s - \frac{2}{3}d^{3/2} + \frac{2}{3}s^{3/2}} + \frac{(1 - s^{1/2})(\frac{2}{3}d^{3/2} - \frac{2}{3}s^{3/2} - \frac{1}{2}d^2 + \frac{1}{2}s^2)}{(d - s - \frac{2}{3}d^{3/2} + \frac{2}{3}s^{3/2})^2} \right] \quad (6\text{-}87)$$

$$C_{ds} = \frac{aqN_d WL}{\phi_{00}} \times \left[\frac{f_4(d, s)}{(d - s - \frac{2}{3}d^{3/2} + \frac{2}{3}s^{3/2})^2} - \frac{2(1 - s^{1/2})\, f_2(d, s)}{(d - s - \frac{2}{3}d^{3/2} + \frac{2}{3}s^{3/2})^3} \right] \quad (6\text{-}88)$$

$$C_{ss} = \frac{aqN_d WL}{\phi_{00}} \times \left[\frac{f_5(d, s)}{(d - s - \frac{2}{3}d^{3/2} + \frac{2}{3}s^{3/2})^2} + \frac{2(1 - s^{1/2}) f_6(d, s)}{(d - s - \frac{2}{3}d^{3/2} + \frac{2}{3}s^{3/2})^3} \right] \quad (6\text{-}89)$$

$$C_{sd} = \frac{aqN_d WL}{\phi_{00}} \times \left[\frac{f_7(d, s)}{(d - s - \frac{2}{3}d^{3/2} + \frac{2}{3}s^{3/2})^2} + \frac{2(1 - d^{1/2}) f_6(d, s)}{(d - s - \frac{2}{3}d^{3/2} + \frac{2}{3}s^{3/2})^3} \right] \quad (6\text{-}90)$$

$$C_{sg} = \frac{aqN_d WL}{\phi_{00}} \times \left[\frac{f_5(d, s) - f_7(d, s)}{(d - s - \frac{2}{3}d^{3/2} + \frac{2}{3}s^{3/2})^2} + \frac{2(d^{1/2} - s^{1/2}) f_6(d, s)}{(d - s - \frac{2}{3}d^{3/2} + \frac{2}{3}s^{3/2})^3} \right] \quad (6\text{-}91)$$

where d and s were given in Eq. (5-33), and

$$f_4(d, s) = f_1(d, s) - f_3(d, s) = \frac{2}{3} d^{3/2} - \frac{1}{2} d^2 - \frac{2}{3} d^{3/2} s^{1/2} + \frac{1}{2} d^2 s^{1/2} - \frac{2}{3} s^{3/2} + \frac{7}{6} s^2 - \frac{1}{2} s^{5/2} \quad (6\text{-}92)$$

$$f_5(d, s) = f_4(d, s) - s^{1/2}\left(1 - s^{1/2}\right) \times \left(d - s - \frac{2}{3} d^{3/2} + \frac{2}{3} s^{3/2}\right) \quad (6\text{-}93)$$

$$f_6(d, s) = \frac{1}{2}\left(\frac{2}{3} d^{3/2} - \frac{2}{3} s^{3/2} - \frac{1}{2} d^2 + \frac{1}{2} s^2\right) \times \left(d - s - \frac{2}{3} d^{3/2} + \frac{2}{3} s^{3/2}\right) - f_2(d, s) \quad (6\text{-}94)$$

$$f_7(d, s) = f_1(d, s) - d^{1/2}(1 - d^{1/2}) \times \left(d - s - \frac{2}{3} d^{3/2} + \frac{2}{3} s^{3/2}\right) \quad (6\text{-}95)$$

Four of the capacitances were shown in Figs. 6-6 and 6-7 for HFETs, and in Figs. 6-11 and 6-12 for MESFETs. We plot the remaining five capacitances for HFETs, as a function of V_{GS} in Fig. 6-18 and of V_{DS} in Fig. 6-19. Similar capacitances for MESFETs are found in Figs. 6-20 and 6-21. As V_{DS} approaches zero, $C_{gs} = C_{gd}$ and $C_{dg} = C_{sg}$. This is expected because the device becomes symmetrical at that bias condition. In saturation, both devices' C_{gd} become zero, since the gate charge becomes insensitive to the drain bias variation as the channel is pinched off.

The transcapacitances C_{sd} and C_{ds} are negative. While it is impossible to have negative capacitances in a regular two-terminal device, it can occur in a three-terminal device. The negative value of C_{ds} simply means that the drain charge Q_D increases in value in response to an increase in the source voltage, while other terminal voltages are maintained constant. Experimental results have confirmed that indeed these two transcapacitances are negative.

We extend the validity of the equivalent circuit of Fig. 6-15 to higher frequencies by incorporating the newly found capacitances. We focus on the small-signal terminal currents $i_d = i_D - I_D$ and $i_g = i_G - I_G$. Starting with Eqs. (6-56) and (6-57) and applying the results of Eqs. (6-68) and (6-69), we write:

$$i_g(t) = C_{gg}\frac{dv_{gs}}{dt} - C_{gd}\frac{dv_{ds}}{dt} \tag{6-96}$$

$$i_d(t) = g_m v_{gs} + g_d v_{ds} - C_{dg}\frac{dv_{gs}}{dt} + C_{dd}\frac{dv_{ds}}{dt} \tag{6-97}$$

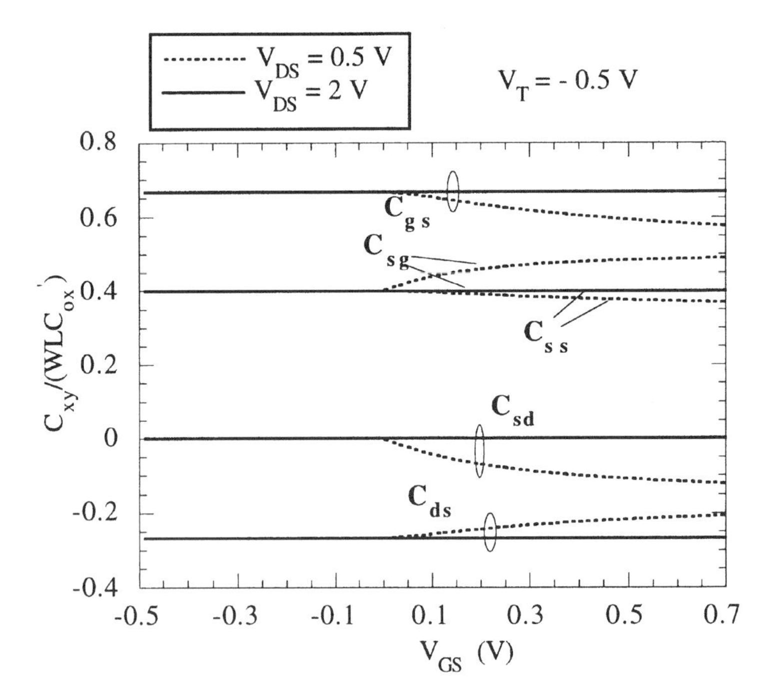

FIGURE 6-18. C_{gs}, C_{ds}, C_{sg}, C_{sd}, and C_{ss} of a HFET as a function of V_{GS}.

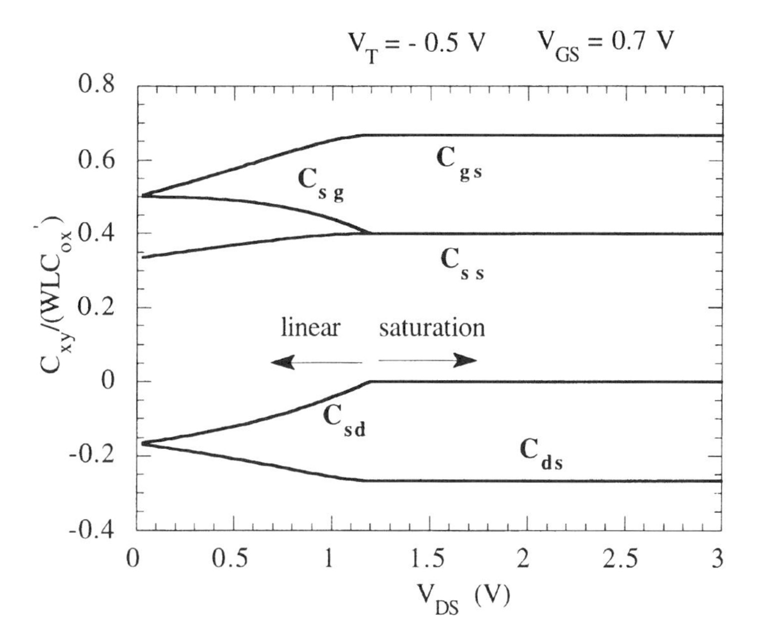

FIGURE 6-19. C_{gs}, C_{ds}, C_{sg}, C_{sd}, and C_{ss} of a HFET as a function of V_{DS}.

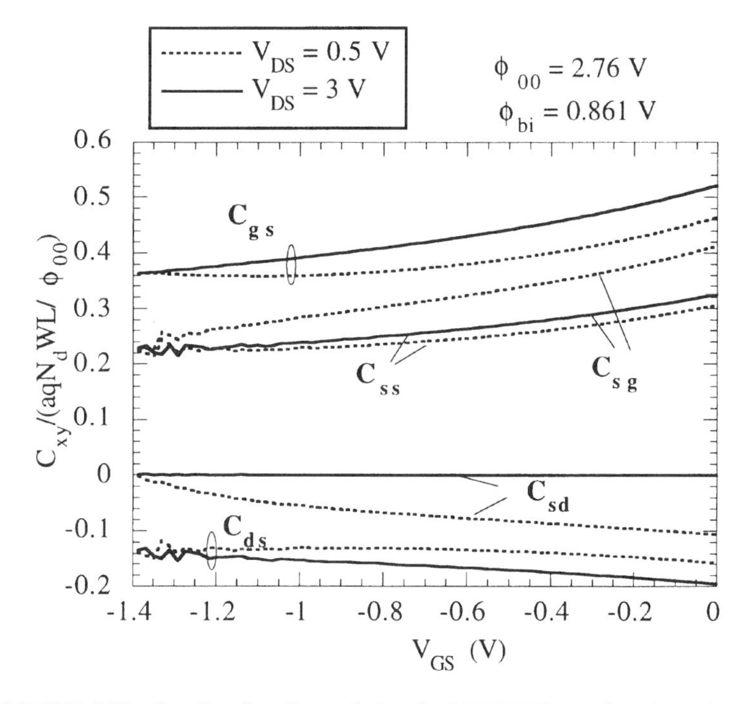

FIGURE 6-20. C_{gs}, C_{ds}, C_{sg}, C_{sd}, and C_{ss} of a MESFET as a function of V_{GS}.

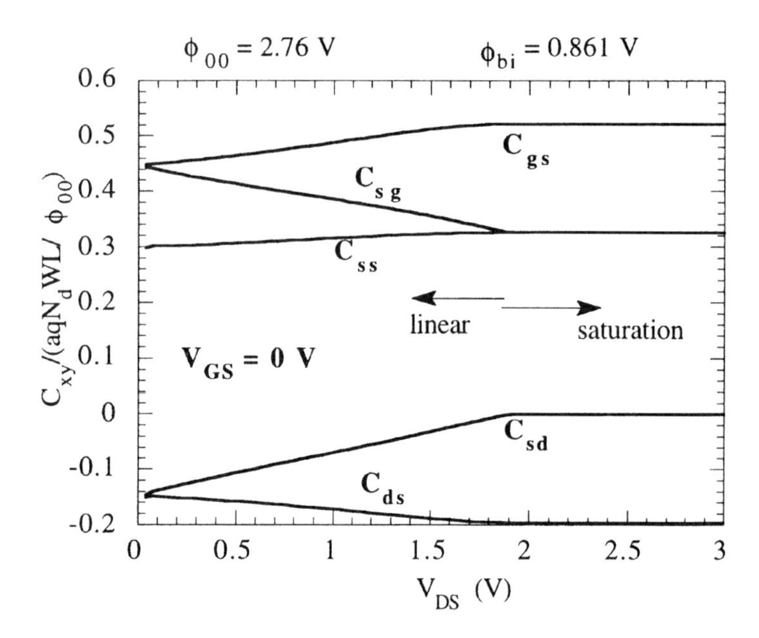

FIGURE 6-21. C_{gs}, C_{ds}, C_{sg}, C_{sd}, and C_{ss} of a MESFET as a function of V_{DS}.

Based on these two equations, the small-signal model shown in Fig. 6-22 is obtained. (This model differs from Fig. 6-14 in that the latter is a large-signal model.) The equivalent circuit applies to FETs in the common-source configuration, because the voltages of Eqs. (6-96) and (6-97) are made with reference to the source. With all voltages referenced to the source, the model is somewhat cumbersome to use. It is desirable, for example, for the majority of the elements in i_d to be made with respect to v_{dg} and v_{ds}, rather than v_{gs}. Similarly, for the gate current, we prefer all elements to be with respect to v_{gs} and v_{gd}, rather than v_{ds}. With these preferences, we rewrite Eqs. (6-96) and (6-97) to reflect the desired modifications:

$$i_g(t) = C_{gg}\frac{dv_{gs}}{dt} - C_{gd}\left(\frac{dv_{gs}}{dt} - \frac{dv_{gd}}{dt}\right) = C_{gs}\frac{dv_{gs}}{dt} + C_{gd}\frac{dv_{gd}}{dt} \tag{6-98}$$

$$\begin{aligned} i_d(t) &= g_m v_{gs} + g_d v_{ds} - C_{dg}\frac{dv_{gs}}{dt} + (C_{sd} + C_{gd})\frac{dv_{ds}}{dt} \\ &= g_m v_{gs} + g_d v_{ds} + C_{sd}\frac{dv_{ds}}{dt} - C_{dg}\frac{dv_{gs}}{dt} + C_{gd}\left(\frac{dv_{gs}}{dt} - \frac{dv_{gd}}{dt}\right) \\ &= g_m v_{gs} + g_d v_{ds} + C_{gd}\frac{dv_{dg}}{dt} + C_{sd}\frac{dv_{ds}}{dt} - C_m\frac{dv_{gs}}{dt} \end{aligned} \tag{6-99}$$

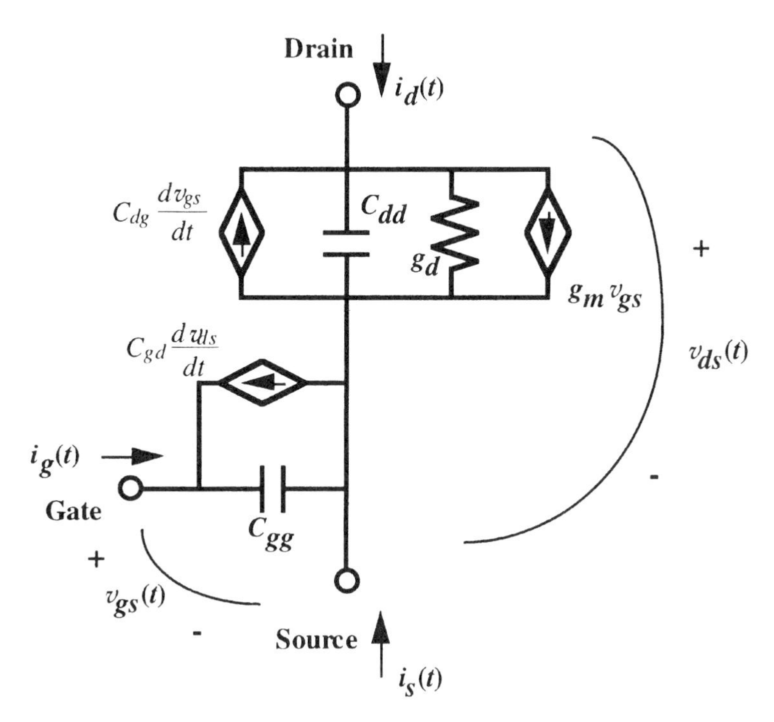

FIGURE 6-22. A small-signal, high-frequency equivalent circuit for a common-source FET. The model is quasi-static.

where

$$C_m \equiv C_{dg} - C_{gd} \tag{6-100}$$

C_m, termed the mutual differential capacitance, is a positive quantity. Based on Eqs. (6-99) and (6-100), we then obtain the small-signal model given by Fig. 6-23.

The equivalent circuit of Fig. 6-23, based ultimately on Eqs. (6-56) and (6-57), implicitly employs the quasi-static assumption. We mentioned that, in the time domain, the quasi-static assumption breaks down when the time scale of interest is smaller than the intrinsic transistor time, $\tau_{\mathrm{tr,QS}}$. We can infer that, in the frequency domain, the quasi-static assumption becomes problematic when the frequency of the small-signal variation exceeds the inverse of the transit time. Although it is not obvious at this moment, the quasi-static assumption embodied in the equivalent circuit results in another deficiency in the model (besides the limitation to a certain frequency). Namely, the circuit of Fig. 6-23 omits a circuit element—the channel resistance, which can be important when the transistor is not heavily turned on. In order to improve the model accuracy (or at least know the error) at high frequencies and to uncover the origin of the channel resistance, we have to drop the quasi-static assumption from the very beginning. We need to determine the y-parameters of the transistor, starting

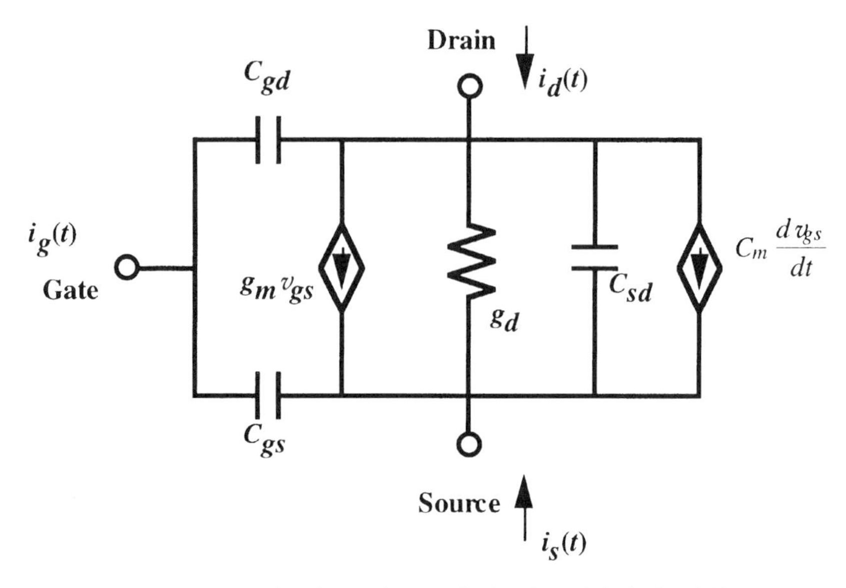

FIGURE 6-23. An alternative small-signal model of Fig. 6-22.

from the two fundamental FET equations: the drift equation and the continuity equation.

Consider a three-terminal FET whose operating point is V_D, V_G, and V_S. On top of these biasing voltages are small-signal biases which are sinusoidal at a radian frequency of ω (equal to$2\pi f$):

$$v_D(t) = V_D + \tilde{v}_d e^{j\omega t} \qquad v_G(t) = V_G + \tilde{v}_g e^{j\omega t} \qquad v_S(t) = V_S + \tilde{v}_s e^{j\omega t} \tag{6-101}$$

Each of the resultant terminal currents consists of a quiescent d.c. value and a small-signal value varying at the same frequency ω (I_S, of course, is equal to $-I_D$):

$$i_D(t) = I_D + \tilde{i}_d e^{j\omega t} \qquad i_G(t) = \tilde{i}_g e^{j\omega t} \qquad i_S(t) = I_S + \tilde{i}_s e^{j\omega t} \tag{6-102}$$

We digress briefly to clarify a point. In § 6-1 and § 6-2, we found that combining the drift and continuity equations forms a second-order nonlinear partial differential equation. We stated that the differential equation has no analytical solution. However, our focus in this section is small-signal operation. When the bias voltages in the time domain can be expressed in a sinusoidal form such as that in Eq. (6-101), then the time dimensionality of the problem can be easily taken care of. Therefore, in small-signal operation, it becomes possible to find the exact solution to the second-order partial differential equation. We are indeed interested in finding the exact solution, and hence, the y-parameters.

From the y-parameters, we can construct an equivalent circuit free of the problems associated with the quasi-static assumption.

A formal definition of the y-parameter is:

$$y_{xy} = \left.\frac{\partial \tilde{i}_x}{\partial \tilde{v}_y}\right|_{\text{node voltages other than } \tilde{v}_y \text{ are set to zero}} \tag{6-103}$$

where x and y refer to any of the for transistor terminals: gate (g), drain (d), and source (s). For example, y_{dg} denotes the ratio of the small-signal current at the drain node with respect to a small-signal voltage at the gate, while the drain and source small-signal voltages are maintained at zero. With this y-parameter definition, we express the terminal small-signal currents as a function of the small-signal bias voltages in a compact form:

$$\begin{bmatrix} \tilde{i}_g \\ \tilde{i}_d \\ \tilde{i}_s \end{bmatrix} = \begin{bmatrix} y_{gg} & y_{gd} & y_{gs} \\ y_{dg} & y_{dd} & y_{ds} \\ y_{sg} & y_{sd} & y_{ss} \end{bmatrix} \begin{bmatrix} \tilde{v}_g \\ \tilde{v}_d \\ \tilde{v}_s \end{bmatrix} \tag{6-104}$$

There are a total of nine y-parameters. Just like the capacitances, only four of them are linearly independent. The remaining five y-parameters can be generated once four parameters are known. Similar to Eqs. (4-73) and (4-75) for the HBTs, we can write:

$$\begin{aligned} &y_{gg} + y_{gd} + y_{gs} = 0 \qquad y_{dg} + y_{dd} + y_{ds} = 0 \qquad y_{sg} + y_{sd} + y_{ss} = 0 \\ &y_{gg} + y_{dg} + y_{sg} = 0 \qquad y_{gd} + y_{dd} + y_{sd} = 0 \qquad y_{gs} + y_{ds} + y_{ss} = 0 \end{aligned} \tag{6-105}$$

In a common-source configuration, all voltages are referenced to the source, which is grounded. The biasing voltages consist of quiescent values as well as the small-signal perturbations. These voltages result in terminal currents which also consist of quiescent values and the small-signal variations. The voltages and currents for a common-source transistor are expressed as:

$$v_{DS}(t) = V_{DS} + \tilde{v}_{ds}e^{j\omega t} \qquad v_{GS}(t) = V_{GS} + \tilde{v}_{gs}e^{j\omega t} \tag{6-106}$$

$$i_D(t) = I_D + \tilde{i}_d e^{j\omega t} \qquad i_G(t) = \tilde{i}_g e^{j\omega t} \tag{6-107}$$

With the gate terminal being port 1, and drain terminal, port 2, the small-signal relationships between the port voltages and currents are:

$$\begin{bmatrix} \tilde{i}_g \\ \tilde{i}_d \end{bmatrix} = \begin{bmatrix} y_{11} & y_{12} \\ y_{21} & y_{22} \end{bmatrix}_s \begin{bmatrix} \tilde{v}_{gs} \\ \tilde{v}_{ds} \end{bmatrix} \tag{6-108}$$

The subscript s is the y-matrix suggests that it applies to the common-source configuration. There are only four y-parameters characterizing a two-port, as opposed to nine y-parameters characterizing a three-port. Comparing Eq. (6-108) with Eq. (6-104), we find the common-source y-parameters to be:

$$\begin{bmatrix} y_{11} & y_{12} \\ y_{21} & y_{22} \end{bmatrix}_s = \begin{bmatrix} y_{gg} & y_{gd} \\ y_{dg} & y_{dd} \end{bmatrix} \tag{6-109}$$

Likewise, in a common-gate configuration in which the source terminal is port 1 and the drain is port 2, the small-signal currents and voltages are related by the following y-parameter matrix:

$$\begin{bmatrix} \tilde{i}_s \\ \tilde{i}_d \end{bmatrix} = \begin{bmatrix} y_{11} & y_{12} \\ y_{21} & y_{22} \end{bmatrix}_g \begin{bmatrix} \tilde{v}_{sg} \\ \tilde{v}_{dg} \end{bmatrix} \tag{6-110}$$

Comparing the above equation to Eq. (6-104), we find the common-gate y-parameters to be:

$$\begin{bmatrix} y_{11} & y_{12} \\ y_{21} & y_{22} \end{bmatrix}_g = \begin{bmatrix} y_{ss} & y_{sd} \\ y_{ds} & y_{dd} \end{bmatrix} \tag{6-111}$$

We briefly go through the procedure to determine the common-source y-parameters of MESFETs in the following. Once the four common-source y-parameters are established, all the nine y-parameters can be identified, through the application of Eq. (6-105). Specifically, the common-gate y-parameters can be realized from the common-source parameters, given by:

$$y_{ss} = y_{gg} + y_{gd} + y_{dg} + y_{dd} \tag{6-112}$$

$$y_{sd} = -y_{gd} - y_{dd} \tag{6-113}$$

$$y_{ds} = -y_{dg} - y_{dd} \tag{6-114}$$

$$y_{dd} = y_{dd} \qquad \text{(unchanged)} \tag{6-115}$$

MESFET is used as the exemplar device to demonstrate the procedure for finding y-parameters. The procedure for finding the y-parameters in HFETs is similar. However, due to the difference in the governing equations (notably the drift equation), the actual implementation of the solution technique is slightly different. The relevant equations for MESFETs are described here, and subsequently, in Appendix A. The equations for HFETs are found in Appendix B.

To find the y-parameters in MESFETs, we start with the two fundamental equations governing MESFETS: the drift [Eq. (5-23)] and the continuity equation [Eq. (5-28)]. For convenience, they are repeated in the following:

$$i_{\mathrm{CH}}(x, t) = qWN_d\mu_n a\left(1 - \sqrt{\frac{\phi_s(x, t)}{\phi_{00}}}\right)\left(-\frac{\partial\phi_s(x, t)}{\partial x}\right) \tag{6-116}$$

$$\frac{\partial i_{\mathrm{CH}}(x)}{\partial x} = aqWN_d\frac{\partial}{\partial t}\sqrt{\frac{\phi_s(x, t)}{\phi_{00}}} \tag{6-117}$$

Just as the currents and voltages which are composed of quiescent values and small-signal perturbations, we can write the surface potential and channel current appearing in the above two equations as

$$\phi_s(x, t) = \phi_{s,\mathrm{dc}}(x) + \tilde{\phi}_{s,\mathrm{ac}}(x)\, e^{j\omega t} \tag{6-118}$$

$$i_{\mathrm{CH}}(x, t) = I_{\mathrm{CH}}(x) + \tilde{i}_{\mathrm{ch}}(x)\, e^{j\omega t} \tag{6-119}$$

Since $\phi_{s,\mathrm{ac}}$, the small-signal perturbation, is much smaller than $\phi_{s,\mathrm{dc}}$, the quiescent value, we can use the Taylor series expansion to approximate an arbitrary function, such as $\phi_s(x, t)$:

$$f\left[\phi_{s,\mathrm{dc}}(x) + \tilde{\phi}_{s,\mathrm{ac}}(x)\, e^{i\omega t}\right] \approx f\left[\psi_{s,\mathrm{dc}}(x)\right] + \left.\frac{\partial f}{\partial\phi_s}\right|_{\phi_s=\phi_{s,\mathrm{dc}}} \times \tilde{\phi}_{s,\mathrm{ac}}(x)\, e^{j\omega t} \tag{6-120}$$

For example, the Taylor expansion for the square root of $\phi_s(x, t)/\phi_{00}$ is:

$$\sqrt{\frac{\phi_{s,\mathrm{dc}}(x) + \tilde{\phi}_{s,\mathrm{ac}}(x)\, e^{j\omega t}}{\phi_{00}}} \approx \sqrt{\frac{\phi_{s,\mathrm{dc}}(x)}{\phi_{00}}} + \frac{1}{2\sqrt{\phi_{00}}\sqrt{\phi_{s,\mathrm{dc}}(x)}} \times \tilde{\phi}_{s,\mathrm{ac}}(x)\, e^{j\omega t} \tag{6-121}$$

We substitute Eqs. (6-118) and (6-119) into Eq. (6-117), the continuity equation. After applying the Taylor approximation and equating the terms which are multiplied with $e^{j\omega t}$, we acquire a differential equation governing the small-signal channel current:

$$\frac{d\tilde{i}_{\mathrm{ch}}}{dx} = -j\omega aq\, WN_d \times \frac{1}{2\sqrt{\phi_{00}}\sqrt{\phi_{s,\mathrm{dc}}(x)}}\tilde{\phi}_{s,\mathrm{ac}}(x) \tag{6-122}$$

We follow the same approach to derive a small-signal equation from the drift equation. The terms containing $e^{j\omega t}$ are given by:

$$\tilde{i}_{\mathrm{ch}} e^{j\omega t} = aqWN_d\mu_n \frac{d}{d\phi_s}\left(1 - \sqrt{\frac{\phi_s}{\phi_{00}}}\right)\Bigg|_{\phi_s=\phi_{s,\mathrm{dc}}} \tilde{\phi}_{s,\mathrm{ac}} e^{j\omega t} \times \left(-\frac{d\phi_{s,\mathrm{dc}}}{dx}\right)$$
$$+ aqWN_d\mu_n\left(1 - \sqrt{\frac{\phi_{s,\mathrm{dc}}}{\phi_{00}}}\right) \times \left(-\frac{d\tilde{\phi}_{s,\mathrm{ac}}}{dx} e^{j\omega t}\right) \quad (6\text{-}123)$$

This equation is simplified to:

$$\tilde{i}_{\mathrm{ch}} = -aqWN_d\mu_n \frac{d}{dx}\left[\left(1 - \sqrt{\frac{\phi_{s,\mathrm{dc}}(x)}{\phi_{00}}}\right)\tilde{\phi}_{s,\mathrm{ac}}(x)\right] \quad (6\text{-}124)$$

Equations (6-122) and (6-124) are combined to form a single differential equation, which governs the small-signal properties of the MESFET:

$$\mu_n \frac{d^2}{dx^2}\left[\left(1 - \sqrt{\frac{\phi_{s,\mathrm{dc}}(x)}{\phi_{00}}}\right)\tilde{\phi}_{s,\mathrm{ac}}(x)\right] = \frac{j\omega}{2\sqrt{\phi_{00}}\sqrt{\phi_{s,\mathrm{dc}}(x)}}\tilde{\phi}_{s,\mathrm{ac}}(x) \quad (6\text{-}125)$$

$\phi_{s,\mathrm{dc}}(x)$ is a known function once the quiescent voltage V_{GS} and V_{DS} are specified (see Eqs. 5-42 and 5-33). It is a complicated function of x, but is nonetheless a known function. The unknown in the above differential equation is $\phi_{s,\mathrm{ac}}(x)$. The boundary conditions for $\phi_{s,\mathrm{ac}}(x)$ are determined by noting that

$$\phi_s(x,t) = \phi_{\mathrm{bi}} - v_{GS}(t) + v_{CS}(x,t) = [\phi_{\mathrm{bi}} - V_{GS} + V_{CS}(x)] - v_{gs}(t) + v_{cs}(x,t) \quad (6\text{-}126)$$

Dropping out the $e^{j\omega t}$ multiplier in all of the time-dependent terms, we obtain:

$$\tilde{\phi}_{s,\mathrm{ac}}(0) = -\tilde{v}_{gs} \qquad \tilde{\phi}_{s,\mathrm{ac}}(L) = -\tilde{v}_{gs} + \tilde{v}_{ds} \quad (6\text{-}127)$$

Therefore, we have a second-order nonlinear differential equation in $\phi_{s,\mathrm{ac}}(x)$, with two boundary conditions specified at the source and the drain, which are related to the applied small-signal voltages. Solving the differential equation (Eq. 6-125) is not trivial, but it can be done, as demonstrated in Appendix A.

After $\phi_{s,ac}(x)$ is determined, the channel current $i_{ch}(x)$ can be established from Eq. (6-124). The drain and source terminal currents are given by:

$$\tilde{i}_s = \tilde{i}_{ch}(x)|_{x=0} \qquad \tilde{i}_d = -\tilde{i}_{ch}(x)|_{x=L} \tag{6-128}$$

After evaluating the drain and source terminal currents, the gate current can then be determined from the requirement of the Kirchoff current law:

$$\tilde{i}_g = -\tilde{i}_d - \tilde{i}_s \tag{6-129}$$

By now we have i_g and i_d as a function of v_{gs} and v_{ds}. The common-source y-parameters y_{gg}, y_{gd}, y_{dg}, and y_{dd} can then be evaluated in accordance with the definition given in Eq. (6-103).

This procedure for finding the y-parameters is straightforward but quite algebraically intensive. Without proper software, the derivation easily takes days and is error-prone. There are generally two techniques to solve the nonlinear differential equation [Eq. (6-125)], both capable of yielding the exact solution. The first is the iteration technique and second involves Bessel functions. In Appendix A, where the MESFET y-parameters are sought, the iteration technique is used. For the sake of variety, the Bessel function approach is used in Appendix B to determine the y-parameters in HFETs. The Bessel solution approach gives the exact solution. Theoretically, the solution obtained by the iteration technique is exact only when the number of iterations used in the solution approaches infinity. The solution given in Appendix A, obtained after two iterations, is nonetheless sufficiently accurate for practical applications.

From the results of Appendix A, we find that, among the various y-parameters, y_{gg} is given by:

$$\begin{aligned} y_{gg} = j\omega \Bigg\{ \frac{qWLN_d a}{\phi_{00}} \Bigg[& \frac{d^{1/2}(1-d^{1/2}) - s^{1/2}(1-s^{1/2})}{d - s - \frac{2}{3}d^{3/2} + \frac{2}{3}s^{3/2}} \\ & + \frac{(d^{1/2} - s^{1/2})(\frac{2}{3}d^{3/2} - \frac{2}{3}s^{3/2} - \frac{1}{2}d^2 + \frac{1}{2}s^2)}{(d - s - \frac{2}{3}d^{3/2} + \frac{2}{3}s^{3/2})^2} \Bigg] \Bigg\} \\ & + \text{2nd-order terms} \end{aligned} \tag{6-130}$$

After spending some time examining the term inside the curly braces, we identify it to be exactly equal to the C_{gg} of Eq. (6-42). Hence, we equate y_{gg} to:

$$y_{gg} = j\omega C_{gg} + \text{2nd-order terms} \tag{6-131}$$

Note that y_{gg} in Eq. (6-130) represents the exact non-quasi-static solution, yet the C_{gg} appearing in Eq. (6-131) was derived in § 6-2, which was based on the quasi-static analysis.

Similarly, we find y_{dd} derived in Appendix A to be:

$$y_{dd} = \frac{W}{L} aq\mu_n N_d[1 - \sqrt{d}] + j\omega \left\{ -2\frac{aqN_dWL}{\phi_{00}} \frac{[(1-d^{1/2})(\frac{2}{3}d^{1/2} - \frac{1}{2}d^2)]}{(d - s - \frac{2}{3}d^{3/2} + \frac{2}{3}s^{3/2})^2} \right.$$
$$+ 2\frac{aqN_dWL}{\phi_{00}}(1 - d^{1/2})\frac{[\frac{4}{15}(d^{5/2} - s^{5/2}) - \frac{7}{18}(d^3 - s^3) + \frac{1}{7}(d^{7/2} - s^{7/2})]}{(d - s - \frac{2}{3}d^{3/2} + \frac{2}{3}s^{3/2})^3}$$
$$\left. + \frac{aqN_dWL}{\phi_{00}} \frac{[d^{1/2}(1 - d^{1/2})]}{(d - s - \frac{2}{3}d^{3/2} + \frac{2}{3}s^{3/2})} \right\} + \text{2nd-order terms} \tag{6-132}$$

Although it is not obvious from the appearance, the term inside the curly braces is identical to the C_{dd} worked out in § 6-2 [Eq. (6-48)]. In addition, the first term of y_{dd} is identical to g_d [Eq. (6-64)]. Therefore, we can say that y_{dd} is equal to:

$$y_{dd} = g_d + j\omega C_{dd} + \text{2nd-order terms} \tag{6-133}$$

The second-order terms in y_{dd} and y_{gg} (as well as other y-parameters) are not identified in Appendix A. We used the iteration technique and went through the iteration twice. With further iteration, these second-order terms can be established, but only after a considerable amount of algebra. If we drop these second-order terms, the simplified y-parameters result. These simplified y-parameters shall be referred to as the *quasi-static y-parameters* because the remaining terms such as g_d, C_{dd}, and C_{gg} are identical to those of § 6-2, which was based on a quasi-static analysis.

We listed only y_{gg} and y_{dd} above. The other two common-source y-parameters can be determined in a similar way. We summarize the quasi-static y-parameters for the common-source configuration in the following:

$$y_{gg,\text{QS}} = j\omega C_{gg} \tag{6-134}$$

$$y_{gd,\text{QS}} = -j\omega C_{gd} \tag{6-135}$$

$$y_{dg,\text{QS}} = g_m - j\omega C_{dg} \tag{6-136}$$

$$y_{dd,\text{QS}} = g_d + j\omega C_{dd} \tag{6-137}$$

Other relationships are also obtained:

$$y_{gs,\text{QS}} = -j\omega C_{gs} \tag{6-138}$$

$$y_{ds,\text{QS}} = -g_d - g_m - j\omega C_{ds} \tag{6-139}$$

$$y_{sg,\text{QS}} = -g_m - j\omega C_{sg} \tag{6-140}$$

$$y_{sd,\text{QS}} = -g_d - j\omega C_{sd} \tag{6-141}$$

$$y_{ss,\text{QS}} = g_d + g_m + j\omega C_{ss} \tag{6-142}$$

Example 6-3:

The HFET studied in Example 6-1 was found to have the following parameters: $V_T = 0.118$ V; $\alpha = 0.476$; $C'_{ox} = 2.91 \times 10^{-7}$ F/cm^2; $C_{gg} = 5.23 \times 10^{-13}$ F; $C_{gd} = 1.31 \times 10^{-13}$ F; $C_{dg} = 1.73 \times 10^{-13}$ F; and $C_{dd} = 1.69 \times 10^{-13}$ F. The device has a gate width of 500 μm and a gate length of 0.25 μm. The mobility is taken to be 4000 cm^2/V-s. The transistor is biased at $V_{GS} = 0.5$ V and $V_{DS} = 0.2$ V. Find the quasi-static common-source y-parameters when the small-signal frequency is 1 GHz.

From Eqs. (6-66) and (6-67):

$$g_m = \frac{500 \times 10^{-4} \cdot 2.91 \times 10^{-7} \cdot 4000}{0.25 \times 10^{-4}} (0.5 - 0.188) \times (1 - 0.476) = 0.38\ 1/\Omega$$

$$g_d = \frac{500 \times 10^{-4} \cdot 2.91 \times 10^{-7} \cdot 4000}{0.25 \times 10^{-4}} (0.5 - 0.188) \times 0.476 = 0.34\ 1/\Omega$$

According to Eqs. (6-134) to (6-137), we have:

$$y_{gg,\mathrm{QS}} = j \cdot 2\pi \cdot 10^9 \cdot 5.23 \times 10^{-13} = j3.28 \times 10^{-3}\ 1/\Omega$$

$$y_{gd,\mathrm{QS}} = -j \cdot 2\pi \cdot 10^9 \cdot 1.31 \times 10^{-13} = -j8.23 \times 10^{-4}\ 1/\Omega$$

$$y_{dg,\mathrm{QS}} = 0.38 - j \cdot 2\pi \cdot 10^9 \cdot 1.73 \times 10^{-13} = 0.38 - j1.09 \times 10^{-3}\ 1/\Omega$$

$$y_{dd,\mathrm{QS}} = 0.34 + j \cdot 2\pi \cdot 10^9 \cdot 1.69 \times 10^{-13} = 0.34 + j1.06 \times 10^{-3}\ 1/\Omega$$

The second-order terms in the y-parameters of MESFETs in Eqs. (6-131) and (6-133) are unknowns because we do not carry out an infinite number of iterations when using the iterative technique in Appendix A. However, the second-order terms in HFETs are identified. In Appendix B, we use a Bessel function approach in solving the governing equations for HFETs. Despite the added degree of complexity, we are able to attain the exact solution of the y-parameters for HFETs. Because these y-parameters contain higher order terms, we shall refer them as the *non-quasi-static y-parameters*, as opposed to the quasi-static y-parameters listed in Eqs. (6-134) to (6-142). It is instructive to examine the y_{gd} expression determined in Appendix B. According to Eqs. (B-50), (B-57), and (B-43),

$$y_{gd,\mathrm{NQS}} = \frac{N_{gd,0} + (j\omega) N_{gd,1} + (j\omega)^2 N_{gd,2} + \cdots + (j\omega)^k N_{gd,k}}{D_0 + (j\omega) D_1 + (j\omega)^2 D_2 + \cdots + (j\omega)^k D_k} \tag{6-143}$$

Because $N_{gd,0}$ is zero and $D_0 = 1$, it can be expressed in the following form:

$$y_{gd,\text{NQS}} = j\omega N_{gd,1} \frac{1 + (j\omega) N_{gd,2}/N_{gd,1} + \cdots}{1 + (j\omega) D_1 + \cdots} \tag{6-144}$$

A comparison of $N_{gd,1}$ of Appendix B with C_{dg} of § 6-1 reveals that they are identical (except by a sign). Hence, we can re-express $y_{dg,\text{NQS}}$ as:

$$y_{gd,\text{NQS}} = -j\omega C_{gd} \frac{1 + (j\omega) N_{gd,2}/N_{gd,1}}{1 + (j\omega) D_1} \tag{6-145}$$

In the above expression, we drop some higher-order terms (but retain some not-so-higher-order terms). The accuracy of keeping various higher-order terms is demonstrated in Fig. 6-24, which plots $y_{dg,\text{NQS}}$ in Eq. (B-50) with $k = 1, 2, 5$, and 10. This figure demonstrates that keeping only up to the $k = 2$ terms is acceptable. Using the $N_{gd,2}$, $N_{gd,1}$, and D_1 expressions given in Appendix B, we can write the $y_{dg,\text{NQS}}$ of Eq. (6-145) in a more convenient form:

$$\begin{aligned} y_{gd,\text{NQS}} &= -j\omega C_{gd} \frac{1 + j(\omega/\omega_0)\tau_3}{1 + j(\omega/\omega_0)\tau_1} \\ &= -j\omega C_{gd} - j\omega C_{gd} \frac{j(\omega/\omega_0)(\tau_3 - \tau_1)}{1 + j(\omega/\omega_0)\tau_1} \end{aligned} \tag{6-146}$$

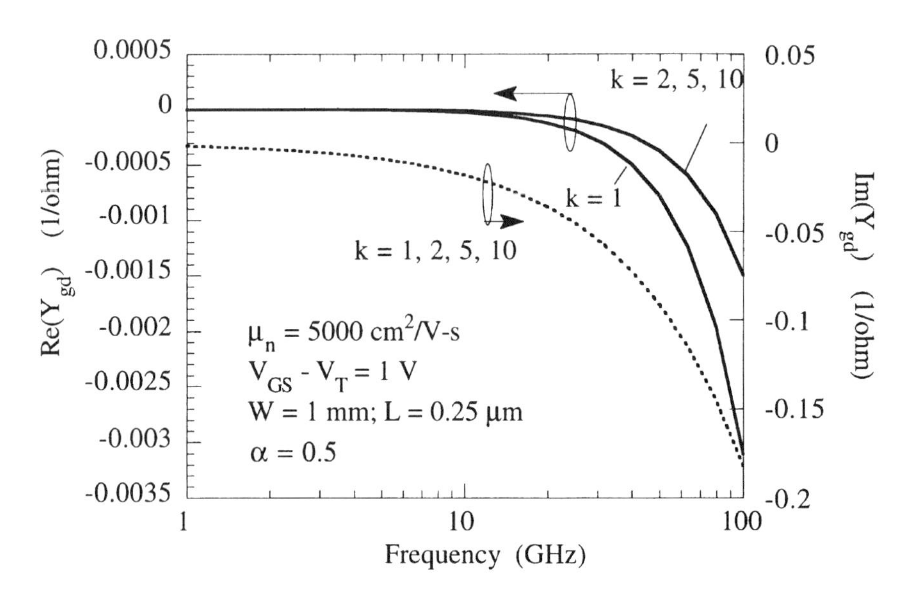

FIGURE 6-24. $y_{dg,\text{NQS}}$ in Eq. (B-50) with $k = 1, 2, 5$, and 10. The accuracy is good when k reaches 2.

where

$$\tau_1 = \frac{4}{15} \frac{1 + 3\alpha + \alpha^2}{(1 + \alpha)^3} \tag{6-147}$$

$$\tau_3 = \frac{1}{15} \frac{5 + 8\alpha + 2\alpha^2}{(1 + \alpha)^2 (2 + \alpha)} \tag{6-148}$$

$$\omega_0 = \frac{\mu_n (V_{GS} - V_T)}{L^2} \tag{6-149}$$

We use the same procedure to write out the other three common-source y-parameters $y_{gg,\mathrm{NQS}}$, $y_{dg,\mathrm{NQS}}$, and $y_{dd,\mathrm{NQS}}$. Together with $y_{gd,\mathrm{NQS}}$ of Eq. (6-146), we can find the remaining five y-parameters. We write all of the nine y-parameters in the following:

$$\begin{aligned} y_{gg,\mathrm{NQS}} = {} & j\omega C_{gg} + j\omega C_{gd} \frac{j(\omega/\omega_0)(\tau_3 - \tau_1)}{1 + j(\omega/\omega_0)\tau_1} \\ & + j\omega C_{gs} \frac{j(\omega/\omega_0)(\tau_2 - \tau_1)}{1 + j(\omega/\omega_0)\tau_1} \end{aligned} \tag{6-150}$$

$$y_{gd,\mathrm{NQS}} = -j\omega C_{gd} - j\omega C_{gd} \frac{j(\omega/\omega_0)(\tau_3 - \tau_1)}{1 + j(\omega/\omega_0)\tau_1} \qquad \text{(same as Eq. 6-146)}$$

$$y_{gs,\mathrm{NQS}} = -j\omega C_{gs} - j\omega C_{gs} \frac{j(\omega/\omega_0)(\tau_2 - \tau_1)}{1 + j(\omega/\omega_0)\tau_1} \tag{6-151}$$

$$y_{dg,\mathrm{NQS}} = g_m \quad j\omega C_{dg} - j\omega C_{gd} \frac{j(\omega/\omega_0)(\tau_3 - \tau_1)}{1 + j(\omega/\omega_0)\tau_1} \tag{6-152}$$

$$y_{dd,\mathrm{NQS}} = g_d + j\omega C_{dd} + j\omega C_{gd} \frac{j(\omega/\omega_0)(\tau_3 - \tau_1)}{1 + j(\omega/\omega_0)\tau_1} \tag{6-153}$$

$$y_{ds,\mathrm{NQS}} = -g_d - g_m - j\omega C_{ds} \tag{6-154}$$

$$y_{sg,\mathrm{NQS}} = -g_m - j\omega C_{sg} - j\omega C_{gs} \frac{j(\omega/\omega_0)(\tau_2 - \tau_1)}{1 + j(\omega/\omega_0)\tau_1} \tag{6-155}$$

$$y_{sd,\mathrm{NQS}} = -g_d - j\omega C_{sd} \tag{6-156}$$

$$y_{ss,\mathrm{NQS}} = g_d + g_m + j\omega C_{ss} + j\omega C_{gs} \frac{j(\omega/\omega_0)(\tau_2 - \tau_1)}{1 + j(\omega/\omega_0)\tau_1} \tag{6-157}$$

where τ_1 was given in Eq. (6-147), τ_3 in Eq. (6-148), and τ_2 by

$$\tau_2 = \frac{1}{15} \frac{2 + 8\alpha + 5\alpha^2}{(1 + \alpha)^2 (1 + 2\alpha)} \tag{6-158}$$

It can be verified that the above nine y-parameters satisfy the relationships imposed in Eq. (6-105).

These non-quasi-static (NQS) y-parameters are applicable to HFETs, not MESFETs. Although the exact forms of the NQS y-parameters in MESFETs are not derived, they are believed to bear similar forms as those given above for the HFETs.

At small frequencies, ω/ω_0 approaches zero. The NQS y-parameters then reduce to the QS (quasi-static) y-parameters given in Eqs. (6-134) to (6-142). Therefore, Eqs. (6-134) to (6-142) apply to both MESFETs as well as HFETs (but the exact forms for C_{gg}, for example, differ in the two types of FETs).

Example 6-4:

Find the non-quasi-static y-parameter $y_{gd,\mathrm{NQS}}$ for the HFET studied in Example 6-3, when the small-signal frequency is 1 GHz. The transistor is biased at $V_{GS} = 0.5$ V and $V_{DS} = 0.2$ V.

According to Example 6-3, the device has a V_T of 0.188 V, and a μ_n of 4000 cm^2/V-s. The channel length is 0.25 μm. Hence, from Eq. (6-149):

$$\omega_0 = \frac{4000(0.5 - 0.188)}{(0.25 \times 10^{-4})^2} = 2.0 \times 10^{12} \text{ rad}$$

With $\alpha = 0.476$, τ_1 and τ_3 are calculated from Eqs. (6-147) and (6-148):

$$\tau_1 = \frac{4}{15} \frac{1 + 3 \cdot 0.476 + 0.476^2}{(1 + 0.476)^3} = 0.67$$

$$\tau_3 = \frac{1}{15} \frac{5 + 8 \cdot 0.476 + 2 \cdot 0.476^2}{(1 + 0.476)^2 \cdot (2 + 0.476)} = 0.11$$

C_{gd} was given in Example 6-3 as equal to 1.31×10^{-13} F. From Eq. (6-146), we have:

$$\begin{aligned} y_{gd,\mathrm{NQS}} = & -j \cdot 2\pi \cdot 10^9 \cdot 1.31 \times 10^{-13} \\ & -j \cdot 2\pi \cdot 10^9 \cdot 1.31 \times 10^{-13} \frac{j[2\pi \cdot 10^9/(2 \times 10^{12})] \cdot 0.11}{1 + j[2\pi \cdot 10^9/(2 \times 10^{12})] \cdot 0.67} \\ = & -1.45 \times 10^{-6} - j8.23 \times 10^{-4} \ 1/\Omega \end{aligned}$$

The $y_{gd,\mathrm{NQS}}$ obtained here can be compared with $y_{gd,\mathrm{QS}}$ obtained in Example 6-3. There is a finite real part in the NQS result.

A useful relationship relates τ_1 to the capacitances (see Problem 7):

$$\tau_1 = \omega_0 \frac{C_{dg} - C_{gd}}{g_m} = \omega_0 \frac{-C_{sd}}{g_d} \tag{6-159}$$

§ 6-4 HYBRID-π MODEL AND CHANNEL RESISTANCE

It is generally accepted that QS y-parameters are a good approximation to the NQS y-parameters when the operating radian frequency (ω) is much lower than ω_0 given in Eq. (6-149). We shall work mostly with the QS y-parameters because they are simpler in form yet capture many important device properties. The occasional need to invoke the more complicated non-quasi-static forms will be mentioned when necessary. We shall from now on drop the subscript QS associated with the quasi-static y-parameters.

To construct the hybrid-π model, we start with the small-signal equivalent circuit of Fig. 4-11. In the common-source configuration, $y_{11} = y_{gg}$, $y_{12} = y_{gd}$, $y_{21} = y_{dg}$ and $y_{22} = y_{dd}$ [see Eq. (6-109)]. From circuit theories, the equivalent circuit of Fig. 4-11 is identical to that of Fig. 4-12, as long as nodes 1′ and 2′ of Fig. 4-11 are the same node. This condition is fulfilled in the common-source configuration, in which both 1′ and 2′ represent the source node of the transistor. The elements required in the equivalent circuit of Fig. 4-12 is given in the following:

$$y_{11} + y_{12} = j\omega C_{gg} - j\omega C_{gd} = j\omega C_{gs} \tag{6-160}$$

$$-y_{12} = j\omega C_{gd} \tag{6-161}$$

$$y_{22} + y_{12} = g_d + j\omega C_{dd} - j\omega C_{gd} = g_d + j\omega C_{sd} \tag{6-162}$$

$$y_{21} - y_{12} = g_m - j\omega C_{dg} + j\omega C_{gd} \tag{6-163}$$

Applying the result of Eq. (6-159), we can rewrite Eq. (6-163) as

$$y_{21} - y_{12} = g_m\left(1 - j\frac{\omega}{\omega_0}\tau_1\right) \tag{6-164}$$

If we were to use y_{dg} and y_{gd} of Appendix B directly and perform the subtraction prior to dropping out the high-frequency-order terms, we would obtain:

$$y_{21} - y_{12} = y_{dg} - y_{gd} = \frac{g_m}{1 + j(\omega/\omega_0)\tau_1} \tag{6-165}$$

This is approximately equal to Eq. (6-164). Indeed, Eq. (6-165) is also a popular form used in the small-signal model. We adopt Eq. (6-164) in the following equivalent-circuit development, so that the NQS y-parameters given in

Eqs. (6-146) to (6-157) retain the property that the sums of all the parameters in a column or a row add up to zero.

Placing the values of Eqs. (6-160), (6-161), (6-162), and (6-164) into Fig. 4-12, we obtain the small-signal model in Fig. 6-25. This is a quasi-static model because the starting y-parameters used in this derivation are the QS y-parameters. This model applies to both HFETs and MESFETs.

The QS small-signal model is a simple model that is capable of achieving high accuracy even at high frequencies. There is one potential drawback in this model, however, in that it does not account for the channel resistance. Figure 6-26 illustrates a physical picture of the transistor operation. We expect the input impedance (between the gate and the source) to contain a resistive part, due to the finite resistance associated with the channel. However, the model of

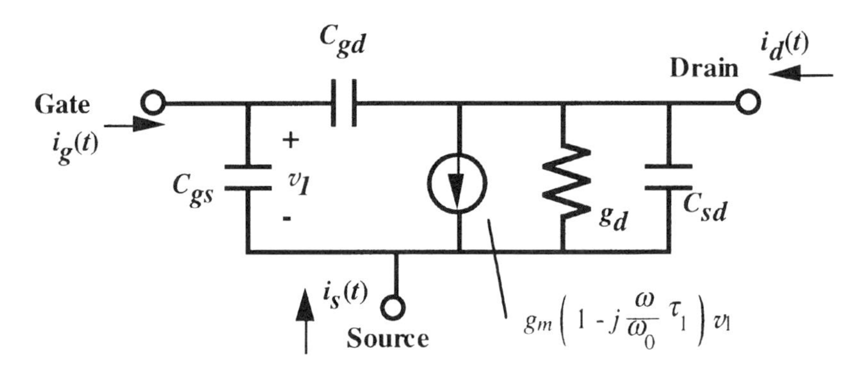

FIGURE 6-25. A hybrid-π small-signal model of a FET. The model is derived from quasi-static y-parameters.

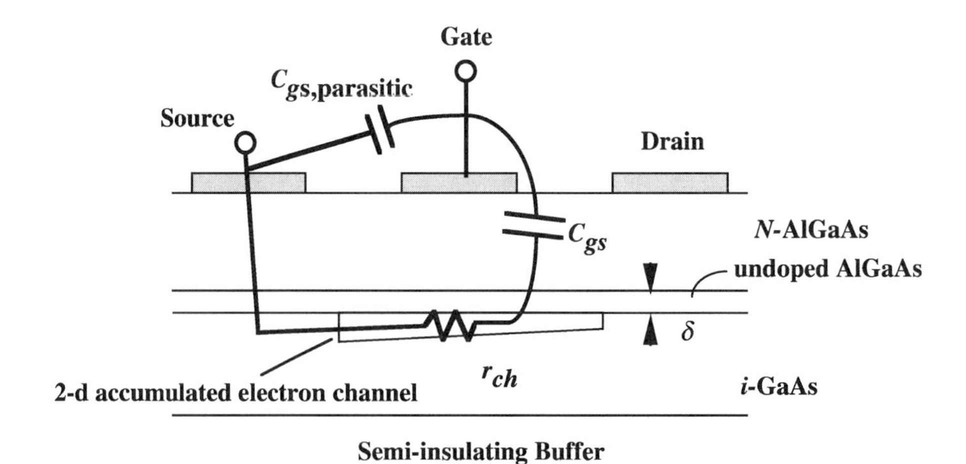

FIGURE 6-26. Schematic input network of a FET, showing the existence of a channel resistance.

Fig. 6-25 reveals that the input impedance is purely imaginary. Only C_{gs} is situated between the gate and the source, without any resistive element. The reason the channel resistance is absent in the QS small-signal model is that a quasi-static analysis assumes the channel charge to respond *instantaneously* to the voltage variation. In reality, the channel resistance acts as an inertia element in the channel, which creates a lag in the response between the external biases and the resulting device currents.

The existence of the channel resistance, not inferable from the QS y-parameters, can be realized from the NQS y-parameters. Let us examine $y_{gs,\mathrm{NQS}}$ in particular. We find that the input impedance looking through the gate to the source (neglecting the drain terminal) is:

$$Z_{\mathrm{in}} = -\frac{1}{y_{gs,\mathrm{NQS}}} = \frac{1}{j\omega C_{gs}} \times \frac{1 + j(\omega/\omega_0)\tau_1}{1 + j(\omega/\omega_0)\tau_2} \tag{6-166}$$

The channel resistance is the real part of the input impedance, given by:

$$r_{\mathrm{ch}} = \mathrm{Re}(Z_{\mathrm{in}}) = \frac{1}{\omega C_{gs}} \times \frac{(\omega/\omega_0)(\tau_1 - \tau_2)}{1 + (\omega/\omega_0)^2\tau_2^2} = \frac{\tau_1 - \tau_2}{\omega_0 C_{gs}} \times \frac{1}{1 + (\omega/\omega_0)^2\,\tau_2^2} \tag{6-167}$$

Equation (6-166) demonstrates that the channel resistance exists even as the frequency approaches zero. We often state that the non-quasi-static effects are important when ω exceeds ω_0. This statement, while true in most situations, is incorrect as far as the channel resistance is concerned. The channel resistance is important even when the frequency is well below ω_0 (or approaches zero).

At frequencies much smaller ω_0, r_{ch} is simplified to:

$$\begin{aligned} r_{\mathrm{ch}} &= \frac{\tau_1 - \tau_2}{\omega_0 C_{gs}} = \frac{1}{\omega_0 WLC'_{\mathrm{ox}}}\,\frac{3\alpha^3 + 15\alpha^2 + 10\alpha + 2}{15(1+\alpha)^3(1+2\alpha)} \times \frac{3}{2}\,\frac{(1+\alpha)^2}{2\alpha + 1} \\ &= \frac{1}{\omega_0 WLC'_{\mathrm{ox}}}\,\frac{3\alpha^3 + 15\alpha^2 + 10\alpha + 2}{10(1+\alpha)\,(1+2\alpha)^2} \end{aligned} \tag{6-168}$$

We substitute ω_0 from Eq. (6-149) into Eq. (6-168). The resulting expression is then in terms of the fundamental device parameters:

$$r_{\mathrm{ch}} = \left[\frac{W}{L}\,\mu_n C'_{\mathrm{ox}}(V_{GS} - V_T)\right]^{-1} \frac{3\alpha^3 + 15\alpha^2 + 10\alpha + 2}{10(1+\alpha)(1+2\alpha)^2} \tag{6-169}$$

Often r_{ch} is expressed in terms of g_m, which was given in Eq. (6-66). Therefore:

$$r_{\mathrm{ch}} = \frac{1}{g_m}\,\frac{(3\alpha^3 + 15\alpha^2 + 10\alpha + 2)(1-\alpha)}{10(1+\alpha)(1+2\alpha)^2} \tag{6-170}$$

Example 6-5:

Find the channel resistance per unit width of a HFET. The transistor is biased in saturation and has a g_m of 400 mS/mm (or 0.4 S/mm).

When the device is in saturation, $\alpha = 0$. Substituting this value into Eq. (6-170), we find:

$$r_{\text{ch}} = \frac{1}{5g_m} = \frac{1}{5 \times 0.4} = 0.5\ \Omega\text{-mm}$$

When the transistor is heavily turned on, g_m is a large number. Hence, the channel resistance has a small value, and the effects of the channel resistance are negligible. However, when the transistor's mutual transconductance is a small value, the channel resistance needs to be inserted in the small-signal model. The channel resistance calculated from Eq. (6-169) as a function of α is plotted in Fig. 6-27.

We develop a NQS small-signal model, using the NQS y-parameters given in Eqs. (6-146) to (6-158):

$$y_{11} - y_{12} = j\omega C_{gs} \frac{1 + j(\omega/\omega_0)\tau_2}{1 + j(\omega/\omega_0)\tau_1} \approx \left(r_{\text{ch}} + \frac{1}{j\omega C_{gs}} \right)^{-1} \tag{6-171}$$

$$-y_{12} = j\omega C_{gd} \frac{1 + j(\omega/\omega_0)\tau_3}{1 + j(\omega/\omega_0)\tau_1} \approx \left(r_{gd} + \frac{1}{j\omega C_{gd}} \right)^{-1} \tag{6-172}$$

$$y_{22} + y_{12} = g_d + j\omega C_{sd} \tag{6-173}$$

$$y_{21} - y_{12} = g_m - j\omega C_{dg} + j\omega C_{gd} = g_m\left(1 - j\frac{\omega}{\omega_0}\tau_1\right) \tag{6-174}$$

The small-signal resistance r_{gd} appearing in $-y_{12}$ can be derived in a similar fashion as r_{ch}. The only difference lies in the usage of τ_3 instead of τ_2, and of C_{gd} instead of C_{gs}.

$$\begin{aligned} r_{gd} &= \frac{\tau_1 - \tau_2}{\omega_0 C_{gd}} = \frac{1}{\omega_0 WLC'_{\text{ox}}} \frac{2\alpha^3 + 10\alpha^2 + 15\alpha + 3}{10\alpha(1+\alpha)(2+\alpha)^2} \\ &= \left[\frac{W}{L}\mu_n C'_{\text{ox}}(V_{GS} - V_T)\right]^{-1} \frac{2\alpha^3 + 10\alpha^2 + 15\alpha + 3}{10\alpha(1+\alpha)(2+\alpha)^2} \end{aligned} \tag{6-175}$$

Its dependence on α is shown in Fig. 6-27. Generally the effects of r_{gd} can be neglected. The most important element in a non-quasi-static model is the introduction of the channel resistance between the gate and the source.

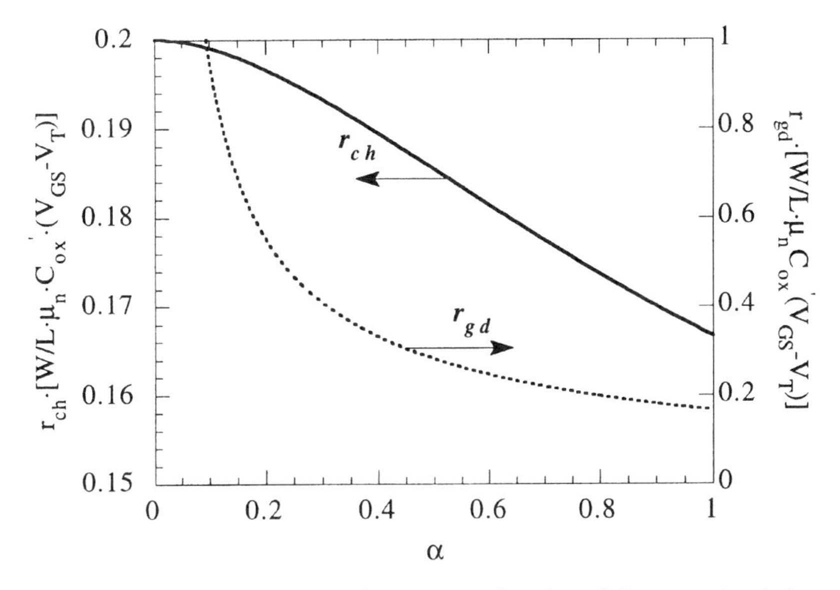

FIGURE 6-27. Channel resistance of HFET as a function of the saturation index, α.

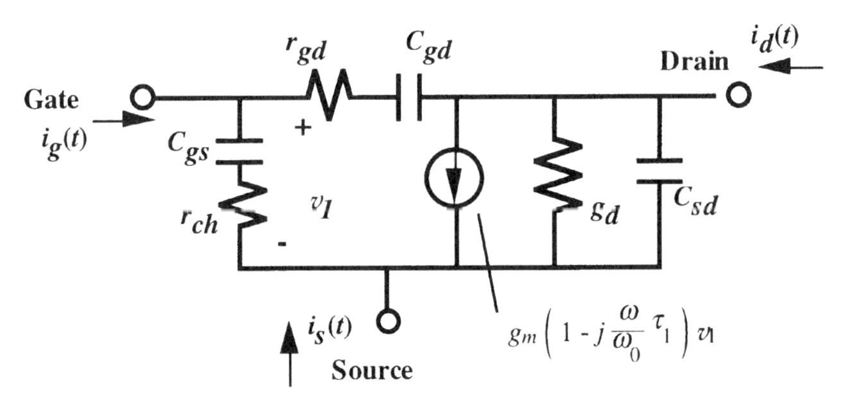

FIGURE 6-28. A non-quasi-static hybrid-π model of a FET accounting for the channel resistance.

Based on Eqs. (6-171) to (6-174), the hybrid-π model accounting for the channel resistance is shown in Fig. 6-28. When the transistor is in saturation, the model simplifies to that of Fig. 6-29.

The small-signal models presented in this section do not include parasitics such as the terminal resistances and the parasitic capacitance between two terminals. These models will be modified in § 6-7 to account for such parasitics.

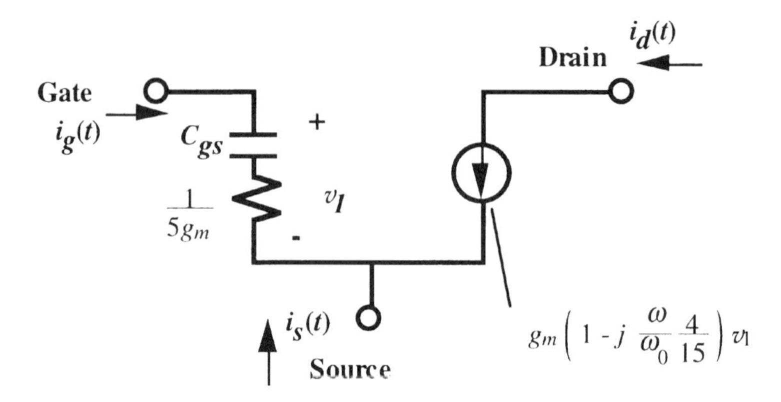

FIGURE 6-29. Simplified model of Fig. 6-28 when the FET operates in saturation.

§ 6-5 GATE RESISTANCE

The gate is made of metal, which is characterized by low resistivity values. Hence, the metal resistance is small and is often neglected in circuit analysis. However, for high-power applications, the gate width can easily be 1000 times that of the gate length. Despite the low metal resistivity, the skinny-finger geometry with a wide gate width still results in a certain amount of resistance. The resistance does not affect the d.c. operation of the transistor—after all, the d.c. gate current is zero. There is no voltage drop associated with the resistance. However, the resistance can have a profound effect in high-frequency operation. There is time-varying current traveling down the gate, which eventually enters the channel through the capacitive coupling such as C_{gs} or C_{gd}. The nonzero transient gate current, upon flowing through the gate resistance, produces a voltage drop which affects the device performance adversely.

An important parameter characterizing the gate resistance is the gate sheet resistance, R_{SHG} (in Ω/square). It is equal to the gate metal's resistivity (ρ_G, in Ω-cm) divided by the metal thickness (Λ_G):

$$R_{SHG} = \frac{\rho_G}{\Lambda_G} \tag{6-176}$$

R_{SHG} is a useful parameter because it is easily measurable from a test structure consisting of a wide metal gate with probe pads at the two ends, as shown in Fig. 6-30. The material underneath the gate structure is designed to be isolated from the active area, so that the current injected from one end of the gate all comes out at the other end. To ensure accurate measurement, two extra voltage-sensing probe pads are attached at the ends of the test

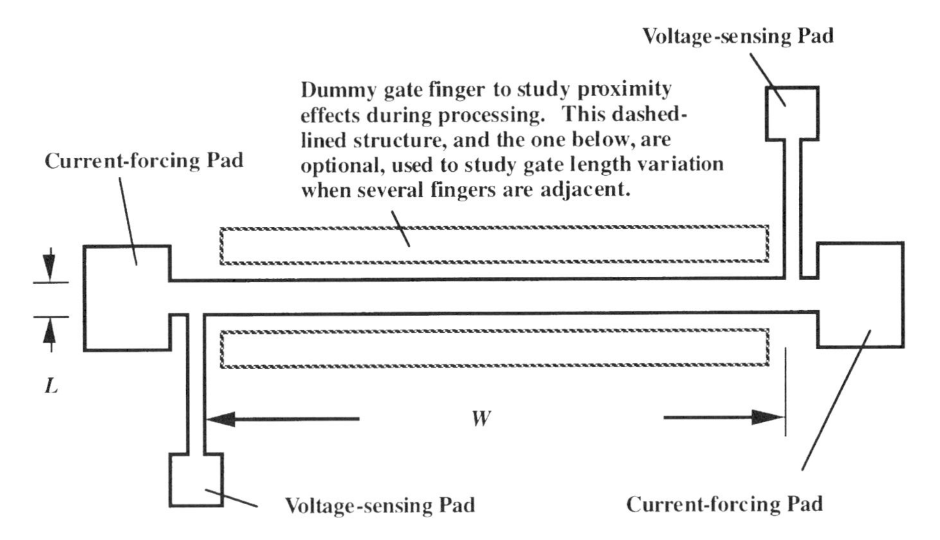

FIGURE 6-30. A test structure used to measure the gate metal sheet resistance.

structure. The gate sheet resistance is related to the measured gate resistance (R_{measure}) by

$$R_{\text{measure}} = R_{SHG} \times \frac{W}{L} \tag{6-177}$$

W and L are the width and the length of the gate metal, respectively. Generally, the gate sheet resistance is about 0.025 Ω/square for a $L = 0.25$ μm Ti/Pt/Au gate with layer thicknesses of 1000/500/4000 Å. Sometimes the gate resistance is expressed in the unit Ω-mm in the FET literature. The resistance is multiplied by millimeters so that the resistance per a given width of the gate can be easily calculated. For example, for an $L = 0.25$ μm gate with a sheet resistance of 0.025 Ω/square, the total gate resistance is 100 Ω when $W = 1$ mm. Alternatively, the gate resistance is expressed as 100 Ω-mm. If the gate width were instead designed to be 2 mm, then the 100 Ω-mm value of the gate-resistance-per-width leads to a gate resistance of 50 Ω.

In order to reduce the gate resistance by increasing the amount of metal cross-sectional area, a T-shaped gate (or T-gate), shown in Fig. 6-31, is routinely used. The length dimension at the top of the T-gate is about three times the channel length L. By increasing the area of conduction, the gate sheet resistance decreases.

When we wrote Eq. (6-177) for the test structure of Fig. 6-30, the current is restricted to the metal, not flowing into the semiconductor. Whatever charges enter through one end come out at the other end. This kind of gate signal flow does not occur in device operation. As shown in Fig. 6-32, the gate charges

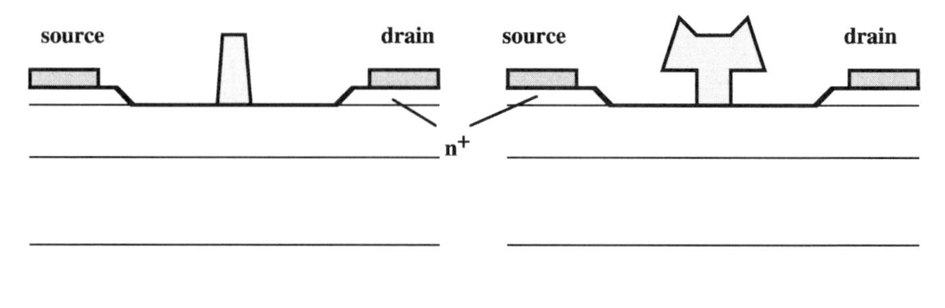

FIGURE 6-31. Conventional gate and T-gate structures used in FETs.

entering one end of the gate do not all come out at the other end. Instead, the gate current is continuously diverted to the channel through the capacitances. At the other end of the gate, there is no more gate current since all of the gate charges have been diverted to the channel. The gate resistance is therefore a distributed resistance, similar to the intrinsic base resistance of the HBTs.

We derive the distributed gate resistance. The gate stripe with a metal height of Λ_G is drawn schematically in Fig. 6-32. The input bias $v_{IN}(t)$ is applied at $x = 0$, but v_{GS} along the gate metal decreases as x increases because of the finite voltage drop across the gate material. Denoting the gate current density at any instant and location as $j_M(x, t)$, we write:

$$v_{GS}(x, t) = v_{IN}(t) - \int_0^x j_M(x', t)\rho_G dx' \tag{6-178}$$

where ρ_G is the metal resistivity, equal to the product of R_{SHG} and Λ_G.

In general, the gate current of a FET without gate resistance was given by Eq. (6-54). The gate current per unit width is:

$$\frac{i_G(t)}{W} = \frac{C_{gg}}{W}\frac{dv_{GS}}{dt} - \frac{C_{gd}}{W}\frac{dv_{DS}}{dt} \tag{6-179}$$

To simplify the analysis, we assume that the output bias v_{DS} stays constant with time. Because the current flowing in the metal at a particular location x will eventually go into the channel through the capacitive couplings, we rewrite Eq. (6-179) as

$$j_M(x, t)\Lambda_G \cdot L = \int_x^W \frac{C_{gg}}{W}\frac{\partial v_{GS}(x', t)}{\partial t} dx' \tag{6-180}$$

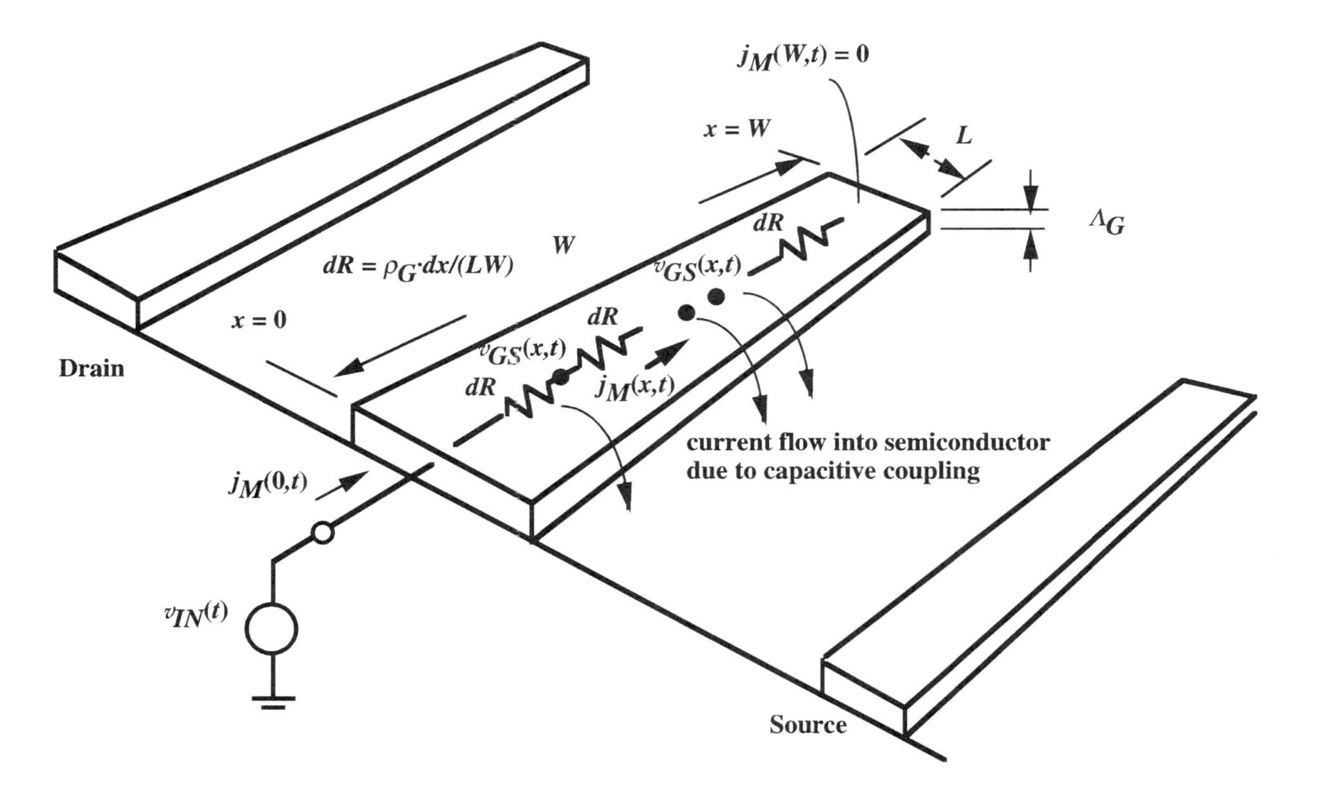

FIGURE 6-32. Schematic diagram depicting the gate structure used in the analysis of distributed gate resistance. The input gate current flows through the metal and diverts into the semiconductor through the gate capacitance.

It is easier to work with the differential forms of Eqs. (6-178) and (6-180). After taking the partial derivative of these equations with respect to x, we obtain:

$$\frac{\partial v_{GS}(x,t)}{\partial x} = -j_M(x,t)\rho_G \tag{6-181}$$

$$\Lambda_G \cdot L\frac{\partial j_M(x,t)}{\partial x} = -\frac{C_{gg}}{W}\frac{\partial v_{GS}(x,t)}{\partial t} \tag{6-182}$$

Although it is possible to solve for the current and voltage under a large-signal excitation [2], here we shall determine the input resistance of the gate structure under a small-signal excitation. It is convenient to convert the differential equations so that only the small-signal quantities appear. To achieve that, we first write out the quiescent component and the small-signal perturbation of several quantities:

$$v_{GS}(x,t) = V_{GS} + \tilde{v}_{gs}(x)e^{j\omega t} = V_{\mathrm{IN}} + \tilde{v}_{gs}(x)e^{j\omega t} \tag{6-183}$$

$$v_{\mathrm{IN}}(x,t) = V_{\mathrm{IN}} + \tilde{v}_{\mathrm{in}}e^{j\omega t} \tag{6-184}$$

$$j_M(x,t) = J_M + \tilde{j}_m(x)e^{j\omega t} = \tilde{j}_m(x)e^{j\omega t} \tag{6-185}$$

The gate current is zero at d.c. Hence, the quiescent value of the metal current density, J_M, is zero. Substituting these quantities into Eqs. (6-181) and (6-182), and equating terms involving $e^{j\omega t}$, we obtain:

$$\frac{\partial \tilde{v}_{gs}(x)}{\partial x} = -\tilde{j}_m\rho_G \tag{6-186}$$

$$\Lambda_G \cdot L\frac{\partial \tilde{j}_m(x)}{\partial x} = -j\omega\frac{C_{gg}}{W}\tilde{v}_{gs}(x) \tag{6-187}$$

These two equations can be combined to form a single differential equation. Together with the boundary conditions appropriate for the small-signal operation under consideration, we have:

$$\frac{\partial^2 \tilde{v}_{gs}(x)}{\partial x^2} = \frac{j\omega}{\gamma}\tilde{v}_{gs}(x) \qquad \text{where } \gamma = \frac{W \cdot L}{R_{SHG} \cdot C_{gg}} \tag{6-188}$$

$$\tilde{v}_{gs}(x=0) = \tilde{v}_{\mathrm{in}} \qquad \left.\frac{\partial \tilde{v}_{gs}}{\partial x}\right|_{x=W} = 0 \tag{6-189}$$

This is a linear second-order differential equation, whose solution can be expressed in hyberbolic sine and cosine functions. After fitting the boundary conditions at $x = 0$ and W, we obtain:

$$\tilde{v}_{gs}(x) = \frac{\tilde{v}_{\mathrm{in}}}{\cosh\sqrt{(\omega/\gamma)}W} \times \cosh\sqrt{\frac{j\omega}{\gamma}}(W-x) \tag{6-190}$$

The current density flowing in the gate metal is obtained by taking the derivative Eq. (6-190), in accordance with Eq. (6-186). We obtain:

$$\tilde{j}_m(x) = \frac{1}{\rho_G} \frac{\tilde{v}_{\text{in}}}{\cosh\sqrt{(j\omega/\gamma)}W} \times \sqrt{(j\omega/\gamma)} \sinh\sqrt{\frac{j\omega}{\gamma}}(W - x) \qquad \text{(6-191)}$$

The input impedance of the gate structure, equal to $v_{\text{in}}/[j_m(x = 0) \times L\Lambda_G]$, is therefore:

$$Z_{\text{in}} = \frac{R_{SHG}}{L}\sqrt{\frac{\gamma}{J\omega}} \coth\sqrt{\frac{j\omega}{\gamma}}W = \frac{1}{j\omega C_{gg}} \times \frac{1 + \sum_{n=1}^{\infty}[1/(2n)!][(j\omega/\gamma)W^2]^n}{1 + \sum_{n=1}^{\infty}[1/(2n+1)!][(j\omega/\gamma)W^2]^n} \qquad \text{(6-192)}$$

Example 6-6:

A HFET is designed with an AlGaAs thickness of 300 Å, and a gate dimension of 0.25 μm × 1 mm. Assume that $\Delta t_b = 70$ Å. The gate metal is characterized by a sheet resistance of 0.025 Ω/square. It is biased in saturation so that $\alpha = 0$. Find the real and imaginary part of the small-signal input impedance of the gate structure, as a function of frequency.

From Eq. (5-121):

$$C'_{ox} = \frac{13.1 \times 8.85 \times 10^{-14}}{(300 + 70) \times 10^{-8}} = 3.13 \times 10^{-7}\ \text{F/cm}^2$$

C_{gg}, in accordance with Eq. (6-16), is:

$$C_{gg} = 0.1 \cdot 0.25 \times 10^{-4} \cdot 3.13 \times 10^{-7} \cdot \frac{2}{3} = 5.22 \times 10^{-13}\ \text{F}$$

With $R_{SHG} = 0.025$ Ω/sq, the diffusion coefficient from Eq. (6-188) is calculated to be:

$$\gamma = \frac{0.1 \cdot 0.25 \times 10^{-4}}{0.025 \cdot 5.22 \times 10^{-13}} = 1.916 \times 10^8\ \text{cm}^2/\text{s}$$

We calculate Z_{in} using Eq. (6-192). If we define $Z_{\text{in}} = R_{\text{in}} + j\omega \cdot C_{\text{in}}$, then R_{in} and C_{in} as a function of frequency is shown in Fig. 6-33. The figure shows that the input resistance is equal to $1/3R_{SHG} \cdot W/L$. Generally, we tend to think that a device's performance degrades over frequency. An intuitive expectation would be that the input resistance increased with frequency, since degradation in

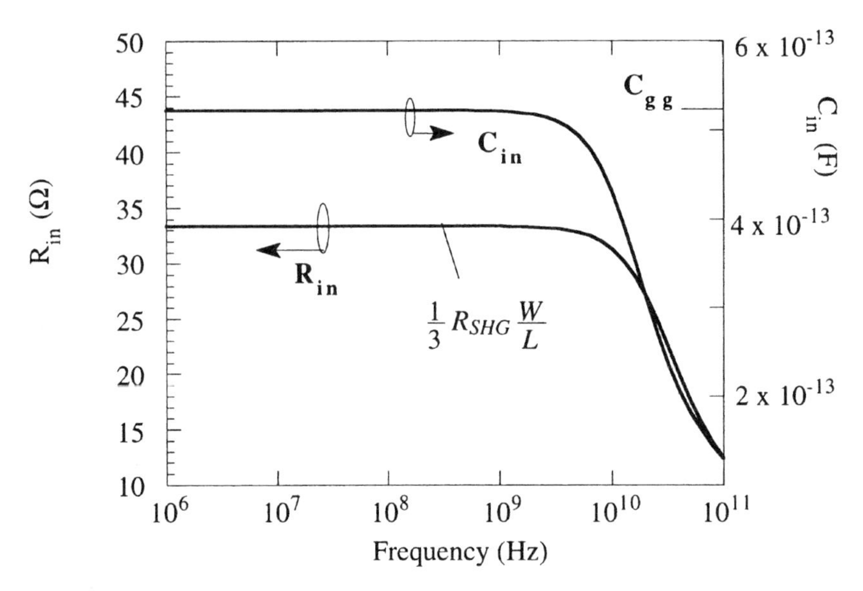

FIGURE 6-33. R_{in} and C_{in} as a function of frequency for the gate structure shown in Fig. 6-32.

device performance was thought to be accompanied by higher resistance. However, the result shown in Fig. 6-33 suggests that the input resistance actually decreases with frequency when ω exceeds a certain high value. This because as the frequency gets higher, less and less signal travels through the entire width of the gate before diverting to the channel. When only a certain portion of the width is accessible to the input signal, the effective width decreases, hence the input resistance decreases. This is similar to the intrinsic base resistance discussed in § 3-7.

Figure 6-33 also shows that the input capacitance has the value C_{gg} when the frequency is low. As frequency increases, the usable portion of the gate width decreases. The input capacitance thus decreases.

From the example, we see that at low frequency, R_G, the real part of the input impedance given in Eq. (6-192) is:

$$R_G = \frac{1}{3} \mathrm{R}_{SHG} \frac{W}{L} \tag{6-193}$$

The 1/3 factor in Eq. (6-193) is absent in R_{measure} of Eq. (6-177). The discrepancy resides in the difference in the current conduction paths in the two structures. In the test structure for which Eq. (6-177) is applicable, the current never leaves the gate metal. The total measured resistance is therefore equal to the sheet resistance times the number of squares in the test structure. In the gate structure of an actual transistor, all of the signals go into one end of the gate structure, but none of them goes out at the other end. These a.c. signals eventually all leave the gate and enter the channel through C_{gg} (in the form of $C_{gg} \cdot dv_{GS}/dt$). Some signals remain in the gate for a longer portion of the gate width before exiting to the channel. Some signals depart sooner. The factor 1/3 accounts for the distributed nature of the current conduction.

If the metal is contacted from both ends, then the gate resistance becomes

$$R_G = \frac{1}{12}\frac{W}{L} R_{SHG} \tag{6-194}$$

One might expect the factor to be 1/6. The resistance decreases by fourfold instead of twofold because of the following. When the gate metal is contacted by two sides, then the location at which j_M equals zero occurs at $x = W/2$ rather than W. The resistance already decreases by twofold due to the width reduction. The overall resistance decreases by another twofold due to the fact that there are two contacts connected in parallel. This factor of 1/12 is similar to that encountered in the intrinsic base resistance of the bipolar transistors (§ 3-7).

Equation (6-193) works for a single-finger device. If the device consists of several fingers, then the overall device gate resistance is given by:

$$R_G = \frac{1}{3}\frac{W}{L} R_{SHG} \times \frac{1}{N} \tag{6-195}$$

Here, W represents the gate width per finger and N is the number of figures.

§ 6-6 SOURCE/DRAIN RESISTANCES AND PARASITIC CAPACITANCES

The gate resistance affects mostly the high-frequency performance of the transistor. There is no gate current at d.c. Hence, there is no d.c. voltage drop across the gate resistance. The source and drain terminal resistances, however, can greatly influence the d.c. characteristics of the transistor. Because all of the current flowing out of the channel flows through the contacts, the source (or drain) resistance is a lumped resistance. This differs from the gate resistance, which is a distributed resistance.

Let us first establish the effects produced by the source and drain resistances. Consider the bias condition shown in Fig. 6-34. Let us call g_m and g_d the measured conductances at nodes extrinsic to the transistor. Conversely, g_{mi} and g_{di} refer to the conductances intrinsic to the device. We find:

$$\begin{aligned}\Delta I_D &= \left.\frac{\partial I_D}{\partial V'_{GS}}\right|_{V'_{DS}} \Delta V'_{GS} + \left.\frac{\partial I_D}{\partial V'_{DS}}\right|_{V'_{GS}} \Delta V'_{DS} \\ &= g_{mi}\Delta V'_{GS} + g_{di}\Delta V'_{DS}\end{aligned} \tag{6-196}$$

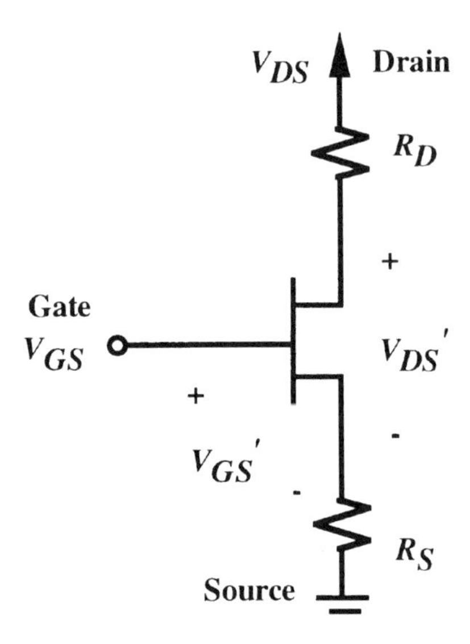

FIGURE 6-34. A FET with finite source and drain resistances under biases of V_{GS} and V_{DS}. The biases appearing in the intrinsic transistor are $V_{GS'}$ and $V_{DS'}$. (From R. S. Chou and D. Antoniads. "Relationship between measured and intrinsic transconductances of FETs". *IEEE Trans. Electron. Dev.* **34,** 448–450, 1987. © IEEE, reprinted with permission)

The extrinsic and the intrinsic bias voltages are related by:

$$V'_{GS} = V_{GS} - R_S I_D \tag{6-197}$$

$$V'_{DS} = V_{DS} - (R_S + R_D) I_D \tag{6-198}$$

Therefore, when V_{DS} is a constant, $\Delta V'_{DS} = -(R_S + R_D)\Delta I_D$ and $\Delta V'_{GS} = \Delta V_{GS} - R_S \Delta I_D$. Substituting these incremental voltage values into Eq. (6-196), we obtain:

$$[1 + g_{mi} R_S + g_{di}(R_S + R_D)]\Delta I_D = g_{mi} \Delta V_{GS} \tag{6-199}$$

$\Delta I_D / \Delta V_{GS}$ is g_m, the measured transconductance at nodes outside of the intrinsic device. Therefore, Eq. (6-199) leads to:

$$g_m = \frac{g_{mi}}{1 + g_{mi} R_S + g_{di}(R_S + R_D)} \tag{6-200}$$

The measured g_m is degraded from the ideal intrinsic-device value of g_{mi} by a factor relating to both R_S and R_D. This expression simplifies to $g_{mi}/(1 + g_{mi} R_S)$ only when $g_{di} = 0$, or in long-channel devices biased in saturation. More generally, such as in a short-channel device, the effects of g_{di} on the mutual transconductance must be considered.

Usually the measured g_m is a known quantity, and g_{mi} is not. It is desirable to infer the intrinsic device transconductance from the measured value, either for the purposes of modeling or process control. In this case, we re-express the above equation in the following manner:

$$g_{mi} = \frac{g_m[1 + g_{di}(R_S + R_D)]}{(1 - g_m R_S)} \tag{6-201}$$

We now consider the situation when V_{GS} is held constant while V_{DS} is allowed to vary. In this case, $\Delta V'_{DS} = \Delta V_{DS} - (R_S + R_D)\Delta I_D$ and $\Delta V'_{GS} = -R_S\Delta I_D$. Substituting these incremental voltages into Eq. (6-196) and applying the definition and $g_d = \Delta I_D/\Delta V_{DS}$, we obtain:

$$g_d = \frac{g_{di}}{1 + g_{mi}R_S + g_{di}(R_S + R_D)} \tag{6-202}$$

The amount of degradation from the intrinsic-device value of g_{di} is determined by the same factor as the degradation in g_m. Sometimes a less accurate derivation leads to $g_d = g_{di}/[1 + g_{di}(R_S + R_D)]$. Such a result is correct only when g_m is small, as in the case when V_{DS} is small (linear region).

The measured value of g_d is often used to extract the intrinsic drain conductance g_{di}. We rearrange the equation to the following:

$$g_{di} = \frac{g_d(1 + g_{mi}R_S)}{1 - g_d(R_S + R_D)} \tag{6-203}$$

Example 6-7:

The intrinsic portion of a $W = 500$ μm FET has a g_{mi} of 600 mS/mm (millisiemens per millimeter) and a g_{di} of 100 mS/mm. The parasitic source and drain resistances are 2 Ω-mm each. Find the measured g_m and g_d.

Because $W = 0.5$ mm, $g_{mi} = 0.3$ S and $g_{di} = 0.05$ S. $R_S = R_D = 4\ \Omega$. According to Eqs. (6-200) and (6-202):

$$g_m = \frac{0.3}{1 + 0.3 \cdot 4 + 0.05(4 + 4)} = 0.115\ 1/\Omega$$

$$g_d = \frac{0.05}{1 + 0.3 \cdot 4 + 0.05(4 + 4)} = 0.019\ 1/\Omega$$

Example 6-7 demonstrates that the overall device g_m and g_d are degraded by the presence of the source-drain resistances. The desire to reduce the terminal resistances eventually leads to the development of modern FET structures (both MESFETs and HFETs) which utilize a cap layer underneath the contacts. Besides reducing the contact resistance at the metal/semiconductor junction, this heavily doped cap layer provides more area for the channel carriers to reach the contacts, as shown in Fig. 6-35*a*. The corresponding equivalent distributed resistive network is shown in Fig. 6-35*b*.

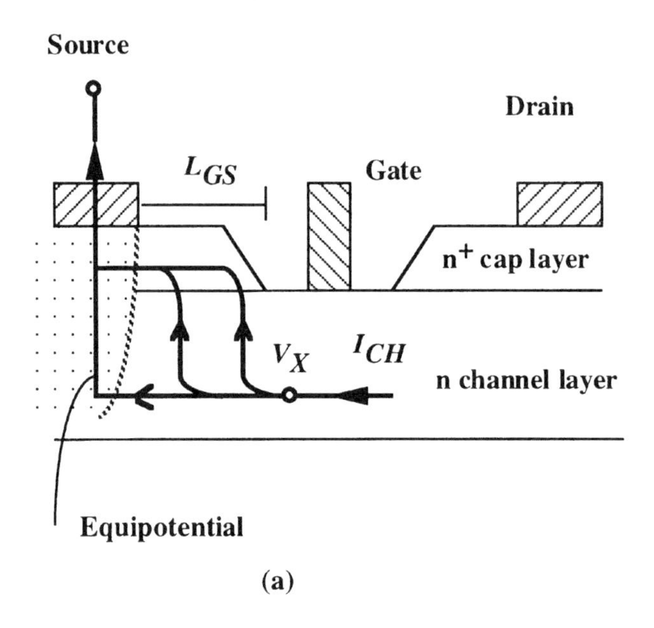

(a)

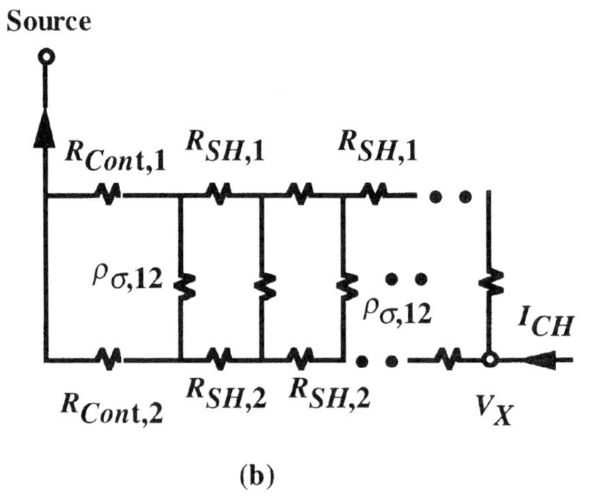

(b)

FIGURE 6-35. (*a*) Schematic diagram showing the current paths through the source area of a FET. (*b*) An equivalent circuit for the current flow through the source.

Due to the two-dimensional current flow through the source and the drain contacts, a theoretical formulation for R_S and R_D in FET devices is more complicated than the terminal resistances in HBTs. The source resistance is defined as the ratio of V_X/I_{CH}, where V_X is shown in Fig. 6-35 and I_{CH} is the current flowing through the contact (generally equal to I_D). An expression for R_S is given by [3]:

$$R_S = \frac{L_{GS}}{W} \times \frac{R_{SH,1} \cdot R_{SH,2}}{R_{SH,1} + R_{SH,2}} + \frac{1}{W} \times \frac{1}{R_{SH,1} + R_{SH,2}} \times \frac{\alpha + \beta \cosh kL_{GS} + \gamma k \sinh kL_{GS}}{(R_{SH,1} + R_{SH,2}) \cosh kL_{GS} + (R_{Cont,1} + R_{Cont,2}) k \sinh kL_{GS}} \tag{6-204}$$

where

$$\alpha = 2R_{SH,2}(R_{SH,1} \cdot R_{Cont,2} - R_{SH,2} \cdot R_{Cont,1}) \tag{6-205}$$

$$\beta = 2R_{SH,2}^2 \cdot R_{Cont,1} + (R_{SH,1}^2 + R_{SH,2}^2) \cdot R_{Cont,2} \tag{6-206}$$

$$\gamma = (R_{SH,1} + R_{SH,2}) R_{Cont,1} R_{Cont,2} + R_{SH,2}^2 \cdot \rho_{\sigma,12} \tag{6-207}$$

$$k = \sqrt{(R_{SH,1} + R_{SH,2}/\rho_{\sigma,12}} \tag{6-208}$$

L_{GS} is the contact spacing between the gate and the source metals (μm); W is the gate width (μm); $R_{SH,1}$ is the sheet resistance of the top cap layer (Ω/sq); $R_{SH,2}$ is the sheet resistance in the access layer (the lower layer) (Ω/sq); $\rho_{\sigma,12}$ is the specific contact resistance between layers 1 and 2 (Ω-cm^2); $R_{Cont,1}$ is the contact resistance per unit width formed between the alloyed metal and layer 1 (Ω-mm); and $R_{Cont,2}$ is the contact resistance per unit width formed between the alloyed metal and layer 2 (Ω-mm). The values for a HFET with a 500 Å, 2×10^{18} cm^{-3} cap layer and a 350 Å, 1.5×10^{18} cm^{-3} $Al_{0.3}Ga_{0.7}As$ barrier layer are [3]: $R_{SH,1} = 216$ Ω/sq; $R_{SH,2} = 1164$ Ω/sq; $\rho_{\sigma,12} = 1.5 \times 10^{-5}$ Ω-cm^2; $R_{Cont,1} = 0.255$ Ω-mm (or 0.0255 Ω-cm); and $R_{Cont,2} = 1.2$ Ω-mm (or 0.12 Ω-cm). If the width is 10 mm, then the $R_{Cont,1}$ and $R_{Cont,2}$ values provided above should be divided by 10, while other parameters are unaffected.

Unlike the gate resistance, the formula for R_S works independent of the number of finger. W in Eq. (6-204) represents the total gate width of the device. It does not matter what the per-finger gate width is.

R_S given in the above expression is in Ω. Whenever that a R_S value is given, it is important to specify the gate width of the device. This is because the source resistance scales linearly with the width. Generally, R_D is taken to be equal to R_S for a FET with symmetrical structures at the two sides of the gate.

Figure 6-36 plots R_S as a function of the gate-to-source spacing. The device is a 1-mm HFET with the aforementioned parameter values, except that $\rho_{\sigma,12}$ is

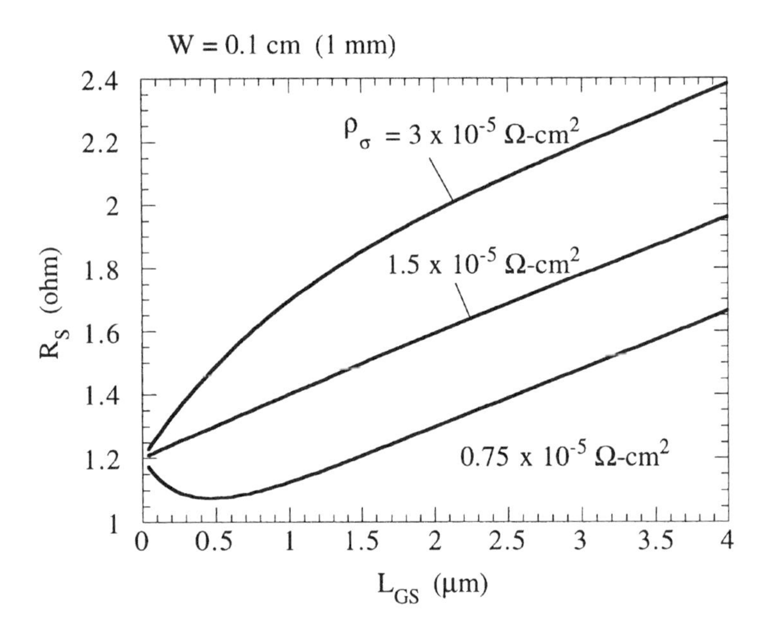

FIGURE 6-36. Source resistance as a function of the gate-to-source spacing for various values of specific contact resistances between the cap and the channel layers. (From P. Roblin, L. Rice, S. Bibyk, and H. Morkoc. "Nonlinear parasitics in MODFET's and MODFET I-V characeristics." *IEEE Trans. Electron. Dev.* **35,** 1207–1214, 1988. © IEEE, reprinted with permission. The figure notations are modified to conform to the symbols used in this book.)

changed from 3, 1.5, to 0.75×10^{-5} Ω-cm^2. In general, we expect the source resistance to increase with L_{GS}. However, if $\rho_{\sigma,12}$ is small, then a lot of the current can be channeled through the low-reistivity cap layer. When the increase in L_{GS} decreases the resistance of the cross-barrier path more than it increases the resistance due to current along the layers, then R_S decreases with L_{GS}.

The R_S formula requires knowledge of several material parameters, such as $R_{\text{SH},1}$ and $\rho_{\sigma,12}$. These parameters need to be determined from special test structures. Once the parameters' values are known, the source resistance is computed. However, there are times when it is desirable to measure directly the source resistance of the device under consideration. There are several measurement techniques [4]. In the following, we describe the *channel-resistance method* to determine the source resistance, applicable to FETs in general (including MOSFETs). To simplify the discussion, we consider symmetrical device with equal gate-source spacing (L_{GS}) and gate-drain spacing (L_{GD}), so that R_S and R_D are identical.

When the gate voltage is high while the drain voltage is small (approaching 0.05 V, for example), then the drain currents can be approximated from Eqs. (6-62) and (6-65):

$$I_D \approx \frac{W}{L} aqN_d\mu_N\left[1 - \sqrt{\frac{\phi_{bi} - V_{GS}}{\phi_{00}}}\right]V_{DS} \qquad \text{(MESFETs)} \qquad (6\text{-}209)$$

$$I_D \approx \frac{W}{L} \mu_n C'_{ox}(V_{GS} - V_T)\, V_{DS} \qquad \text{(HFETs)} \qquad (6\text{-}210)$$

For the MESFET equation, we took advantage of the Taylor expansion that $f(x_0 + \Delta x) \approx f(x_0) + df/dx \times \Delta x$. Therefore, when V_{DS} is small:

$$\left(\frac{\phi_{bi} - V_{GS} + V_{DS}}{\phi_{00}}\right)^{3/2} \approx \left(\frac{\phi_{bi} - V_{GS}}{\phi_{00}}\right)^{3/2} + \frac{3}{2}\left(\frac{\phi_{bi} - V_{GS}}{\phi_{00}}\right)^{1/2} \times \frac{V_{DS}}{\phi_{00}} \qquad (6\text{-}211)$$

For the HFET equation, we simply neglected the term $V_{DS}^2/2$ in comparison to V_{DS}.

The drain current in either device is linearly proportional to the drain voltage when V_{DS} is small. Therefore, when a FET operates under the linear region, the transistor presents a constant channel resistance (R_{CH}), given by:

$$R_{CH} \approx \frac{L}{W}\frac{1}{aqN_d\mu_n}\left[1 - \sqrt{\frac{\phi_i - V_{GS}}{\phi_{00}}}\right]^{-1} \qquad \text{(MESFETs)} \qquad (6\text{-}212)$$

$$R_{CH} \approx \frac{L}{W}\frac{1}{\mu_n C'_{ox}(V_{GS} - V_T)} \qquad \text{(HFETS)} \qquad (6\text{-}213)$$

This channel resistance (R_{CH}) is unrelated to the non-quasi-static channel resistance (r_{ch}) of the small-signal model discussed in § 6-4. There should be no confusion between them because they appear in quite dissimilar scenarios.

The measured total resistance (R_T) at the nodes extrinsic to the transistor is equal to the sum of the source-drain resistances, as well as the channel resistance. That is:

$$R_T = R_S + R_{CH}(V_{GS}, L) + R_D \qquad (6\text{-}214)$$

We specifically point out that the channel resistance is a function of the gate-to-source bias and the channel length. By varying several devices with different L's but biasing them at the same V_{GS}, we then expect to reveal the R_T variation with L, as shown schematically in Fig. 6-37. By extrapolating the R_T values to $L = 0$, we obtain the y-axis intercept as the sum of the source and the drain resistance. Since they are identical in a symmetrical structure, $R_S = R_D =$ half of the intercept value.

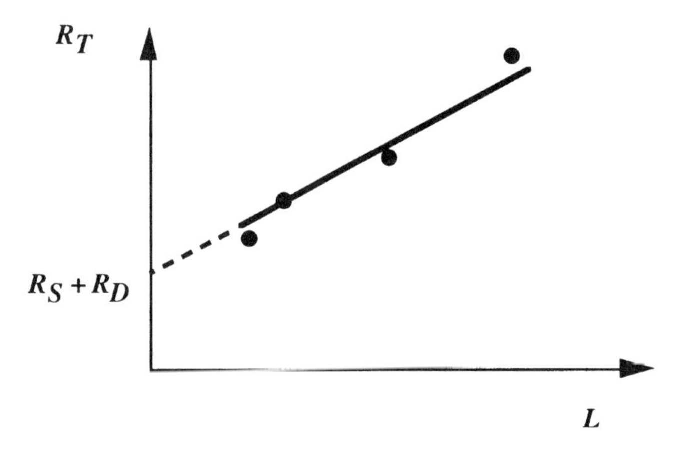

FIGURE 6-37. Total transistor resistances of transistors with various channel lengths. The extrapolated resistance at $L = 0$ is the sum of the source and the drain resistances.

If only one device is available for the resistance measurement, it is still possible to extract R_S and R_D. We need only measure the device at various V_{GS}'s. By plotting the measured resistance as a function of $1/(V_{GS} - V_T)$, we again obtain a line whose slope is proportional to R_{CH}. The source resistance is half of the extrapolated value at the y-axis intercept.

Besides parasitic resistances, there are parasitic capacitances in FETs, as illustrated in Fig. 6-38. Let us examine the parasitic capacitance between the drain and the source electrodes first. Because the drain-to-source spacing is much smaller than the substrate thickness, we can consider the substrate to be infinitely thick and the metal thickness to be infinitely thin. Using the conformal mapping technique elaborated in Appendix C, the capacitance between two plates of metal on the same plane can be determined. With ϵ_s and ϵ_0 being the dielectric constants for the GaAs semiconductor and air, respectively, we write:

$$C_{ds,p} = C_{sd,p} = \frac{\epsilon_s + \epsilon_0}{2} W \times \frac{K\left(\sqrt{1 - k_{ds}^2}\right)}{K(k_{ds})} \qquad k_{ds} = \frac{L_{DS}}{L_{DS} + 2L_S} \tag{6-215}$$

where $K(k)$ is the complete elliptic integral of the first kind, defined by

$$K(k) = \int_0^1 \frac{dw}{(1 - w^2)^{1/2}(1 - k^2 w^2)^{1/2}} \tag{6-216}$$

This integral has been evaluated numerically, with results tabulated in mathematical handbooks.

The capacitance symbol ($C_{ds,p}$) contains a subscript p, signifying that this is a parasitic component. The capacitance depends solely on the geometric

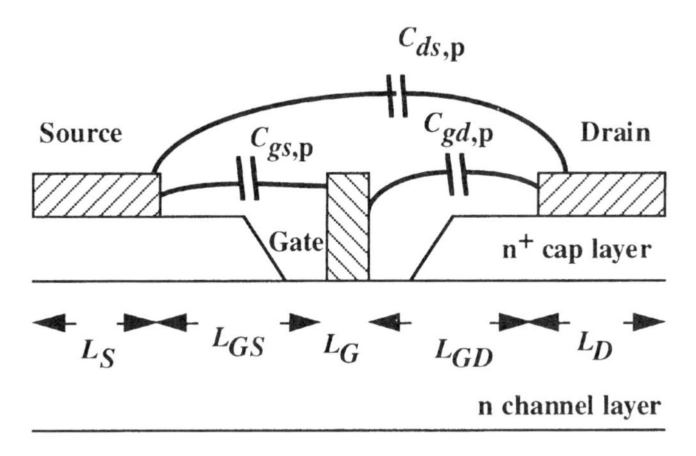

FIGURE 6-38. Locations of parasitic capacitances in FETs.

information between the source and the drain. No transistor action is involved in the formula. This parasitic capacitance can be treated as a two-terminal parallel-plate capacitance, in that the incremental drain charge due to source voltage variation is identical to the incremental source charge due to drain voltage variation. Therefore, $C_{ds,p} = C_{sd,p}$. This kind of equality does not apply to the transistor capacitances ($C_{ds} \neq C_{sd}$).

In the derivation of $C_{ds,p}$ in Eq. (6-216), L_S is assumed to be equal to L_D. As is, the equation is already complicated. In Appendix C, we list the formulas for $C_{dg,p}$, which is equal to $C_{gd,p}$, and $C_{sg,p}$, which is equal to $C_{gs,p}$. These formulas are even more complicated than Eq. (6-216), mainly because $L_G \neq L_D$ and $L_G \neq L_S$. We will not list them here.

$C_{gs,p}$ (or $C_{sg,p}$) is usually not important in normal operation. When the transistor is turned on and operates in saturation, C_{gs} has some large values. $C_{gs,p}$ is usually insignificant in comparison. Therefore, the total gate-to-source capacitance, $C_{gs,t}$, is composed mostly of C_{gs}. (Note: a capacitance without an additional subscript such as t or p means that it is an intrinsic capacitance belonging to the intrinsic portion of the transistor.) While $C_{gs,p}$ is unimportant, $C_{gd,p}$ is. This is because, when the transistor is in saturation, C_{gd} is approximately zero. The overall gate-to-drain capacitance, $C_{gd,t}$, consists exclusively of the parasitic component.

§ 6-7 CUTOFF AND MAXIMUM OSCILLATION FREQUENCIES

The y-parameters of Eqs. (6-134) to (6-142) are those of the intrinsic transistor. An actual device consists of series parasitic resistances at the terminals as well as the parasitic capacitances. Let us focus first on the effects brought about by the parasitic capacitances. Because the parasitic capacitances are all in parallel with their intrinsic counterpart, we can add the parasitics directly to their intrinsic values. We write all nine of the total device capacitances, although only four of them are linearly independent:

$$C_{gg,t} = C_{gg} + C_{gd,p} + C_{gs,p} \tag{6-217}$$

$$C_{gd,t} = C_{gd} + C_{gd,p} \tag{6-218}$$

$$C_{gs,t} = C_{gs} + C_{gs,p} \tag{6-219}$$

$$C_{dg,t} = C_{dg} + C_{dg,p} \tag{6-220}$$

$$C_{dd,t} = C_{dd} + C_{dg,p} + C_{ds,p} \tag{6-221}$$

$$C_{ds,t} = C_{ds} + C_{ds,p} \tag{6-222}$$

$$C_{sg,t} = C_{sg} + C_{sg,p} \tag{6-223}$$

$$C_{sd,t} = C_{sd} + C_{sd,p} \tag{6-224}$$

$$C_{ss,t} = C_{ss} + C_{sg,p} + C_{sd,p} \tag{6-225}$$

where C_{xy}'s without the t and p subscripts denote the intrinsic values, given in § 6-1 for HFETs and in § 6-2 for MESFETs.

When the parasitic capacitances are included, the y-parameters for a FET in the common-source configuration are given in the following, according to Eqs. (6-109) and (6-134) to (6-137):

$$[y]_s = \begin{bmatrix} y_{gg} & y_{gd} \\ y_{dg} & y_{dd} \end{bmatrix} = \begin{bmatrix} j\omega C_{gg,t} & -j\omega C_{gd,t} \\ g_m - j\omega C_{dg,t} & g_d + j\omega C_{dd,t} \end{bmatrix} \tag{6-226}$$

We proceed to find the cutoff frequency and the maximum oscillation frequency, without considering the parasitic resistances (only the parasitic capacitances are accounted for). Because f_T characterizes the switching speed and $f_{\max}$ measures the amount of power gain, f_T is more critical to digital circuit design while $f_{\max}$ is the more pertinent to r.f. applications.

The cutoff frequency is the frequency at which the magnitude of the forward current gain (h_{21}) decreases to unity. The forward current gain is given by $h_{21} = y_{21}/y_{11}$. Therefore:

$$h_{21} = \frac{g_m - j\omega C_{dg,t}}{j\omega C_{gg,t}} \quad \text{(neglecting parasitic resistances)} \tag{6-227}$$

When $\omega = \omega_T$, $|h_{21}| = 1$. Using this condition, we find ω_T to be:

$$\omega_T = \frac{g_m}{\sqrt{C_{gg,t}^2 - C_{dg,t}^2}} \tag{6-228}$$

It is seen that the imaginary part in the numerator of h_{21}, $j\omega C_{dg,t}$, increases the value of ω_T. A conservative estimation of ω_T is obtained by neglecting the extra phase contributed by this imaginary part. In this case, ω_T is written as

$$\omega_T = 2\pi f_T = \frac{g_m}{C_{gg,t}} \tag{6-229}$$

The cutoff frequency f_T (in Hz) is the frequency corresponding to the radian frequency ω_T (in rad/s).

Because g_m is inversely proportional to L [see Eq. (6-63) or (6-66)] while $C_{gg,t}$ is proportional to L [see Eq. (6-42) or (6-16)], ω_T is proportional to L^{-2}. As each generation of FET shrinks the gate length, f_T is bound to increase. However, it is not always certain that the maximum oscillation frequency will necessarily increase as L shrinks. This is because $f_{\max}$ depends, much more so than f_T, on the values of the parasitics. Note that in short-channel devices, g_m can become relatively independent of L. In this case, the cutoff frequency improvement by shrinking the gate length is not as dramatic.

We proceed to find the maximum oscillation frequency, which is the frequency at which the unilateral power gain (U) is equal to 1. The unilateral power gain is given by

$$U = \frac{|y_{21} - y_{12}|^2}{4[\mathrm{Re}(y_{11})\mathrm{Re}(y_{22}) - \mathrm{Re}(y_{12})\mathrm{Re}(y_{21})]} \tag{6-230}$$

$$= \frac{|z_{21} - z_{12}|^2}{4[\mathrm{Re}(z_{11})\mathrm{Re}(z_{22}) - \mathrm{Re}(z_{12})\mathrm{Re}(z_{21})]} \tag{6-231}$$

We substitute the y-parameters from Eq. (6-226) into Eq. (6-230). Because $\mathrm{Re}(y_{11}) = 0$ and $\mathrm{Re}(y_{12}) = 0$, U is equal to infinity. This is clearly an overestimation of what the device can deliver. This error can be corrected by two considerations. First, the y-parameters of Eq. (6-226) are quasi-static y-parameters. When the NQS y-parameters are used, we find that there is a real part in the expression for y_{11}, related to the channel resistance [r_{ch} of Eq. (6-170)]. Inclusion of such a resistance prevents U from reaching a nonphysical value. Another correction, which is by far more important, is that there are terminal resistances in an actual device. The device is better represented by the equivalent circuit of Fig. 6-39, which includes a lumped resistance connected in series at each of the three terminals. The gate resistance expression is given in Eq. (6-193). The source and drain resistances are given by Eq. (6-204) (R_D is obtained from the R_S expression as well). To establish the r.f. properties of the overall transistor, we shall now use the equivalent circuit of Fig. 6-39 to calculate the forward current gain and the unilateral power gain.

Equation (6-226) is the y-parameter matrix of the intrinsic device. It has not yet incorporated the parasitic resistances. To obtain the small-signal parameters of the overall device, we first convert the intrinsic y-parameters to z-parameters.

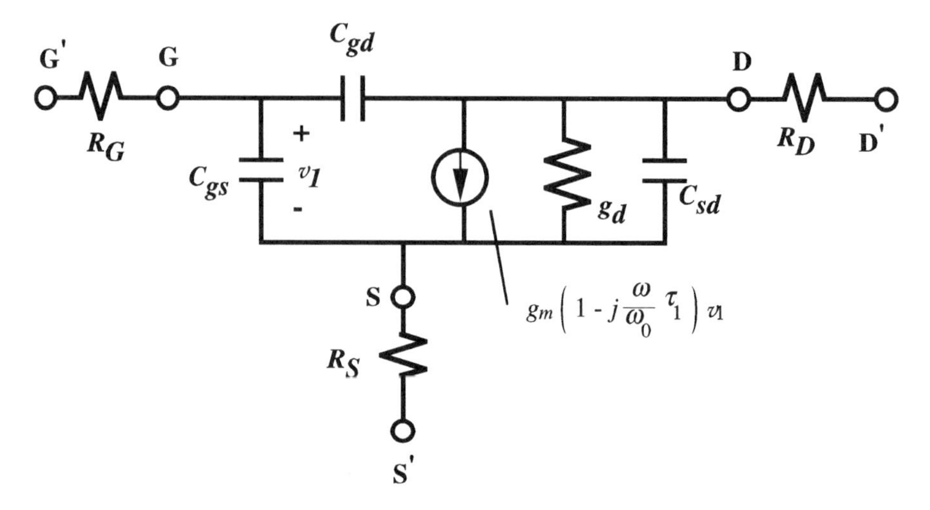

FIGURE 6-39. Equivalent small-signal circuit of realistic FETs including the parasitic resistances.

The conversion is performed with the formula shown in Fig. 4-17. Next, the series resistances are appended to the $[z]$ matrix. The final $[z']$ matrix for the overall transistor is given as:

$$[z'] = [z] + \begin{bmatrix} R_G + R_S & R_S \\ R_S & R_D + R_S \end{bmatrix} = \begin{bmatrix} R_G + R_S + \dfrac{y_{22}}{\Delta y} & R_S - \dfrac{y_{12}}{\Delta y} \\ R_S - \dfrac{y_{21}}{\Delta y} & R_D + R_S + \dfrac{y_{11}}{\Delta y} \end{bmatrix} \tag{6-232}$$

where Δy is

$$\Delta y = y_{11}y_{22} - y_{12}y_{21} = j\omega(g_d C_{gg,t} + g_m C_{gd,t}) + \omega^2(C_{gd,t}C_{dg,t} - C_{gg,t}C_{dd,t}) \tag{6-233}$$

According to Fig. 4-17, h_{21} is equal to $-z_{21}/z_{22}$. Or, working with the $[z']$ matrix given in Eq. (6-232), h'_{21} of the primed two-port of Fig. 6-39 is $-z'_{21}/z'_{22}$. After a lengthy derivation and neglecting the second- and higher-order terms of ω, we find the forward current gain of the overall transistor to be:

$$h'_{21} = -\frac{g_m - j\omega(R_S C_{gg,t} g_d + R_S C_{gd,t} g_m + g_m)}{j\omega[C_{gg,t} + (R_S + R_D)C_{gd,t} g_d + (R_S + R_D)C_{gd,t} g_m]} \tag{6-234}$$

We omit the $j\omega$ term in the numerator of Eq. (6-234) to acquire a simple expression for the cutoff frequency. Substituting $\omega = 2\pi f_T$ into the equation and setting the condition $|h'_{21}| = 1$, we find f_T as:

$$\frac{1}{2\pi f_T} = \frac{C_{gg,t}}{g_m} + \frac{C_{gg,t}}{g_m}(R_S + R_D)\, g_d + (R_S + R_D)\, C_{gd,t} \tag{6-235}$$

In the ideal case where the drain and source resistances are zero, Eq. (6-235) reduces to the well-known (but approximate) expression given in Eq. (6-229).

Applying the expression in Eq. (6-231), we obtain the unilateral power gain as:

$$U = \frac{|(y_{21} - y_{12})/\Delta y|^2}{4[\mathrm{Re}(R_S + R_G + y_{22}/\Delta y)\mathrm{Re}(R_S + R_D + y_{11}/\Delta y) - \mathrm{Re}(R_S - y_{12}/\Delta y)\mathrm{Re}(R_S - y_{21}/\Delta y)]} \tag{6-236}$$

After some algebraic manipulations, the unilateral gain can be expressed as:

$$\begin{aligned}\frac{1}{U} = \frac{4\omega^2}{g_m^2}\Big\{ & R_G[(R_S + R_D)(C_{gg,t}g_d + C_{gd,t}g_m)^2 + C_{gg,t}(C_{gg,t}g_d + C_{gd,t}g_m)]. \\ & + R_D[R_S(C_{gg,t}g_d + C_{gd,t}g_m)^2 + C_{gd,t}(C_{dg,t}g_d + C_{dd,t}g_m)] \\ & + R_S[g_m(C_{gg,t} - C_{gd,t})(C_{gd,t} - C_{dd,t}) + g_d(C_{gg,t} - C_{gd,t})(C_{gg,t} - C_{dg,t})]\Big\}\end{aligned} \tag{6-237}$$

We neglect the latter two terms in Eq. (6-237) to establish a manageable equation for the maximum oscillation frequency. Replacing U by 1 and ω by $2\pi \cdot f_{\max}$, we find:

$$f_{\max} = \sqrt{\frac{f_T}{8\pi R_G C_{gd,t}\left[1 + \left(\dfrac{2\pi f_T}{C_{gd,t}}\right)\Psi\right]}} \tag{6-238}$$

where Ψ is

$$\Psi = (R_D + R_S)\frac{C_{gg,t}^2 g_d^2}{g_m^2} + (R_D + R_S)\frac{C_{gg,t}C_{gd,t}g_d}{g_m} + \frac{C_{gg,t}^2 g_d}{g_m^2} \tag{6-239}$$

§ 6-8 EXEMPLAR CALCULATION OF f_T AND f_{max}

We consider a HFET whose gate width is 0.6 mm and the gate length is 0.3 μm. The gate sheet resistance is 0.03 Ω/sq. The source and drain resistances per unit width are equal, each given by 1.2 Ω-mm. The HFET has a C'_{ox} of 3×10^{-7} F/cm^2 and a V_T of 0.3 V. The electron mobility in the channel is 4000 cm^2/ V-s. It is biased at $V_{GS} = 0.5$ V and $V_{DS} = 3$ V. We shall assume that the parasitic capacitance between the gate and the drain (and between the gate and the source) is 1/10 of the intrinsic device's C_{gg}. The parasitic capacitance between the drain and the source is assumed to be 1/20 of the intrinsic device's C_{gg}. We attempt to evaluate the cutoff and maximum oscillation frequencies of the device.

Let us evaluate the saturation index α first. According to Eq. (5-130):

$$V_{DS,\text{sat}} = 0.5 - 0.3 = 0.2 \text{ V}$$

Because $V_{DS} > V_{DS,\text{sat}}$, α determined from Eq. (5-132) is zero. With $W \times L = 600 \times 0.3 \times 10^{-8}$ cm^2, we find the intrinsic capacitances from Eqs. (6-16), (6-17), (6-27), and (6-28):

$$C_{gg} = 600 \times 0.3 \times 10^{-8} \cdot 3 \times 10^{-7} \times \frac{2}{3}\frac{1 + 4.0 + 0^2}{(1 + 0)^2} = 3.6 \times 10^{-13} \text{ F}$$

$$C_{gd} = 600 \times 0.3 \times 10^{-8} \cdot 3 \times 10^{-7} \times \frac{2}{3}\frac{2.0 + 0^2}{(1 + 0)^2} = 0 \text{ F}$$

$$C_{dg} = 600 \times 0.25 \times 10^{-8} \cdot 3 \times 10^{-7} \times \frac{2}{15}\frac{2 + 14.0 + 11.0^2 + 3.0^3}{(1 + 0)^3} = 1.2 \times 10^{-13} \text{ F}$$

$$C_{dd} = 600 \times 0.25 \times 10^{-8} \cdot 3 \times 10^{-7} \times \frac{2}{15}\frac{8.0 + 9.0^2 + 3.0^3}{(1 + 0)^2} = 0 \text{ F}$$

The parasitic capacitances described above are: $C_{gd,p} = C_{dg,p} = C_{gs,p} = C_{sg,p} = 0.1 \times C_{gg} = 3.6 \times 10^{-14}$ F; and $C_{ds,p} = C_{sd,p} = 0.05 \times C_{gg} = 1.8 \times 10^{-14}$ F. According to Eqs. (6-217), (6-218), (6-220), and (6-221), the total capacitances are:

$$C_{gg,t} = 3.6 \times 10^{-13} + 3.6 \times 10^{-14} + 3.6 \times 10^{-14} = 4.32 \times 10^{-13} \text{ F}$$

$$C_{gd,t} = 0 + 3.6 \times 10^{-14} = 3.6 \times 10^{-14} \text{ F}$$

$$C_{dg,t} = 1.2 \times 10^{-13} + 3.6 \times 10^{-14} = 1.56 \times 10^{-13} \text{ F}$$

$$C_{dd,t} = 0 + 3.6 \times 10^{-14} + 3.6 \times 10^{-14} = 7.2 \times 10^{-14} \text{ F}$$

The transconductance g_m and the conductance g_d are found from Eqs. (6-66) and (6-67):

$$g_m = \frac{600 \times 10^4 \cdot 3 \times 10^{-7} \cdot 4000}{0.3 \times 10^{-4}}(0.5 - 0.3) \times (1 - 0) = 0.48\ 1/\Omega$$

$$g_d = \frac{600 \times 10^{-4} \cdot 3 \times 10^{-7} \cdot 4000}{0.3 \times 10^{-4}}(0.5 - 0.3) \times 0 = 0\ 1/\Omega$$

Because the width is 0.6 mm, $R_D = R_S = 1.2\ \Omega\text{-mm}/0.6\ \text{mm} = 2\ \Omega$. The gate resistance is given in Eq. (6-193):

$$R_G = \frac{1}{3} \times 0.03 \times \frac{600 \times 10^{-4}}{0.3 \times 10^{-4}} = 20\ \Omega$$

According to Eq. (6-235), we have:

$$\frac{1}{2\pi f_T} = \frac{4.32 \times 10^{-13}}{0.48} + \frac{4.32 \times 10^{-13}}{0.48}(2 + 2) \cdot 0 + (2 + 2) \cdot 3.6 \times 10^{-14}$$
$$= 1.044 \times 10^{-12}\ \text{s}$$

Therefore, $f_T = 150$ GHz. To find the maximum oscillation frequency, we determine the parameter Ψ from Eq. (6-239): it turns out all three terms of Ψ are multiplied by g_d, which is zero in this calculation. Therefore, $\Psi = 0$. From Eq. (6-238), we then have:

$$f_{\max} = \sqrt{\frac{f_T}{8\pi R_G C_{gd,t}}} = \sqrt{\frac{150 \times 10^9}{8\pi \cdot 20 \cdot 3.6 \times 10^{-14}}} = 9.1 \times 10^{10}\ \text{Hz}$$

So, $f_{\max}$ is 91 GHz. The maximum oscillation frequency is limited by the amount of the gate resistance. If the device's gate width is reduced so that R_G is smaller, then $f_{\max}$ increases.

REFERENCES

1. Liu, W., Bowen, C., and Chang, M. (1996). "A CAD-compatible non-quasi-static MOSFET model." *IEEE Int. Electron Dev. Meeting*, 151–154.
2. Liu, W., and Chang, M. (1998). "Transistor transient studies including transcapacitive current and distributive gate resistance for inverter circuits." *IEEE Trans. Circuits Sys. I.*, **45,** 416–422.
3. Feuer, M. (1985). "Two-layer model for source resistance in selectively doped heterojunction transistors." *IEEE Trans. Electron Dev.* **32,** 7–11. The exprcssion is modified somewhat to conform with the notation used in this book.

4. del Alamo, J., and Azzam, W. (1989). "A floating-gate transmission-line model technique for measuring source resistance in heterostructure field-effect transistors." *IEEE Trans. Electron Dev.* **36,** 2386–2393. This reference describes several resistance measurement techniques.

PROBLEMS

1. Start with Eq. (6-4). Multiply both sides by $(L - x)$ and then carry out the integration from $x = 0$ to $x = L$. Show that with the quasi-static approximation and the source current equation given by Eq. (6-29), the source charge is given by:

$$Q_S = -\frac{WC'_{\text{ox}}}{L}\int_0^L (L - x)U_{\text{CH}}dx$$

2. The $Al_{0.35}Ga_{0.65}As$ barrier layer of a FET has a thickness of 280 Å. The undoped portion of the AlGaAs layer adjacent to the channel is 30 Å. The doped portion adjacent to the gate is doped at 8×10^{17} cm^{-3}. Assume that the barrier height between the gate metal and the AlGaAs barrier layer to be 1.0 eV. The device has a gate width of 500 μm and a gate length of 0.25 μm. $V_{GS} = 0.3$ V and $V_{DS} = 2$ V.

a. What is the threshold voltage.

b. Find C_{gg}, C_{gd}, C_{dg}, and C_{dd}.

c. Find g_m and g_d.

d. Find the four common-source y-parameters.

3. Derive the drain charge expression in MESFET, given by Eq. (6-44).

a. The first step is to write down the differential equation governing the MESFET dynamics. Show that, after combining Eqs. (5-23) and (5-28), the equation is

$$\mu_n \frac{\partial}{\partial x}\left[\left(1 - \sqrt{\frac{\phi_s(x.t)}{\phi_{00}}}\right)\frac{\partial \phi_s(x.t)}{\partial x}\right] = \frac{\partial}{\partial t}\sqrt{\frac{\phi_s(x.t)}{\phi_{00}}}$$

b. Follow the steps in § 6-1, similar to Eqs. (6-20) to (6-24). During the step of integration by parts, the left integral can be evaluated with $u = x$ so that $du = dx$, and dv is $\partial/\partial x$ of the square-bracket term, so that v is exactly equal to that enclosed in the square brackets of the equation in part **a.**

4. A MESFET has the following parameters: $\phi_B = 0.9$ eV; $a = 1500$ Å; $N_d = 2 \times 10^{17}$ cm^{-3}; $W = 200$ μm; and $L = 0.5$ μm. The transistor is biased at $V_{GS} = 0$ V and $V_{DS} = 3$ V.

a. Find the parameters d and s.

b. Find C_{gg}, C_{gd}, C_{dg}, and C_{dd}.

c. Find g_m and g_d.

d. Find the four common-source y-parameters.

5. True or false:

a. It is known that $y_{sd,\text{QS}}$ as $-g_d - j\omega C_{sd}$. We therefore can write $y_{ds,\text{QS}} = -g_d - j\omega C_{ds}$.

b. As V_{DS} approaches zero, $C_{gs} = C_{gd}$, for both MESFETs and HFETs.

c. The remaining five capacitances of a three-terminal FET can be determined once C_{gg}, C_{gd}, C_{ds}, and C_{ss} are known.

d. In a FET without parasitic terminal resistances, the input impedance is always imaginary.

e. Consider an ideal transistor without leakage current in the gate. The small-signal current gain h_{21} is infinite because the gate current is zero.

f. The gate charge is negative in MESFETs and is positive in HFETs. Therefore, C_{gg} in one of the FETs is positive and, in the other FET, negative.

6. The drain of the FET in Fig. 6-1 is connected to a load capacitor C_L. Because C_L is initially charged to a high value of V_{DD}, v_{DS} is a large value at $t = 0$. As v_{GS} increases, the drain current flows to the source, discharging the capacitor. v_{DS} then decreases with time. For simplicity, we shall assume that during the entire transient, v_{DS} remains above $v_{DS,\text{sat}}$, so that $\alpha = 0$. This assumption becomes problematic only as v_{DS} decreases near zero.

a. Is Eq. (6-57) or Eq. (6-97) appropriate for the situation under consideration?

b. Show that at $t \geq 0$, v_{DS} is governed by the following equation:

$$-C_L \frac{dv_{DS}}{dt} = \frac{W\mu_n C'_{\text{ox}}}{2L}(v_{GS} - V_T)^2 + C_{dd,t}\frac{dv_{DS}}{dt} - C_{dg,t}\frac{dv_{GS}}{dt}$$

c. Solve for $v_{DS}(t)$.

7. Use C_{dg} of Eq. (6-27), C_{gd} of Eq. (6-17), ω_0 of Eq. (6-149), and g_m of Eq. (6-66), show that τ_1 of Eq. (6-147) satisfies the relationship given by Eq. (6-159).

8. Show that as the frequency approaches zero, the real part of Z_{in} in Eq. (6-192) is given by R_G of Eq. (6-193).

9. What is wrong with the small-signal model shown in Fig. 6-40, which is meant to apply for all bias conditions? Under what bias condition will this model be correct?

10. The y-parameters of a FET are: $y_{gg} = j4 \times 10^{-3}\ \Omega^{-1}$; $y_{gd} = -j3 \times 10^{-3}\ \Omega^{-1}$; $y_{dg} = 0.4 - j1 \times 10^{-3}\ \Omega^{-1}$; and $y_{dd} = 0.1 + j9.5 \times 10^{-4}\ \Omega^{-1}$. The terminal resistances are $R_S = R_D - 4\ \Omega$ and $R_G - 8\ \Omega$. Find current gain h_{21} and the unilateral power gain.

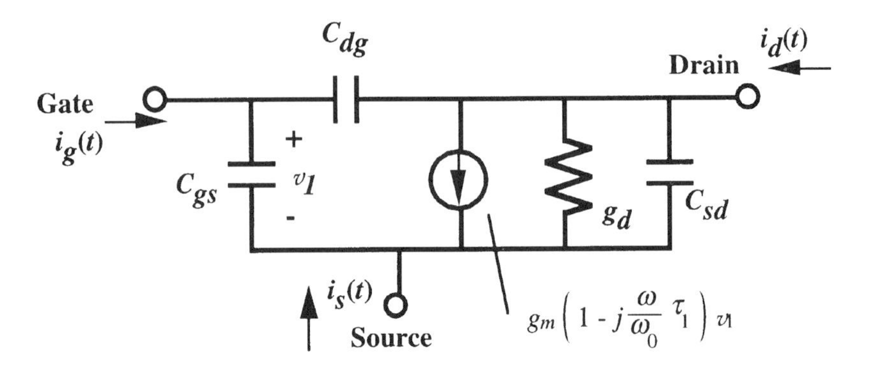

FIGURE 6-40. A problematic equivalent circuit analyzed in Problem 9.

11. Show that as $V_{DS} \rightarrow 0$, the non-quasi-static channel resistance is given by:

$$r_{\text{ch}} \rightarrow \frac{1}{6\, g_m} \frac{V_{DS}}{V_{DS,\,\text{sat}}}$$

12. The gate metal of a MESFET has a sheet resistance of 0.03 Ω/sq. The gate length is 0.5 μm and the overall device width is 1 mm.

- **a.** Find the device gate resistance if the device consists of only one finger; hence, the finger has a width of 1 mm.
- **b.** Find the device gate resistance if the device consists of 10 fingers which are joined at one side. Each finger in this structure has a width of 0.1 mm.
- **c.** Find the device gate resistance if the device consists of 5 fingers which are joined at two sides. Each finger in this structure has a width of 0.2 mm.

13. Consider an eight-finger HFET with $0.29 \times 32\ \mu\text{m}^2$ per finger area. $R_{SHG} = 5\ \Omega/\text{sq}$; $R_S = 1\ \Omega$; $R_D = 3\ \Omega$; AlGaAs barrier thickness = 300 Å; $\alpha = 0$; $g_d = 5.73$ mS, $g_m = 70$ mS. Find the cutoff and the maximum oscillation frequencies. Assume that the gate-to-drain and the gate-to-source capacitances is 10% of WLC'_{ox}.

14. A FET has exactly the same total area as the transistor in Problem 13. However, this FET has a wide gate width, consisting of only one finger. Find the cutoff and the maximum oscillation frequencies.

15. Sometimes, when the transistor is potentially unstable, the maximum stable gain (MSG) is used as a figure of merit. MSG is given by y_{21}/y_{12}. Express MSG in terms of the device parameters, with and without the parasitic capacitances.

CHAPTER 7

TRANSISTOR FABRICATION AND DEVICE COMPARISON

§ 7-1 HBT D.C. FABRICATION

There are countless ways of fabricating an HBT. However, they can be grouped into two general categories: d.c. fabrication and r.f. fabrication. Depending on the specific analysis to be performed on the transistor, one of the two processes is chosen and the associated mask is designed. We describe these processes in some detail, noting that all processes described in the following lead to an *emitter-up* configuration for the transistors. This means that, if the substrate is at the bottom, then the adjacent epitaxial layer is the collector, followed by the base layer. The emitter layer sits on the top, forming a structure such as that shown in Fig. 4-26. *Collector-up* HBTs, in contrast, are the HBTs whose collector layer is at the top while the emitter layer is adjacent to the substrate. The apparent advantage for collector-up structure is that there is no extrinsic base–collector capacitance, in the same manner that there is no extrinsic base–emitter capacitance in the emitter-up geometry. The removal of the extra base–collector capacitance should increase the devices's high-frequency performance. The transistor should attain higher values of f_{max}, even though there exists the "extrinsic" base-emitter capacitance. From time to time the collector-up HBTs are reported; however, the emitter-up structure is the dominant HBT technology.

The d.c. process is used whenever possible because it is the simple and not time-consuming. It is used in a qualifying stage whose sole purpose is to verify that the epitaxial layers are grown within specification. It is also used in the studies of d.c. properties of the HBTs, such as to determine the surface passivation characteristics or the breakdown characteristics. Besides its simplicity and time-saving feature, the d.c. process avoids many undesirable device characteristics associated with the r.f. processes to be described later. For example, the d.c. process does not use ion implantation to isolate devices, thus leakage currents associated with the implantation damage in the base–emitter and the base–collector junction are eliminated.

A schematic mask layout used for this process is shown in Fig. 7-1. The schematic cross-sectional view of an HBT going through this process is shown in Fig. 7-2. The process starts with evaporating a metal layer that is then selectively lifted off using the "EMIT" mask. The liftoff process starts by spinning a layer of photoresist. Typically the positive photoresist is used because the negative photoresist generally results in worse resolution and is more difficult to remove. Once the emitter pattern is defined and subsequent photolithography steps, such as baking and developing, are followed through, an opening in the photoresist results. It is important for a successful liftoff process that an overhang edge structure exists on the top of the photoresist profile, as shown in Fig. 7-3. Such overhang is possible with commercially available photoresist through some baking and preexposure steps. As evidenced in Fig. 7-3, the overhang prevents the evaporated metal from forming a continuous layer at the opening boundaries, allowing for an easier liftoff in which the acetone is used to dissolve the photoresist and remove the metal on top of the photoresist.

The choice of emitter metal depends somewhat on the epitaxial layer structure of the wafer. If the top layer is GaAs, then it is mandatory to have AuGe/Ni/Au (or any similar materials that allow alloying) as the emitter metal. The maximum

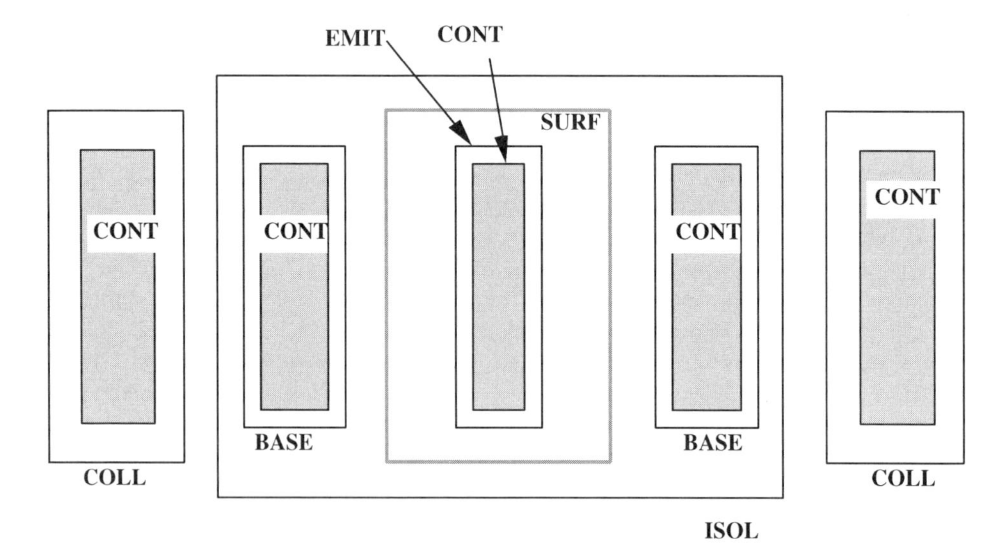

FIGURE 7-1. Mask layout for the d.c. process. (From W. Liu, "Microwave and d.c. characterizations of Npn and Pnp HBTs." Ph.D. dissertation, Stanford University, Stanford, CA, 1991.)

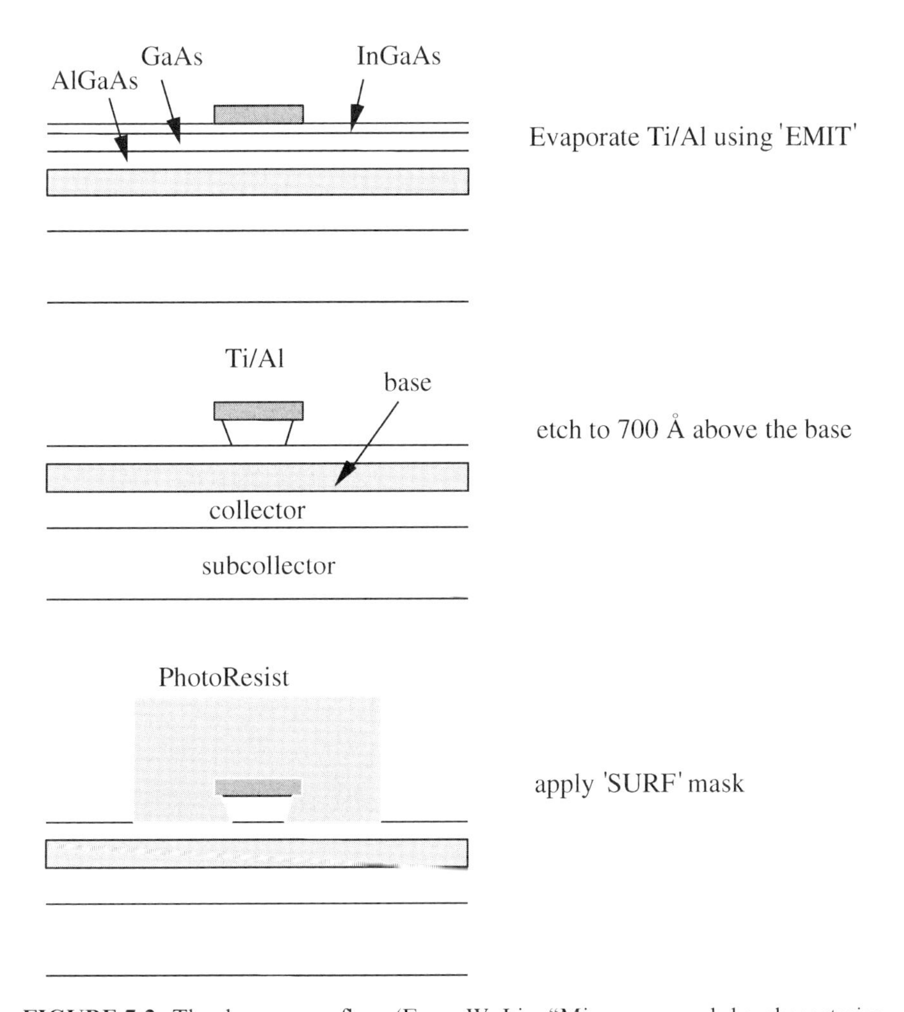

FIGURE 7-2. The d.c. process flow. (From W. Liu, "Microwave and d.c. characterizations of Npn and Pnp HBTs". Ph.D. dissertation, Stanford University, Stanford, CA, 1991.) (Continued on next page).

practical doping of *n*-type GaAs is about 5×10^{18} cm^{-3}, which is too small to allow for a nonalloyed ohmic contact. However, with AuGe/Ni/Au as the contacting metal and a subsequent alloying at typically 440°C for 30 s, an ohmic contact whose contact sheet resistivity on the order of 10^{-6} Ω-cm^2 is routinely obtainable. In addition to the *n*-type GaAs layer, another popular choice of top layer is a lattice-mismatched InGaAs layer first discussed in § 1-4. An InGaAs layer has the advantage that it is easily doped to more than 2×10^{19} cm^{-3} when the indium mole fraction exceeds 50%. The availability of the heavy doping concentration allows for the emitter contact to be ohmic without alloying. For wafers with an InGaAs top layer, the emitter contact can be just about any

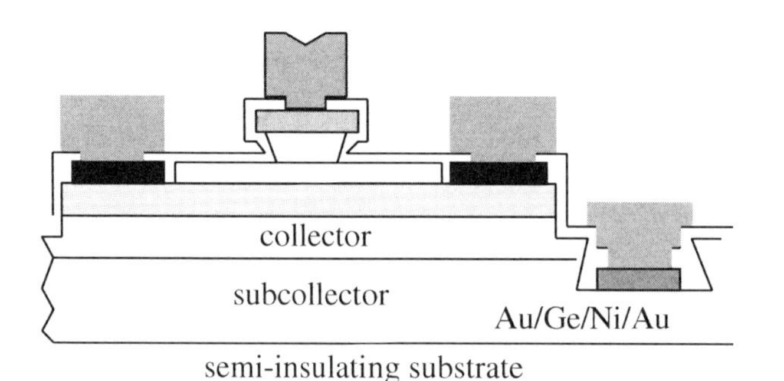

deposite nitride
define contact hole using
'CONT'
RIE nitride

apply 'PAD' pattern
evaporate pad metal
liftoff

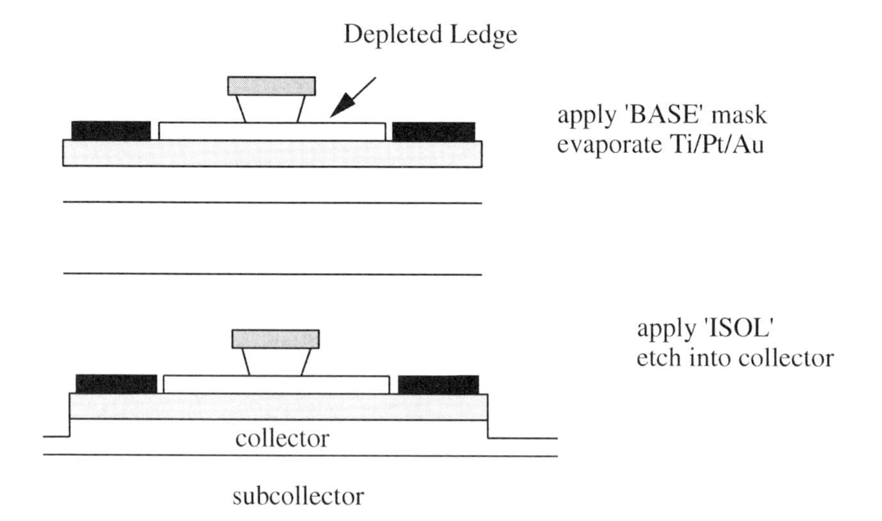

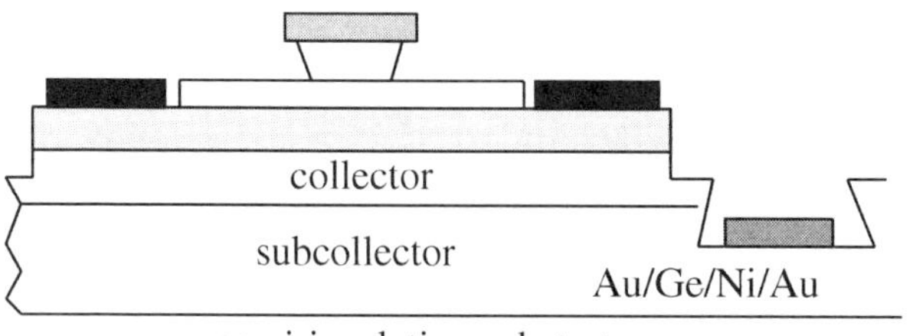

apply 'COLL'
etch to subcollector
form collector contact

FIGURE 7-2. *(Continued)*

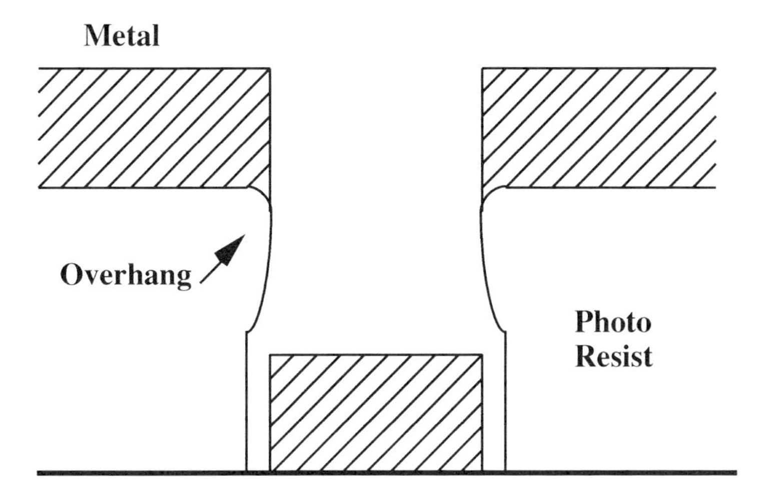

FIGURE 7-3. Overhang structure nearby the opening of a pattern.

metal. A typical choice is Ti/Au. Titanium is known for its ability to hold on to the semiconductor material. It is used as the intermediate layer between the semiconductor and the low-resistivity Au layer to enhance the sticking of the emitter metal, ensuring that the contact remains intact during various processes, such as metal liftoff and wet etching. A variation of the Ti/Au metal is Ti/Pt/Au. Platinum is inserted in between because it prevents Au from diffusing through Ti and entering the semiconductor, thus eliminating possible long-term reliability problems. The potential drawback of platinum is its high resistivity, although a Ti/Pt/Au contact often exhibits the same order of contact resistivity as the Ti/Au contact. Considering that the goal of d.c. process is to fabricate devices for short-term measurements, we think that Ti/Pt/Au is not necessary and Ti/Au suffices. Whether Ti/Pt/Au or Ti/Au is used perhaps does not result in any significant difference. However, a derivation of this type of nonalloyed contact, Pt/Au, should be discouraged. At times the platinum reacts with the InGaAs layer, even converting the normally shiny golden color of the Au top layer to white. When this happens (typically occur after some alloying, such as during the collector contact formation), the emitter metal becomes highly resistive.

Of course, even AuGe/Ni/Au can be used to contact the InGaAs top layer, with or without alloying. However, the latter is found to result in a higher resistivity than Ti/Au when the top layer is InGaAs. This is perhaps relating to some kind of reaction between the emitter metals and the indium atoms in the InGaAs layer.

Once the emitter contact is formed, it is subsequently used as a natural mask, allowing for the wet-etch removal of the regions other than the emitter mesas, as shown in Fig. 7-2. The etching process includes an initial etch of a H_3PO_4:H_2O_2:H_2O solution to remove the top InGaAs cap layer and a portion of the

GaAs layer, followed by a $NH_4OH:H_2O_2$ solution to expose the AlGaAs emitter layer. We describe a little about each of the etching solutions. A solution that etches GaAs, such as the first aforementioned etching solution, contains two essential ingredients: an *oxidizing agent* and a *complexing agent*. Often the solution also contains a certain portion of water or methanol, mainly for dissolving the reaction products and for diluting the solution in order to slow down the etching rate. The etching process begins when the oxidizing agent oxidizes the GaAs material, converting the top layer into some form of oxide. The predominant oxidizing agent used for III–V materials is H_2O_2. The oxide layer just formed is insoluble in the solution, however. In the second part of an etching process, the oxide is made soluble by the complex formation enabled by the complexing (solubilizing) agent. Either acid or base can be used as the complexing agent, although most often acid is chosen. Some commonly used acids include, in the order of acidity, H_2SO_4, HCl, and H_3PO_4. Because of the strong acidity of sulphuric acid, the etch rate is fast if the solution is not diluted sufficiently with water. For example, a 8:1:1 $H_2SO_4:H_2O_2:H_2O$ etches GaAs at a rate of more than 1 μm/min, which is too fast for HBT device processing. Such a solution, however, is often used to prepare molecular beam epitaxy (MBE) or metalorganic chemical vapor deposition (MOCVD) growth, rapidly removing the surface GaAs layer before exposing the fresh GaAs layer underneath and loading into the growth chamber. The use of sulphuric acid is also tricky in that the acid is *exothermic*, releasing heat when water is added to the solution. If the solution is just mixed without cooling down to a certain temperature, the etch rate, which increases exponentially with temperature (in agreement with the Arrhenius law), can be dramatically higher than expected. For an etch rate of about 1000 Å/min, a good etching solution is 3:1:50 (by volume) $H_3PO_4:H_2O_2:H_2O$ solution at room temperature.

Theories have been proposed about the wet-etching chemistry that allows further understanding of the process. For example, the etching process can be categorized into *reaction-rate-limited* or *diffusion-limited*, depending on the behaviour of the oxidizing agent. When the solution is viscous and the amount of H_2O_2 is small, the diffusion of H_2O_2 in the etching solution to reach the solution–oxide interface does not keep up with the adsorption of the H_2O_2 on the same interface. This is then a diffusion-limited etch. Conversely, when the absorption of the H_2O_2 limits the amount of etching, it is a reaction-rate-limited etch. Such a classification allows a more in-depth intuition about several wet etching phenomena, such as the stirring dependence of the etch, the mask etch trenching wherein the etch rate is faster on the GaAs adjacent to the mask edge, and the general etch rate as a function of concentration of the oxidizing and complexing agents. Despite these theories, the prevalent practice in device fabrication is still to experiment with several wet etching solutions, determine the etch profiles through scanning electron microscope (SEM), and lock in on a particular wet etching solution to be used repeatedly in the whole process.

Usually an etching solution that etches GaAs etches AlGaAs and InGaAs as well. The etch rates are usually slightly faster than in GaAs, depending on the crystalline quality of the grown material as well as the mole fractions of aluminum and indium. However, the second solution applied to expose the AlGaAs layer, a $NH_4OH:H_2O_2$ solution, can be made to etch GaAs, but not AlGaAs or InGaAs. First, the solution needs to have an exceptionally high volume ratio of H_2O_2 to NH_4OH, on the order of 200 to 1. (When mixed in this kind of proportion, the solution is referred to as a *superoxol solution.*) Second, the *selectivity* that measures the ability of the etchant to remove the top layer compared to the layer underneath depends on the aluminum and indium concentrations. For a typical HBT whose aluminum mole fraction is 30% to 35%, this 200:1 etch is highly selective in removing GaAs but stops at the AlGaAs layer. Therefore, after this selective etching step, the designed 700-Å AlGaAs emitter layer is exposed to air. This layer is, to a first approximation, fully depleted because of the combination of surface Fermi level pinning on the top and being the more lightly doped layer in the base–emitter junction below. The depleted AlGaAs forms a protection ledge, which suppresses the surface recombination in HBTs. This depletion ledge is important. It allows us to analyze the other base current components of the HBT independent of the surface effects at the extrinsic base surface.

To contact the base, we must etch away this depleted AlGaAs layer to reach the *p*-type base layer. However, prior to indiscriminately etching all of this AlGaAs layer, the "SURF" mask is used to protect a narrow depletion ledge that surrounds the emitter mesa and prevents it from being removed. Once the photoresist (PR) is defined to protect the not-to-be-etched AlGaAs, the 3:1:50 solution is used to etch down to the base surface. This etch is somewhat critical, since the 3:1:50 solution is not a selective etching solution. However, in a d.c. *Npn* HBT, the base thickness design is typically 1000 Å. This base layer is thick enough so that etching completely past the base layer can be prevented, as long as the etch rate of the solution is calibrated and some care is taken. A method to estimate whether the base layer is reached is to measure the breakdown voltage at the surface, with two clean probes touching on wafer surface and a bias applied between them. If the base layer is not yet reached, the surface breakdown voltage is on the order of 10 V since the doping level in the emitter is light. As the base layer is approached, the breakdown voltage decreases but remains at some finite value. When the base layer is finally reached, the two probes measure an ohmic characteristic due to the fact that the base layer is heavily doped. This method will fail if the base layer is doped lightly, such as levels of about 5×10^{18} cm^{-3}.

After this base contact etch, the base contact region is defined with the "BASE" mask, and a base contact metal of Ti/Pt/Au is deposited by electron-beam (e-beam) evaporation. Again, titanium is used to provide a good adhesion, and platinum is used to prevent Au from diffusing through the titanium and entering the semiconductor. Since reliability is hardly a concern in devices fabricated to allow examination of their d.c. properties, Ti/Au can be used as well. The active device region is then defined with the "ISOL" mask. The area outside this masked region is etched past the base layer and into the collector layer. If this isolation etch does not take place, the base–collector junction would have a very large area, essentially equalling to the area of the wafer. Even though the leakage current density through a reverse-biased base–collector junction is small, when multiplied by a large area, the overall leakage current can be comparable to the desired signal current. An excessive leakage current can cause the base current to become negative, resulting in a negative current gain and rendering the device useless.

As far as device isolation is concerned, the requirement for the isolation etch is to etch past the base layer so that the physical base–collector junction area is defined by the "ISOL" mask. It is entirely possible to etch all the way to the subcollector layer, removing both the base and the collector layers. This has the advantage that the collector metal can be evaporated immediately after the collector pattern "COLL" is made. When the isolation etch stops short of the subcollector layer and is somewhere in the collector layer, a wet etch is initiated to reach the subcollector layer before evaporating the collector metal. In either case, without the luxury of a heavily doped InGaAs layer (which is not lattice-matched to the GaAs substrate), the choice of the collector metal is most often based on the AuGe/Ni/Au system. An alloying is required for the collector contact.

The collector pattern shown in Fig. 7-1 consists of two stripes placed at each side of the base contacts. If the collector pattern instead forms a rectangular frame that encloses the device, then the "ISOL" pattern is not needed. That is, right after the base metal liftoff, the collector pattern is defined. During the etch to reach the subcollector, the base–collector junction area is defined along the process. Another method of defining, the collector is to throw out the "COLL" pattern altogether (but "ISOL" must be used in this case). When a n^+ substrate is used instead of a semiinsulating substrate, then the contact metal of Au/Ge/Ni/Au can be evaporated at the backside of the wafer, without any patterning.

After the collector contact is formed, a nitride layer is deposited on the entire wafer. The contact holes are defined with the "CONT" mask, and the exposed nitride is removed with a C_2F_6 (Freon 116) plasma. We note that, because the d.c. process is not used to fabricate devices for high-frequency performance, the emitter width in the d.c. mask does not have to be as narrow as those of high-frequency transistors. The dimension can be as wide as 4 μm. With such a dimension, it is easy to open up a 2-μm-wide contact hole on the emitter and still allow for a reasonable 1-μm misalignment tolerance. The same logic applies to the base contact width. The base contact width can be made wide because we are not interested in the device's high-frequency performance. When the base contact width is at least 4 μm, the contact hole on the base contacts can be opened just as easily. Finally, the interconnect metal of Ti/Au is evaporated and lifted off. The mask, "PAD," which is used to define the interconnect metal location, is not shown in Fig. 7-1. It should be large enough to allow direct probing. A pad of size 75×75 μm^2 is large enough comfortably to accommodate a probe. If more test structures per die are required, a smaller pad size of 50×50 μm^2 is tolerable.

§ 7-2 HBT R.F. FABRICATION

In many ways the mask layout and the fabrication process are very different between an r.f. transistor and a d.c. transistor. For an r.f. transistor, the need to reduce the contact pad shunt capacitances and conductances dictates the use of a semi-insulating substrate. The use of a simple backside collector contact requiring the use of a n^+ substrate is not possible for an r.f. transistor. More important, both the emitter width and base contact width for high-frequency transistors must be necessarily narrow, on the order of 2 μm or less. Without a stepper as the lithographical tool, it is difficult to align the emitter contact hole directly on top of the emitter mesa (at least not easily). The r.f. process, as well as the mask that goes along, thus calls for a totally different processing methodology from that of § 7-1. An r.f. mask layout suitable for a single-finger HBT is shown in Fig. 7-4. The relative position of the device with respect to the contact pads is shown in Fig. 7-5. Note that the area of contact pads is much larger than that of the device. These pads in such an arrangement are designed for common-emitter configuration measurements, with the emitters connected to the two outer pads and the base and collectors connected the central pads. A standard on-wafer high-frequency probe can be thought to consist of three subprobes, as shown in Fig. 7-6. It is called a *GSG probe*, meaning that the three subprobes are arranged in order, to connect to ground, signal, and ground, respectively. Since in the mask layout the emitter is connected to the outer two pads, the emitter is automatically grounded during the high-frequency measurements. The signal from one GSG probe is connected to the base. The signal from another GSG probe is connected to the collector. The distance between the ground subprobe and the signal subprobe is fixed by the probe manufacturer, usually being 125 μm or 150 μm. Therefore, it is critical during a mask layout to draw out proper distances for the pads.

The sequence of the r.f. process is delineated in Fig. 7-7. The process begins by depositing 7000 Å of silicon dioxide. The first mask level, "ISOL," is then defined, and 1.5 μm of aluminum is evaporated and lifted off. The wafer is sent to a reactive ion etch (RIE) chamber. With the aluminum as a mask protecting the silicon dioxide directly underneath, the oxide outside of the "ISOL" pattern is removed, resulting in a profile shown in Fig. 7-7*a*. A dual proton implantation is applied to define the active device region. The implantation schedule consists of a dose of $2 \times 10^{15}/\text{cm}^2$ at 200 keV, followed by a dose of $2 \times 10^{15}/\text{cm}^2$ at 70 keV. The substrates are slightly tilted during implantation to avoid channeling. For a transistor whose epitaxial layers are not excessively thick, this implantation schedule is sufficient to make the area unprotected by the aluminum/oxide electrically inactive. If somehow the transistor has a collector or a subcollector thickness on the order of 1.5 μm, the protons do not penetrate far enough inside the substrate to isolate the device effectively. Other types of high-energy implants will be necessary.

A shallow etching is applied right after the implantation so that a trail mark of the active device area is left. This facilitates the alignment of future mask levels to this first mask. Subsequently, the masking materials of aluminum and oxide

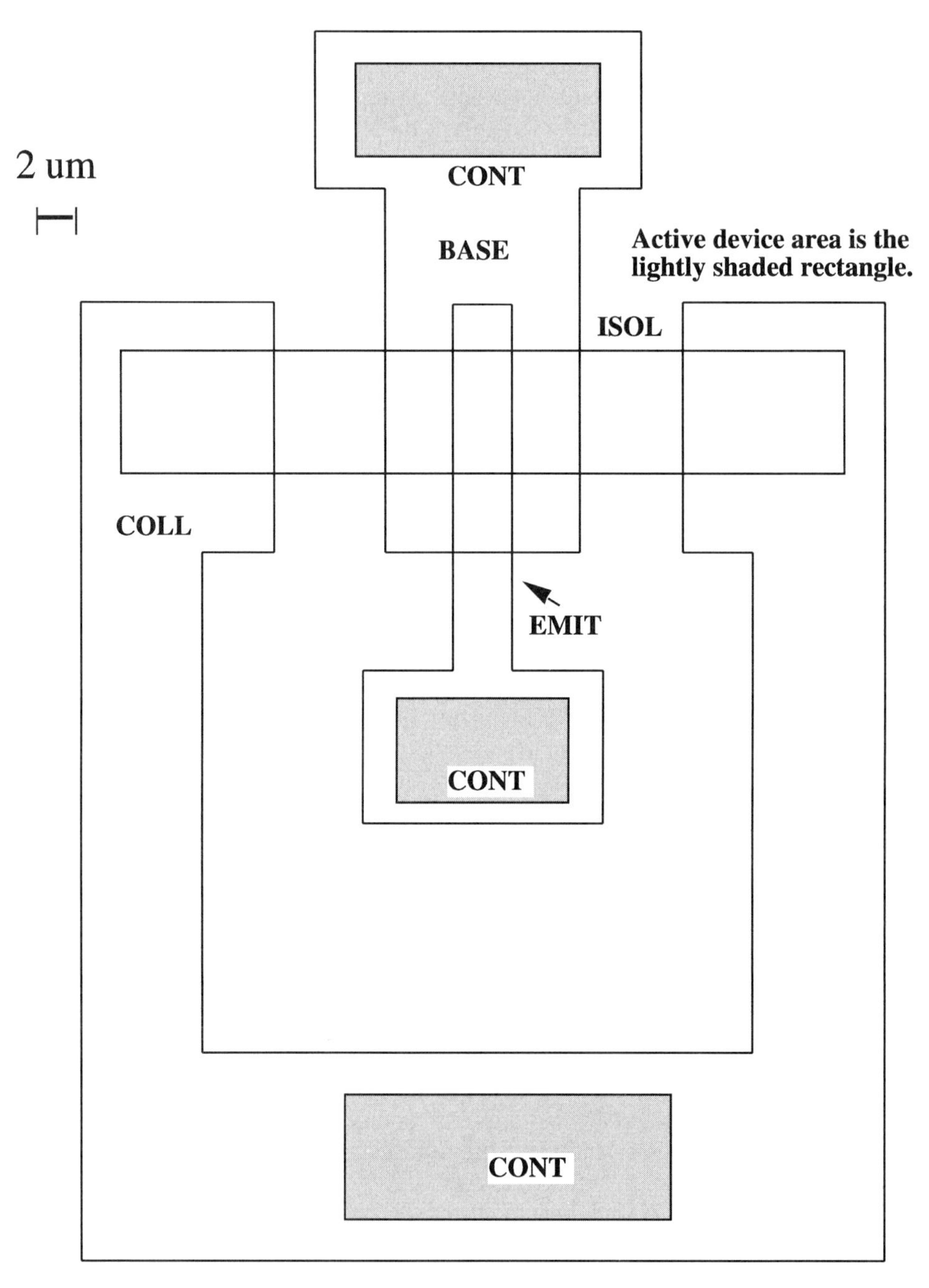

FIGURE 7-4. Mask layout for the r.f. process. (From W. Liu, "Microwave and d.c. characterizations of Npn and Pnp HBTs." Ph.D. dissertation, Stanford University, Stanford, CA, 1991.)

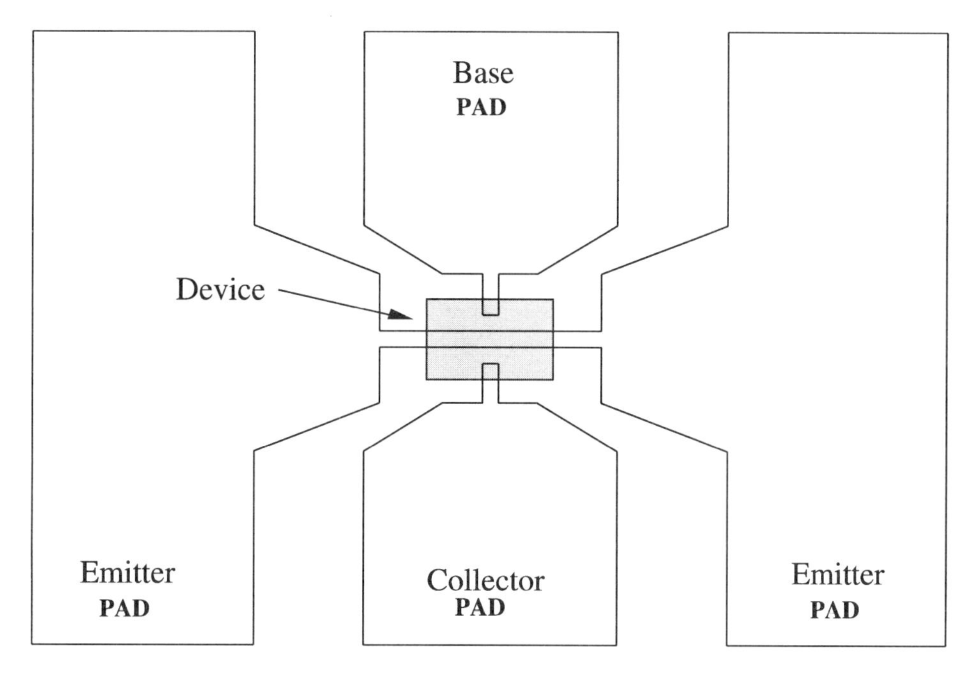

FIGURE 7-5. Relative position of the device with respect to the contact pads. (From W. Liu, "Microwave and d.c. characterizations of Npn and Pnp HBTs." Ph.D. dissertation, Stanford University, Standford, CA, 1991.)

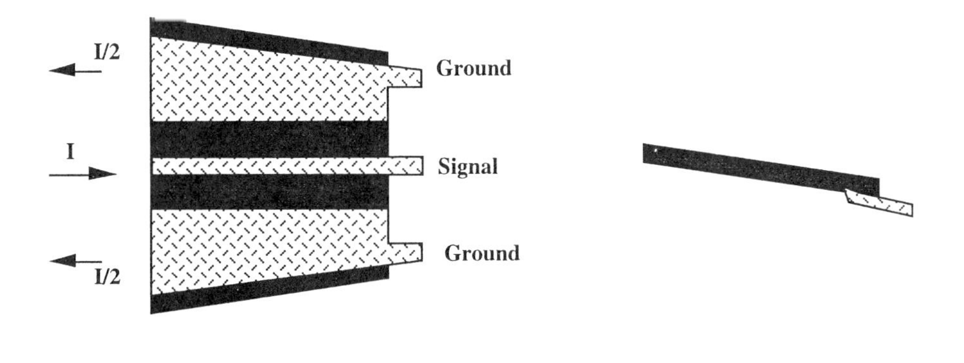

FIGURE 7-6. Standard on-wafer high-frequency probe. The two ground pins are shorted internally. (From D. Costa, Microwave and low-frequency noise characterization of Npn AlGaAs/GaAs heterojunction bipolar transistors. Ph.D. dissertaion, Stanford University, Stanford, CA, 1991. Reprinted with permission.)

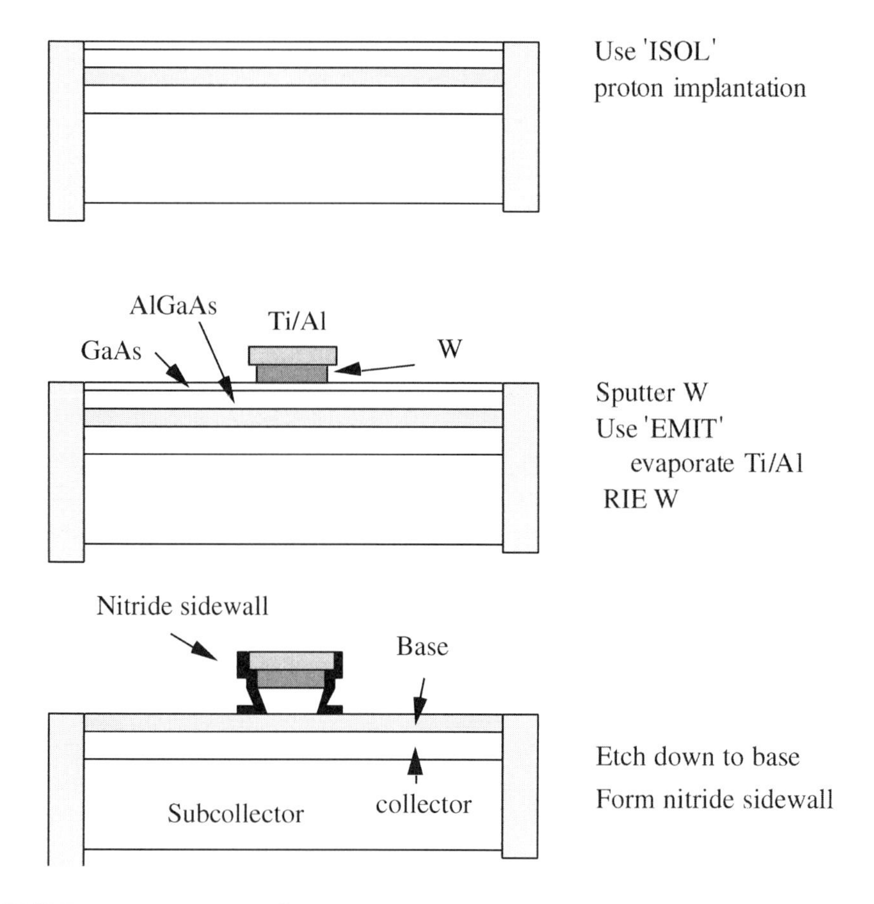

FIGURE 7-7. R.F. process flow. (From W. Liu, "Microwave and d.c. characterizations of Npn and Pnp HBTs." Ph.D. Dissertation, Stanford University, Stanford, CA, 1991.)

are removed using a 1:1 H_3PO_4:H_2O solution heated to 70°C and a buffered hydroflouric acid, respectively. The removal of the masking material re-exposes the fresh InGaAs's top surface. After a short surface cleaning with a dip in an acidic solution, a tungsten (W) layer is sputtered onto the wafer. The "EMIT" mask is used to define the emitter, and Ti/Al is evaporated and lifted off. The W not underneath the emitter is then removed using a RIE with C_2F_6/SF_6. Tungsten, being a refractory metal, has a stable property that it does not react with the semiconductor underneath. If W is not used and the titanium from the emitter metal is allowed to contact the InGaAs directly, long-term reliability problems can arise as titanium reacts with indium. Other refractory metal suitable for HBT fabrication includes tungsten silicide (WSi) and titanium tungsten (TiW).

After the W outside the emitter mesa is removed, the 3:1:50 H_3PO_4:H_2O_2:H_2O solution is used to etch down to the base. It should be

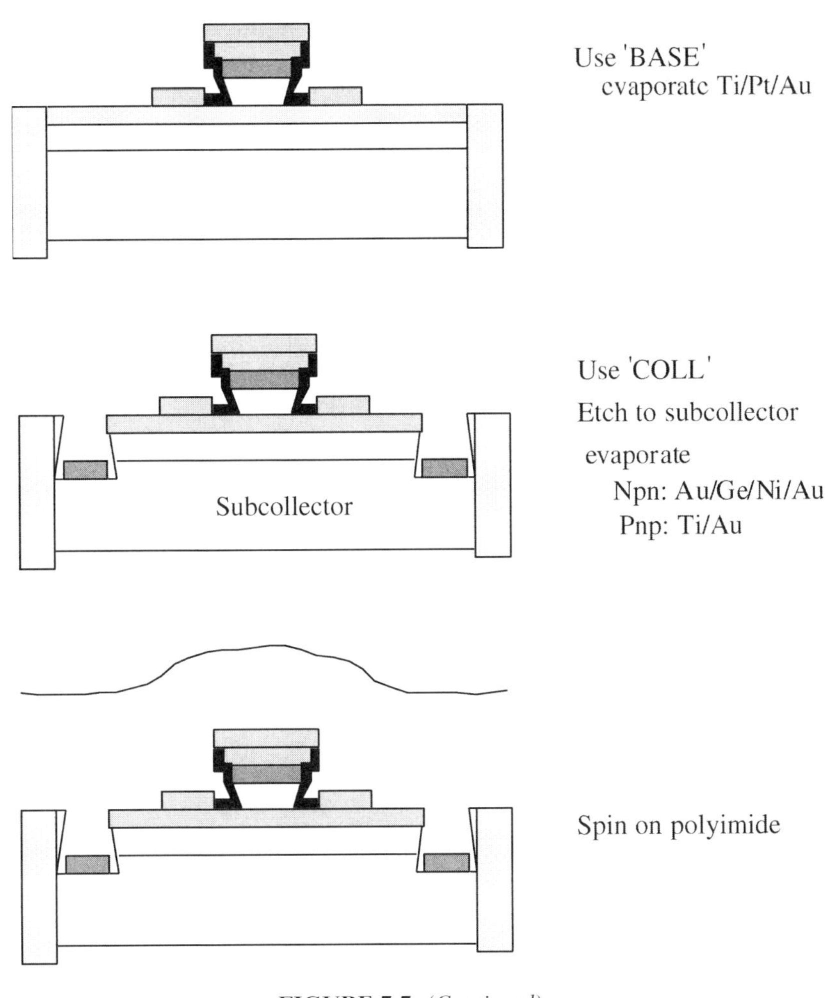

FIGURE 7-7. *(Continued)*

noted that there is no need for a "SURF" mask in the r.f. process because the extrinsic base surface recombination does not significantly affect HBT high-frequency performance. (From the standpoint of reliability, passivation ledge is desirable. It can be fabricated in a similar fashion described for the d.c. process.) The etching distance is roughly 1500 Å and the entire etch is not a selective process. The 3:1:50 solution etches GaAs just as effectively as it etches AlGaAs. As described in the d.c. process, some care in calibrating the etching solution is required to etch down successfully right to the top portion of the base layer. After etching down to the base, a layer of silicon nitride is deposited by plasma-enchanced chemical vapor deposition (PECVD). The nitride deposition is conformal, forming a layer of nitride everywhere, including on the vertical edges. Immediately after the nitride deposition, the nitride is then etched by RIE. Since the vertical electric field is set up in the RIE chamber such that the chemical species removing the nitride on average moves only in the vertical direction, the sidewall nitride covering the sides of the emitter mesas are untouched while a majority of nitride layer that lies flat on the wafer is removed. The nitride sidewall is used to improve the yield of the device, dramatically cutting down the possible electrical short between the emitter contact and the base contact which is to be deposited shortly.

The step after the nitride sidewall formation is the "BASE" lithography. As shown in the mask layout, there is no separation between the "BASE" and the "EMIT" levels. The base metal during the evaporation therefore partly lands on the emitter mesa, although some of it lands on the base layer as described. The process has the advantage that, even if the "BASE" level is somewhat misaligned with the "EMIT" level, the distance between the emitter and base contacts remains unchanged and is at the minimum value set by the thickness of the nitride sidewall. The latter advantage is especially critical to the high-frequency performance since minimizing such a distance reduces the extrinsic base resistance, as well as the extrinsic base–collector capacitance. In this manner, the base contact is said to be *self-aligned* to the emitter. Typically, Ti/Pt/Au is used as the choice of the base metal.

This place serves as a good place to discuss the orientation of the wafer. As mentioned in § 1-1, most GaAs wafer substrates are in the [100] direction so that alternative layers of Group III atoms and Group V atoms are grown on top. For substrates in such an orientation, the major flat is usually perpendicular to the $[01\bar{1}]$ direction, and the minor flat is perpendicular to the [011] direction. The emitter pattern in the stripe geometry can be placed in parallel with either the major flat or the minor flat (Fig. 7-8). It is a GaAs material property that the wet etching in certain crystallographic planes occurs faster in preference to some others. This preferential (or anisotropic) etching property results in different etch profiles. The mesa edge can bend inward or outward, depending on the crystalline direction. When the emitter stripe is parallel to the major flat, the resulting emitter mesa is narrower at the bottom than at the top. The converse is true when the emitter stripe is parallel to the minor flat. This difference has a major implication in the device yield if the base metal is self-aligned to the emitter. If the etch edge extends outward, a portion of the base contact metal lands on the emitter layer, connecting it to the base layer. This is not a direct shorting, as would be in the case if a nitride sidewall were absent and somehow during a not-so-perfect liftoff a shred of metal connected the base metal to the emitter metal or W. Therefore, the connection between the emitter layer and the base metal is sometimes called a *semi-shorting*. The device would still work, but not properly. Figure 7-9 shows the I-V characteristics typical of a semi-shorted device. It seems that the device has a very small Early voltage, with the collector current increasing at a large slope as V_{CE} increases. It is also characterized by a small current gain at low base currents. Sometimes at high base current level, the collector current exhibits a snap-back behaviour that can be captured in

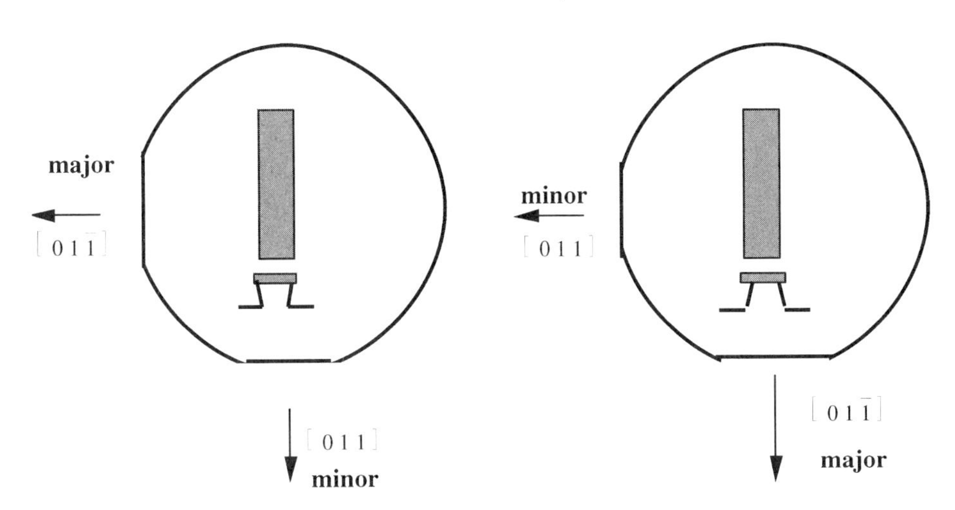

FIGURE 7-8. Etch profiles at different crystalline orientations. (From W. Liu, E. Beam, T. Kim, and A. Khatibzadeh. "Recent developments in GaInP/GaAs heterojunction bipolar transistors." In M. F. Chang, ed., *Current Trends Heterojunction Bipolar Transistors*. Singapore: World Scientific. © World Scientific, reprinted with permission.)

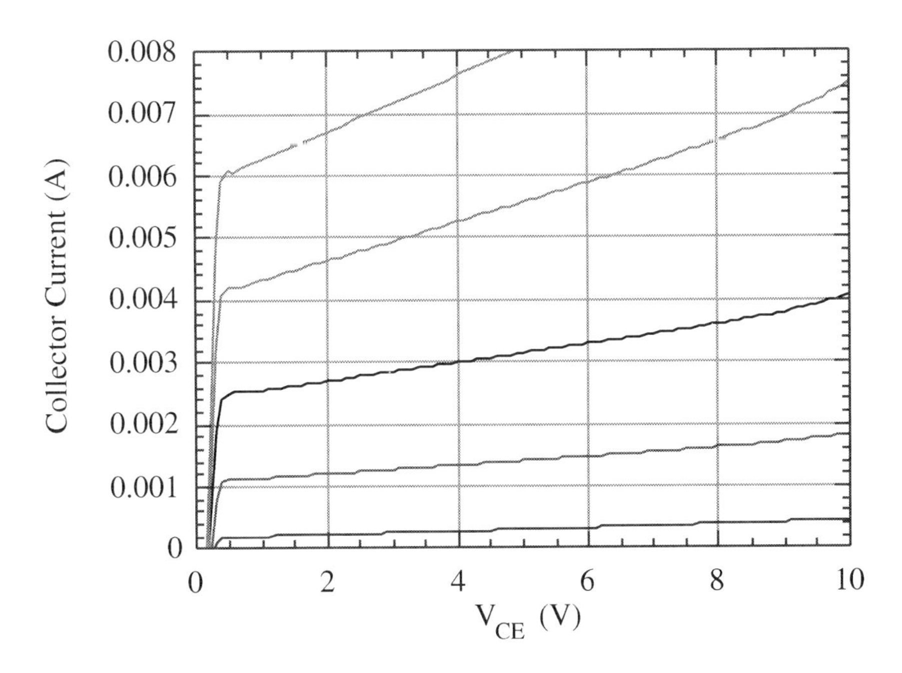

FIGURE 7-9. I-V characteristics of a semi-shorted device.

a curve tracer. Soon after the snap-back (not shown in the figure), a further increase of input power destroys the device. The semi-shorting problem is avoided if the emitter stripe is in parallel with the major flat. In such an orientation, the etch profile slants inward, rather than outward, thus reducing the chance that the base metal lands on to part of the emitter. Wafer orientation is not an issue for d.c. processing. With a non-self-aligned technology, there is a significant amount of spacing separating the emitter and the base contacts.

After the base metal formation, the collector is defined and the contact metal of Au/Ge/Ni/Au is deposited. After the contact is alloyed, a polyimide layer is spun to planarize the device. The contact holes are then defined and Ti/Au is evaporated to contact the different electrodes.

§ 7-3 FET FABRICATION

There are generally no distinct "d.c." and "r.f." processes for FETs, as there are for HBTs. The FETs are fabricated in a given sequence, with the end goal of producing r.f.-functional devices. However, d.c. testing of the FET devices is performed intermittently during the processing steps. If the devices reveal problematic characteristics, the wafer is then scrapped. Otherwise, the process continues until it yield r.f. devices.

A typical FET fabrication process is shown in Fig. 7-10. The associated mask levels are drawn in Fig. 7-11. The FET fabrication starts with a definition of the active device area, which can be defined by implanting ions at the area outside of the "ISOL" mask, as is often done for HBTs (see § 7-2). Alternatively, the active device area can be formed by etching the material outside the active area. This isolation technique, called *mesa etch*, is a feasible option for FETs because the channel layer is near the top of the surface. A photolithography is performed so that photoresist remains on the area enclosed by the "ISOL" pattern. The photoresist then serves as a natural mask for the subsequent etching, which etches past the active channel layer and reaches the semiinsulating buffer layer. Isolated islands (mesas) with preserved active layers are formed. Because the exposed buffer layer is on the order of 1000–2000 Å below the top surface, the resulting wafer surface is fairly planar. Mesa etch is not usually used for HBTs, since the buffer layer below the subcollector can be 2 μm below the top surface.

The next step is contact formation, in which both the source and the drain metal contacts are placed on top of the n^+ cap layer. The photoresist is formed on the wafer with the "S/D" mask, allowing the metal to be deposited on the source and drain sides of the transistor. Generally the contacts are alloyed, with a schedule (temperature and duration) appropriate for the particular composition of the deposited AuGe/Ni/Au metal.

The "RCSS" (recess-etch) mask is used to remove the n^+ cap layer on top of the gate area. Removing this layer increases the breakdown voltage. The feature length of this mask is usually between 1 and 2 μm, so the usual optical lithography can be used in this step. Sometimes, however, in the interest of reducing

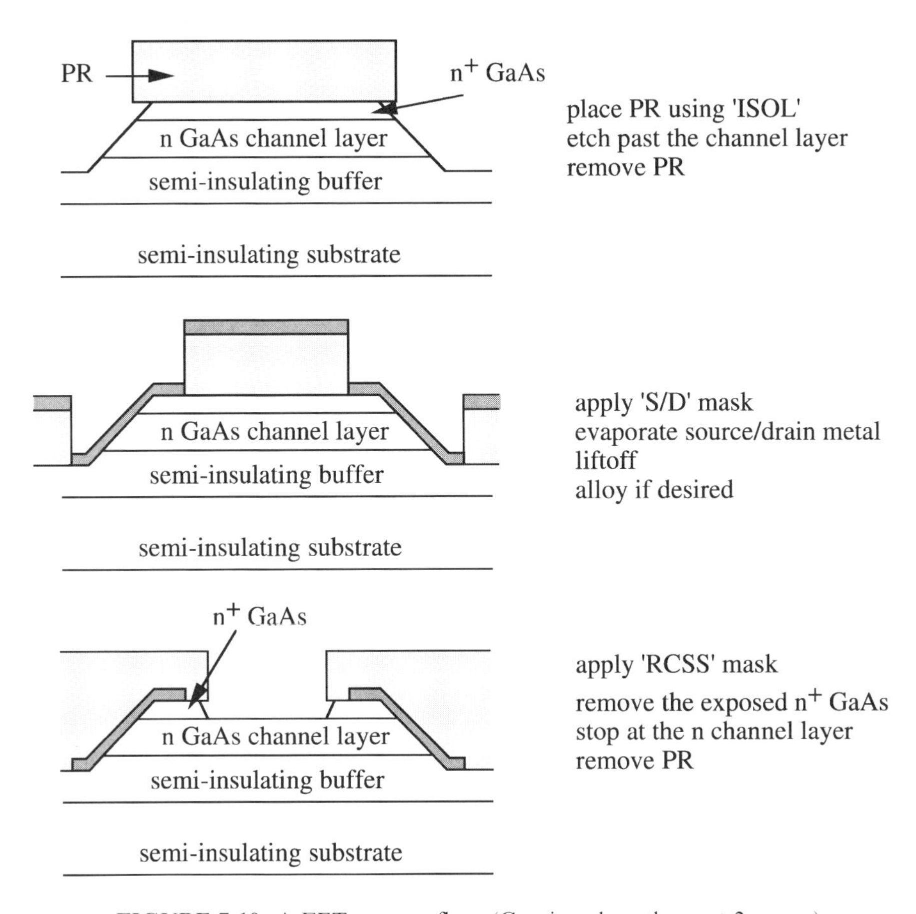

FIGURE 7-10. A FET process flow. (Continued on the next 2 pages.)

the source and drain resistances, fine feature below 1 μm is used. Therefore, occasionally this mask level is performed with e-beam lithography. Once the opening is defined by lithography, wet etching is proceeded. The depth of this etch, unlike the isolation etch, needs to be controlled carefully. If the etch does not reach the top of the channel, a portion of the n^+ cap layer still remains, a fact which results in a low breakdown voltage. If the etch depth penetrates into the channel layer too much, the threshold voltage also becomes off the target. In addition, as the available channel charges decrease from overetching, the output current is reduced. This recess etch is sometimes referred to as the *first recess etch*. There can be a second recess etch, which is also aimed at improving the breakdown voltage.

The next step, gate formation, requires the use of e-beam lithography if the gate length is on the order of 0.5 μm or shorter. Previously, when conventional optical lithography was used, the photoresist was spun on the wafer first. For

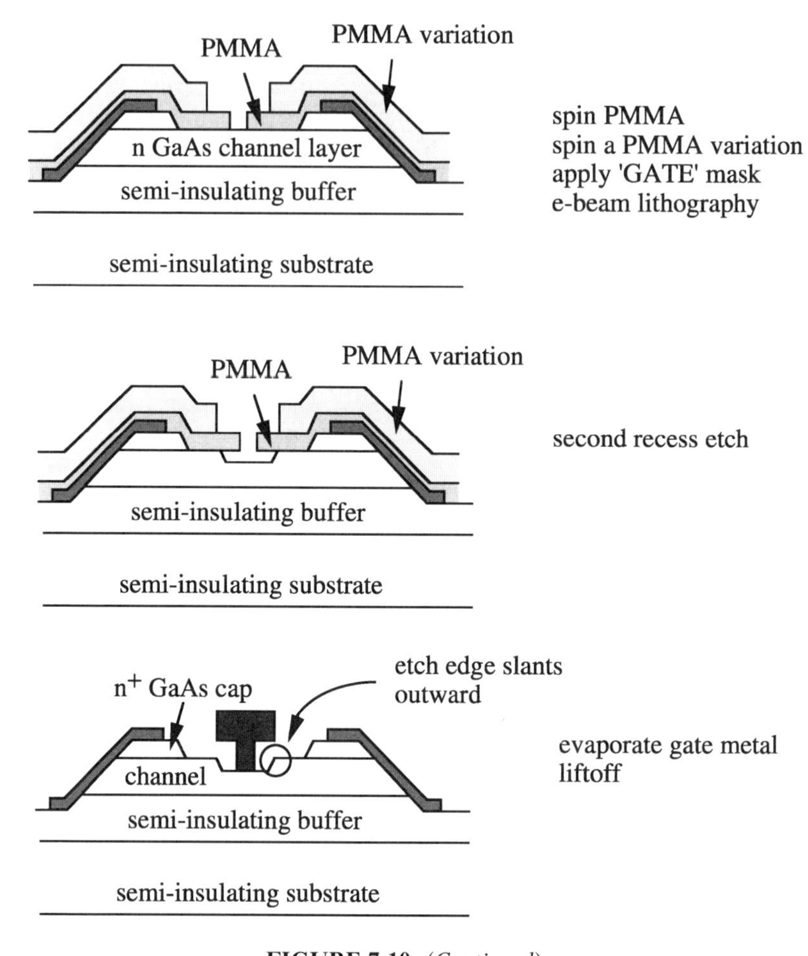

FIGURE 7-10. (*Continued*)

e-beam lithography, the appropriate material to be spun is poly(methyl methacrylate), or PMMA. As mentioned in §6-5, a T-gate with extra metal cross section reduces the parasitic gate resistance. If the gate metal is evaporated with only one layer of PMMA on the wafer, the resulting metal cros section will be rectangular at best, and is often triangular with a tapered end at the top (see Fig. 6-31). The tapered cross section increases the gate resistance dramatically. In order to form T-gate, a second layer of photoresist material related to PMMA needs to be spun on. This second layer has higher light sensitivity compared to the PMMA beneath. After the lithographical exposure and resist development, a larger opening is made at the top layer than at the lower layer. The gate metal is deposited and lifted off. A typical gate metal consists of 1000 Å Ti, 500 Å Pt, and 4000 Å Au.

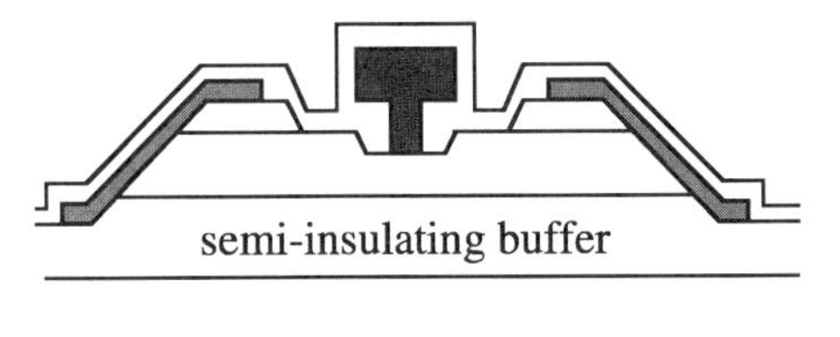

device passivation --
PECVD nitride

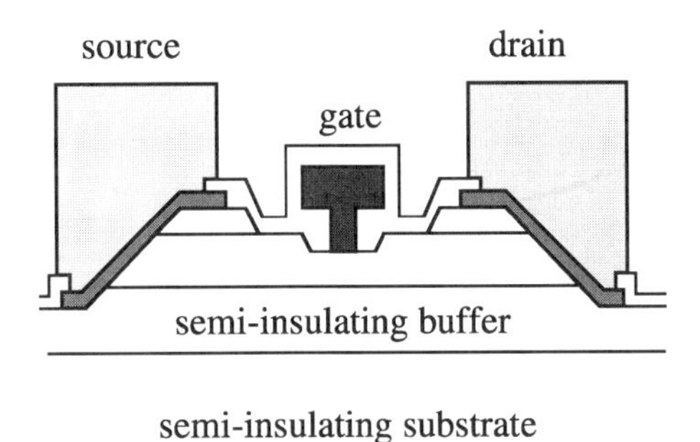

apply 'CONT' mask
RIE ntiride through pattern
remove PR

apply 'PAD' mask
evaporate interconnect metal
liftoff

FIGURE 7-10. (*Continued*)

Right before the gate metal deposition but after the resist opening, an optimal etch is sometimes performed. This etch, taking place locally exactly where the gate metal will land, is called the *second recess etch*. This is perhaps the most critical step in establishing high performance in HFETs. The second recess etch creates a desired device cross section, with a corner between the drain contact and the gate contact. The edge of the slope, as indicated by a circle in Fig. 7-10, is slanted outward toward the source/drain contacts, rather than inward. This particular geometric shape has been found to be effective in increasing the breakdown voltage. As noted in § 7-2, the etch profile, whether the slope is slanted outward or inward, depends on the crystallographical direction. Since it is the outward slope which is desirable, the gate finger in FETs is placed parallel to the minor flat (right portion of Fig. 7-8). This is the opposite to HBTs, in which the emitter finger is placed parallel to the major flat. Nonetheless, due to other concerns, sometimes the FET fingers are placed parallel to the *major flat*.

Especially in HFETs, the second recess etch serves another purpose besides increasing the breakdown voltage. In order to let the gate exert a great influence on the charges in the channel behind the barrier layer, it is desirable to place the gate as close to the channel as possible. Placing it too close, however, can deplete the charge completely. There is an optimum distance such that the mutual transconductance is at a maximum while maintaining a high operating current. In a research environment, the amount of etch can be decided *in situ*. After an

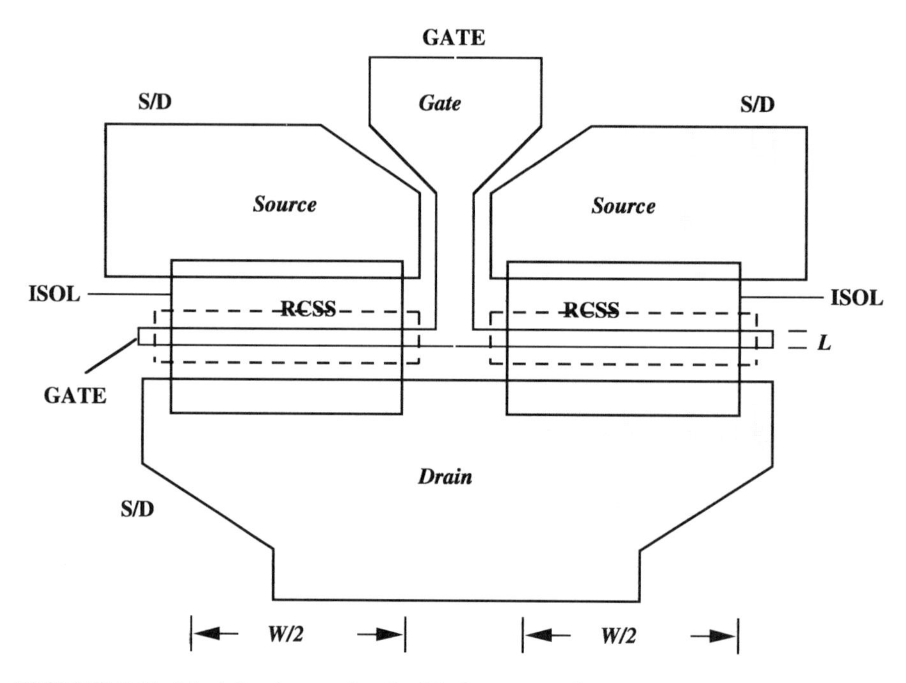

FIGURE 7-11. Mask levels associated with the process flow described in Fig. 7-10. The levels "CONT" and "PAD" are not shown.

incremental amount of the etch is performed, the current of the thus-far-ungated HFETs is measured between the evaporated source and drain contacts. Once the measured current reaches some predetermined value from calibration, the second recess etch is stopped and the gate metal is evaporated. This kind of processing technique usually achieves the proper amount of etching, and hence, superior device performance. However, this *in situ* etching technique should not be adopted in a production environment. A great challenge for FET production is to control the second recess etch to a level within the performance boundary while producing good yield.

After the gate metal evaporation and liftoff, the bulk of the device is finished. At this point, the device is unpassivated, meaning that the device's surface layer is exposed to air. The exposure is a reliability concern, since prolonged exposure results in oxidation, which increases surface depletion and modifies the device characteristics. To passivate the device, a nitride layer of usually 2000 Å is deposited through plasma enhanced chemical vapor deposition (PECVD). We caution that in FET terminology, passivation means covering the exposed GaAs or AlGaAs layers to prevent oxidation. Depositing a nitride layer all over the wafer constitutes device passivation. In HBTs, however, the nitride layer on GaAs does not reduce the surface recombination velocity on the GaAs interface. It is thus not considered a passivation layer. Passivation in HBTs means the use of an AlGaAs ledge which is grown epitaxially on top of the GaAs layer. This way, the interface surface velocity is reduced significantly, resulting in a smaller amount of surface recombination. The passivation of the extrinsic base surface in HBTs is especially critical in increasing the device lifetime [1].

The nitride layer, as shown in Fig. 7-10, is *conformal*. That is, the deposited nitride follows the features of the wafer surface prior to the deposition. If it is desirable to planarize the wafer surface, polyimide can be used.

Now that the nitride layer covers the entire wafer, it is desirable to selectively open holes to contact the source, drain, and gate metals. The "CONT" mask level is used for this purpose. With the resist openings right on the top of the terminal metals, a reactive ion etch (RIE) is used to remove the expose the nitride. After the bottom metal is exposed, the photoresist is removed. A follow-on photolithographical step using the "PAD" mask is performed. This level allows us to evaporate a thick interconnect metal to connect the terminal contacts to the external world.

§ 7-4 HBT AND FET COMPARISON

There are many differences between HBTs and FETs. Some of them relate purely to terminology. Figures 7-12*a* and 7-12*b* illustrate the I-V characteristics of common-emitter HBTs and common-source FETs, respectively. There are two distinctive regions in each set of I-V characteristics. In one, the collector or (drain current) increases rapidly with V_{CE} (or V_{DS}). This region is called the *saturation* region for HBTs, but it is the *linear* region for FETs. The I-V plots also reveal the second region where the currents are relatively constant with the voltage. This region is referred to as the *linear* region for HBTs, but is named the *saturation* region for FETs.

It is somewhat unfortunate that the names are exactly reversed for HBTs and FETs. However, the names are so chosen to reflect the device operation in each

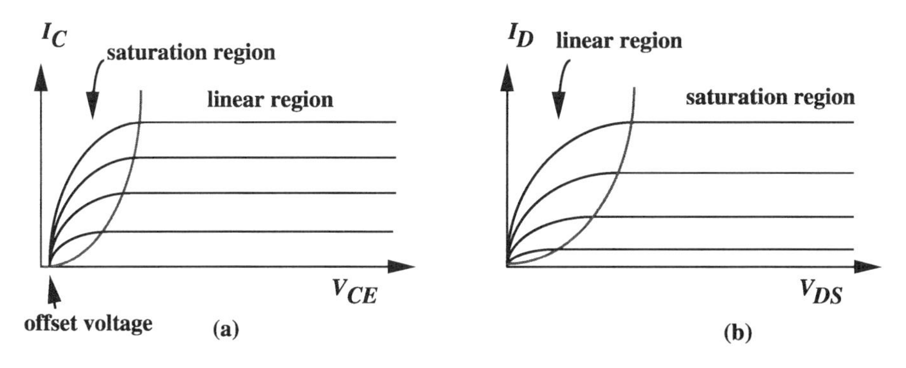

FIGURE 7-12. I-V characteristics of: (*a*) common-emitter HBT; (*b*) common-source FET.

of the region. When V_{CE} is small, both the base–emitter and base–collector junctions are forward-biased. The saturation region in HBTs is so named because, in this situation, the base is saturated with minority carriers injected from both junctions. The linear region in FETs is so named because when V_{DS} is small, the drain current increases roughly linearly with V_{DS}. This fact is observed in Eqs. (6-209) and (6-210) for MESFETs and HFETs. We proceed to discuss the reasons for the naming of the region where the currents are relatively constant with the bias voltage. In HBTs, this region is called the linear region because the current gain is constant. The output collector current increases linearly with the input base current. This behavior is unlike the saturation region, where the collector current has a weak dependence on the base current. In FETs, this region is called the saturation region because the current saturates to a relatively constant value. In long-channel devices, the current saturation is a result of pinchoff of the channel charges at the drain end. In short-channel devices, the current saturation is due to velocity saturation.

In HBTs, the point at which $I_C = 0$ does not occur right at $V_{CE} = 0$ (Fig. 7-12*a*). There is an offset voltage before the collector current changes sign. The offset voltage is a finite value because the base–emitter and the base–collector junctions are not symmetrical. The base–collector junction turns on at a slightly smaller bias voltage. In contrast, for FETs, the drain current is precisely zero when $V_{DS} = 0$. When $V_{DS} \neq 0$, a field is established to cause the available charges to drift, and hence, a finite current. Occasionally the threshold voltage of a FET is referred to as the offset voltage. In this sense, the offset voltage in HBTs and the offset voltage in FETs refer to totally different quantities.

The above differences in terminology come mainly from the differences in the physical operation of the devices. The following difference, however, is likely due to historical reasons. The critical dimension (on the order of 2 μm) in HBTs is called the emitter *width*. The dimension perpendicular to it (on the order of 30–60 μm), is called the emitter length. For FETs, the critical dimension (on the order of 0.25 μm) is referred to as the gate *length*, leaving *gate width* to refer to the larger dimension, on the order of 10 μm to 1 mm. We have followed these conventions in previous chapters, avoiding annoying statements like "the gate width is long."

Many parameters quantifying the device performance are expressed per unit length for FETs and per unit area in HBTs. For example, the power density in FETs is expressed in W/mm. It is in W/μm^2 in HBTs. The discrepancy lies ultimately in the fact that the FET is a horizontal device while the HBT is a vertical device. For HBTs, the entire emitter area conducts current (neglecting the emitter crowding effects). The wider the emitter width, and/or the longer the emitter length, the higher the output power that is delivered. Therefore, the power density is expressed by the power divided by the emitter area. When a practical HBT epitaxial structure is demonstrated to give a certain power, we expect to achieve twice the power when the emitter area is doubled, through increase in either the emitter width or the length. For FETs, however, the current is proportional only to the gate width, not the gate length. The power density is thus expressed by the power divided by the gate width. We expect the power of a FET to double when the gate width is doubled, but not when the gate length is doubled. The power density in W/mm is therefore a good measure to compare the device performances of various devices. Occasionally, however, HBT power performance in the unit W/mm has been reported, mainly to facilitate the comparison between HBTs and FETs.

Another terminology confusion relates to the meaning of passivation. As pointed in § 5-8, passivation in FET devices can be as simple as depositing a nitride layer, covering the exposed GaAs or AlGaAs layer. In HBTs, surface passivation generally refers to the use of AlGaAs ledge to passivate the extrinsic base surface.

Now that the terminology issues have been pointed out, we are ready to explore the difference in the HBT and FET performance. The following points are generally considered the advantages of HBTs.

- HBTs are vertical devices, while FETs are horizontal devices. The key distances governing the electron transit time in HBTs are established by epitaxial growth rather than by lithography. Therefore, high-speed performance is obtainable with conventional lithography. In contrast, the critical dimension in FETs, the gate length, is defined with submicron lithography such as electron-beam lithography. The requirement on the lithographical equipment is more stringent for FET processing. For HBTs, f_T and f_{max} of 100–200 GHz are obtained with 1–3 μm lithography. To achieve similar high-frequency performance, the gate dimension in FETs are typically 0.2–0.5 μm.
- The mutual transconductance in HBTs is directly proportional to the collector current, due to the exponential dependence of the collector current on the base–emitter bias [Eq. (3-2)]. For a typical emitter width of 2 μm and a operating current density of 5×10^4 A/cm^2, g_e approaches 3.9×10^4 S/mm. For FETs, we shall consider HFETs, which generally enjoy higher values of transconductance than MESFETs. When the thickness of the depleted layer between the metal gate and the accumulated channel is approximately 200 Å, the maximum intrinsic transconductance can be as high as 600, 800, and 1500 mS/mm in AlGaAs/GaAs, AlGaAs/InGaAs/GaAs, and In AlAs/InGaAs/InAlAs/InP device structures. These values are far smaller than that achievable with HBTs, since the FET's drain increase only linearly (if velocity saturates) or quadratically (if long channel) with the input voltage. The high transconductance values in HBTs allow them to be operated with small input voltage swings and fast charging of load capacitances in ICs.
- The whole emitter area of HBTs conducts current. In a horizontal device such as an FET, the effective device area is the product of the channel thickness and the gate width. Because the channel thickness cannot be arbitrarily increased, the FET's current-handling capability is dramatically less than that

of an HBT on a per-unit distance basis. The following examples calculate the required emitter length in HBTs and gate width in FETs to deliver a current level of 1 A. In HBTs, a collector current density of 5×10^4 A/cm^2 is routinely achievable. With a 2-μm emitter width, the required emitter length is 1000 μm. In HFETs with quantum well designs, the two-dimensional electron carrier density in the channel is typically equal to 2×10^{12} cm^{-2}. The electron saturation velocity is about 1×10^7 cm/s. In order to deliver 1 A, the gate width needs to be about 3125 μm. Because the output power of a transistor depends directly on the maximum operating current level, a FET amplifier generally occupies three times the device area of a HBT amplifier.

- The breakdown voltage of HBT is not determined by surface effects. It is straightforward to design the required epitaxial structure based on a given breakdown voltage criterion. For FETs, the breakdown voltage is strong function of the surface treatment during device processing.
- The turn-on voltage of HBTs is determined by the energy gap of the base and the emitter materials. HBTs are *well matched*; that is, two HBTs near each other have nearly identical turn-on characteristics. The matching property is critical to circuit blocks such as a differential amplifier in analog applications or a comparator in A/D converter applications. In contrast, the threshold voltage in FETs is determined by the doping level and the thickness of the channel layer. The uniformity of the threshold voltage on a wafer depends on the uniformity of the doping level as well as the epitaxial thickness. Control of the threshold voltage is especially difficult when the channel layer thickness is thin.
- Both MOCVD and MBE can be used to grow the HBT epitaxial layers. Control over the layer thickness and doping level on the order of 10% is sufficient to result in reproducible HBT performances. A background doping concentration on the order of 5×10^{15} cm^{-3} is adequate for most applications. For FETs, the requirement on the layer thickness and doping level is more stringent. Only MBE is routinely used to grow FET structures.
- Surface effects are not an important concern in HBTs, whose main device action occurs well below the top surface, in the bulk of the device. The current conduction in FETs, however, takes place near the wafer surface. The device operation is more susceptible to the dynamics of surface traps. Consequently, the $1/f$ noise is higher in FETs than in HBTs.
- III-V FETs are naturally depletion mode devices, with a negative threshold voltage. A negative gate bias is required to shut off the FET to its standby mode, while a positive drain bias is needed to turn on the device. Hence, a FET circuit requires bias voltage of opposite polarities. This directly contradicts a system requirement of reducing circuit components by having just a single battery source, especially in mobile phone applications. Moreover, shutting off the FET device completely is not easy. There is a finite standby current even with a certain negative gate bias. Therefore, some sort of current switch circuitry (usually made by silicon PMOS) is required to turn the current off.

Sometimes an enhancement-mode FET, one whose threshold voltage is positive, is made to remove the above shortcomings. It is hoped that, with a positive V_T, the FET can be shut off with a zero bias, so that the negative power supply is no longer crucial. However, in III-V FETs, the threshold voltage must be smaller than the metal/semiconductor junction built-in potential (ϕ_{bi}) to prevent forward-biasing the metal/semiconductor junction. It is found that, at least for MESFETs, these devices really cannot be shut off completely by zero bias although the threshold voltage is claimed to be positive. The current switch is still needed after all.
HBTs, in normal operation, use positive biases for both the base–emitter electrodes and the collector–emitter electrodes. The issues of a negative power supply do not exist.

- Both HBTs and FETs have a critical processing step. It is the base etch (or etch to the top of the base layer) for the HBTs, and the gate recess etch for the HFETs. The base layer in HBTs is about 800–1000 Å. There is some latitude in the base etch without significantly affecting the device performance. An overetch of 200 Å into the base layer still leaves plenty of thickness for base contact formation. The gate recess etch in FETs, in contrast, demands much tighter control. Assuming that the epitaxial layers are grown uniformly, the gate recess etch needs to be within 20 Å to achieve a 25-mV variation in the threshold voltage across the wafer. (For power amplifier applications in which the threshold voltage requirement is more relaxed, 100-Å control is sometimes tolerated.) Some FET structures incorporate etch stop layers, relying selective etching to achieve uniformity during the gate recess etch. For HBTs, because of the more relaxed requirement on the etching precision, a nonselective timed etch can be used in production.
- HBTs are said to exhibit better linearity characteristics than FETs. In mobile communication applications, a power amplifier is required to meet a certain linearity requirement while delivering a certain large-signal power. Generally, it is easier for HBTs to achieve the specification without special device design.

The above lists the advantages of HBTs over FETs. Most of the advantages are fundamental, originating from the ways the devices are operated. There are, however, practical advantages that FETs enjoy over the HBTs. In many applications, especially for telecommunication, both HBT and FET parts are widely used. We list the properties which are considered to be to the FETs' advantage in the following:

- HBTs have a vertical device structure. The epitaxial layers need to be etched at various places to access the base and collector layers underneath the top emitter layer. There are more processing steps for the HBTs than for FETs, although the overall cycle time can be equivalent due to the use of e-beam lithography for FETs. There is, however, a more important drawback associated with these layer-access etches. The resultant HBT device structure is most often nonplanar, with the emitter, base, and collector contacts being placed at different distances away from the top. The nonplanarity is especially severe in power HBTs, whose collector thickness is on the order of 1 μm. The nonplanarity does not preclude the monolithic integration of HBTs, but it can create a yield problem in production, due to significant amounts of metal step coverage. In contrast, all of the FET terminals are placed roughly at the top. The resultant planar structure facilitates the metal routing and simplifies processing.

- FETs do not exhibit thermal instability problems. The critical parameters which make up the drain current of FET devices, such as the mobility and the saturation velocity, all decrease with temperature. Therefore, there is an intrinsic negative feedback mechanism to stabilize the drain current. When the drain current increases, the self-heating increases the junction temperature, which in turn lowers the carrier velocity. There is no current hogging in a multi-finger FET device.

  HBTs, by virtue of being bipolar devices, can have thermal instability if precaution is not exercised. As discussed in § 3-10, the thermal instability in multifinger HBTs results in the collapse of current gain, in which the collector current exhibits a sudden decrease when V_{CE} increases. Thermal instablilty in HBTs (or collapse of current gain) can be avoided with proper ballasting as well as other techniques. Unconditional thermal instability has been achieved in state-of-the-art power amplifiers [1]. Nonetheless, the potential of having thermal instability is a disadvantage for HBTs, because it complicates the device design.

- The design processes involved in fabricating FETs are simpler than HBT. The most difficult step in FET fabrication, gate formation, was resolved the moment e-beam lithography equipment became available. The mask layouts for various patterns such as isolation are straightforward due to its planar structure. For HBTs, the mask layout design is more elaborate. Therefore, FETs are thought to be easier to fabricate.
- Although HBTs enjoy low $1/f$ noises at low frequency, FETs enjoy significantly better noise performance at high frequencies. The figure of merit of high-frequency noise is the noise figure. In HBTs, the noise figure is high because the shot noise is proportional to the operating current. The FETs' noise figure is determined by the thermal noise, and it generally has a low value.
- When the requirement on performance is not stringent, ion-implanted MESFET can be used. The active layer of the ion-implanted MESFET is made through ion-implantation. There is no epitaxial layer, unlike the HBTs and HFETs. This reduces the cost, since epitaxy typically costs about $1000 per wafer. The use of bare GaAs wafer as the starting material also reduces the cycle time involved in epitaxial growth.

- There is excess charge storage in the base of HBTs when the devices operate in saturation, in which both the emitter–base and the base–collector junctions are forward-biased. It takes a finite amount of time to discharge these storage charges before the devices back away from the saturation region and enter the forward-active mode. The charge storage problem is absent in FETs.

Often the decision to implement a product in either HBT or FET does not rest on whichever device is physically able to deliver superior performance. In fact, for many applications, both devices are more than capable of meeting the design requirements. Therefore, many other issues, such as time-to-market, availability of a process, or simply historical reasons, can become the dominant factors in making the decision.

REFERENCE

1. Liu, W. (1998). *Handbook of III-V Heterojunction Bipolar Transistors*, New York: Wiley.

PROBLEMS

1. Devise a d.c. HBT process such that a transistor can be fabricated with three masks. Can this process be used to fabricate a small device?

2. Devise a d.c. HBT process such that the transistor can be fabricated with only two mask levels.

3. A $2 \times 400\,\mu m^2$ HBT delivers a output power of 2 W. What is the power density in W/μm^2 and in W/mm?

APPENDIX A

MESFET *y*-PARAMETERS

The differential equation governing the small-signal properties was given by Eq. (6-125). We express it in a slightly different form:

$$\frac{d^2}{dx^2}\left[g(\phi_{s,\mathrm{dc}})\,\tilde{\phi}_{s,\mathrm{ac}}\right] = \frac{j\omega aqWN_d}{2\sqrt{\phi_{00}}\sqrt{\phi_{s,\mathrm{dc}}(x)}}\,\tilde{\phi}_{s,\mathrm{ac}}(x) \tag{A-1}$$

where

$$g(\phi_{s,\mathrm{dc}}) = aqWN_d\mu_n\left(1-\sqrt{\frac{\phi_{s,\mathrm{dc}}}{\phi_{00}}}\right) = g_0\left(1-\sqrt{\frac{\phi_{s,\mathrm{dc}}}{\phi_{00}}}\right) \tag{A-2}$$

$$g_0 = aqWN_d\mu_n \tag{A-3}$$

The appropriate boundary conditions were given by Eq. (6-127). Despite that $\tilde{\phi}_{s,\mathrm{ac}}$ is the unknown of the differential equation, it is not the ultimate quantity we are interested in. We need to determine the terminal currents, from which the two-port y-parameters of the transistor can be established. The relationship between the small-signal terminal currents and $\tilde{\phi}_{s,\mathrm{ac}}$ can be found from Eqs. (6-128) and (6-124). We express them in terms of $g(\phi_{s,\mathrm{dc}})$ just introduced:

$$\tilde{i}_s = -\frac{d}{dx}\left[g(\phi_{s,\mathrm{dc}})\,\tilde{\phi}_{s,\mathrm{ac}}\right]\Big|_{x=0} \tag{A-4}$$

$$\tilde{i}_d = +\frac{d}{dx}\left[g(\phi_{s,\mathrm{dc}})\,\tilde{\phi}_{s,\mathrm{ac}}\right]\Big|_{x=L} \tag{A-5}$$

$$\tilde{i}_g = -\tilde{i}_d - \tilde{i}_s \tag{A-6}$$

We adopt a series expansion approach to solve Eq. (A-1). Further, since we are interested primarily in the first-order frequency terms (the quasi-static terms), we will expand the variables only to their $j\omega$ terms. Therefore, we write:

$$\tilde{\phi}_{s,\mathrm{ac}}(x) \approx \tilde{\phi}_{s,0}(x) + j\omega\tilde{\phi}_{s,1}(x) \tag{A-7}$$

Both $\tilde{\phi}_{s,0}$ and $\tilde{\phi}_{s,1}$ are unknown. They are determined by substituting Eq. (A-7) into Eq. (A-1):

$$\frac{d^2}{dx^2}[g(\phi_{s,\mathrm{dc}})\,\tilde{\phi}_{s,0}] = 0 \qquad \tilde{\phi}_{s,0}(0) = -\,\tilde{v}_{gs} \qquad \tilde{\phi}_{s,0}(L) = -\,\tilde{v}_{gs} + \tilde{v}_{ds} \tag{A-8}$$

$$\frac{d^2}{dx^2}[g(\phi_{s,\mathrm{dc}})\,\tilde{\phi}_{s,1}] = \frac{aqWN_d}{2\sqrt{\phi_{00}}\sqrt{\phi_{s,\mathrm{dc}}(x)}}\,\tilde{\phi}_{s,0}(x) \qquad \tilde{\phi}_{s,1}(0) = 0 \qquad \tilde{\phi}_{s,1}(L) = 0 \tag{A-9}$$

Similar to the first-order approximation made to $\tilde{\phi}_{s,\mathrm{ac}}$, we approximate the terminal small-signal currents as:

$$\tilde{i}_g = y_{gg}\tilde{v}_{gs} + y_{gd}\,\tilde{v}_{ds} \approx \tilde{i}_{g,0} + j\omega\tilde{i}_{g,1} \tag{A-10}$$

$$\tilde{i}_d = y_{dg}\tilde{v}_{gs} + y_{dd}\,\tilde{v}_{ds} \approx \tilde{i}_{d,0} + j\omega\tilde{i}_{d,1} \tag{A-11}$$

$$\tilde{i}_s \approx \tilde{i}_{s,0} + j\omega\tilde{i}_{s,1} \tag{A-12}$$

To determine y_{11} and y_{21}, we first consider only the presence of perturbation in v_{gs}. The perturbation v_{ds} is assumed to be zero first. It will be considered later when we determine y_{12} and y_{22}. The solution to Eq. (A-8) involves a straightforward integration:

$$g(\phi_{s,\mathrm{dc}})\tilde{\phi}_{s,0}(x) = \left(A\frac{x}{L} + B\right)\tilde{v}_{gs} \tag{A-13}$$

$$A = g[\phi_{s,\mathrm{dc}}(0)] - g[\phi_{s,\mathrm{dc}}(L)] = g_0(1 - \sqrt{s}) - g_0(1 - \sqrt{d}) = g_0(\sqrt{d} - \sqrt{s}) \tag{A-14}$$

$$B = -\,g[\phi_{s,\mathrm{dc}}(0)] = -\,g_0(1 - \sqrt{s}) \tag{A-15}$$

d and s were given in Eq. (5-33).

Applying the definition of the source and drain current, as well as the first-order approximations delineated in Eq. (A-10), we obtain:

$$\tilde{i}_{s,0} = -\frac{d}{dx}[g(\phi_{s,\mathrm{dc}})\tilde{\phi}_{s,0}]|_{x=0} = -\frac{A}{L}\,\tilde{v}_{gs} \tag{A-16}$$

$$\tilde{i}_{d,0} = \frac{d}{dx}[g(\phi_{s,\mathrm{dc}})\tilde{\phi}_{s,0}]|_{x=L} = \frac{A}{L}\,\tilde{v}_{gs} \tag{A-17}$$

$$\tilde{i}_{g,0} = -\,\tilde{i}_{d,0} - \tilde{i}_{s,0} = 0 \tag{A-18}$$

After finding $\tilde{\phi}_{s,0}(x)$ from Eq. (A-13), we substitute it into Eq. (A-9) to determine $\tilde{i}_{s,1}(x)$. We recall that, from Eqs. (5-29) and (5-31),

$$I_D = qW\mu_n N_d a\left(1 - \sqrt{\frac{\phi_{s,\mathrm{dc}}}{\phi_{00}}}\right)\frac{d\phi_{s,\mathrm{dc}}}{dx} = g(\phi_{s,\mathrm{dc}})\frac{d\phi_{s,\mathrm{dc}}}{dx} \tag{A-19}$$

We combine Eq. (A-19) with Eq. (A-13) to express $\tilde{\phi}_{s,0}(x)$ as

$$\tilde{\phi}_{s,0}(x) = (A\frac{x}{L} + B)\tilde{v}_{gs}\frac{1}{I_D}\frac{d\phi_{s,\mathrm{dc}}}{dx} \tag{A-20}$$

Therefore, the differential equation governing $\tilde{\phi}_{s,1}(x)$ [from Eq. (A-9)] becomes:

$$\frac{d^2}{dx^2}[g(\phi_{s,\mathrm{dc}})\tilde{\phi}_{s,1}] = \frac{aqWN_d}{2\sqrt{\phi_{00}}\sqrt{\phi_{s,\mathrm{dc}}(x)}} \times \left(A\frac{x}{L} + B\right)\frac{\tilde{v}_{gs}}{I_D}\frac{d\phi_{s,\mathrm{dc}}}{dx} \tag{A-21}$$

Integrating the differential equation once from $x' = 0$ to $x' = x$, and invoking the boundary condition at the source, we obtain:

$$\frac{d}{dx}[g(\phi_{s,\mathrm{dc}})\tilde{\phi}_{s,1}] = K + aqWN_d\frac{\tilde{v}_{gs}}{I_D} \times\left[\left(A\frac{x}{L} + B\right)u^{1/2} - B\,s^{1/2} - \frac{Ag_0\phi_{00}}{LI_D}\left(\frac{2}{3}u^{3/2} - \frac{2}{3}s^{3/2} - \frac{1}{2}u^2 + \frac{1}{2}s^2\right)\right] \tag{A-22}$$

where $u(x)$ was given in Eq. (5-33) and K denotes

$$K - \frac{d}{dx}[g(\phi_{s,\mathrm{dc}})\tilde{\phi}_{s,1}]|_{x=0}. \tag{A-23}$$

Noting that the unit for $\tilde{\phi}_{s,1}$ is V-s while that for $\tilde{\phi}_{s,0}$ is V, we find that the units on both sides of Eq. (A-22) are identical.

We integrate Eq. (A-22) from $x' = 0$ to $x' = x$, again using the identity of Eq. (A-19):

$$\begin{aligned} g(\phi_{s,\mathrm{dc}})\tilde{\phi}_{s,1} = Kx &+ \frac{aqWN_d}{I_D^2}\tilde{v}_{gs}g_0\phi_{00}\left[\left(A\frac{x}{L} + B\right)\left(\frac{2}{3}u^{3/2} - \frac{1}{2}u^2\right) - B\left(\frac{2}{3}s^{3/2} - \frac{1}{2}s^2\right)\right] \\ &- \frac{aqWN_d}{I_D}\tilde{v}_{gs}\left[B\,s^{1/2}x - \frac{ag_0\phi_{00}}{I_D L}\left(\frac{2}{3}s^{3/2} - \frac{1}{2}s^2\right)x\right] \\ &- 2A'\frac{aqWN_d g_0^2\phi_{00}^2}{I_D^3 L}\tilde{v}_{gs}\left[\frac{4}{15}(u^{5/2} - s^{5/2}) - \frac{7}{8}(u^3 - s^3) + \frac{1}{7}(u^{7/2} - s^{7/2})\right] \end{aligned} \tag{A-24}$$

The constant K is established from the remaining boundary condition at the drain, that $\tilde{\phi}_{s,1}(L)$ is equal to zero. Furthermore, we substitute A and B in terms of g_0, s, and d:

$$\begin{aligned}
\frac{K}{\tilde{v}_{gs}} &= \frac{aqN_dWL}{\phi_{00}}\left(\frac{g_0\phi_{00}}{I_DL}\right)^2\left[(1-d^{1/2})\left(\frac{2}{3}d^{3/2}-\frac{1}{2}d^2+\frac{2}{3}s^{3/2}-\frac{1}{2}s^2\right)\right] \\
&\quad -2\frac{aqN_dWL}{\phi_{00}}\left(\frac{g_0\phi_{00}}{I_DL}\right)^2\left[(1-s^{1/2})\left(\frac{2}{3}s^{3/2}-\frac{1}{2}s^2\right)\right] \\
&\quad -\frac{aqN_dWL}{\phi_{00}}\left(\frac{g_0\phi_{00}}{I_DL}\right)[s^{1/2}(1-s^{1/2})] \\
&\quad +2\frac{aqN_dWL}{\phi_{00}}\left(\frac{g_0\phi_{00}}{I_DL}\right)^3(d^{1/2}-s^{1/2}) \\
&\quad \times\left[\frac{4}{15}(d^{5/2}-s^{5/2})-\frac{7}{18}(d^3-s^3)+\frac{1}{7}(d^{7/2}-s^{7/2})\right]
\end{aligned} \tag{A-25}$$

With K given in Eq. (A-25), we have a complete solution of $\tilde{\phi}_{s,1}(x)$ which is given in Eq. (A-24). However, we are interested in the small-signal terminal currents. Therefore, we need to carry the derivation one more step. From the definitions of Eqs. (A-4) to (A-6):

$$\tilde{i}_{s,1} = -\frac{d}{dx}[g(\phi_{s,\mathrm{dc}})\tilde{\phi}_{s,1}]|_{x=0} = -K \tag{A-26}$$

$$\begin{aligned}
\tilde{i}_{d,1} &= \frac{d}{dx}[g(\phi_{s,\mathrm{dc}})\tilde{\phi}_{s,1}]|_{x=L} \\
&= K - \frac{aqN_dWL}{\phi_{00}}\tilde{v}_{gs}\left(\frac{g_0\phi_{00}}{I_DL}\right)[d^{1/2}(1-d^{1/2})-s^{1/2}(1-s^{1/2})] \\
&\quad -\frac{aqN_dWL}{\phi_{00}}\tilde{v}_{gs}\left(\frac{g_0\phi_{00}}{I_DL}\right)^2\left[(d^{1/2}-s^{1/2})\left(\frac{2}{3}d^{3/2}-\frac{2}{3}s^{3/2}-\frac{1}{2}d^2+\frac{1}{2}s^2\right)\right]
\end{aligned} \tag{A-27}$$

$$\begin{aligned}
\tilde{i}_{g,1} &= -\tilde{i}_{d,1}-\tilde{i}_{s,1} \\
&= \frac{aqN_dWL}{\phi_{00}}\tilde{v}_{gs}\left(\frac{g_0\phi_{00}}{I_DL}\right)[d^{1/2}(1-d^{1/2})-s^{1/2}(1-s^{1/2})] \\
&\quad +\frac{aqN_dWL}{\phi_{00}}\tilde{v}_{gs}\left(\frac{g_0\phi_{00}}{I_DL}\right)^2\left[(d^{1/2}-s^{1/2})\left(\frac{2}{3}d^{3/2}-\frac{2}{3}s^{3/2}-\frac{1}{2}d^2+\frac{1}{2}s^2\right)\right]
\end{aligned} \tag{A-28}$$

In the above derivation, a convenient relationship has been used:

$$\frac{du}{dx} = \frac{I_D}{g_0\phi_{00}}\frac{1}{1-u^{1/2}} \tag{A-29}$$

We are now ready to determine y_{gg} and y_{dg}. According to Eq. (A-10), y_{gg} is:

$$y_{gg} = \frac{\tilde{i}_{g,0} + j\omega\tilde{i}_{g,1}}{\tilde{v}_{gs}}$$

$$= j\omega \times \left\{ \frac{aqN_d WL}{\phi_{00}} \left(\frac{g_0\phi_{00}}{I_D L} \right) [d^{1/2}(1 - d^{1/2}) - s^{1/2}(1 - s^{1/2})] \right.$$

$$\left. + \frac{aqWN_d WL}{\phi_{00}} \left(\frac{g_0\phi_{00}}{I_D L} \right)^2 \left[(d^{1/2} - s^{1/2}) \left(\frac{2}{3} d^{3/2} - \frac{2}{3} s^{3/2} - \frac{1}{2} d^2 + \frac{1}{2} s^2 \right) \right] \right\} \quad \text{(A-30)}$$

y_{gg} is purely imaginary because $\tilde{i}_{g,0}$ is zero. Similarly. we find y_{dg} from:

$$y_{dg} = \frac{\tilde{i}_{d,0} + j\omega\tilde{i}_{d,1}}{\tilde{v}_{gs}} = \frac{g_0}{L}(\sqrt{d} - \sqrt{s}) - y_{gg} + \frac{K}{\tilde{v}_{gs}} \quad \text{(A-31)}$$

where y_{gg} and K/v_{gs} were given in Eqs. (A-30) and (A-25), respectively.

There are a total of four y-parameters to characterize a two-port. We have found y_{gg} and y_{dg}. To establish the remaining two y-parameters, we consider only small-signal excitation in the drain and make $\tilde{v}_{gs} = 0$. Re-solving Eq. (A-8), except with the boundary condition that $\tilde{\phi}_{s,0}(0) = 0$ and that $\tilde{\phi}_{s,0}(L) = \tilde{v}_{ds}$, we obtain a solution for $\tilde{\phi}_{s,0}(x)$:

$$g(\phi_{s,\text{dc}})\tilde{\phi}_{s,0} = \frac{A'}{L} x\tilde{v}_{ds} \quad \text{(A-32)}$$

$$A' = g[\phi_{s,\text{dc}}(L)] = g_0\,(1 - \sqrt{d}) \quad \text{(A-33)}$$

Applying the definition of the source and drain currents [Eqs. (A-4) to (A-6)], as well as the first-order approximations delineated in Eq. (A-10), we obtain:

$$\tilde{i}_{s,0} = -\frac{d}{dx}[g(\phi_{s,\text{dc}})\tilde{\phi}_{s,0}]|_{x=0} = -\frac{A'}{L}\tilde{v}_{ds} \quad \text{(A-34)}$$

$$\tilde{i}_{d,0} = \frac{d}{dx}[g(\phi_{s,\text{dc}})\tilde{\phi}_{s,0}]|_{x=L} = \frac{A'}{L}\tilde{v}_{ds} \quad \text{(A-35)}$$

$$\tilde{i}_{g,0} = -\tilde{i}_{d,0} - \tilde{i}_{s,0} = 0 \quad \text{(A-36)}$$

Having found $\tilde{\phi}_{s,0}(x)$ from Eq. (A-32), we can substitute it into Eq. (A-9) to rewrite the differential equation for $\tilde{\phi}_{s,1}(x)$:

$$\frac{d^2}{dx^2}[g(\phi_{s,\text{dc}})\tilde{\phi}_{s,1}] = \frac{aqWN_d}{2\sqrt{\phi_{00}}\sqrt{\phi_{s,\text{dc}}(x)}} \times \left(A' \frac{x}{L} \right) \frac{\tilde{v}_{ds}}{I_D} \frac{d\phi_{s,\text{dc}}}{dx} \quad \text{(A-37)}$$

Integrating the differential equation once from $x' = 0$ to $x' = x$ and invoking the boundary condition at the source, we obtain:

$$\frac{d}{dx}[g(\phi_{s,\mathrm{dc}})\tilde{\phi}_{s,1}] = K' + aqWN_d\frac{\tilde{v}_{ds}}{I_D} \times\left[\frac{A'x}{L}u^{1/2} - \frac{A'g_0\phi_{00}}{LI_D}\left(\frac{2}{3}u^{3/2} - \frac{2}{3}s^{3/2} - \frac{1}{2}u^2 + \frac{1}{2}s^2\right)\right] \quad \text{(A-38)}$$

where K' is given by

$$K' = \frac{d}{dx}[g(\phi_{s,\mathrm{dc}})\tilde{\phi}_{s,1}]|_{x=0} \quad \text{(A-39)}$$

Again using the identity of Eq. (A-19) in the integration of Eq. (A-37) from $x' = 0$ to $x' = x$, we obtain:

$$g(\phi_{s,\mathrm{dc}})\tilde{\phi}_{s,1} = K'x + \frac{aqWN_d g_0\phi_{00}}{I_D^2}\tilde{v}_{ds}\left[\frac{A'x}{L}\left(\frac{2}{3}u^{3/2} - \frac{1}{2}u^2 + \frac{2}{3}s^{3/2} - \frac{1}{2}s^2\right)\right] - 2A'\frac{aqWN_d g_0^2\phi_{00}^2}{I_D^3 L}\tilde{v}_{ds}\left[\frac{4}{15}(u^{5/2} - s^{5/2}) - \frac{7}{18}(u^3 - s^3) + \frac{1}{7}(u^{7/2} - s^{7/2})\right] \quad \text{(A-40)}$$

The constant K' is established from the remaining boundary condition at the drain, that $\tilde{\phi}_{s,1}(L)$ is equal to zero. Furthermore, we substitute A' in terms of g_0, d, and s:

$$\frac{K'}{\tilde{v}_{ds}} = -\frac{aqN_dWL}{\phi_{00}}\left(\frac{g_0\phi_{00}}{I_DL}\right)^2\left[(1 - d^{1/2})\left(\frac{2}{3}d^{3/2} - \frac{1}{2}d^2 + \frac{2}{3}s^{3/2} - \frac{1}{2}s^2\right)\right] + 2\frac{aqN_dWL}{\phi_{00}}\left(\frac{g_0\phi_{00}}{I_DL}\right)^3(1 - d^{1/2})\left[\frac{4}{15}(d^{5/2} - s^{5/2}) - \frac{7}{18}(d^3 - s^3) + \frac{1}{7}(d^{7/2} - s^{7/2})\right] \quad \text{(A-41)}$$

Thus far, both $\tilde{\phi}_{s,0}(x)$ and $\tilde{\phi}_{s,1}(x)$ are found. However, in the interest of finding y_{gd} and y_{dd}, we need to proceed to determine the small-signal terminal currents. The terminal currents are given by:

$$\tilde{i}_{s,1} = -\frac{d}{dx}[g(\phi_{s,\mathrm{dc}})\tilde{\phi}_{s,1}]|_{x=0} = -K' \quad \text{(A-42)}$$

$$\tilde{i}_{d,1} = \frac{d}{dx}[g(\phi_{s,\mathrm{dc}})\tilde{\phi}_{s,1}]|_{x=L} = K' + \frac{aqN_dWL}{\phi_{00}}\tilde{v}_{ds}\left(\frac{g_0\phi_{00}}{I_DL}\right)[d^{1/2}(1 - d^{1/2})] - \frac{aqN_dWL}{\phi_{00}}\tilde{v}_{ds}\left(\frac{g_0\phi_{00}}{I_DL}\right)^2(1 - d^{1/2})\left(\frac{2}{3}d^{3/2} - \frac{2}{3}s^{3/2} - \frac{1}{2}d^2 + \frac{1}{2}s^2\right) \quad \text{(A-43)}$$

$$\tilde{i}_{g,1} = -\tilde{i}_{d,1} - \tilde{i}_{s,1}$$
$$= -\frac{aqN_dWL}{\phi_{00}}\tilde{v}_{ds}\left(\frac{g_0\phi_{00}}{I_DL}\right)[d^{1/2}(1-d^{1/2})]$$
$$+\frac{aqN_dWL}{\phi_{00}}\tilde{v}_{ds}\left(\frac{g_0\phi_{00}}{I_DL}\right)^2(1-d^{1/2})\left(\frac{2}{3}d^{3/2}-\frac{2}{3}s^{3/2}-\frac{1}{2}d^2+\frac{1}{2}s^2\right) \quad \text{(A-44)}$$

We are now ready to determine y_{gd} and y_{dd}. From Eq. (A-11), we find:

$$y_{dd} = \frac{\tilde{i}_{d,0} + j\omega\tilde{i}_{d,1}}{\tilde{v}_{ds}}$$
$$= \frac{g_0}{L}(1-\sqrt{d} + j\omega \times \left\{-2\frac{aqN_dWL}{\phi_{00}}\left(\frac{g_0\phi_{00}}{I_DL}\right)^2\right.$$
$$\times\left[(1-d^{1/2})\left(\frac{2}{3}d^{1/2}-\frac{1}{2}\mathrm{d}^2\right)\right]$$
$$+2\frac{aqN_dWL}{\phi_{00}}\left(\frac{g_0\phi_{00}}{I_DL}\right)^3(1-d^{1/2})$$
$$\times\left[\frac{4}{15}(d^{5/2}-s^{5/2})-\frac{7}{18}(d^3-s^3)+\frac{1}{7}(d^{7/2}-s^{7/2})\right]$$
$$\left.+\frac{aqN_dWL}{\phi_{00}}\left(\frac{g_0\phi_{00}}{I_DL}\right)[d^{1/2}(1-d^{1/2})]\right\} \quad \text{(A-45)}$$

Similarly, we find y_{gd} to be:

$$y_{gd} = \frac{\tilde{i}_{g,0} + j\omega\tilde{i}_{g,1}}{\tilde{v}_{ds}}$$
$$= -j\omega \times \left\{\frac{aqN_dWL}{\phi_{00}}\left(\frac{g_0\phi_{00}}{I_DL}\right)[d^{1/2}(1-d^{1/2})]\right.$$
$$\left.-\frac{aqN_dWL}{\phi_{00}}\left(\frac{g_0\phi_{00}}{I_DL}\right)^2\left[(1-d^{1/2})\left(\frac{2}{3}d^{3/2}-\frac{2}{3}s^{3/2}-\frac{1}{2}d^2+\frac{1}{2}s^2\right)\right]\right\} \quad \text{(A-46)}$$

The terminal currents all contain the term $g_0\phi_{00}/(I_DL)$. With g_0 From Eq. (A-3) and I_D from Eq. (5-40), we find this particular term unitless, being equal to:

$$\frac{g_0\phi_{00}}{I_DL} = \frac{1}{d - s - \frac{2}{3}d^{3/2} + \frac{2}{3}s^{3/2}} \quad \text{(A-47)}$$

In summary, the common-source y-parameters, up to the $(j\omega)$ term, are established. y_{gg}, y_{gd}, y_{dg}, and y_{dd} are given by Eqs. (A-30), (A-46), (A-31), and (A-45), respectively.

APPENDIX B

HFET y-PARAMETERS

The charge control model for the HFET is nearly identical to that for the MOSFET. The main difference lies in that there is a bulk terminal in the MOSFET, with the substrate material doped in the medium range. Many of the expressions in MOSFET bear a constant δ [1], typically on the order of 0.2, to reflect the nonzero influence of the bulk voltage on the channel charge (body effects). In HFET, the bulk is nearly semiinsulating, and there is no bulk terminal. If the MOSFET equations have been derived, then the corresponding HFET equations can be written by setting δ to zero.

The analysis carried out here assumes that the mobility is a constant. As will be clear shortly, even with this simple model, the derivation of y-parameters is quite challenging. The velocity saturation is not considered.

Our starting equations are the continuity equation and the drift equation given by Eqs. (5-135) and (5-136). Suppose that the transistor is biased at V_{GS} and V_{DS} with sinusoidal perturbations of v_{gs} and v_{ds}. Then:

$$v_{GS}(t) = V_{GS} + \tilde{v}_{Gs}e^{j\omega t} \tag{B-1}$$

$$v_{CS}(x, t) = V_{CS}(x) + \tilde{v}_{cs}(x)e^{j\omega t} \tag{B-2}$$

$$v_{CS}(x = 0, t) = 0 \qquad v_{CS}(x = L, t) = V_{DS} + \tilde{v}_{ds}e^{j\omega t} \tag{B-3}$$

We can express $u_{CH}(x, t)$ and $i_{CH}(x, t)$ as

$$u_{CH}(x, t) = U_{CH}(x) + \tilde{u}_{ch}(x)e^{j\omega t} \tag{B-4}$$

$$i_{CH}(x, t) = I_{CH} + \tilde{i}_{ch}(x)e^{j\omega t} \tag{B-5}$$

$I_{CH}(x)$ is equal to the negative of I_D, which was given in Eq. (5-139). $U_{CH}(x)$ is given in Eq. (5-137). The small-signal channel potential is given by:

$$\tilde{u}_{ch}(x) = \tilde{v}_{gs} + \tilde{v}_{cs}(x) \tag{B-6}$$

Substituting these small-signal quantities into Eqs. (5-135) and (5-136), we find the terms involving with $e^{j\omega t}$ give rise to the following differential equations:

$$\frac{\partial \tilde{i}_{ch}(x)}{\partial x} = j\omega W C'_{ox} \tilde{u}_{CH}(x) \tag{B-7}$$

$$\tilde{i}_{ch}(x) = W C'_{ox} \mu_n \frac{d}{dx} [U_{CH}(x) \cdot \tilde{u}_{ch}(x)] \tag{B-8}$$

According to Eq. (5-139), we have:

$$\frac{d}{dx} = \frac{I_D}{W \mu_n C'_{ox}} \times \frac{1}{U_{CH}(x)} \frac{d}{dU_{CH}} \tag{B-9}$$

Substituting Eq. (B-9) into Eq. (B-8), we obtain:

$$\tilde{i}_{ch}(x) = \frac{I_D}{U_{CH}(x)} \frac{d}{dU_{CH}} [U_{CH}(x) \cdot \tilde{u}_{ch}(x)] \tag{B-10}$$

Similarly, substituting Eq. (B-9) into Eq. (B-7), we obtain:

$$\frac{d\tilde{i}_{ch}}{dU_{CH}} = j\omega \frac{(W C'_{ox})^2 \mu_n}{I_D} U_{CH} \tilde{u}_{CH} \tag{B-11}$$

Taking the derive of Eq. (B-11) and then utilizing the result of Eq. (B-10), we find that the master equation to be solved is:

$$\frac{d^2 \tilde{i}_{ch}}{dU_{CH}^2} - jDU_{CH} \tilde{i}_{ch} = 0 \tag{B-12}$$

where D is given by

$$D = \omega \mu_n \left(\frac{W C'_{ox}}{I_D} \right)^2 \tag{B-13}$$

Equation (B-12) is a cylindrical differential equation satisfying the form $x^2 y'' + (1 - 2a)xy' - ((b^2 c^2 x^{2c}) + (c^2 p^2 - a^2))y = 0$, with $a = \frac{1}{2}$, $c = \frac{3}{2}$, $b = \frac{2}{3}$ times the square root of (jD), and $p = \frac{1}{3}$. With these coefficients, the solution of Eq. (B-12) is given by [2]:

$$\tilde{i}_{ch}(U_{CH}) = C'_1 \sqrt{U_{CH}} \hat{I}_{1/3} \left(\frac{2}{3} \sqrt{jD} U_{CH}^{3/2} \right) + C'_2 \sqrt{U_{CH}} \hat{K}_{1/3} \left(\frac{2}{3} \sqrt{jD} U_{CH}^{3/2} \right) \tag{B-14}$$

where $\hat{I}$ and $\hat{K}$ are the modified Bessel functions. A mathematical identity states that the general solution written above, with $p \neq$ integer, can be alternatively expressed by:

$$\tilde{i}_{ch}(U_{CH}) = C_1 \sqrt{U_{CH}} \hat{I}_{1/3} \left(\frac{2}{3} \sqrt{jD} U_{CH}^{3/2} \right) + C_2 \sqrt{U_{CH}} \hat{I}_{-1/3} \left(\frac{2}{3} \sqrt{jD} U_{CH}^{3/2} \right) \tag{B-15}$$

The square root of j is equal to $\exp(j\pi/4)$. One mathematical definition is helpful:

$$\hat{I}_n(\sqrt{j}x) \equiv \text{Ber}_n(x) + j\,\text{Bei}_n(x) \tag{B-16}$$

where $\text{Ber}_n(x)$ and $\text{Bei}_n(x)$ are derivative functions of the Bessel function. They are given by:

$$\text{Ber}_n(x) = \sum_{k=0}^{\infty} \frac{(x/2)^{2k+n}}{k!\,\Gamma(n+k+1)} \cos\left(\frac{3n}{4} + \frac{k}{2}\right)\pi \tag{B-17}$$

$$\text{Bei}_n(x) = \sum_{k=0}^{\infty} \frac{(x/2)^{2k+n}}{k!\,\Gamma(n+k+1)} \sin\left(\frac{3n}{4} + \frac{k}{2}\right)\pi \tag{B-18}$$

With this information, Eq. (B-15) reduces to:

$$\begin{aligned}\tilde{i}_{\text{ch}}(x) = {} & k_1 U_{\text{CH}}(x)\,[1+j]\left[1 + j\frac{DU_{\text{CH}}^3(x)}{12} - \frac{D^2U_{\text{CH}}^6(x)}{504} - j\frac{D^3U_{\text{CH}}^9(x)}{45{,}360} + \cdots\right] \\ & + k_2[1-j]\left[1 + j\frac{DU_{\text{CH}}^3(x)}{6} - \frac{D^2U_{\text{CH}}^6(x)}{180} - j\frac{D^3U_{\text{CH}}^9(x)}{12{,}960} + \cdots\right]\end{aligned} \tag{B-19}$$

k_1 and k_2 are unknown constants. They need to be determined from the boundary conditions that:

$$\tilde{u}_{\text{ch}}(0) = \tilde{v}_{gs} \qquad \tilde{u}_{\text{ch}}(L) = \tilde{v}_{gs} - \tilde{v}_{ds} \tag{B-20}$$

Thus far, the solution is in terms of $i_{\text{ch}}(x)$. We need to determine $u_{\text{ch}}(x)$ somehow. According to Eq. (B-11), u_{ch} is equal to:

$$\tilde{u}_{\text{ch}} = \frac{1}{jDI_D} \times \frac{1}{U_{\text{CH}}} \frac{d\tilde{i}_{\text{ch}}}{dU_{\text{CH}}} \equiv \frac{1}{jDI_D} \times [k_1 F_1(U_{\text{CH}}) + k_2 F_2(U_{\text{CH}})] \tag{B-21}$$

where, according to Eq. (B-19), $F_1(U_{CH})$ and $F_2(U_{\text{CH}})$ are given by:

$$F_1(U_{\text{CH}}) = (1+j) \times \left(\frac{1}{U_{\text{CH}}} + j\frac{1}{3}DU_{\text{CH}}^2 - \frac{1}{72}D^2U_{\text{CH}}^5 - j\frac{1}{4536}D^3U_{\text{CH}}^8 + \cdots\right) \tag{B-22}$$

$$F_2(U_{\text{CH}}) = (1-j) \times \left(j\frac{1}{2}DU_{\text{CH}} - \frac{1}{72}D^2U_{\text{CH}}^4 - j\frac{1}{1440}D^3U_{\text{CH}}^7 + \cdots\right) \tag{B-23}$$

The unit of $F_1(x)$ is V^{-1} and the unit of $F_2(x)$ is V^{-2}.

We plug the boundary conditions given in Eq. (B-20) into Eq. (B-21). With $F_1(0)$ denoting $F_1[U_{\text{CH}}(x=0)]$, $F_1(L) = F_1[U_{\text{CH}}(x=L)]$, and likewise for F_2's, we can write

k_1 and k_2 as:

$$k_1 = \frac{\tilde{v}_{gs}F_2(L) - (\tilde{v}_{gs} - \tilde{v}_{ds})\ F_2(0)}{F_1(0)F_2(L) - F_1(L)F_2(0)} \times jDI_D \tag{B-24}$$

$$k_2 = \frac{\tilde{v}_{gs}F_1(L) - (\tilde{v}_{gs} - \tilde{v}_{ds})\ F_1(0)}{F_1(L)F_2(0) - F_1(0)F_2(L)} \times jDI_D \tag{B-25}$$

Some normalizing variables are introduced to simplify equations:

$$\omega_0 = \frac{\mu_n(V_{GS} - V_T)}{L^2} \tag{B-26}$$

$$D = \frac{\omega}{\omega_0} \times \frac{4}{U_{CH}^3(0)(1 - \alpha^2)^2} \tag{B-27}$$

$$G = \frac{W}{L}\ \mu_n C'_{ox}(V_{GS} - V_T) \tag{B-28}$$

It can be verified that:

$$\frac{G}{\omega_0} \equiv WLC'_{ox} \tag{B-29}$$

With these normalization variables, Eqs. (B-22) and (B-23) are rewritten as:

$$F_1(x) = \frac{1+j}{U_{CH}(x)} \times \left[1 + j\frac{4}{3}\frac{\omega}{\omega_0}\frac{1}{(1-\alpha^2)^2}\frac{U_{CH}^3(x)}{U_{CH}^3(0)} - \frac{16}{72}\left(\frac{\omega}{\omega_0}\right)^2 \frac{1}{(1-\alpha^2)^4}\frac{U_{CH}^6(x)}{U_{CH}^6(0)} - j\frac{64}{4536}\left(\frac{\omega}{\omega_0}\right)^3 \frac{1}{(1-\alpha^2)^6}\frac{U_{CH}^9(x)}{U_{CH}^9(0)} + \cdots \right] \tag{B-30}$$

$$F_2(x) = \frac{1-j}{U_{CH}^2(x)} \times \left[j\frac{4}{2}\frac{\omega}{\omega_0}\frac{1}{(1-\alpha^2)^2}\frac{U_{CH}^3(x)}{U_{CH}^3(0)} - \frac{16}{30}\left(\frac{\omega}{\omega_0}\right)^2 \frac{1}{(1-\alpha^2)^4}\frac{U_{CH}^6(x)}{U_{CH}^6(0)} - j\frac{64}{1440}\left(\frac{\omega}{\omega_0}\right)^3 \frac{1}{(1-\alpha^2)^6}\frac{U_{CH}^9(x)}{U_{CH}^9(0)} + \cdots \right] \tag{B-31}$$

For the convenience of the following derivation, we shall define X_1, X_2, X_3, X_4 such that:

$$F_2(0) = \frac{1-j}{U_{CH}^2(0)} \times X_1 \qquad F_2(L) = \frac{1-j}{U_{CH}^2(L)} \times X_4 \tag{B-32}$$

$$F_1(0) = \frac{1+j}{U_{CH}(0)} \times X_2 \qquad F_1(L) = \frac{1+j}{U_{CH}(L)} \times X_3 \tag{B-33}$$

Up to now, we have expanded the series of the Bessel functions. From here on, we shall write in concise notation. It can be verified by comparing Eqs. (B-32) and (B-33) to

Eqs. (B-30) and (B-31) that:

$$X_1 = \sum_{n=1}^{\infty}\left(\frac{j\omega}{\omega_0}\right)^n \frac{1}{(1-\alpha^2)^{2n}} \times \frac{4^n}{(n-1)!\,3^{n-1}\prod_{i=1}^{n} 3i-1} \tag{B-34}$$

$$X_2 = 1 + \sum_{n=1}^{\infty}\left(\frac{j\omega}{\omega_0}\right)^n \frac{\alpha^{3n}}{(1-\alpha^2)^{2n}} \times \frac{4^n}{n!\,3^n\prod_{i=1}^{n-1} 3i+1} \tag{B-35}$$

$$X_3 = 1 + \sum_{n=1}^{\infty}\left(\frac{j\omega}{\omega_0}\right)^n \frac{1}{(1-\alpha^2)^{2n}} \times \frac{4^n}{n!3^n\prod_{i=1}^{n-1} 3i+1} \tag{B-36}$$

$$X_4 = \sum_{n=1}^{\infty}\left(\frac{j\omega}{\omega_0}\right)^n \frac{\alpha^{3n}}{(1-\alpha^2)^{2n}} \times \frac{4^n}{(n-1)!\,3^{n-1}\prod_{i=1}^{n} 3i-1} \tag{B-37}$$

We further define Z_1 and Z_2 such that the channel current of Eq. (B-19) can be written as:

$$\tilde{i}_{\mathrm{ch}}(x) = k_1\, U_{\mathrm{CH}}(x)(1+j)Z_1 + k_2(1-j)Z_2 \tag{B-38}$$

Therefore, Z_1 and Z_2 are:

$$Z_1 = 1 + \sum_{n=1}^{\infty}\left(\frac{j\omega}{\omega_0}\right)^n \frac{\alpha^{3n}}{(1-\alpha^2)^{2n}} \times \frac{4^n}{n!\,3^n\prod_{i=1}^{n} 3i+1} \tag{B-39}$$

$$Z_2 = 1 + \sum_{n=1}^{\infty}\left(\frac{j\omega}{\omega_0}\right)^n \frac{\alpha^{3n}}{(1-\alpha^2)^{2n}} \times \frac{4^n}{n!\,3^n\prod_{i=1}^{n} 3i-1} \tag{B-40}$$

The small-signal drain current is simply the channel current evaluated at $x = L$. Hence, $i_d = i_{\mathrm{ch}}(x = L)$, after some algebra, is equal to:

$$\begin{aligned}\tilde{i}_d &= \frac{j(\omega/\omega_0)}{\alpha X_1 X_2 - X_3 X_4}\frac{2\alpha}{1-\alpha^2} \times G \times \{[\alpha^2(\tilde{v}_{gs} - \tilde{v}_{ds})X_1 - \tilde{v}_{gs}X_4]Z_1 \\ &\quad + [\tilde{v}_{gs}X_2 - (\tilde{v}_{gs} - \tilde{v}_{ds})X_3]Z_2\} \\ &= \frac{G}{D(j\omega)} \times \{[\alpha^2(\tilde{v}_{gs} - \tilde{v}_{ds})X_1 - \tilde{v}_{gs}X_4]Z_1 + [\tilde{v}_{gs}X_2 - (\tilde{v}_{gs} - \tilde{v}_{ds})X_3]Z_2\}\end{aligned} \tag{B-41}$$

where $D(j\omega)$ is given by:

$$D(j\omega) = D_0 + (j\omega)D_1 + (j\omega)^2 D_2 + \cdots \tag{B-42}$$

$D_0 = 1$ and $D_n(j\omega)$ are:

$$\begin{aligned}D_n = \omega_0^{-n}\frac{4^{n+1}}{2(1-\alpha^2)^{2n+1}} \times \Bigg[&\frac{1-\alpha^{3n+2}}{n!\,3^n\prod_{i=1}^{n+1} 3i-1} \\ &+ \sum_{m=1}^{n}\frac{\alpha^{3n-3m+3} - \alpha^{3m-1}}{(m-1)!\,(n-m+1)!\,3^n\prod_{i=1}^{m} 3i - 1\prod_{i=1}^{n-m} 3i+1}\Bigg]\end{aligned} \tag{B-43}$$

The source current can be similarly obtained, except that i_s is $i_{\rm ch}$ evaluated at $x = 0$. We shall not write out the expression here.

Now that the small-signal drain and source currents are determined, we want to find the gate current to complete the device analysis. There are two ways to find the gate current. The first is to equate i_g to $-i_d - i_s$. This was the approach used in the derivation of MESFET y-parameters in Appendix A. In a four-terminal device such as a MOSFET, this approach would not work. With the extra bulk terminal, the information of i_d and i_s are not sufficient to establish i_g, since the current through the bulk node remains unknown. Although HFET is a three-terminal device and we could as well adopt the said approach to find i_g, we will establish the gate current with the following second approach. We first determine the total gate charge, which is negative of the total charge in the channel. The gate current is then set equal to the time derivative of the gate charge. We note that because the channel charge of a HFET is negative, the gate charge is positive. This is the opposite of the situation in a MESFET.

In a small-signal operation, it is useful to equate the instantaneous gate charge (q_G) to the biasing gate charge (Q_G) plus the small-signal component (q_g). We shall denote $\tilde{q}_g$ as the phasor of q_g, which, in accordance with Eq. (6-11), is equal to:

$$\tilde{q}_g = W \int_0^L C'_{\rm ox}\, \tilde{u}_{\rm ch}\, dx \tag{B-44}$$

The small-signal gate current is the time derivative of the gate charge:

$$\tilde{i}_g = \frac{d\tilde{q}_g}{dt} = j\omega C'_{\rm ox} WL \times \frac{1}{L}\int_0^L \tilde{u}_{\rm ch}\, dx \tag{B-45}$$

Substituting $u_{\rm ch}$, k_1, and k_2 into the above expression, we find i_g after some algebra as:

$$\begin{aligned}
\tilde{i}_g = {} & j\omega C'_{\rm ox} WL \frac{(1-j)U_{\rm CH}(0)}{2X_3X_4 - 2\alpha X_1 X_2}\left[\tilde{v}_{gs}X_4 - \alpha^2(\tilde{v}_{gs} - \tilde{v}_{ds})X_1\right] \\
& \times (1+j)\int_0^L \frac{1}{L}\left(\frac{1}{U_{\rm CH}(x)} + j\frac{D}{3}U_{\rm CH}^2 - \frac{D^2}{72}U_{\rm CH}^5 - j\frac{D^3}{4536}U_{\rm CH}^8 + \cdots\right)dx \\
& + j\omega C'_{\rm ox} WL \frac{(1+j)U_{\rm CH}^2(0)}{2X_1X_2 - 2X_3X_4/\alpha}\left[\tilde{v}_{gs}X_2 - \alpha(\tilde{v}_{gs} - \tilde{v}_{ds})X_3\right] \\
& \times (1-j)\int_0^L \frac{1}{L}\left(j\frac{D}{2}U_{\rm CH}(x) - \frac{D^2}{30}U_{\rm CH}^4 - j\frac{D^3}{1440}U_{\rm CH}^7 + \cdots\right)dx
\end{aligned} \tag{B-46}$$

We shall define the first integral in the i_g expression as Y_1 and the second, Y_2. $U_{\rm CH}(x)$ itself was given in Eq. (5-142); it is a square-root function of position. Carrying out the integrations, we find that Y_1 and Y_2 in general can be written as the following series expansions:

$$Y_1 = \frac{2}{1+\alpha} + \sum_{n=1}^{\infty}\left(\frac{j\omega}{\omega_0}\right)^n \frac{1}{(1-\alpha^2)^{2n}} \times \frac{4^n}{n!\,3^n\prod_{i=1}^{n-1} 3i+1} \times \frac{2(\alpha^{3n} + \alpha^{3n-1} + \cdots + 1)}{(3n+1)(1+\alpha)} \tag{B-47}$$

$$Y_2 = \sum_{n=1}^{\infty}\left(\frac{j\omega}{\omega_0}\right)^n \frac{1}{(1-\alpha^2)^{2n}} \times \frac{4^n}{(n-1)!\,3^{n-1}\prod_{i=1}^{n} 3i-1} \times \frac{2(\alpha^{3n-1} + \alpha^{3n-2} + \cdots + 1)}{3n(1+\alpha)} \tag{B-48}$$

After some simplification, we find:

$$\tilde{i}_g = \frac{C'_{ox}WL}{D(j\omega)} \frac{\omega_0(1-\alpha^2)}{2\omega} \times \{[\alpha^2(\tilde{v}_{gs} - \tilde{v}_{ds})\, X_1 - \tilde{v}_{gs}X_4]Y_1$$
$$+ [\alpha\tilde{v}_{gs}X_2 - \alpha^2(\tilde{v}_{gs} - \tilde{v}_{ds})X_3]Y_2\} \tag{B-49}$$

With i_d given in Eq. (B-41) and i_g in Eq. (B-49), we are ready to find y_{gg}, y_{gd}, y_{dg}, and y_{dd}. We express these y-parameters in the following forms:

$$y_{gg} = \frac{N_{gg}(j\omega)}{D(j\omega)} \qquad y_{gd} = \frac{N_{gd}(j\omega)}{D(j\omega)} \qquad y_{dg} = \frac{N_{dg}(j\omega)}{D(j\omega)} \qquad y_{dd} = \frac{N_{dd}(j\omega)}{D(j\omega)} \tag{B-50}$$

where $D(j\omega)$ was given in Eq. (B-42), and the numerator terms are:

$$N_{gg}(j\omega) = N_{gg,0} + (j\omega)\, N_{gg,1} + (j\omega)^2\, N_{gg,2} + \cdots \tag{B-51}$$

$$N_{gd}(j\omega) = N_{gd,0} + (j\omega)\, N_{gd,1} + (j\omega)^2\, N_{gd,2} + \cdots \tag{B-52}$$

$$N_{dg}(j\omega) = N_{dg,0} + (j\omega)\, N_{dg,1} + (j\omega)^2\, N_{dg,2} + \cdots \tag{B-53}$$

$$N_{dd}(j\omega) = N_{dd,0} + (j\omega)\, N_{dd,1} + (j\omega)^2\, N_{dd,2} + \cdots \tag{B-54}$$

$$N_{gg,0} = 0 \qquad N_{gd,0} = 0 \quad N_{dg,0} = \alpha G = g_d \qquad N_{dg,0} = (1-\alpha)G = g_m \quad D_0 = 1 \tag{B-55}$$

The following is for $n \geq 1$ [for D_n, see Eq. (B-43)]:

$$\frac{N_{gg,n}}{WLC'_{ox}} = \frac{\omega_0^{1-n}4^n}{(1-\alpha^2)^{2n-1}(1+\alpha)\,3^{n-1}} \left\{\frac{1}{(n-1)!\prod_{i=1}^{n} 3i-1}\left[\alpha - \alpha^{3n-1}\right.\right.$$
$$\left. + \frac{(1-\alpha)(\alpha^{3n-1} + \alpha^{3n-2} + \cdots + 1)}{3n}\right]$$
$$+ \sum_{m=1}^{n-1} \frac{\alpha - \alpha^{3m-1}}{(m-1)!\,(n-m)!\prod_{i=1}^{m} 3i-1\prod_{i=1}^{n-m-1} 3i+1}$$
$$\times \frac{\alpha^{3n-3m} + \alpha^{3n-3m-1} + \cdots + 1}{3n-3m+1}$$
$$\left. + \frac{(\alpha^{3m} - \alpha)(\alpha^{3n-3m-1} + \alpha^{3n-3m-2} + \cdots + 1)}{m!\,(n-m-1)!\,(3n-3m)\prod_{i=1}^{m-1} 3i+1\prod_{i=1}^{n-m} 3i-1}\right\} \tag{B-56}$$

$$\frac{N_{gd,\mathrm{n}}}{WLC'_{ox}} = \frac{\omega_0^{1-n}4^n\alpha}{(1-\alpha^2)^{2n-1}(1+\alpha)}\left[\frac{1}{(n-1)!\,3^{n-1}\prod_{i=1}^{n}3i-1}\right.$$

$$\times\left(\frac{\alpha^{3n-1}+\alpha^{3n-2}+\cdots+1}{3n}-1\right)$$

$$+\sum_{m=1}^{n-1}\frac{\alpha^{3m-1}+\alpha^{3m-2}+\cdots+1}{m!\,(n-m)!\,3^n\prod_{i=1}^{m}3i-1\prod_{i=1}^{n-m-1}3i+1}$$

$$\left.-\frac{\alpha^{3m}+\alpha^{3m-1}+\cdots+1}{m!\,(n-m-1)!\,3^{n-1}(3m+1)\prod_{i=1}^{m-1}3i+1\prod_{i=1}^{n-m}3i-1}\right] \tag{B-57}$$

$$\frac{N_{dg,\mathrm{n}}}{WLC'_{ox}} = \frac{\omega_0^{1-n}4^n}{(1-\alpha^2)^{2n}}\left[\frac{\alpha^2-\alpha^{3n}}{(n-1)!\,3^{n-1}\prod_{i=1}^{n}3i-1}+\frac{\alpha^{3n}-\alpha}{n!\,3^n\prod_{i=1}^{n-1}3i+1}\right.$$

$$+\frac{(1-\alpha)\,\alpha^{3n}}{n!\,3^n\prod_{i=1}^{n}3i-1}$$

$$+\sum_{m=1}^{n-1}\frac{(\alpha^2-\alpha^{3m})\cdot\alpha^{3n-3m}}{(m-1)!\,(n-m)!\,3^{n-1}\prod_{i=1}^{m}3i-1\prod_{i=1}^{n-m}3i+1}$$

$$\left.+\frac{(\alpha^{3m}-\alpha)\cdot\alpha^{3n-3m}}{m!\,(n-m)!\,3^n\prod_{i=1}^{m-1}3i+1\prod_{i=1}^{n-m}3i-1}\right] \tag{B-58}$$

$$\frac{N_{dd,\mathrm{n}}}{WLC'_{ox}} = \frac{\omega_0^{1-n}4^n}{(1-\alpha^2)^{2n}}\left[\frac{\alpha}{n!\,3^n\prod_{i=1}^{n-1}3i+1}+\frac{\alpha^{3n+1}}{n!\,3^n\prod_{i=1}^{n}3i-1}\right.$$

$$-\frac{\alpha^2}{(n-1)!\,3^{n-1}\prod_{i=1}^{n}3i-1}$$

$$+\sum_{m=1}^{n-1}\frac{\alpha^{3n-3m+1}}{m!\,(n-m)!\,3^n\prod_{i=1}^{m-1}3i+1\prod_{i=1}^{n-m}3i-1}$$

$$\left.-\frac{\alpha^{3n-3m+2}}{(m-1)!\,(n-m)!\,3^{n-1}\prod_{i=1}^{m}3i-1\prod_{i=1}^{n-m}3i+1}\right] \tag{B-59}$$

It is convenient to express some of these coefficients explicitly, up to $n = 2$. The terms with $n = 0$ were described in Eq. (B-55). The following are the terms associated with $n = 1$ and 2:

$$D_1 = \frac{4}{15}\frac{1}{\omega_0}\frac{\alpha^2+3\alpha+1}{(1+\alpha)^3} \tag{B-60}$$

$$D_2 = \frac{1}{45}\frac{1}{\omega_0^2}\frac{\alpha^2+4\alpha+1}{(1+\alpha)^4} \tag{B-61}$$

$$N_{gg,1} = WLC'_{ox}\frac{2}{3}\frac{\alpha^2+4\alpha+1}{(1+\alpha)^2} \tag{B-62}$$

$$N_{gg,2} = \frac{WLC'_{ox}}{\omega_0}\frac{2}{45}\frac{2\alpha^2+11\alpha+2}{(1+\alpha)^3} \tag{B-63}$$

$$N_{gd,1} = -WLC'_{ox}\frac{2}{3}\frac{\alpha(\alpha+2)}{(1+\alpha)^2} \tag{B-64}$$

$$N_{gd,2} = -\frac{WLC'_{ox}}{\omega_0}\frac{2}{45}\frac{\alpha(2\alpha^2+8\alpha+5)}{(1+\alpha)^4} \tag{B-65}$$

$$N_{dg,1} = -WLC'_{ox}\frac{2}{3}\frac{\alpha(\alpha+2)}{(1+\alpha)^2} \tag{B-66}$$

$$N_{dg,2} = -\frac{WLC'_{ox}}{\omega_0}\frac{2}{45}\frac{\alpha(2\alpha^2+8\alpha+5)}{(1+\alpha)^4} \tag{B-67}$$

$$N_{dd,1} = WLC'_{ox}\frac{2}{3}\frac{\alpha(\alpha+2)}{(1+\alpha)^2} \tag{B-68}$$

$$N_{dd,2} = \frac{WLC'_{ox}}{\omega_0}\frac{2}{45}\frac{\alpha(2\alpha^2+8\alpha+5)}{(1+\alpha)^4} \tag{B-69}$$

REFERENCES

1. Tsividis, Y. (1987). *Operation and Modeling of the MOS Transistor*, New York: McGraw-Hill.
2. Liu, W. (1998). *Handbook of III-V Heterojunction Bipolar Transistors*. New York: Wiley, p. 860.

APPENDIX C

PARASITIC CAPACITANCES

Due to the complexity in appearance, we did not list the expressions of $C_{gd,p}$ and $C_{gs,p}$ when we first discussed the parasitic capacitances in § 6-6. We shall show the derivation of all the parasitic capacitances in this Appendix. We caution that these formulas differ somewhat from those used in a classic, authoritative paper [1]. For completeness, we present the parasitic capacitance formula used in the reference in the following:

$$C_{gs,p} = C_{sg,p} = (\epsilon_s + \epsilon_0)W \times \frac{K(\sqrt{1-k_{gs}^2})}{K(k_{gs})} \qquad k_{gs} = \sqrt{\frac{L_{GS}}{L_{GS}+L}} \tag{C-1}$$

$$C_{gd,p} = C_{dg,p} = (\epsilon_s + \epsilon_0)W \times \frac{K(\sqrt{1-k_{gd}^2})}{K(k_{gd})} \qquad k_{gd} = \sqrt{\frac{L_{GD}}{L_{GD}+L}} \tag{C-2}$$

$$C_{ds,p} = C_{sd,p} = (\epsilon_s + \epsilon_0)W \times \frac{K(\sqrt{1-k_{ds}^2})}{K(k_{ds})} \qquad k_{ds} = \sqrt{\frac{(2L_S+L_{DS})L_{DS}}{(L_S+L_{DS})^2}} \tag{C-3}$$

where $K(k)$ is the complete elliptic integral of the first kind, given in Eq. (6-216).

We first discuss the derivation of $C_{ds,p}$, which was given in Eq. (6-215). The equation is based on the conformal mapping analysis—a power tool in the study of two-dimensional problems (but not in three-dimensional problems). Since the width of the electrodes, W, in practical devices is much larger than the length dimension such as L, L_{GD}, etc., conformal mapping can be used. In order to be concise, we leverage on the results of Ref. [2] as much as possible. Many of the following terms, unspecified in this book, are found in the reference.

We apply the Schwarz-Christoffel transformation to the upper half of the w-plane (Fig. C-1a) to a rectangle in the z-plane (Fig. C-1b). The points A ($w = -1/k$), B ($w = -1$), C ($w = 1$), and D ($w = 1/k$) are mapped to A′ ($z = -K + jK'$),

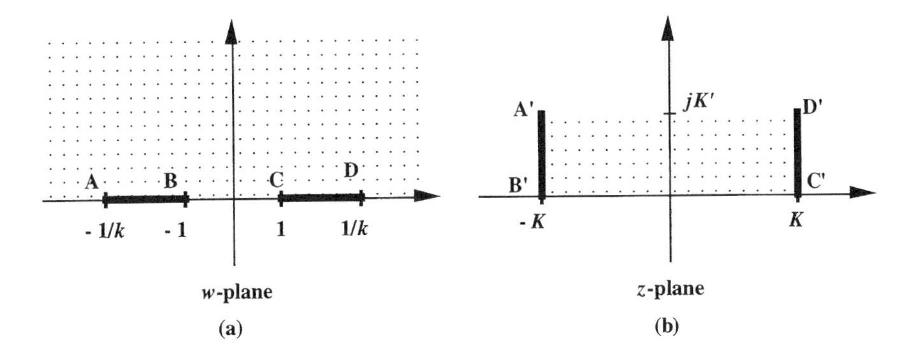

FIGURE C-1. (*a*) *w*-plane; two metal plates of infinitely thin thickness are located at AB and CD. (*b*) *z*-plane; the two transformed metal plates are at A′B′ and C′D′. The Schwarz-Christoffel transformation transforms the upper half of the *w*-plane to a rectangle in the *z*-plane.

B′ ($z = -K$), C′ ($z = K$), and D′ ($z = K + jK'$), respectively, where K and K', the notation used in Ref. [2], are given by:

$$K = \int_0^1 \frac{dw}{(1 - w^2)^{1/2}(1 - k^2 w^2)^{1/2}} \tag{C-4}$$

$$K' = \int_0^1 \frac{dw}{(1 - w^2)^{1/2}(1 - k'^2 w^2)^{1/2}} = \frac{1}{j}\int_1^{1/k} \frac{dw}{(1 - w^2)^{1/2}(1 - k^2 w^2)^{1/2}} \tag{C-5}$$

where

$$k' = \sqrt{1 - k^2} \tag{C-6}$$

It is clear that:

$$K'(k) = K(\sqrt{1 - k^2}) \tag{C-7}$$

Because an electrode (shown as a dark line in Fig. C-1) on the *w*-plane is from $w = 1$ to $w = 1/k$ whereas it is from $x = L_{DS}/2$ to $L_S + L_{DS}/2$ in real device geometry, k is equal to:

$$k = \frac{L_{DS}}{L_{DS} + 2L_S} \tag{C-8}$$

Reference [2] is concerned mostly with the resistance between the two electrodes in the *w*-plane. Due to the principle that the resistance is invariant under conformal transformation, the resistance in the *w*-plane is identical to the resistance in the *z*-plane. We note that Fig. C-1 is only a two-dimensional representation of the sample. The sample has a width equal to W extending in the dimension perpendicular to the page. The electrodes in the real *w*-plane are the segments AB and CD. The transformed electrodes are the segments A′B′ and C′D′ in the *z*-plane. Imagine that a voltage V_0 is applied at C′D′ and $-V_0$ is applied at A′D′; then the resistance in the *z*-plane is equal to:

$$R = \frac{2K(k)}{K'(k)} \frac{\rho}{W} \tag{C-9}$$

The resistance equation is identical to that found in Ref. [2]. Here, however, we are more interested in the capacitance between electrode A′B′ and C′D′. The capacitance follows the parallel-capacitor formula, given by:

$$C = \frac{\epsilon_0}{2K(k)} K'(k) W \tag{C-10}$$

It is interesting to note that the RC product is equal to $\epsilon_0 \cdot \rho$, which is a well-known static relationship in electromagnetics [3].

Due to the invariance of capacitance in conformal transformation, the capacitance in the real w-plane is also given by Eq. (C-10). The equation developed thus far considers only the top half of the w-plane. In a real problem, the bottom half of the plane also contributes to capacitance, as indicated in Fig. 6-38. Unlike the upper half, whose dielectric constant is ϵ_0, the bottom half, made of GaAs, has a dielectric constant of ϵ_s. The total capacitance is therefore:

$$C = \frac{\epsilon_0 + \epsilon_s}{2} \times \frac{K'(k)}{K(k)} W \tag{C-11}$$

When $\epsilon_0 = \epsilon_s$, the above capacitance formula becomes that mentioned in Ref. [4]. For either a MESFET or a HFET, we write:

$$C_{ds,p} = C_{sd,p} = \frac{(\epsilon_s + \epsilon_0)}{2} W \times \frac{K(\sqrt{1 - k_{ds}^2})}{K(k_{ds})} \qquad k_{ds} = \frac{L_{sd}}{L_{sd} + 2L_s} \tag{C-12}$$

The derivation for $C_{sg,p}$ and $C_{dg,p}$ is slightly complicated since the gate length L is not identical to the source contact length L_S (or the drain contact length L_D). Due to the unequalness in length, the electrodes EF in the w-plane are transformed to E′F′ on the z-plane, as shown in Fig. C-2. Although we know that A′ is located at $z = -K + jK'$ and B′ is at $-K$, we do not know the coordinates of E′ and F′. Well, the coordinate on the z-plane corresponding to point E is

$$z = \int_0^1 \frac{dw}{(1 - w^2)^{1/2}(1 - k^2 w^2)^{1/2}} + \int_1^{1/e} \frac{dw}{(1 - w^2)^{1/2}(1 - k^2 w^2)^{1/2}} = K + jE' \tag{C-13}$$

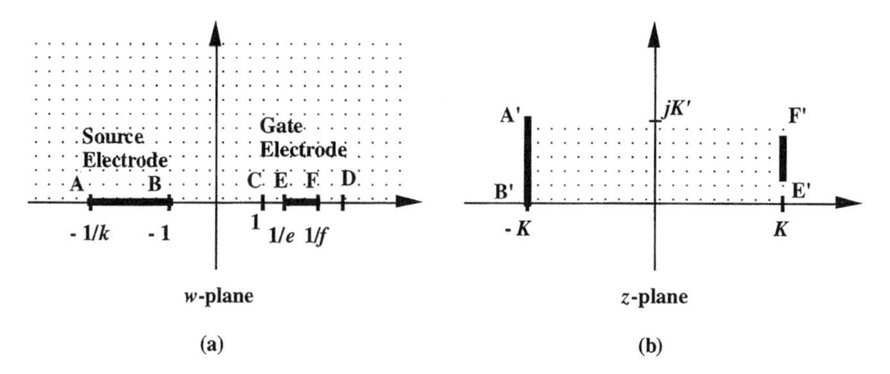

FIGURE C-2. (*a*) w-plane; two metal plates of infinitely thin thickness are located at AB and EF. The distances AB and EF are different. (*b*) z-plane; the two transformed metal plates are at A′B′ and E′F′. The Schwarz-Christoffel transformation transforms the upper half of the w-plane to a rectangle in the z-plane.

where e is identified on Fig. C-2, and E' is

$$E' = \frac{1}{j}\int_1^{1/e} \frac{dw}{(1-w^2)^{1/2}(1-k^2w^2)^{1/2}} \tag{C-14}$$

Similarly, we can write the z-coordinate of transformed point F as $z = K + jF'$, where F' is

$$F' = \frac{1}{j}\int_1^{1/f} \frac{dw}{(1-w^2)^{1/2}(1-k^2w^2)^{1/2}} \tag{93}$$

We have chosen to leave E' and F' in their integral forms, rather than in terms of elliptic functions such as $\mathrm{sn}\,\chi$ and $\mathrm{cn}\,\chi$ [2]. The functions can be quite bewildering for people unfamiliar with conformal transformation.

If we can determine the capacitance of Fig. C-2*b*, with one electrode being E′F′ and another, A′B′, then we can know the capacitance between EF and CD in the w-plane, which is the actual problem under consideration. To determine the capacitance between E′F′ and A′B′, we first consider the case of Fig. C-3, with the upper plate extending from $-\infty$ to point E′, and the bottom plate from $-\infty$ to $+\infty$. The distance separating these two plates is d. Following the strategy used in Chapter 7 of Ref. [2], we find that the capacitance is equal to:

$$C = \frac{\epsilon_0}{d} W \times \left(\text{top plate length} + \frac{d}{\pi} + \frac{4d}{\pi}\right) \tag{C-16}$$

The first term is just the ideal capacitance between the top plate and the bottom plate if the fringing field at the side can be neglected. The two additional terms account for the additional charges due to the presence of the fringing field. The additional length term d/π accounts for the extra charge induced inside the top plate, whereas the term $4d/\pi$ accounts for the extra charge induced on the top of the top plate. The distance d corresponds to $2K(k)$ of the previous discussion. The formula basically assumes that the top plate (which is E′F′) is much longer than d. If not, then we can just drop the additional fringing terms altogether.

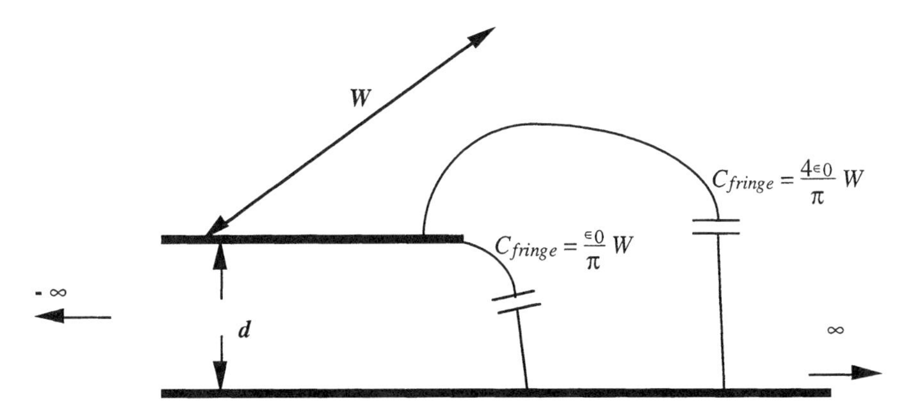

FIGURE C-3. A semi-idealized capacitance structure (the plates extends from minus infinity to a certain location) considered by Eq. (C-16).

The above concerns the situation when the top plate is between $-\infty$ and E′. A similar formula can be derived for a top plate running from F′ to ∞. From the invariance principle of capacitance under transformation, the capacitance in the w-plane between EF and AB is thus found. Finally, note that the problem considered so far encompasses only the top half of the w-plane. Once we consider the bottom half of the w-plane as well, we obtain the capacitance as

$$C_{gs,p} = C_{sg,p} = \begin{cases} \dfrac{\epsilon_0 + \epsilon_s}{2K(k_{gs})} W \times \left[(E' - F') + \dfrac{20K(k_{gs})}{\pi} \right] & \text{if } E' - F' \gg K(k_{gs}) \\ \dfrac{\epsilon_0 + \epsilon_s}{2K(k_{gs})} W \times (E' - F') & \text{if } E' - F' \ll K(k_{gs}) \end{cases} \tag{C-17}$$

where E' and F' were given in Eqs. (C-14) and (C-15), respectively, and k_{gs} is given by:

$$k_{gs} = \frac{L_{GS} - (L_S - L)/2}{L_{GS} - (L_S - L)/2 + 2L_S} \tag{C-18}$$

Similarly, the gate-drain parasitic capacitance is given by:

$$C_{gd,p} = C_{dg,p} = \begin{cases} \dfrac{\epsilon_0 + \epsilon_s}{2K(k_{gd})} W \times \left[(E' - F') + \dfrac{20K(k_{gd})}{\pi} \right] & \text{if } E' - F' \gg K(k_{gd}) \\ \dfrac{\epsilon_0 + \epsilon_s}{2K(k_{gd})} W \times (E' - F') & \text{if } E' - F' \ll K(k_{gd}) \end{cases} \tag{C-19}$$

where k_{gd} is:

$$k_{gd} = \frac{L_{GD} - (L_D - L)/8}{L_{GD} - (L_D - L)/8 + 2L_D} \tag{C-20}$$

REFERENCES

1. Pucel, R., Haus, H., and Statz, H. (1975). "Signal and noise properties of gallium arsenide microwave field-effect transistors." In Marton, L., ed., *Advances in Electronics and Electron Physics*. New York: Academic Press, **38,** 195–265.
2. Gibbs, W. (1958). *Conformal Transformations in Electrical Engineering*. London: Chapman & Hall.
3. See Cheng, D. (1988). *Field and Wave Electronmagnetics*. New York: Addison-Wesley.
4. Smythe, W. (1950). *Static and Dynamic Electricity*. New York: McGraw-Hill, problem 58 of Chap. IV, p. 109.

APPENDIX D

UNIVERSAL CONSTANTS AND UNITS

Speed of light	c	3×10^{10} cm/s
Free electron mass	m_0	9.11×10^{-31} kg
Electric charge	q	1.6×10^{-19} C
Permittivity of free space	ϵ_0	8.85×10^{-14} F/cm
Permeability of the free space	μ_0	1.256×10^{-8} H/cm
Boltzmann's constant	k	8.62×10^{-5} eV/°C
Plank's constant	h	6.63×10^{-34} J-s

1 Å = 10^{-8} cm
1 μm = 10^{-4} cm
1 mil ≈ 25 μm
1 eV = 1.6×10^{-19} J
0°C = 273 K

APPENDIX E

SEMICONDUCTOR MATERIAL PARAMETERS AT ROOM TEMPERATURE

	GaAs	$Al_{0.3}GaAs$	$Ga_{0.51}InP$	InP	$In_{0.53}GaAs$	$In_{0.52}AlAs$	Si
ϵ_r	13.1	~12.2	~11.9	12.6	~14.0	~12.5	11.9
N_c (cm^{-3})	4.7×10^{17}	—	—	5.8×10^{17}	$\sim 2.8 \times 10^{17}$	—	2.8×10^{19}
N_v (cm^{-3})	7.0×10^{18}	—	—	1.0×10^{19}	$\sim 6.0 \times 10^{18}$	—	1.04×10^{19}
n_i (cm^{-3})	1.79×10^{6}	—	—	1.05×10^{7}	6.31×10^{11}	—	1.45×10^{10}
E_g (eV)	1.424 (direct)	1.80 (direct)	1.86 (direct)	1.35 (direct)	0.75 (direct)	1.46 (direct)	1.12 (indirect)
$E_{\Gamma-L}$ (eV)	0.28	0.10	0.15	0.61	0.55	0.54	—
m_e^*/m_0	0.067	0.092	0.099	0.078	0.041	0.074	0.26
k_{th}(W/cm°C)	0.46	0.12	0.05	0.68	0.05	—	1.5
s (cm/s)	1×10^{6}	1×10^{6}	—	—	1×10^{3}	—	1×10^{3}
τ_n(s)	1×10^{-9}	—	—	—	—	—	1×10^{-6}
μ_n (cm^2/V-s)	4000	1400	1000	3200	7000	900	800
μ_p (cm^2/V-s)	250	130	75	150	300	180	400
v_{sat} (cm/s)	8×10^{6}	—	4.4×10^{6}	1.5×10^{7}	7×10^{6}	6×10^{6}	8×10^{6}

Note: When the parameter value of a material (such as $Al_{0.3}Ga_{0.7}As$) is obtained by linearly extrapolating between two extreme materials (such as AlAs and GaAs), an approximate sign is placed in front. All of the values are room-temperature values. Electron mobilities are taken at $N_d = 1 \times 20^{17}$ cm^{-3}. All mobility values are majority carrier values.

APPENDIX F

HETEROJUNCTION PARAMETERS AT ROOM TEMPERATURE

	$Al_{0.3}Ga_{0.7}As$/ GaAs	$Ga_{0.51}In_{0.49}P$/ GaAs	InP/ $In_{0.53}Ga_{0.47}As$	$In_{0.52}Al_{0.48}As$/ $In_{0.53}Ga_{0.47}A$	Si/ $Si_{0.8}Ge_{0.2}$
ΔE_g (eV)	0.37	Ordered: 0.43 Disordered: 0.46	0.60	0.71	Unstrained: 0.078 Strained: 0.165
ΔE_c (eV)	0.24	Ordered: 0.03 Disordered: 0.22	0.23	0.50	
ΔE_v (eV)	0.13	Ordered: 0.40 Disordered: 0.24	0.37	0.21	
s (cm/s)	103	2–200			

INDEX